AEPINUS'S
ESSAY ON THE THEORY OF
ELECTRICITY
AND
MAGNETISM

AEPINUS'S
ESSAY ON THE THEORY OF
ELECTRICITY
AND
MAGNETISM

INTRODUCTORY MONOGRAPH
AND NOTES BY
R. W. HOME

TRANSLATION BY
P. J. CONNOR

PRINCETON UNIVERSITY PRESS
PRINCETON, NEW JERSEY

WILLIAM MADISON RANDALL LIBRARY UNC AT WILMINGTON

COPYRIGHT © 1979 BY PRINCETON UNIVERSITY PRESS
PUBLISHED BY PRINCETON UNIVERSITY PRESS, PRINCETON, NEW JERSEY
IN THE UNITED KINGDOM: PRINCETON UNIVERSITY PRESS, GUILDFORD,
SURREY

ALL RIGHTS RESERVED
LIBRARY OF CONGRESS CATALOGING IN PUBLICATION DATA WILL BE
FOUND ON THE LAST PRINTED PAGE OF THIS BOOK

PUBLICATION OF THIS BOOK HAS BEEN AIDED BY THE WHITNEY DARROW
PUBLICATION RESERVE FUND OF PRINCETON UNIVERSITY PRESS
THIS BOOK HAS BEEN COMPOSED IN LINOTYPE BASKERVILLE

CLOTHBOUND EDITIONS OF PRINCETON UNIVERSITY PRESS BOOKS
ARE PRINTED ON ACID-FREE PAPER, AND BINDING MATERIALS ARE
CHOSEN FOR STRENGTH AND DURABILITY.

PRINTED IN THE UNITED STATES OF AMERICA BY PRINCETON
UNIVERSITY PRESS, PRINCETON, NEW JERSEY

UXORI DEDICAVIT
UTERQUE SUAE

CONTENTS

	ILLUSTRATIONS	viii
	PREFACE	ix
	ABBREVIATIONS	xiii
	CHRONOLOGY	xiv

INTRODUCTION

One	Biographical Outline	3
Two	The Electrical Background	65
Three	Electricity in the *Essay*	107
Four	Magnetism	137
Five	A Place in History	189

AN ESSAY ON THE THEORY OF ELECTRICITY AND MAGNETISM BY F. U. T. AEPINUS

	DEDICATION	227
	PREFACE	233
	INTRODUCTION	237
I.	General Principles of the Theory of Electricity and Magnetism	241
II.	Concerning Electrical and Magnetic Attraction and Repulsion	304
III.	Concerning the Communication of Electricity and Magnetism	350
IV.	Concerning Certain Phenomena of Bodies Immersed in Electric and Magnetic Vortices, and concerning the Magnetism of the Earth	392
	DISSERTATION I. An Explanation of a Certain Phenomenon of the Leyden Jar Discovered by the Celebrated Richmann	453
	DISSERTATION II. An Explanation of a Certain Paradoxical Magnetic Phenomenon	469
	APPENDIX: Annotated Bibliography of Aepinus's Published Writings	479
	BIBLIOGRAPHY OF SECONDARY WORKS CITED	499
	INDEX	505

ILLUSTRATIONS

1. The Leyden experiment — 72
2. Aepinus's notebook diagram illustrating the continuity of the force law — 135
3. Descartes' diagram showing how grooved particles are produced — 141
4. Musschenbroek's attempted determination of the law of magnetic action — 172
5. Aepinus's notebook illustration of a possible method of measuring magnetic forces — 187
6. Facsimile Plate I from the *Essay* (Figs. I-XXIV) — 313
7. Corrected version of Aepinus's Fig. XXXIX — 347
8. Facsimile Plate II from the *Essay* (Figs. XXV–XLIII) — 357
9. Facsimile Plate III from the *Essay* (Figs. XLIV–LXIII) — 381
10. Facsimile Plate IV from the *Essay* (Figs. LXIV–LXXVI) — 391
11. Diagram to clarify a mathematical argument of Aepinus — 404
12. Facsimile Plate V from the *Essay* (Figs. LXXVII–XCVI) — 411
13. Facsimile Plate VI from the *Essay* (Figs. XCVII–CXIV) — 449
14. Facsimile Plate VII from the *Essay* (Figs. CXV–CXXV) — 477

PREFACE

Aepinus's *Essay on the Theory of Electricity and Magnetism*, first published in St. Petersburg in 1759, was one of the outstanding achievements of eighteenth-century physics. Previously, the sciences of electricity and magnetism had always been investigated in a wholly qualitative and non-mathematical way. In Aepinus's hands they acquired for the first time something like their modern rigorous and highly mathematical form.

Moreover, in successfully bringing these subjects within the reach of mathematical analysis, Aepinus spearheaded an important general development in physics. In the half-century or so following the appearance of his book, most of the other branches of traditional experimental physics were also successfully mathematized for the first time. These later developments were, for the most part, the achievement of the French school headed by Laplace and Coulomb. But Coulomb in particular drew heavily upon Aepinus's work, and so, too, did mathematically inclined natural philosophers elsewhere, such as the Edinburgh professor John Robison. A new breed of mathematical physicist emerged at this period. Aepinus's book came to be seen by members of this ever more influential group as the starting point of modern electrical and magnetic science, and, more generally, as an excellent example of the kind of physics they themselves hoped to achieve.

It is not simply as a precursor nor even as an inspirer of important later developments, however, that Aepinus's work is of interest. It was also the climax of a fascinating but insufficiently studied chain of development that saw the sciences of electricity and magnetism advance from a primitive state, where various patently ad hoc mechanisms were put forward to account for an extremely limited range of known phenomena, into full-blown studies in their own right. In the process, and partly as a direct result of this development, men's views of the nature and limitations of physical inquiry underwent a series of profound changes that proved of lasting significance.

Despite its historical importance, Aepinus's *Essay* has in recent years slipped almost totally from view. Only in the Soviet Union, where a Russian translation was published some years ago, has it continued to attract a significant readership. Its failure to capture the attention of historians elsewhere is doubtless due in part to the fact that copies of the book are now scarce. It is also symptomatic, however, of a general lack of interest by historians of science in developments in physics in the eighteenth century. Too often the period has been seen, most unfairly, as an uneventful interlude between two great eras of scientific discovery. By making Aepinus's book available in English for the first time, and by setting the work in context and showing why it subsequently enjoyed such high esteem, I hope to create a better appreciation not only of Aepinus's own contribution but of some of the central problems and achievements of eighteenth-century science as a whole.

No attempt is made to give a full account of eighteenth-century developments in the study of electricity and magnetism. Yet the dearth of previous studies, especially in the case of magnetism, has necessitated a fairly extensive treatment. I have also felt it necessary to re-assess the contribution of Aepinus's great predecessor, Benjamin Franklin, to the science of electricity. The American left certain vital questions unanswered and the answers to others in a less than satisfactory state. To say this is not to belittle the importance of Franklin's work: far from it. But until we recognize the open-endedness of his contribution, we are bound to misunderstand the central concerns of the next generation of electrical theorists, and of Aepinus more than anyone else. Franklin did not bring about a once-and-for-all revolution in electrical theory. On the contrary, the theory evolved continuously throughout most of the century. While Franklin's writings mark one milestone along the way, Aepinus's *Essay* marks another; in it, the American's incompletely worked out ideas were for the first time rendered precise and fully consistent with each other.

A notable feature of the story that follows is its international character. Throughout the eighteenth century, electricity was everywhere, for scientist and dilettante alike, a popular subject for investigation. In no other field did the "republic of learning" more completely transcend national boundaries. When we take up the work of Aepinus, a German who went to live in the capital of the Tsars, we recognize at once that, whatever the standing of France and Britain as the leading scientific nations of the day, vital developments also took place elsewhere, and in due course had their impact at the center. The historian of eighteenth-century physics ignores events east of the Rhine at his peril.

Though Aepinus's book was published in St. Petersburg, his thinking seems not to have been influenced by contemporary ideas within the Russian Academy of Sciences. On the contrary, he had already developed his novel approach to the theory of electricity and magnetism during his period in Berlin as a member of Frederick the Great's

PREFACE

Royal Academy of Sciences, and he arrived in St. Petersburg with his leading doctrines settled. Furthermore, though Aepinus was highly esteemed by his fellow academicians in the northern capital, the views set out in his *Essay* seem not to have attracted any particular support either in the Academy or within the wider Russian intellectual community of the day. So far as the history of science in Russia is concerned, therefore, Aepinus's book stands as an isolated achievement, a fact that is clearly reflected in the general lack of attention this discussion will give to the electrical and magnetic investigations of his Russian predecessors and contemporaries. Nevertheless, in the years following the publication of his book, Aepinus himself came, one way and another, to occupy a noteworthy place in the government and intellectual life of his adopted country. Within the limits imposed by the sparseness of available evidence, I have traced his later career, and to this extent this study may contribute to the general intellectual history of eighteenth-century Russia.

This study would not have been possible without financial support from the Australian Research Grants Committee and from various funds within the University of Melbourne. Equally important have been the generous sabbatical leave provisions, traditional until recently in Australian universities, which have made it possible for me to visit crucial archival repositories in various parts of Europe.

No undertaking of this kind could hope to succeed without the cooperation of librarians and archivists around the world. Among the many who have assisted in the course of this project, I must single out for special thanks the Director of the University Archives at Rostock University, Dr. B. Wandt; successive Librarians of the Royal Society of London, Mr. I. Kaye and Mr. N. H. Robinson; the Librarian of the American Philosophical Society, Dr. Whitfield J. Bell, Jr.; the staff of the Leningrad branch of the Archives Department of the Academy of Sciences of the U.S.S.R.; and the staff of the Archives Department of the Academy of Sciences of the D.D.R. Above all, I must thank the Reference Department of the Baillieu Library, University of Melbourne, and especially Miss E. Garran and Mr. P. Singleton, for their unfailing patience and resourcefulness in the face of one seemingly impossible request after another.

I am grateful to the following institutions for allowing me to quote unpublished materials from their collections: the Archives Department of the Academy of Sciences of the U.S.S.R.; the American Philosophical Society Library; the Beinecke Rare Book and Manuscript Library, Yale University; the British Library; the Biblioteca Nazionale Braidense, Milan; the Institution of Electrical Engineers, London;

the Royal Medical Society, Edinburgh; and the Royal Society of London.

I have been fortunate to have had from the very beginning of the project the services of Mrs. Hazel Maxian as research assistant. Her knowledge of European languages has been a priceless resource on which I have greatly depended. In addition, my prose has regularly benefited under her critical eye for the niceties of English usage. My colleagues in the Department of History and Philosophy of Science at the University of Melbourne, though subjected often enough to my outpourings, have been remarkably tolerant and have, into the bargain, made a great number of very helpful suggestions. So, too, has Dr. K. R. Hutchison, to whom I likewise offer my grateful thanks. Such errors as remain are, of course, entirely my responsibility.

Finally, I would like to express my personal thanks to Peter Connor for agreeing to take part in the project in the first place, and for his patience and forbearance since then. He must often have wondered whether it was worth the effort. I hope that, in the end, he will feel that it was.

R.W.H.
October 1977.

ABBREVIATIONS

The following bibliographical abbreviations are used throughout:

AAN	Akademiya Nauk S.S.S.R. : Arkhiv, Leningradskoe otdelenie.
HAS	*Histoire de l'Académie Royale des Sciences [de Paris], avec les mémoires de mathématique et de physique.*
Mém. Acad. Berlin	*Histoire de l'Académie Royale des Sciences et des Belles-Lettres de Berlin, avec les mémoires.*
Phil. Trans.	*Philosophical Transactions of the Royal Society of London.*
Teoriya	Aepinus, *Teoriya elektrichestva i magnetizma,* ed. Ya. G. Dorfman, Moscow/Leningrad, 1951.

CHRONOLOGY

1724	13 December. Born at Rostock.
1740	Commences studies at Rostock University.
1744–46	*Wanderjahre.*
1747	5 February. Proclaimed Doctor of Philosophy at Rostock University.
1747–55	*Privatdozent* attached to Rostock University.
1755	17 April. Member of Berlin Academy of Sciences.
1756	Experiments with tourmaline crystal.
1757	10 May. Arrives in St. Petersburg.
1759	Beginning of dispute with Lomonosov.
	Appointed tutor to Grand Duchess Catherine.
	8 December. Publication of *Essay on the Theory of Electricity and Magnetism*.
1760	Appointed Director of St. Petersburg Academy's Observatory.
1761	Appointed Director of Studies at Imperial Corps of Noble Cadets.
	Transit of Venus.
1762	Catherine becomes Empress after successful coup against her husband.
	Publication of *Recueil de différents mémoires sur la tourmaline*.
1765	Appointed tutor to Tsarevich Paul Petrovich. Relinquishes regular academic duties.
	First duties at College of Foreign Affairs.
	Death of Lomonosov.
1770	President of St. Petersburg Free Economic Society.
1782	Appointed to Education Commission.
1783	Knighthood in Order of St. Anne.
1784	Invents achromatic microscope.
1798	Retires to Dorpat.
1802	10 August. Dies.

INTRODUCTION

CHAPTER ONE
Biographical Outline

Franz Ulrich Theodosius Aepinus [1] was born in Rostock, the chief town of the north German Duchy of Mecklenburg, on 13 December 1724, the second son and fifth and last child of Franz Albert Aepinus, the highly esteemed professor of theology in the local university.

In days gone by, Rostock had been an exceedingly prosperous trading center, enjoying all the advantages that a fine harbor and membership of the powerful Baltic trading alliance, the Hanseatic League, could bring. By the eighteenth century, however, and the years of Aepinus's childhood, both its population and its economic power had been considerably reduced. The chief cause of this had been the Thirty Years' War, which had left the whole of Mecklenburg, like so many other parts of Germany, in a state of utter devastation. But a second time, too, during the so-called Great Northern War between Sweden and Russia (1699–1721), the duchy had suffered dreadfully as contending armies fought across its territory. By the time the troops eventually departed, the population of the countryside had been decimated, and that of Rostock itself had dropped from its sixteenth-century peak of perhaps 14,000 to a mere 8,000 or so; agriculture had been reduced to a very low level throughout the province, and the economic life of the towns was almost at a standstill. Recovery was slow, and was not helped by a protracted dispute between Karl Leopold, the then Duke of Mecklenburg-Schwerin, and the Mecklenburg Estates. This led to several interventions by the imperial authorities (including the dispatch at one point of a Hanoverian army to the province and, later, the actual removal of the Duke from his office) before it was eventually settled in the 1730s. In the countryside, the survival of serfdom on a wide scale continued to keep the bulk of the populace in thrall. (As late as 1766, Thomas Nugent still found

---

[1] Over the past hundred years, scholars have shown considerable uncertainty over Aepinus's given names. Some have written "Udalric" rather than "Ulrich"; others have had "Theodor" instead of "Theodosius"; some have gratuitously added "Maria" to the list. However, the principal eighteenth-century documents, including the baptismal record, agree with what is given here. There has also been a tendency to spell Aepinus's surname with a diphthong "Æ." Aepinus himself, however, did not do so.

much to criticize in regard to the condition of the peasantry, and also concerning the depressed state of agriculture in general.)[2] Recovery was quicker in the towns, but even in Rostock itself it was mid-century before the many waste spaces created by a fire as long ago as the 1670s were rebuilt with houses.[3]

Despite the downturn in economic affairs, life in Rostock in the early years of the eighteenth century continued much as it had done for hundreds of years. The town had always been dominated culturally by a small but tightly knit group comprising the merchant community, the professional and university people, and the local civic and court officials; and while this group was now perhaps not as large as it had once been, its members, including university professors such as the elder Aepinus, retained their superior social position. Their families were invariably well-educated and enjoyed for the most part a very comfortable if seldom over-luxurious standard of living. Such was certainly the case with the Aepinus family.

Rostock University had been founded in 1419. For the first three hundred years of its existence, its excellent reputation had attracted students and teachers from the whole of north Germany and the Baltic region. Early in the eighteenth century, however, at about the time the elder Aepinus was first appointed a professor, the university's standing suffered a sharp decline under the impact of the Northern War and then the disruptions brought about by the Duke's dispute with his Estates. During the century that followed, it remained an institution of merely local importance; almost all its professors and most of its students were Mecklenburgers, Rostock itself providing a large proportion of the total. Fully one quarter of those who held professorial office in the university during the course of the eighteenth century were themselves the sons of professors, and the fathers of most of the others were lesser academics attached to the university, or officials or merchants based in the town. Frequently the various families involved were further related by marriage.[4]

The Aepinus family was typical of this well-defined little social group. The father, Franz Albert Aepinus, had fitted quite naturally into it, for, although he was not a native of the town, his family had occupied a succession of official appointments in one or other of the north German principalities for many generations past, and his

[2] Thomas Nugent, *Travels through Germany . . . With a Particular Account of the Courts of Mecklenburg*, London, 1768, I, 278; II, 44–45.

[3] G. Kohfeldt, *Rostocker Professoren und Studenten im 18. Jahrhundert*, Rostock, 1919, p. 7.

[4] Ibid., pp. 6–14.

father had held a prominent administrative position in the district of Stargard, in Mecklenburg-Güstrow.⁵ The family proudly traced its ancestry to Johann Aepinus (1499–1553), one of Luther's leading supporters in north Germany during the Reformation, and the man who had changed the family name from the traditional Hoch or Hoek, deriving the new one from the Greek αἰπεινος ("lofty").⁶

Franz Albert Aepinus was a dominant figure in the life of Rostock University for almost half a century. Born at Wanzke, in Mecklenburg, in 1673, he studied at Rostock and then at Jena for some years before returning to graduate as *Magister* in the philosophical faculty of his home university in 1696. Thereafter he taught at Rostock as a *Privatdozent* (that is, as a private lecturer associated with the university) for a number of years. In 1710 he was awarded the degree of Doctor of Theology, and in 1712 was appointed *professor extraordinarius* of logic.⁷ In 1721, he succeeded to the chair of theology, which he retained until his death in 1750. Over the years he was also active in the administration of the university, serving as rector nine times and as dean of the faculty on no fewer than twenty-one occasions. At the same time he was appointed to a series of higher positions in the Church, culminating in his nomination in 1733 to the bench of the important ecclesiastical court based on Rostock. In this office, as in the others that he held from time to time, he performed his duties with considerable distinction. He was, it seems, a very hard worker. He was also a popular teacher, and a throng of students regularly gathered to hear his lectures. Many of these he also provided with board and lodging. On doctrinal matters he defended the tenets of orthodox Lutheranism against the inroads of both pietism and mysticism, and later of rationalism as well. He maintained the traditional approach in his philosophical teaching, too, in opposition to the new ideas associated with the name of Christian Wolff that were then sweeping the universities of Germany. In defense of his views he published a large number of addresses and dissertations, as well as several popular textbooks.⁸

⁵ Hermann Becker, ed., *Sacrum piis manibus et exequiis viri illustris atque summe reverendi Domini Franc. Alberti Æpini . . . sanctum,* Rostock, 1750, pp. IV–VI.

⁶ Ibid., p. v.

⁷ The *Allgemeine deutsche Biographie,* Leipzig, 1875–1912, I, 128, says that F. A. Aepinus became headmaster of the cathedral school in Ratzeburg, another important Mecklenburg town, in 1709. In fact, however, while he was offered the position, he did not take it up, but recommended another person for it (Becker, *Sacrum piis manibus et exequiis . . .* , p. XI).

⁸ Becker, *Sacrum piis manibus et exequiis;* Kohfeldt, *Rostocker Professoren,* pp. 26–27. A full listing of the elder Aepinus's publications is given in Johann Christian

In September 1715, Franz Albert Aepinus married Agnesa Dörcksen, the daughter of an official of the local ducal court. Three children of the marriage survived childhood (one daughter died in infancy, another at the age of five), and the careers mapped out for them well illustrate the close-knit nature of the Rostock academic community of the day. The two sons, Franz Ulrich Theodosius and his older brother Angelius Johann Daniel Aepinus, were both clearly pointed toward academic careers from the start. Angelius received his early education at home from a succession of private tutors; in 1732, at the age of fourteen, he was admitted to the university; in 1739 he graduated *Magister* and then taught as a *Privatdozent* until his appointment in 1746 to the position he occupied for the rest of his life, professor of rhetoric.[9] When he chose a wife, Angelius Aepinus settled on Anna Christiana Eschenbach, a member of a family that contributed three professors to the university in the course of the century. Franz Ulrich Aepinus's education was very similar to that provided for his older brother. He, too, then taught in the university as a *Privatdozent* for a number of years and would undoubtedly have succeeded to a professorship in due course had he not moved on to higher things elsewhere. So far as is known, he never married. His surviving sister, Francisca Agnesa Beata Aepinus, completes the pattern; in 1744 she married Johann Peter Becker, a pastor in one of the local churches and for a time a lecturer in mathematics in the university, son of the long-time (1697–1753) professor of mathematics Peter Becker, and member of a family that, like the Eschenbachs, could boast other Rostock professors as well during the century.

Within this sedate provincial environment, then, Aepinus grew up. Along with his older brother and sister, he was reared in conditions of some material comfort—the family could afford to employ a manservant in addition to the mandatory serving girl [10]—but their father neither approved nor permitted extravagance or luxury. On the other hand, he did believe in providing his children with the best education available in the town, and therefore employed a succession of private tutors for them. His procedure here was again typical of the eighteenth-century Rostock professoriate. Most of the other professors, too, chose to have their families receive most if not all of their early

---

Burgmann, *Memoriae posteritatis perennaturae, insigne gloriae monimentum, quod sibi, dum vixit, statuit ipse, Vir maxime venerabilis . . . Dominus Franc. Albert. Aepinus . . .*, Rostock, 1750.

[9] J. C. Koppe, *Jetztlebendes gelehrtes Mecklenburg*, Rostock/Leipzig, 1783–84, I, 1–9; article "Aepinus, Angelius Johann Daniel."

[10] Kohfeldt, *Rostocker Professoren*, p. 144.

education at home at the hands of private tutors rather than at the town school. These tutors were usually students or newly graduated *Magisters* in need of some extra income; occasionally even one of the less well-paid professors might be thus employed. The name of only one of Franz Ulrich Aepinus's tutors has come down to us, that of a *Magister* called Dankert who subsequently became headmaster of the town school at Neubrandenburg, sixty miles southeast of Rostock. Of the rest, we are told merely that they were "skilled" (*geschickt*).[11] Aepinus matriculated into the university in October 1736, two months before his twelfth birthday.[12] However, this was only a formality (arranged perhaps for sentimental reasons connected with the fact that his father was serving as rector during that particular term), and it was not until 1740 that he actually began attending lectures.[13] The father had evidently cherished hopes that his son would follow in his footsteps and become a theologian, for in the original entry in the university's *Matrikelbuch* the boy was described as a theological student. As time went by, though, it became clear that his true vocation was not theology but mathematics. This presented something of a problem, because mathematics was as always taught in the junior faculty of the university, the philosophical faculty, and Aepinus's father believed strongly that his son should not be content with an enrollment at this level, but should enroll in one of the three higher faculties—theology, medicine, or law—as well. A choice clearly had to be made, and the lad chose medicine since it was the area where the work was most closely related to his mathematical and scientific interests. His father was satisfied, and the entry in the *Matrikelbuch* was altered accordingly.[14]

Mathematics remained, however, the young Aepinus's great love; indeed, we are told that, even before he began his university studies, he had worked his way conscientiously through Christian Wolff's introductory Latin text on the subject, his *Elementa*.[15] Unfortunately, this report is not precise enough for us to be certain which book it was that Aepinus studied so diligently. A list of Wolff's publications covering the period [16] suggests that the text in question must have been the *Elementa matheseos universae*. This first appeared as a two-

[11] Koppe, *Jetztlebendes gelehrtes Mecklenburg*, I, 10; article "Aepinus, Franz Ulrich Theodosius."
[12] A. Hofmeister, *Die Matrikel des Universität Rostock*, Rostock, 1904, IV, 194.
[13] Koppe, I, 10.
[14] Ibid.; also Hofmeister, IV, 194.
[15] Koppe, I, 10.
[16] J. H. Zedler, ed., *Grosses vollständiges Universal-Lexicon aller Wissenschaften und Künste*, Leipzig/Halle, 1732–50, LVIII, cols. 549–677.

volume work (1713–1715), but subsequently it was republished in a considerably expanded five-volume edition (1730–1741), by which time it embraced not only the elements of mathematics per se but also mechanics, optics, astronomy, geography, architecture, and several other subjects. It seems unlikely that Aepinus could have worked his way through such a bulky compendium at so early an age; thus we may conclude either that he used the earlier, unexpanded edition of the book or that the report refers to only the first few chapters of the complete work, those dealing with the elements of mathematics proper. Either of these alternatives would have been enough to make Aepinus a most unusual student in an age when the school curriculum leading up to university entrance generally amounted to imparting a working knowledge of Latin and the catechism and very little else.

At the university, Aepinus pursued his mathematical interests chiefly under the guidance of the *Magister* L. F. Weiss rather than under the professor of mathematics, Peter Becker, by then a very old man who no longer lectured on a regular basis (though he did take an interest in Aepinus and helped him from time to time with his studies).[17] In philosophy, Aepinus attended the lectures of his father and his older brother, the former of whom, at least, would certainly have approached his subject from the traditional point of view. New ideas were abroad, however, most of them associated with the name of Wolff; and since the tutor Weiss was one of those mainly responsible for introducing Wolff's doctrines into Rostock,[18] Aepinus undoubtedly acquired from him a knowledge of the most recent philosophical trends in addition to his mathematics.

The lectures he attended in the medical faculty constituted Aepinus's formal introduction to the study of natural science. Later in life his own scientific work was characterized by a vigorous Newtonian empiricism, and it is therefore perhaps significant that the medical professor from whom he gained most of his early knowledge, G. C. Detharding (1699–1784), had not only learned his medicine in the bracingly empirical air of Holland but had also spent some time in England and had actually met Newton while he was there.[19] However, the extent and nature of Detharding's influence on Aepinus is unknown, as indeed is the influence of his other teachers in Rostock, since no information has come down to us about these matters.

[17] Koppe, I, 10. In the university's *Catalogus lectionum* for 1742, it is expressly stated that Professor Becker no longer gave lectures, but that his son would offer them in his stead.
[18] Kohfeldt, *Rostocker Professoren*, p. 47.
[19] Ibid., p. 43.

In an earlier age it had been the custom for young men to spend only part of their student days at their home university before embarking on an extended study tour to the principal centers of European intellectual life. By the eighteenth century this practice was greatly weakened, and when Rostock students of the period went on tour they usually did so on a much more restricted basis, visiting only one or two middle and north German schools and some of the more important libraries and museums in the area. When Aepinus's turn came, he embarked on a tour that was altogether typical for the time. Jena was his chief port of call, as it had been for his father and for his older brother, and as it was for most of his Rostock contemporaries. He matriculated into the university there on 25 April 1744 [20] after traveling directly from Rostock, and the only further travels he undertook before returning home two and a half years later were a number of brief trips to such nearby centers as Weimar and Erfurt, and one slightly longer journey to Halle.[21]

At Jena, Aepinus spent most of his time following the lectures on mathematics, physics, and the various medical sciences of the well-known teacher G. E. Hamberger (1697–1755), for many years one of the principal ornaments of the university and at that particular period its professor of botany, anatomy, and surgery. Two thick books of notes taken by Aepinus from Hamberger's lectures have survived from this period, one dated 1744, the other 1745.[22] Unfortunately these deal not with physics and mathematics, but with the subjects of his medical course, namely physiology, pathology, chemistry both theoretical and experimental, and materia medica. They are thus not particularly informative about what turned out to be the central features of Aepinus's scientific education. However, some interesting points emerge.

The notes from Hamberger's physiology lectures, for example, adopt a consistently iatromechanical line. They are thorough but wear a slightly dated appearance, the authorities cited generally being late seventeenth- rather than eighteenth-century figures. Aepinus obviously sensed this weakness at the time, and he remedied the situation by using in conjunction with his lecture notes an up-to-date (indeed, *the* most up-to-date) textbook on the subject, Albrecht von Haller's annotated edition of Herman Boerhaave's lectures on physiology, which had been published not long before under the title *Praelectiones academicae, in proprias institutiones rei medicae*

---

[20] Personal communication from the Director, Universitätsbibliothek Jena.
[21] Koppe, I, 11.
[22] AAN, raz. v, op. э–7, nos. 22, 23.

(Göttingen, 1739–1744). If Boerhaave's opinion differed from that of Hamberger on a particular point, Aepinus recorded this in his notes and added a reference to the appropriate page of Boerhaave's work.

Boerhaave's influence is not so evident in the chemistry notes, despite the esteem in which his textbook on that subject was held at the time. More surprisingly, perhaps, the opinions of the renowned early eighteenth-century German chemist G. E. Stahl appear to have been entirely without influence, despite Jena's proximity to Halle, where Stahl taught. Phlogiston, Stahl's most notorious contribution to chemical theory, is indeed mentioned at one point, but the general approach adopted is far removed from Stahl's traditionalist emphasis on the qualitative features of reactions. Rather, a mechanistic chemistry is displayed in which the phenomena are consistently discussed from a physical point of view, in terms of the particles thought to be involved. Often it is a supposed difference of specific gravity between one kind of particle and another that is presented as the crucial factor in the situation, although calcination (to take one important example) is seen as a purely mechanical conversion of a non-pulverizable body into a pulverizable one. The general tone of the notes is more reminiscent of the previous generation of chemists, men such as Robert Boyle, than it is of the most important chemists of Hamberger's own day; and this is the case even though more recent discoveries such as Geoffroy's table of affinities are presented in due order as the discussion proceeds. Nevertheless, a large number of individual reactions are also described in the standard "recipe-book" fashion of the time, with much less emphasis on the mechanisms that might be involved than is the case in the earlier, more theoretical sections of the notes. From this part of the course, Aepinus would have received a good grounding in chemical practice, while from the earlier sections he would have become thoroughly familiar with a fairly representative version of the physicalist approach to chemical theory so widespread among the philosophers (as distinct from the practicing chemists) of his day.

But the contributions that Aepinus subsequently made to science were chiefly in mathematical physics, not in chemistry or physiology. What kind of training did he receive at Jena in the field that he was later to make his own? In the absence of his physics lecture notes, we turn to Hamberger's published textbook on the subject,[23] since this would almost certainly have been the text used in his classes. Even if this was not the case, Aepinus soon afterward used Hamberger's book as the text in one of the first courses he taught following his return

[23] G. E. Hamberger, *Elementa physices, methodo mathematica in usum auditorii conscripta,* Jena, 1727; 3rd ed., Jena, 1741.

to Rostock.[24] Somehow or other, then, he must have become well versed in its contents during his time away.

Hamberger's book is a long and comprehensive introduction to the physics of his day. I have indicated already that Aepinus's later scientific work was strongly Newtonian in outlook, so it is interesting to find that, even though Newton is cited a number of times in Hamberger's first chapter, "De Motu," the foundations of mechanics are in fact set out here in a straightforwardly Leibnizian, not Newtonian, form. For example, Hamberger asserts at the outset that all space is filled with matter. As usual with those who uphold this doctrine, he then invokes antiperistasis to explain how motion is nevertheless possible. Different bodies can have different specific weights because not all matter gravitates, and the proportion of gravitating to non-gravitating matter can vary from one body to another. (Newton, on the other hand, was an atomist, and believed that differences in specific weight arose from the varying proportions of matter to void in bodies.) Most revealing of all, the characteristics of any given body are determined not only by the sizes and shapes of its parts but also by the internal force (*vis*) with which these parts are endowed; "the nature of body," Hamberger tells us, "consists in the disposition of things (*entia*) to determined effects," and "the corporeal universe is a complex of *entia* endowed with force." [25]

It was of course the notion, displayed here so prominently, of force as a dispositional property inherent in a body that was the most distinguishing feature of Leibniz's dynamics. Newton, by contrast, held that matter was essentially inert, as Descartes had supposed it to be; a body could change its motion if a force was impressed *on* it by something else, but it was simply incorrect, in his view, to speak of the force *of* a body in the way that Leibniz did. Hamberger discussed the Englishman's ideas at one point, reporting (correctly) how he ascribed change of motion in bodies to the action of impressed forces, whether of the impulsive or continuously acting variety. But from this he felt able to conclude immediately, in typically Leibnizian fashion, "therefore an impressed force is not a new force translated into a body, but only a new determination, through occasional causes of motion, of that [force] which was created by God for the body when he created the body"; and in a scholium he added that "indeed it cannot even be thought that force can go from one body into another." [26]

[24] F.U.T. Aepinus, *De curvis, in quibus corpora gravitate naturali agitata . . . descendunt . . .* , Rostock, 1747, p. [2].
[25] Hamberger, *Elementa physices* (1741), p. 9.
[26] Ibid., p. 15.

Other Leibnizian orthodoxies can be cited. Space and time are defined, for example, in relational rather than absolute terms. The principle of sufficient reason is presented, in the form "nothing can be or come to be, in this place or in that, without a sufficient reason, and vice versa." [27] Of course Hamberger comes down on Leibniz's side in the well-known *vis viva* controversy. Nevertheless, as the work proceeds through Chapter II, on the equilibrium of fluids and solids, and Chapter III, on the cohesion of bodies, a gradual shift in emphasis can be detected. Particularly in the discussion of cohesion, Hamberger has more and more frequent recourse to experimental evidence to support his arguments. This trend continues in the later chapters of the book, which deal in turn with the physical aspects of fire, air, water, and earth, and the various combinations of these—under which headings Hamberger manages to deal with all the usual branches of physical inquiry—and finally "the system of the world." In large measure, then, the book is a treatise of experimental physics, and despite its Leibnizian orientation its approach is similar to the texts being published at about the same period in Holland and in England by 'sGravesande, Musschenbroek, and Desaguliers; [28] it could in fact be justly described as the German equivalent of these. Hamberger is careful to acknowledge his sources, and we find frequent references to all the best-known experimentalists of the age—Musschenbroek, 'sGravesande, Desaguliers, Sturm, Hartsoeker, Newton (especially his *Opticks*), Boyle, Mariotte, Réaumur, and so on. In all, Hamberger's presentation is thorough, wide-ranging, and for the most part reasonably up-to-date. Any student who worked his way conscientiously through the book would have gained a broad acquaintance with all the chief questions of eighteenth-century physics.

Aepinus probably used the third edition of Hamberger's text, published in 1741. If so he would have been confronted at the outset with a long preface, new to this edition, in which Hamberger set out his views on scientific method. Here, as with the introductory chapters of the text itself, there is a striking contrast between the views Aepinus would hold in later life and those of his teacher. Aepinus would become the person chiefly responsible for the formulation of electrical

[27] Ibid., p. 12: "Nihil hic vel illic tale esse aut fieri potest, sine ratione sufficiente, et contra."

[28] W. J. 'sGravesande, *Physices elementa mathematica, experimentis confirmata* . . . , Leyden, 1720. Pieter van Musschenbroek, *Epitome experimentorum physicomathematicorum in usus academicos,* Leyden, 1726; 2nd ed. entitled *Elementa physicae,* Leyden, 1734. J. T. Desaguliers, *A Course of Experimental Philosophy* . . . , London, 1734-44.

and magnetic theory on the basis of Newtonian-style forces acting in some wholly unexplained manner at a distance between particles of matter. By contrast, Hamberger's preface aimed precisely at demonstrating the absurdity of having any recourse to forces acting at a distance in scientific theorizing! All such forces between bodies, Hamberger argued, must be due to the action of some intermediary body; otherwise the notion would be self-contradictory, because to say that true action occurred at a distance would be to say that an effect could have a proximate cause (for that was what a force was, in Hamberger's view) that was not proximate.[29] If some material agency was necessarily involved as an intermediary, Hamberger went on, we would be better advised to have recourse to this in our explanations from the start, even in total ignorance of the way it brought about the effect in question, than we would be to invoke some kind of action at a distance that we *know* cannot occur.[30]

When he came to consider the "system of the world," Hamberger followed his own advice and attributed the force holding the planets in their orbits about the sun to a centripetal pressure exerted on them by the ambient aether which, he supposed, filled all the parts of space not occupied by ordinary bodies.[31] Interestingly, however, Hamberger did not invoke a Cartesian-style aethereal vortex to account for this centripetal pressure, but relied instead simply on density gradients. When God created the world, Hamberger suggested, He might have arranged things in such a way that the aether was denser (and hence, in Hamberger's terms, endowed with greater innate force) nearer the center than it was further out, and if so, this would suffice to explain the effect. Elaborating a little, Hamberger even attributed the earth's daily rotation on its axis to an "aether drag" deriving ultimately from this same density gradient![32] Apart from additions such as this, however, Hamberger's aethereal mechanism bore a certain resemblance to the one invoked by Newton himself in the queries that he added to the second (1717) English edition of his *Opticks*[33] (except that, for Newton, the density of the aether was supposed to increase, not decrease, with increasing distance from a center of gravitational force). However, where Newton spoke of the "elasticity" of the aether, Ham-

[29] Hamberger, *Elementa physices* (1741), Preface, p. 53.
[30] Ibid., Preface, pp. 63, 69.
[31] Ibid., pp. 577–80.
[32] Ibid., pp. 580–81.
[33] Isaac Newton, *Opticks, or a Treatise of the Reflections, Refractions, Inflections and Colours of Light*, New York, 1952 (based on 4th ed., London, 1730), pp. 347–54. See especially Query 21 (pp. 350–51) for the explanation of gravity in terms of varying aethereal densities.

berger used the term "innate force," which to his Leibnizian taste was methodologically much more acceptable.

Hamberger would therefore have tended to impose upon Aepinus's early education as a physicist a basically Leibnizian outlook, but one that carried with it considerable emphasis on experimental investigation. His, however, was not the only influence on Aepinus's training as a scientist. At least as important was the influence of the famous Swiss mathematician Leonhard Euler (1707–1783).

Euler was at this period a dominant figure in the newly reformed Prussian Royal Academy of Sciences and Belles-Lettres in Berlin, having moved there in 1741 from St. Petersburg, where he had first established his reputation. We cannot be certain when Aepinus first became acquainted with his work, nor is it known which of his writings most influenced the young man. However, in one of Aepinus's earliest publications, dating from his period as a *Privatdozent* at Rostock, he refers quite casually [34] to the development of certain mathematical paradoxes concerning the infinite in Euler's treatise *Mechanica* (1735). Hence by December 1751 at the latest, when Aepinus wrote the tract in which these remarks appear, he had become thoroughly familiar with that epoch-making work. Not long afterward, Euler came to exert a stronger influence on the young Rostocker than he could ever have done through his writings alone, for Aepinus moved to Berlin in March 1755 to work in the Academy of Sciences, and he spent the next two years in close, almost daily, personal contact with the master. For some years after that, during Aepinus's early years in St. Petersburg, the two occasionally corresponded.

Euler's influence, both through his writings and in person, must have been wholly beneficial to the young Aepinus. Euler would have strongly reinforced Aepinus's confidence in the value and importance of detailed mathematical studies, but at the same time he would have done nothing to dampen any enthusiasm for experimental investigation that might have been aroused by Hamberger's lectures. Euler was, however, resolutely opposed to the Leibnizian approach to science. Not for a moment would Euler accept that matter could possess an internal dynamism of the kind postulated by the Leibnizians. As a devout Calvinist, he viewed such a notion as theologically unacceptable, and as a scientist, he regarded it as simply mistaken.[35] He

[34] Aepinus, *Demonstrationes primariarum quarundam aequationibus algebraicis competentium proprietatum*, Rostock, 1752, p. 11.

[35] Euler's convictions in this regard are clearly revealed in many places throughout his writings. To cite but one example, see his *Letters to a German Princess*, trans. Henry Hunter, London, 1802, I, 292–307.

insisted over and over again that the chief characteristic of matter was its *inertia,* whereby it would remain, in the absence of external agencies, in whatever state it happened to be, whether this was one of rest or of uniform motion in a straight line. To change this state, Euler maintained, some *external* force had to act on the body, and the business of mechanics was to study the connections between these forces and the changes they produced.[36]

In large measure, then, Euler's approach to science was much akin to Newton's; he based his whole system of mechanics on an explicit notion of forces acting on bodies that were themselves inert, in the same way that Newton had done. Given this, there is a second respect in which his influence on Aepinus would have been extremely important. Euler shared none of Hamberger's reservations about invoking forces that acted at a distance in his physics. True, he agreed with Hamberger (and with the followers of Descartes) that such forces must in every case derive ultimately from some material cause, presumably some action of the extremely subtle aether that in his opinion filled all the spaces surrounding bodies. But in the absence of any knowledge of the way these causal mechanisms operated, Euler was perfectly prepared to talk in terms of the forces themselves. Aepinus was happy to follow his example, and indeed to do so with rather more licence than Euler himself would allow. A few years later he developed a theory of electricity and magnetism based on a multiplicity of action-at-a-distance forces whose supposed joint occurrence in nature made it extremely implausible that there was some underlying aethereal mechanism giving rise to them all. When Aepinus's theory was published, Euler criticized the inclusion in it of "arbitrary" forces acting at a distance;[37] but it was presumably only to these explanatory difficulties that he alluded, since Aepinus's procedure in other respects conformed closely to what he himself had done elsewhere. It was in fact almost certainly Euler rather than anyone else who had first inspired Aepinus to approach nature in the way he did.

One final aspect of Aepinus's scientific upbringing remains to be considered, and that is the training he received in mathematics. Very little direct information has come down to us in this regard beyond the problematic assertion that he worked his way through Wolff's *Elementa* at an early age. We know too that most of his early lessons in Rostock were at the hands of the *Magister* L. F. Weiss, and that

[36] Euler, *Opera omnia,* ser. 2, v, 112.
[37] Euler to G. F. Müller, 10 January 1761; A. P. Yushkevich et al., eds., *Die Berliner und die Petersburger Akademie der Wissenschaften im Briefwechsel Leonhard Eulers,* Berlin, 1959–76, I, 166.

the old professor, Peter Becker, also took an interest in his progress. At Jena, Aepinus studied mathematics under Hamberger as well as physics and the subjects of his medical course. Following his return to Rostock, he studied a series of advanced texts at the same time as he earned a living teaching the elements of the subject to a number of private pupils.[38] As his text for these lessons he chose J. A. Segner's *Elementa arithmetices et geometriae in usus auditorum* (Göttingen, 1739), but this is an elementary work and tells us nothing about the extent or the sources of Aepinus's own knowledge at this time.[39] To obtain such information, our best recourse is the not altogether reliable one of seeing which works Aepinus cited in his early publications.

Not surprisingly, it was to Wolff's work that Aepinus referred most frequently in his first writings. When he did so, he referred invariably to Wolff's "Elem. analys. finit. & infinit.," that is, the chapter on analysis from the *Elementa matheseos universae*. As time went by, however, Aepinus's work quickly began to show the benefit of the course of advanced reading he had undertaken, and by 1754, when he published a discourse on the concept of negative numbers (a favorite topic in eighteenth-century mathematics texts) as an advertisement for the lectures he was to give in the coming term, he was able to refer in familiar terms to most of the leading authors of the age.[40] Among these, Leibniz, whom Aepinus described unequivocally as the inventor of the calculus, enjoyed pride of place (no mention was made of Newton's contribution to this subject nor of the disgraceful disputes over priorities in this regard that had so wracked the mathematical literature of the previous generation). In connection with the calculus, however, Aepinus did emphasize Leibniz's failure to establish his system upon sufficiently rigorous foundations before simply asserting that the conceptual difficulties involved could be made to vanish if they were approached by means of the Archimedian method of quadratures and the method of exhaustion built upon this by Gregory of St. Vincent.[41] Among the works Aepinus mentioned once he turned to the particular theme of his discourse were such classics as the

[38] Koppe, *Jetztlebendes gelehrtes Mecklenburg*, I, 10–11.

[39] Aepinus, *De curvis, in quibus corpora . . . descendunt . . .* , p. [2]. A set of notes taken from Aepinus's lectures on Segner's text in 1752 by his most important pupil, Johan Carl Wilcke, is preserved in Stockholm, along with two further sets of notes from Aepinus's lectures, one on the geometry of conic sections, the other on algebraic equations of higher degree (Kungl. Vetenskapsakademiens Bibliotek, Stockholm; MS Wilcke, J. C. 20, 17, 21, respectively).

[40] Aepinus, *Commentatio de notione quantitatis negativae*, Rostock, 1754.

[41] Ibid., p. 4.

Geometry of Descartes and the *Arithmetica infinitorum* of John Wallis, together with L'Hospital's *Analyse des infiniment petits* (1696), a paper by Varignon in the *Mémoires* of the Paris Academy of Sciences,[42] Guido Grandi's *De infinitis infinitorum* (1710), Fontenelle's *Élémens de la géométrie de l'infini* (1727), Newton's *Arithmetica universalis* (in the Leyden edition of 1732), and several lesser works. Elsewhere in Aepinus's early writings, as we have noted already, Euler's *Mechanica* was cited, and so too was Daniel Bernoulli's *Hydrodynamica* (1738),[43] as well as papers by Maupertuis and Clairaut on the solving of certain classes of differential equations.[44] Granted that a single citation does not prove that Aepinus had made a close study of the whole of any particular work, it is nevertheless clear simply from the extent of this list and the importance of the works he cited, as well as from the evident familiarity with which he mentioned individual ones among them, that he had by the early 1750s at the latest become thoroughly immersed in the mathematical literature of his day. If there is a surprising omission from this list, it is perhaps Euler's classic *Introductio in analysin infinitorum* (1748), although this can no doubt be explained by the fact that the book had been published only a short time before. Aepinus did cite this work in another paper that he composed just a little later in his career, and in a way that makes it clear that he had, by then at least, worked his way through it very carefully.[45] By that time, however, he had spent two years working in close contact with the master himself, an experience that must have been of even greater significance for his mathematical development than it was for his physics.

This account of Aepinus's scientific education has of necessity taken us somewhat away from a strictly chronological record of his career.

[42] Varignon, "Sur les espaces plus qu'infinis de M. Wallis," *HAS*, 1706, Mém. pp. 15–23.

[43] Cited in Aepinus, *Meditationes de causa et indole febrium intermittentium*, Rostock, 1748, p. XII.

[44] Maupertuis, "Sur la séparation des indéterminées dans les équations différentielles," *HAS*, 1731, Mém. pp. 103–109; Clairaut, "Sur l'intégration ou la construction des équations différentielles du premier ordre," ibid., 1740, Mém. pp. 293–323. Both papers were cited in Aepinus, *De integratione, et separatione variabilium, in aequationibus differentialibus, duas variabiles continentibus, commentatio*, Rostock, 1755, pp. XII and VI, respectively.

[45] Aepinus, "De functionum algebraicarum integrarum factoribus trinomialibus realibus commentatio," *Novi commentarii Academiae Scientiarum Imperialis Petropolitanae*, VIII, 1760–61, 182. The paper was actually submitted to the St. Petersburg Academy rather earlier than its date of publication would indicate, in December 1758 (*Protokoly zasedanii konferentsii Imperatorskoĭ Akademii Nauk s 1725 po 1803 goda*, St. Petersburg, 1897–1911, II, 418).

INTRODUCTION

To pick up the thread of that story again, Aepinus arrived in Jena in April 1744. He returned home to Rostock for the start of Michaelmas term in 1746, and presented himself forthwith for examination for the degree of Doctor of Philosophy, the essential prerequisite to teaching in the University as a *Magister*.[46] In conjunction with this examination, a candidate was normally required to prepare a thesis and defend it publicly, but in Aepinus's case, as in some others from around this period, this procedure was not followed. Instead, once Aepinus had successfully completed the examination and had thereby satisfied the members of the faculty of his competence, he was proclaimed Doctor of Philosophy immediately, on 5 February 1747, and given his licence to teach, without ever submitting a thesis. He formally graduated two months later, on 17 April 1747, with all the usual pomp and ceremony and amid all the festivities that normally marked such occasions. Simultaneously Aepinus presented to the faculty for approval a "programma" announcing the course of lectures he intended to give in the following term, but designed chiefly to display the talents he would be bringing to those lectures. The faculty's approval was forthcoming, and the work was promptly published under the title *De curvis, in quibus corpora gravitate naturali agitata, ea lege descendunt, ut quantitatem descensus metiatur quaevis potestas temporis.*[47]

For the next few years, Aepinus lectured regularly in the university on mathematics and natural philosophy. At the same time he continued his own mathematical studies in depth. He also pursued the study of modern languages in concert with his older brother (in whose house he lived after the death of his father in 1750).[48] Later in life he showed himself able to read and write French fluently, so this was evidently (as might have been expected in any case) one of the languages he studied. He probably learned to read English at this time, too, if not necessarily to speak it. Soon afterward, when he and his student Wilcke were carrying out their important electrical researches during their sojourn in Berlin in the mid-1750s, they used the English original of Benjamin Franklin's *Experiments and Observations on Electricity, made at Philadelphia in America* rather than the French translation,[49]

---

[46] Koppe, *Jetztlebendes gelehrtes Mecklenburg*, I, 11.

[47] Universitätsarchiv, Rostock; Rostock University, Philosophical Faculty, *Protocollum* for Easter term, 1747. Aepinus's pamphlet has often been said in the secondary literature to have been his doctoral thesis, but this is incorrect.

[48] Koppe, *Jetztlebendes gelehrtes Mecklenburg*, I, 11.

[49] Wilcke's German edition (1758) of Franklin's book was explicitly stated to be "translated from the English."

while only a little later again, Aepinus managed to sustain a quite successful correspondence with the English electrical investigator Benjamin Wilson, even though the latter probably wrote all his letters in his native tongue. Certainly in Rostock Aepinus would have had the opportunity to learn English, because this, along with French and Italian, was one of the languages taught by the man under whom he and his brother studied, one J. C. Schreiber.[50]

Besides these activities, Aepinus also spent some time during his first year of teaching preparing himself for the Rostock medical faculty's doctoral examination and composing a thesis to defend in conjunction with this. He did so, not because he had any desire to practice medicine, but simply to please his ailing father and to demonstrate publicly that he had not neglected his studies in this field during his time away at Jena. He successfully defended his thesis in April 1748, but, perhaps to save himself the considerable expense involved in what would have been, for him, a quite useless ceremonial, he never formally took out his degree; hence he continued to be characterized as "medical candidate" in his formal publications in Rostock even after this date. Nor did he ever undertake the course of clinical training in a hospital that would have been necessary before he could have practiced medicine at all successfully. His thesis, like his medical course as a whole, was highly theoretical in its approach; it amounted merely to a straightforward extension of Boerhaave's overall "fluids-and-fibers" physiological theory to account for the symptoms associated with "intermittent fever" (i.e. malaria).[51]

Aepinus was apparently a successful teacher, since he attracted a number of talented students to his classes. The most notable of these was Johan Carl Wilcke (1732–1796).[52] Wilcke's father, Samuel Wilcke, had been for some years during his student days in Rostock one of the tutors employed by Professor Aepinus to look after the education of his elder son, Angelius. Subsequently, he had been appointed pastor of the German church in Stockholm, and his son Johan Carl was brought up in the Swedish capital. The latter began his university studies in Sweden, at Uppsala, early in 1750, but eighteen months later he transferred to Rostock. His father's intention was that he study under Angelius Aepinus, by now professor of rhetoric, in preparation for a career in the Church. In Rostock, Wilcke lived in Angelius

[50] Kohfeldt, *Rostocker Professoren,* p. 89.

[51] Aepinus, *Meditationes de causa et indole febrium intermittentium,* Rostock, 1748.

[52] For what follows I have drawn heavily upon C. W. Oseen, *Johan Carl Wilcke, Experimental-Fysiker,* Uppsala, 1939.

Aepinus's house, and there he quickly became friends with his fellow-lodger, the professor's mathematically inclined younger brother. For the next two years, until he transferred to the university at Göttingen, Wilcke attended Franz Ulrich Aepinus's classes regularly, as well as those of his host. To his father's dismay, he soon abandoned all plans for a life in the Church, and resolved instead upon a career in science.

Aepinus's success as a teacher was undoubtedly due in part to that orderliness and clarity of thought that is found in all his publications, including those from this early period. It was also due in part to his having taken the trouble to acquire a good supply of scientific and mathematical instruments, for demonstration purposes as well as for use in his own research.[53] In addition, Aepinus was careful to keep his name before the public by means of a judiciously spaced series of publications, some of these being separate pamphlets, others being articles in a journal edited locally by his brother Angelius, the Rostock *Gelehrte Nachrichten*. The majority of these publications dealt with mathematical subjects of one kind or another, but, beginning in 1753, they also began to reflect the interest in observational astronomy that Aepinus developed at this period—an interest that soon afterward was to lead to a radical transformation in his whole way of life.

Aepinus's astronomical studies began modestly enough, with a regular program of observations undertaken in conjunction with Peter Becker from an observatory set up in the latter's house, the pastor's house attached to St. Jacob's Church, next door to the university.[54] (Aepinus's sister and her husband, Professor Becker's son Johann Peter Becker, who had succeeded his father as pastor at the church, would also have been living in the house at this period.) Most of Aepinus's observations were quite routine, with one major exception. On 6 May 1753, one of Mercury's relatively rare transits across the face of the sun occurred, and Aepinus went to some lengths to observe this as accurately as possible from a specially prepared position in the tower of St. Jacob's Church. The results he obtained, though somewhat restricted by cloud, were excellent, and he quickly made them known by publishing them in his brother's journal and by passing them on to both Euler in Berlin and J.-N. Delisle in Paris.[55] Probably he also

[53] Koppe, *Jetztlebendes gelehrtes Mecklenburg*, I, 11.
[54] Ibid., p. 12.
[55] Aepinus, "[Nachricht vom Durchgang des Merkurs]," [*Rostock*] *Gelehrte Nachrichten*, 1753, pp. 210–13, 489–92. Euler's copy is preserved, along with his other surviving correspondence from Aepinus, in AAN, fond 136, op. 2, no. 5. Aepinus's letter to Delisle, and a copy of Delisle's reply, are at the Observatoire, Paris, MS B. 1.7, nos. 248 and 251, respectively. Aepinus's response to Delisle's letter is at Archives de France, MS Marine 2 JJ 68.

sent a copy to Göttingen, where Tobias Mayer, the greatest German astronomer of the age, was at that time a professor; not long afterward, at any rate, both Mayer and J. A. Segner, one of Mayer's colleagues and the man whose mathematics textbook Aepinus used in his classes, seem to have been thoroughly familiar with his work. Indeed, Aepinus seems to have been corresponding regularly with both Mayer and Segner at this period, and both of them clearly held a high opinion of his talents as an astronomer.[56]

Aepinus's anxiety to make his observations known is but one of a number of signs of his growing ambitions during the 1750s. Evidently he felt that he had been a *Privatdozent* long enough, and that he should now be seeking a more established position in academic life. In the previous year, for example, when he published a short dissertation on a mathematical theme, he dedicated it to the Swedish Royal Academy of Sciences, even though he had not previously had any contact with that body; and he was then careful to send ten copies of the work to Stockholm, together with a flowery covering letter in explanation.[57] The Swedish Academy was not a body of salaried professional scientists like the academies in Paris, Berlin and St. Petersburg. Instead, like the Royal Society of London, it was very much an association of amateurs. In bringing himself to the Academy's attention, therefore, Aepinus could not have had in mind any possibility of being offered employment in Stockholm. Rather, prompted perhaps by Wilcke (who had then but recently arrived from the Swedish capital), he probably simply hoped to be offered foreign membership in the Academy. Any such international recognition of his worth would have greatly improved his prospects of finding a paying position closer to home. If Aepinus was thinking along these lines, his hopes were in vain, for the Academy made no response to his gesture.

Nothing daunted, Aepinus voyaged soon afterward to nearby Greifswald, ostensibly to get to know the professors in the university there and to inspect the scientific instruments they had at their disposal,[58] but in reality, one suspects, to make himself known (unsuccessfully as it turned out) as a candidate for any professorships that became vacant there. When Wilcke moved to Göttingen in 1753 he would certainly

[56] Mayer to Euler, 23 February 1755; Eric G. Forbes, ed., *The Euler-Mayer Correspondence*, London, 1971, p. 97. Segner to Euler, 27 November 1754; AAN, fond 136, op. 2, no. 3, l. 427.

[57] Aepinus, *Demonstrationes*, pp. 1–4. Aepinus to Pehr Wargentin and to the Swedish Academy collectively, 27 February 1752; Kungl. Vetenskapsakademiens Bibliotek, Stockholm: MS Wargentin, P., Brev; and Berg. brevs. 7: 717.

[58] Koppe, *Jetztlebendes gelehrtes Mecklenburg*, I, 12.

have carried news of Aepinus's talents with him. Soon, either through Wilcke's efforts on his behalf or on his own initiative, Aepinus had established a correspondence with both Segner and Mayer there. But his best prospects for a position would still have lain in Rostock, and he must have been bitterly disappointed when he did not succeed to Peter Becker's chair of mathematics following the latter's death in November 1753; the town council had previously promised the first chair that fell vacant to somebody else.

Aepinus did not remain disappointed for very long. Only twelve months later his prospects of academic success multiplied to an extent he could scarcely have dreamed possible. For some time past, the Berlin Academy of Sciences had been searching for a new professor of astronomy. Euler had been negotiating on the Academy's behalf throughout most of 1753 and 1754 with Tobias Mayer in Göttingen, and for much of that time his efforts seemed certain to be crowned with success. Mayer had every intention of moving, and even went so far as to promise his position at Göttingen to Aepinus.[59] In the end, though, in October 1754, he decided to stay where he was, and the Academy was forced to look elsewhere for its astronomer. Now Euler's eye, too, fell on Aepinus, who promptly received an invitation to visit Berlin. He accepted at once and the visit took place less than a month later, in December 1754. Euler was apparently pleased by what he saw, and arrangements for Aepinus to join the Academy were completed before the end of the visit. The young Rostocker returned home shortly before Christmas to complete his teaching obligations for the term and to wind up his other affairs in the town,[60] and, though this took a little longer than he had originally expected, he was ready to leave again soon after presenting a farewell dissertation in the university auditorium early in March 1755.[61] He was in Berlin before the end of the month and, once some initial difficulties with housing had been overcome,[62] was ready to begin work. On 10 April he attended a meeting of the Academy as a visitor. Euler, acting on behalf of the absent president, Maupertuis, formally proposed him for membership, and after the statutory week's delay, he was admitted "en qualité

---

[59] Forbes, *Euler-Mayer Correspondence*, pp. 64–93. Also C. L. Scheidt to J. D. Michaelis, 23 September 1754; Niedersächische Staats- und Universitätsbibliothek Göttingen, Handschriftenabteilung, Cod. MS. philos. 157. (I am indebted to Dr. Forbes for this reference.)

[60] These details are drawn from a series of letters from this period from Aepinus to Euler, now preserved in Leningrad; AAN, fond 136, op. 2, no. 5.

[61] Aepinus, *De integratione, et separatione variabilium*.

[62] Aepinus to Euler, 28 March 1755; AAN, fond 136, op. 2, no. 5, l. 15.

de Membre ordinaire dans la Classe de Mathématiques, et d'Astronome de l'Académie." [63]

As the Academy's official astronomer Aepinus was expected to devote part of his time to preparing the astronomical calendar the Academy published each year. Moreover, this was a duty he could not afford to neglect, because the Academy relied for much of its income on the proceeds from the sale of its calendars. Nor did he want for suitable equipment to do the job, because when Maupertuis had taken over as president some years previously, one of his earliest concerns had been to establish this side of the Academy's work on a sound footing. When Aepinus took charge of the observatory, he signed an inventory of the instruments placed in his charge. From this we learn that in 1755 the Academy owned eleven telescopes of one kind or another (several of these being provided with micrometers), twelve quadrants, five globes, six clocks, and a large number of less important items of observatory furniture.[64] Both Maupertuis and Euler believed, as did Tobias Mayer, that the observatory was by then well enough equipped to make significant contributions to the progress of astronomy.[65] One of Aepinus's first tasks in Berlin was to put the observatory's instruments in order, for they had lain unused for some time prior to his arrival. The competence with which he did this greatly pleased his fellow academicians, particularly Euler, who wrote to Mayer in most satisfied terms about the standard of what was being done.[66] Aepinus himself clearly enjoyed this kind of work, and over the years he went on to publish a number of articles suggesting ways in which various pieces of optical apparatus might be improved.

Apart from his duties in the observatory, Aepinus's new position made few formal demands upon him. In Rostock he would have had to devote much of his time to teaching in order to make a living, but in Berlin he was paid 500 thaler a year—which, even allowing for the higher cost of living, made him rather better off than he had been before—simply to do research. He had also moved to a much more stimulating intellectual environment. In such conditions, and with his talents, his work quickly flourished, and within two years, he was

[63] E. Winter, ed., *Die Registres der Berliner Akademie der Wissenschaften 1746–1766*, Berlin, 1957, pp. 211–12.

[64] Archiv der Akademie der Wissenschaften der D.D.R., Berlin; Abschnitt I, Abth. xiv, no. 27, pp. 42–47.

[65] Maupertuis to Frederick the Great, undated memorandum, probably (to judge from its contents) composed in 1749; A. Le Sueur, ed., *Maupertuis et ses correspondants*, Montreuil-sur-Mer, 1896, p. 85. Euler to Mayer, 17 May 1753, Forbes, *Euler-Mayer Correspondence*, p. 68. Mayer to Euler, 7 September 1754, ibid., p. 92.

[66] Euler to Mayer, 27 May 1755, Forbes, *Euler-Mayer Correspondence*, p. 98.

making contributions of fundamental and lasting significance to science.

The new intellectual environment was especially important, and this was dominated by the presence of Euler. Aepinus's links with the master became close; he took to dining regularly at Euler's house and established a firm friendship with the Euler children, particularly with the eldest son Johann Albrecht Euler, a talented young man of about his own age who had himself been admitted to membership of the Academy just a few months earlier. Aepinus was, in fact, only one of a number of bright young men dedicated to science who gathered around the elder Euler at this period. From Switzerland came the mathematician Louis Bertrand, from Russia the future academicians Stepan Rumovskiĭ and Semyon Kotel'nikov, from Göttingen Aepinus's friend and student Wilcke. Together with Euler and his children, and joined perhaps by others such as the mineralogist Johann Gottlob Lehmann and the young academicians Johann Bernard Merian and Johann Jakob Huber, they formed an enthusiastic *Tischgesellschaft* (the word used by Aepinus to describe the group) which met regularly at Euler's house for meals and conversation about things scientific.[67]

At the time Aepinus joined the group, a leading topic of conversation would undoubtedly have been electricity, for young Johann Albrecht Euler had only a short time before put the finishing touches to an essay he had entered in the St. Petersburg Academy's prize competition on that subject. (Euler *père* had transmitted the essay to St. Petersburg on his son's behalf on 20/31 December 1754.)[68] And we know that the father, too, was interested, because when it was announced some months later that his son's entry had won the prize, he confessed that the original ideas upon which the paper was based had been his, but that, not being sure whether he was eligible to enter the competition, he had passed these on to his son to be worked out in the necessary detail.[69] These were the years just after Benjamin

[67] Aepinus to Euler, 14 February 1755, AAN, fond 136, op. 2, no. 5, l. 13. That such a group gathered around Euler is not well known. Its existence and composition have been determined largely from the correspondence of his son, now preserved in the Academy of Sciences archives in Leningrad (especially fond 1, op. 3, no. 46). For Lehmann's possible membership of the group, see Oseen, *Johan Carl Wilcke*, pp. 35–36.

[68] Euler to Müller, 20/31 December 1754, Yushkevich et al., *Die Berliner und die Petersburger Akademie*, I, 72.

[69] Euler to Müller, 26 September 1755, ibid., p. 92. On the basis of this letter, a number of recent authors have tended, I believe wrongly, to attribute the prize-winning essay as a whole to Leonhard Euler rather than to his son. For a discussion

Franklin's famous identification of lightning as a large-scale electrical discharge, and electricity was a favorite subject with scientists everywhere. Aepinus, so far as we know, had not been much interested in it before his arrival in Berlin (though it had certainly been brought to his attention in one of the chapters of Hamberger's textbook), but evidently he now became an enthusiastic experimenter, encouraged, we may be sure, by the Eulers. So too did Wilcke, who arrived in the autumn of 1755. The two men shared a house, and no doubt did much of their experimenting together; Wilcke in fact later testified to the help he had received from Aepinus,[70] while for his part the latter, in giving a circumstantial account of his discoveries, made clear his indebtedness to "a friend" who, the context reveals, was plainly Wilcke.[71] By early 1757, Aepinus's researches were well advanced, and in a paper he read to the Berlin Academy in March of that year, he displayed an extremely high level of sophistication in both electrical theory and experiment.[72] A few months later it was Wilcke's turn to parade his knowledge of such things when he presented a long and important doctoral thesis on electricity to the Philosophical Faculty at Rostock.[73] Nor would the two friends have been left to pursue their research on their own, for the younger Euler also continued to work on the subject. He kept in close touch with what the others were doing, and later frankly admitted the powerful impact Aepinus's work, in particular, had had on his thinking.[74] Electrical questions were also no doubt frequently aired over meals at Euler's among the group as a whole.

It may have been on one such occasion toward the end of 1756 that Aepinus's friend, the mineralogist Lehmann, drew his attention to the peculiar ability a small crystal in his possession had of drawing ashes to itself when placed on a burning coal.[75] It may even have been Lehmann who suggested that the crystal, now called "tourmaline," ought to be investigated electrically. In any event, when Aepinus took

---

of this point, see R. W. Home, "On two supposed works by Leonhard Euler on electricity," *Archives internationales d'histoire des sciences*, xxv, 1975, 3–7.

[70] Wilcke, *De electricitatibus contrariis*, Rostock, 1757, p. 93.

[71] Below, p. 283; cf. Wilcke, *De electricitatibus contrariis*, pp. 83–88.

[72] Aepinus, "Mémoire concernant quelques nouvelles expériences électriques remarquables," *Mém. Acad. Berlin*, xii, 1756, 105–21.

[73] Wilcke, *De electricitatibus contrariis*.

[74] J. A. Euler, "Recherches sur la cause physique de l'électricité," *Mém. Acad. Berlin*, xiii, 1757, 125.

[75] Aepinus, *Mém. Acad. Berlin*, xii, 1756, 107. The dating is provided below, p. 237; but cf. Wilcke, "Geschichte des Tourmalin," *Abhandl. Königl. Schwed. Akad. Wiss.*, 1766, p. 99, where a slightly later date is indicated.

the matter up, he found at the first try that the effect was an electrical one.[76] More importantly, he quickly recognized that he was dealing with something very different from the usual kinds of electrical phenomena. The tourmaline crystal became electrified simply by being warmed; none of the usual rubbing was required. What is more, he found that when a tourmaline was warmed, it did not become electrical in the usual manner; instead of acquiring an overall charge, the crystal acquired opposite electrical charges on two opposite faces.[77] Aepinus was struck immediately by the analogy between an electrified tourmaline and a magnetized piece of iron, and this in turn led him to wonder whether useful parallels could be drawn between the actual processes involved in magnetizing the iron and electrifying the tourmaline. Armed with this new insight into the workings of nature, Aepinus at once began the series of magnetic researches that was to lead him to a revolutionary new theory of magnetism, the one that he would set out in his great *Essay on the Theory of Electricity and Magnetism* of 1759.

Aepinus had by this time clearly become an extremely skillful experimenter. The electrical properties of the tourmaline crystal are complex, being under most circumstances a combination of pyroelectric and piezoelectric effects—both of which are dipolar in character—with frictional and other charges usually superimposed as well. What is worse, the contributions from the two dipolar effects vary in a perplexing manner according to the circumstances of the experiment, to the extent that at times they will reinforce each other, while at other times they will be in opposition. Aepinus did not succeed altogether in sorting out the situation, but he did get a long way toward a satisfactory resolution of it, and this in itself was a remarkable achievement.

Aepinus reported the results of his tourmaline experiments in a paper he read to the Academy in March 1757. This was, however, one of the last meetings of that body that he attended, for he had in the meantime accepted a position in the Academy of Sciences in St. Petersburg, and he left soon afterward for the north. It has been suggested that the onset of the Seven Years' War had something to do with Aepinus's departure.[78] However, he had made up his mind that life

[76] Aepinus nowhere mentions that Lehmann suggested that electricity might be involved; indeed, by his choice of phrase (*Mém. Acad. Berlin*, XII, 1756, 107) he indicated quite strongly that the idea was his own. The suggestion that Lehmann was responsible for introducing electricity into the discussion is due to Wilcke (*Abhandl. Königl. Schwed. Akad. Wiss.*, 1766, pp. 99–100).

[77] Aepinus, *Mém. Acad. Berlin*, XII, 1756, 110ff.

[78] *Teoriya*, p. 461; also H. Pupke, *Die Naturwissenschaften*, XXXVII, 1950, 49.

in Berlin was not for him nearly twelve months earlier, long before the outbreak of war and only a year after he first joined the Academy.[79] From then on, he had attended meetings less and less regularly.[80] Evidently it had not taken him very long to become disenchanted with the situation in the Prussian capital. In the absence of definite evidence, we can only speculate as to what made him feel this way, but it must have been a powerful aversion indeed to override the attractions of living and working in close and friendly contact with Euler. Most likely his thoroughly German soul was repelled by the highly Frenchified atmosphere of Frederick's Academy. This could make life difficult even for Euler from time to time, and, indeed, was one of the factors that caused him to return to St. Petersburg a few years later. Its effect on a much more junior German member of the Academy could well have been oppressive.

Once he had made up his mind to leave Berlin, Aepinus apparently began making cautious inquiries about jobs elsewhere. Segner in Göttingen knew as early as April 1756 that he wanted to move,[81] and by June word reached St. Petersburg, where the historian G. F. Müller, the secretary of the Academic Conference of the Imperial Academy of Sciences, reacted promptly. The St. Petersburg Academy had been seeking a new professor of physics ever since the tragic death of G. W. Richmann while investigating the electricity of thunderclouds almost three years earlier. Now Müller immediately made overtures to Aepinus through the latter's brother-in-law, a member of the Eschenbach family from Rostock and at that time a merchant in St. Petersburg. At the same time Müller sought the views of Euler and of Rumovskiĭ and Kotel'nikov (both of whom returned to Russia from Berlin at about this time) as to Aepinus's suitability for the position. All three replied extremely favorably. The negotiations with Aepinus himself proceeded smoothly: by October agreement was reached, and he was formally appointed professor of physics in the St. Petersburg Academy at a salary of 860 roubles a year. In addition he was promised a sum of 200 roubles to cover his removal expenses, and an unconditional release after five years' service should he wish that when the time came.[82]

[79] Segner to Euler, April 1756, AAN, fond 136, op. 2, no. 4, l. 7.

[80] Winter, *Die Registres der Berliner Akademie der Wissenschaften*, pp. 211–31.

[81] See above, n. 79.

[82] Aepinus's reply to Müller's initial inquiry has survived (Aepinus to Müller, 9 July 1756, AAN, fond 21, op. 3, no. 3, l. 2), and enables us to date the start of the negotiations fairly precisely. Allowing for the usual delays in the mails, Müller must have contacted Aepinus's brother-in-law at least two weeks before Aepinus wrote his reply, that is, no later than about 26 June 1756 (N.S.). Given this, we, like Dorfman (*Teoriya*, p. 483), can interpret the more widely known exchange between

With his appointment thus settled, all that remained was for Aepinus to obtain his release from Berlin. This, however, was easier said than done. The king had been furious when one of Aepinus's predecessors, A. N. Grischow, had abandoned his position and departed surreptitiously for St. Petersburg, and he proved very reluctant to see another of his astronomers leave so soon after having been appointed to his post. It took considerable effort on the part of Euler (acting as director of the Academy in the absence of Maupertuis, who was ill) before permission was finally granted.[83] By the time this was forthcoming, Aepinus was busily engaged on his tourmaline experiments, so he decided further to delay his departure by some weeks until these were completed [84]—a wise precaution, since tourmaline was at that time a very rare stone (occurring, so far as was known, only on certain beaches in Ceylon), and for all Aepinus knew, there might not have been any samples at all in St. Petersburg for him to experiment with.

Not until late March did Aepinus finally leave Berlin. He then spent a month in Rostock among his family and friends before proceeding to Lübeck to take ship. A brief stop at the island fortress of Kronstadt guarding the approaches to the Russian capital enabled him to deliver some letters on Euler's behalf,[85] and finally, on 10 May 1757, he arrived in the city where henceforth he would make his home. Two days later he attended the first of many meetings at the Academy of Sciences.[86]

The St. Petersburg Academy of Sciences had been founded some

Müller and Euler as the seeking and obtaining of a report on Aepinus's work, rather than as the preliminary to Müller's direct negotiations with Aepinus that it has sometimes been taken to be (Müller to Euler, 22 June 1756 O.S. [3 July 1756 N.S.], and Euler's reply, 20/31 July 1756; Yushkevich et al., *Die Berliner und die Petersburger Akademie*, I, 116, 119). Both Müller's final report recommending Aepinus's appointment and the initial contract that Aepinus signed with the St. Petersburg Academy have also been preserved (AAN, fond 21, op. 1, no. 25, and fond 3, op. 1, no. 700, l. 184–85, respectively). Nothing in any of these documents supports the suggestion that has sometimes been made (*Teoriya*, pp. 482–83; also Heilbron, "Aepinus," *Dictionary of Scientific Biography*, New York, 1970–78, I, 66–68) that the St. Petersburg Academy deliberately sought to replace Richmann with someone who would be able to continue the tradition of electrical inquiry that he and his fellow academician M. V. Lomonosov had begun.

[83] Euler to Müller, 9 October 1756, and idem., 16 April 1757, Yushkevich et al., *Die Berliner und die Petersburger Akademie*, I, 128, 140.

[84] Aepinus to Müller, 7 April 1757, AAN, fond 21, op. 3, no. 3, l. 4.

[85] Aepinus to Euler, 8 June 1757, AAN, fond 136, op. 2, no. 5, l. 19.

[86] *Protokoly*, II, 380. For the date of Aepinus's arrival in St. Petersburg, see *Teoriya*, p. 541.

thirty years earlier, in 1725, by Tsar Peter the Great as part of his remarkable attempt to transform his realm all at once into a modern European state. Until then, Russian attitudes toward the natural world had remained firmly rooted in traditional learning; the new approach that had developed in western Europe during the so-called "scientific revolution" of the previous hundred years or so had penetrated hardly at all. Peter, determined as he was to introduce the new ideas into his empire, had had to staff his new Academy entirely with foreigners, mostly Germans, rather than with native-born Russians. Things had changed only very gradually in the years that followed, and when Aepinus joined the Academy, it remained the only center of scientific activity in the country, and in its membership it was still rather a German institution than a Russian one.

In certain respects the Academy had actually slipped backward during the intervening years. Peter had been remarkably successful in persuading a number of first-rate scientists to join the first group of academicians to take up residence in St. Petersburg, but some of these had died, while others had returned to their homelands once they had served out the terms of their contracts. For the most part the replacements had not been men of the same caliber as those who had left. On the other hand, there had been a slow but steady increase in the Russian element within the Academy, so that the institution was no longer as cut off from Russian society as it had been initially. The day-to-day administrative affairs were handled in Russian by Russian clerks. Skilled Russian mechanics staffed the Academy's workshop and also its printery, one of the first in the country to operate outside the control of the Church. An active group of translators rendered a large number of scientific and technical works into Russian, and in the process played an important part in the development and enrichment of the Russian written language. The first native-born Russians were appointed to professorships within the Academy in the early 1740s, and from then on their numbers, while remaining small, increased steadily. So too did their influence on academic affairs.

Of particular importance for both the subsequent history of the Academy and Aepinus's personal career within it was the character of one of the earliest native-born Russian professors, Mikhail Vasil'evich Lomonosov (1711–1765). Brought up near Kholmogory in the far north of Russia on the shores of the White Sea, Lomonosov succeeded against great odds in acquiring as a youth something of an education. Then by a fortunate chance he came to the attention of the authorities in the new Academy of Sciences in St. Petersburg, and in 1736 he was sent at the Academy's expense to Marburg, in Germany, to study under the great Christian Wolff himself. Following his return to Russia five years later, Lomonosov became a junior member of the Academy's staff. In 1745 he was promoted to the ranks of the academicians with the title Professor of Chemistry.

Lomonosov was a man of extraordinary talents, and in the years that followed his appointment to the Academy he was active in a number of widely differing fields. In addition to pursuing a distinguished scientific career, he quickly established himself as one of the leading Russian poets of his generation. As a grammarian he occupies a prominent place in the history of the Russian written language. As an historian he engaged in an important debate over the origins of the Russian state. As a pioneering industrialist he established a glass factory outside St. Petersburg in which, for perhaps the first time in Russia, an attempt was made to marry the new knowledge of the scientists to the techniques of the artisan. As an educationalist, he played an important part in persuading the government to establish a university in Moscow during the 1750s, and he also devoted much effort to improving the sadly neglected educational institutions (a secondary school and a university) attached to the Academy itself in St. Petersburg. His activities extended into several other fields as well.

Lomonosov, however, was not merely a man blessed with remarkable talents, he was also an extraordinarily volatile individual, full of violent temper and passionate enthusiasms. As a result, his presence in the Academy had a powerful and often abrasive impact on the other members, not least upon Aepinus. Lomonosov's fierce national pride, for example, led him to resent bitterly both the continuing German dominance within the Academy and the patronizing attitude many of the German professors adopted toward their Russian hosts. When provoked he never hesitated to give free rein to his criticisms on this score. In addition, he felt very strongly that the Academy should be more responsive to the clear and pressing needs of contemporary Russian society, but his efforts to bring this about inevitably exacerbated still further his relations with his fellow academicians, who for the most part wanted only to pursue their individual research interests undisturbed.

Shortly before Aepinus's arrival in St. Petersburg in 1757, Lomonosov was appointed as one of the councillors in the Academy's chancellery, the administrative apparatus that, under the Academy's charter, exercised a considerable measure of control over the academic professors. In this new position he was able to press his views concerning the working of the Academy more vigorously than ever. The powers granted to the chancellery, and their ruthless exploitation by the long-time secretary Johann Schumacher, had already led to much discontent within the Academy. Following Lomonosov's appointment, relations between the chancellery and the academicians became even worse. Lomonosov's passions and enthusiasms became constant thorns in the flesh of the other professors. He argued with them, he insulted them, he bullied them. In response, the other academicians came to demand more and more insistently that the Academy's structure be reformed: "the Academy will then take on quite another appearance," wrote Stepan Rumovskiĭ, "the members of the Academy will have no chicanery to fear, and the despotic power to which Mr. Lomonosov aspires will be reduced to nothing."[87] This wish was eventually granted in 1765; the chancellery not only lost all its powers over the academicians, it was abolished altogether. But by the time this happened, Lomonosov had died, and the Academy was about to enter on a more tranquil phase in its history.

It was into a turbulent environment, then, that Aepinus entered when he took up his appointment in St. Petersburg. For a time, however, he managed to avoid entanglement in the arguments raging around him. On at least one occasion he acted as peacemaker, and thereby saved the Academy acute embarrassment, following an ill-conceived trip to the Baltic coast by the astronomer Grischow (the man who had so enraged Frederick of Prussia a few years earlier) to try to establish his pretended discovery of a remarkably variability in the Earth's gravitational acceleration.[88]

During this period Aepinus threw himself into his research with great vigor, hindered only by a bout of illness that laid him low for some weeks soon after his arrival in the Russian capital. His formal academic duties were not at all onerous, and most of his time was his own.[89] His new post offered Aepinus particular scope for his experimental work, since it gave him access to—indeed, the supervision of—the excellent physics "cabinet" that had been built up by his predecessor in the chair of physics, G. W. Richmann. Most notably, from his point of view, Aepinus found himself heir to a first-rate set of electrical apparatus, Richmann having been particularly interested in electricity for a number of years before his death. As well as the collection of apparatus, Aepinus had at his disposal the comprehensive workshop facilities that the Academy had built up over the years.

[87] Rumovskiĭ to J. A. Euler, 22 November/2 December 1764, AAN, fond 1, op. 3, no. 46, 1. 274–75.

[88] *Protokoly*, II, 387. Cf. Aepinus to L. Euler, 30 September 1757, AAN, fond 136, op. 2, no. 5, l. 21.

[89] Aepinus to L. Euler, 30 September 1757, AAN, fond 136, op. 2, no. 5, l. 21.

In short, in St. Petersburg the opportunities for research were first-class, and Aepinus, then at the height of his powers, was quick to take advantage of them.

Aepinus's first concern was to pursue further the electrical and magnetic research he had begun so successfully in Berlin. At that time, he and Wilcke had devoted much effort to exploring in detail one of the crucial components of Benjamin Franklin's theory of electricity, the idea that there are two contrary modes of electrification possible in nature (these being called by Franklin "plus" and "minus"). Both men had quickly become firm adherents of Franklin's views, and Aepinus had used the theory to great effect in sorting out the complex electrical properties of the tourmaline. Now, in St. Petersburg, Aepinus turned again to an elaboration of the Franklinian doctrine. Very early in his investigations, his mathematician's eye had enabled him to see that the theory, as Franklin had left it, contained a number of serious flaws. These Aepinus now set out to eliminate; more precisely, he successfully demonstrated that, with the flaws eliminated, the theory agreed even more exactly with experiment than it had done previously. In addition, Aepinus explored still further the analogy he had perceived between magnetism and electricity. Finally, Aepinus saw that Franklin's theory left a great deal unsaid about such wider questions as the connections between electricity and the structure of matter in general. These questions clearly needed answering, and Aepinus applied himself to them diligently. He recorded his efforts, though in disappointingly fragmentary fashion, in a notebook that has survived from this period.[90] He evidently found the results he obtained unsatisfactory, however, because he never referred to them in his published work.

Aepinus by no means confined himself during this period to his electrical and magnetic investigations. Both his notebook and his published writings make it clear that he was active in a number of fields. Astronomical questions, for example, continued to attract his attention. Many pages of his notebook are devoted to extracts from "de la Caille Astron." (presumably the Abbé La Caille's *Leçons élémentaires d'astronomie géométrique et physique*, Paris, 1746; 2nd ed. 1755). In addition, a number of jottings relate to such matters as the forthcoming transit of Venus across the face of the sun, the possibility that the moon might have an atmosphere, and Halley's prediction that the great comet of 1682 would reappear some time in the late 1750s. When a bright comet duly appeared in September 1757,

[90] AAN, raz. v, op. 3–7, no. 1. See below, pp. 131–35.

Aepinus became most excited,[91] but his calculations soon showed that this was not the one predicted by Halley. He therefore contented himself with composing an article on the nature of comets in general, and on Halley's prediction in particular, for the popular-style Russian-language journal entitled *Yezhemesyachnyya sochineniya k pol'ze i uveseleniyu sluzhashchiya* (*Monthly Works of Benefit and Amusement*) that the Academy published each month under Müller's editorship.[92]

Aepinus also maintained the interest in scientific instrumentation that had so impressed Euler. In his notebook, for example, much space is devoted to the design of telescopes, the use of micrometers in optical instruments, and the like. From the minute books of the Academic Conference we learn that he was working on other kinds of instruments as well, for in September 1757 he presented to the Academy a description of a new "pyrometer" he had designed for measuring the thermal expansion and contraction of metals.[93] For some reason this was never published, but a number of his articles on the improvement of optical instruments and on various subjective factors in relation to optics did subsequently appear in the Academy's *Novi commentarii*. In one of these articles, Aepinus proposed some particularly important modifications that would enable a solar microscope to be used to observe opaque as well as transparent objects. His suggestions on this score had been inspired, as Aepinus himself freely admitted, by his knowledge of the earlier work of Euler and the famous German microscopist Johann Nathanael Lieberkühn on the same subject, but the ideas he put forward were very much his own. A few years later, in the early 1770s, Aepinus's ideas were taken up by the English instrument maker Benjamin Martin, and through him they eventually became part of the standard design for this very popular instrument.[94]

Nor did Aepinus neglect his mathematical interests. In November 1757 he submitted a paper to the Academy of Sciences on the binomial

[91] Aepinus to Euler, 30 September 1757, AAN, fond 136, op. 2, no. 5, ll. 20–21.

[92] Aepinus, "Razmyshleniya o vozvrate komet, s kratkim izvestiyem o nyne yavivsheĭsya komete," *Yezhemesyachnyya sochineniya*, October 1757, pp. 329–48.

[93] *Protokoly*, II, 391. Aepinus's manuscript has been preserved (AAN, raz. I, op. 2, no. 43) and has been discussed at some length by V. L. Chenakal in his article, "Problema metallicheskoga termometra v fizika XVIII v. i Lomonosov," pp. 96–135 in *Lomonosov: sbornik stateĭ i materialov*, VI, Moscow/Leningrad, 1965.

[94] Aepinus, "Emendatio microscopii solaris," *Novi commentarii Academiae Scientiarum Imperialis Petropolitanae*, IX, 1762–63, 316–25. For a discussion of Aepinus's proposals and their significance, see S. L. Sobol', *Istoriya mikroskopa i mikroskopicheskikh issledovaniĭ v Rossii v XVIII veke*, Moscow/Leningrad, 1949, pp. 230–34.

theorem, that famous theorem that lay behind so much of eighteenth-century mathematical analysis, and that had already attracted Aepinus's attention earlier in his career.[95] A few months later he sent to Euler a paper based on the principle of least action, for presentation to the Academy in Berlin.[96] However, like his paper on his new "pyrometer," this was never published, and unfortunately the manuscript has now disappeared, so we do not know his views on this controversial subject. A few months later again, Aepinus presented another mathematically oriented paper to the St. Petersburg Academy, this time on an algebraical theme.[97]

In all, therefore, it is apparent that Aepinus embarked on a wide-ranging program of important scientific researches as soon as he arrived in the Russian capital. His talents obviously made an impression on his colleagues, and within a year of his arrival he was appointed to read one of the papers at the forthcoming public meeting of the Academy to be attended by the Empress Elisabeth and the assembled dignitaries of her court. Since Aepinus was by now well advanced with his electrical and magnetic investigations, he decided to use these as the basis of his presentation on this occasion. More specifically, since his experiments had confirmed the belief first prompted by his work on the tourmaline that a close analogy could be drawn between the phenomena of magnetism and those of electricity, he decided to use the occasion to launch this idea into the world. Given the nature of his audience, he could not afford to make his paper too technical, and he therefore limited himself to demonstrating that each of the principal magnetic phenomena had a parallel in the electrical realm. Only at the end of his address did Aepinus outline briefly the main points of the radically new theory, modeled on Franklin's theory of electricity, that he had devised to account for them. And yet, despite its limited scope (perhaps, indeed, because of it), his essay evidently attracted a great deal of interest, for it quickly appeared in at least three separate German editions besides the Latin and Russian versions that were published initially in St. Petersburg.[98]

---

[95] Aepinus, "Demonstratio generalis theorematis Newtoniani de binomio ad potentiam indefinitam elevando," *Novi commentarii*, VIII, 1760–61, 169–80. Cf. his earlier note, "Demonstratio theorematis binomialis," [*Rostock*] *Gelehrte Nachrichten*, 1752, p. 136.

[96] Winter, *Die Registres der Berliner Akademie der Wissenschaften*, p. 240. Cf. Aepinus to Euler, 28 February 1758, AAN, fond 136, op. 2, no. 5, l. 22.

[97] Aepinus, "De functionum algebraicarum integrarum factoribus trinomialibus realibus commentatio," *Novi commentarii*, VIII, 1760–61, 181–88.

[98] See Appendix, items **10A–10E**.

An extended review appeared soon afterward in the important and influential *Göttingische Anzeigen von gelehrten Sachen*.[99]

Aepinus's interest in magnetism was not limited to his novel theoretical ideas on the subject. He also attempted to sort out the complexities of the magnetic phenomena themselves, and (as of course we might have expected, given his general interest in scientific instrumentation) to apply his discoveries to practical affairs. In both these respects, as well as in the theoretical realm, he enjoyed a great deal of success. Most of his results were reported for the first time in 1759 in his *Essay on the Theory of Electricity and Magnetism*, and will be discussed at length later. But Aepinus also sought to make the practical benefits that could flow from his research more widely known, and to this end he contributed two articles on magnetic themes to Müller's journal, *Monthly Works of Benefit and Amusement*.[100] In one of these he described some methods he had devised for preparing magnetized needles that would perform more reliably than the ones normally used in nautical compasses. In the other article he revealed a new method of increasing the strength of natural magnets (this too being of interest to navigators, whose equipment customarily included a magnet that could be used from time to time to re-magnetize their compass needles). Aepinus also sent papers on these subjects to Germany. One was subsequently published in the *Hamburgisches Magazin*[101]—a suitable home for it, surely, in view of Hamburg's importance as a seaport—and another appeared in the *Acta* of the newly formed Society of Sciences in Erfurt.[102] In connection with the latter, Aepinus was elected to membership of the Erfurt Society, an honor that is recorded on the title page of many of his later publications.

During the winter of 1758–1759 Aepinus's work was interrupted by a protracted illness;[103] nevertheless by the following June the completed manuscript of his *Essay on the Theory of Electricity and Magnetism*

---

[99] *Göttingische Anzeigen von gelehrten Sachen*, 1759, pp. 116–19.

[100] This was by now appearing with a slightly modified title, namely *Sochineniya i perevody, k pol'ze i uveseleniyu sluzhashchiya (Works and Translations of Benefit and Amusement)*. Aepinus's articles appeared in the issues for November 1758 and January 1759.

[101] Aepinus, "Abhandlung über einige neue Verbesserungen der Magnetnadel und des Seecompasses," *Hamburgisches Magazin*, XXIV, 1760, 563–84.

[102] Aepinus, "Descriptio acuum magneticarum, noviter inventarum, quae vulgaribus praestantiores sunt, atque artificii, vires magnetum naturalium insigniter augendi," *Acta academiae electoralis Moguntinae scientiarum quae Erfurti est*, II. 1761, 255–72.

[103] See below, p. 233.

was ready for presentation to the St. Petersburg Academy for publication.[104] A week later, after two of his fellow academicians had reported favorably on it, the manuscript was sent to the printery with a request that printing be expedited as much as possible; the result of the Academy's previously announced prize competition on the construction of artificial magnets was due to be made public early in September, and, since Aepinus had treated the same topic at some length in his book, his colleagues were anxious that his account should if possible be in print before the announcement was made.[105] In the event, the academicians' hopes were not realized. In September, with the prize announcement made and printing still unfinished, Aepinus submitted two further short dissertations to be included in his book as appendices,[106] and it was early December before copies of the completed work finally reached the Academy's bookshop.[107] Because his work had thus been overtaken by events, Aepinus included in his preface a graceful acknowledgement of the priority of the author of the winning entry in the competition (Antheaulme, *Syndic des Tontines* in Paris) in the invention of the improved method of magnetizing steel bars described at length in his book—and this despite the fact that, as is clear from the dates involved, he had developed the method entirely independently of Antheaulme.

The year 1759 was a fateful one for Aepinus, for in addition to seeing his master work published, two other major developments occurred. First, a bitter dispute erupted between Aepinus and Lomonosov that was to drag on for several years; and second, the year marked the beginning of Aepinus's career at court, a career that within a few years was to bring his scientific activities to an end. The immediate effect of this latter development was to divert Aepinus from his electrical and magnetic researches just when these had reached their most productive stage. As a result, his further contributions to these subjects—subjects that he had by now so clearly made his own—were not nearly as substantial as they might otherwise have been.

So far as his disagreements with Lomonosov are concerned, Aepinus has generally been treated somewhat cavalierly by historians. Most

---

[104] *Protokoly*, II, 428.

[105] Ibid.

[106] Ibid., p. 436. See below, pp. 453–78.

[107] Personal communication from Mr. D. E. Bertel's, Archives Department, Academy of Sciences of the U.S.S.R. A number of recent authors, apparently following the lead of Poggendorff in his *Biographisch-Literarisches Handwörterbuch zur Geschichte der exacten Wissenschaften*, I, Leipzig, 1863, cols. 14–15, have stated that Aepinus's book was published in Rostock. This is certainly wrong.

accounts have been based almost entirely on Lomonosov's contributions to the proceedings, and these have tended to be accepted uncritically as though their author were an impartial observer and not one of the parties to the dispute. The affair thus merits more attention here than it might otherwise have received, in order to redress the balance a little. But it is also of interest in its own right for the insight it affords into Aepinus's scientific activities during this period and into the atmosphere in which he was working in St. Petersburg.

When Aepinus joined the St. Petersburg Academy, he managed for a time to avoid becoming embroiled in the arguments raging around him. Lomonosov himself later freely acknowledged the friendliness Aepinus showed toward him at this period,[108] and an entry in Aepinus's notebook reveals that on at least one occasion Aepinus actually visited Lomonosov in his home.[109] But this happy state of affairs did not last, and in April 1758 a first hint of discord appeared between the two men.

On this occasion, Lomonosov reported to the Academy that Aepinus had communicated to him some criticisms of his paper proposing the development of a "night-vision" telescope, that is, one that would enable dimly lit objects to be seen more clearly.[110] Aepinus regarded Lomonosov's plan as impracticable; indeed, on what seemed to be well-established theoretical grounds, as impossible. Lomonosov rejected Aepinus's arguments, and suggested that the perfecting of his scheme should be announced by the Academy as the subject for a competition. The other academicians requested that he first make Aepinus's criticisms available to them, which Lomonosov promised to do. Since the matter was taken no further, it appears that Aepinus's arguments prevailed.

This, however, was not the end of the affair. Lomonosov remained unconvinced by Aepinus's arguments, and at a session of the Academy in June 1759, he raised the matter again by demonstrating a new telescope that his friend and protector Ivan Shuvalov had just received from England. Its function, Lomonosov asserted, was the same as that of the instrument he had proposed.[111]

Lomonosov's apparent triumph was short-lived. Two weeks later, Aepinus delivered a devastating critique of his claims; and on this occasion he was not content simply to set out the theoretical difficulties

[108] P. S. Bilyarskiĭ, *Materialy dlya biografii Lomonosova*, St. Petersburg, 1865, p. 074.
[109] AAN, raz. v, op. э-7, no. 1, l. 20.
[110] *Protokoly*, II, 406.
[111] Ibid., p. 430.

confronting Lomonosov. He went on to report how, when he had tested the newly arrived instrument, he had found its performance in dull light to be not even as good as that of an ordinary reflecting telescope. Upon dissecting it, he said, he had found it to be constructed in a "completely bizarre manner" using twice as many lenses as necessary; it thus betrayed at once, he continued, how it was that its maker had been able to make such fantastic claims for it—he was totally ignorant of even the first principles of optics. Aepinus asked pointedly why the English learned journals, usually so eager to defend the priority claims of British inventors, had remained utterly silent about the new device. He provided his own answer in the most insulting terms: "Without doubt the unfortunate effort of an ignoramus is seen in England as unworthy of discussion." [112] Finally, he suggested that, once Lomonosov had been given an opportunity to prepare a reply to his report, both the report and the reply should be sent to the Académie Royale des Sciences in Paris for judgment, with the authors' names suppressed to insure impartiality.[113] At this point, however, Lomonosov seems finally to have given up the struggle, for there is no record of his having taken up Aepinus's challenge.[114]

Aepinus's report was couched in extremely blunt and undiplomatic language, and Lomonosov certainly resented it bitterly; in fact, in a memorandum written some time afterward, he identified it as the starting point of all his subsequent quarrels with its author.[115] Yet a tone of such exasperation pervades Aepinus's document that it is diffi-

[112] AAN, raz. v, op. 9–7, no. 11.
[113] *Protokoly*, II, 432.
[114] A convenient account of Lomonosov's design for his telescope has been provided by Valentin Boss, *Newton and Russia: the Early Influence, 1698–1796*, Cambridge, Mass., 1972, pp. 207–10. S. I. Vavilov has argued (in *Lomonosov: Sbornik statei i materialov*, II, Moscow/Leningrad, 1946, 74–80, and elsewhere) that the main issue in this dispute lay between Lomonosov's great experimental insight, which enabled him to glimpse a truth not again recognized until only a few years ago, and Aepinus's over-rigid adherence to a theory of optics that we now know to have been incorrect. Not enough evidence has survived to enable us to reach a definite conclusion about this (most notably, no trace has been found of Aepinus's chief critique of Lomonosov's ideas), but it is doubtful whether Lomonosov really carried his experiments far enough to justify Vavilov's argument. Otherwise it would be extremely difficult to understand, in the light of Aepinus's criticisms of it, how Lomonosov could have waxed so enthusiastic over the telescope Shuvalov received from England. Surely he would have recognized its failings for himself had he got very far with his own experiments? Boss has suggested (*Newton and Russia*, pp. 209–10) that Shuvalov's telescope was one of John Dollond's newly invented achromatic refractors. Aepinus's remark about the multiplicity of lenses in it is consistent with this.
[115] Quoted by P. P. Pekarskiĭ, *Istoriya Imperatorskoĭ Akademii Nauk v Peterburge*, St. Petersburg, 1870–73, II, 649.

cult not to see it as the culmination of a series of increasingly acrimonious exchanges on the subject between the two men, rather than as the starting point of their dispute. What is more, other points of disagreement between them had arisen, some stated, some not, by the time Aepinus delivered his report. Less than a month earlier, Aepinus had submitted to the Academy for approval the manuscript of his *Essay on the Theory of Electricity and Magnetism*. For many years previously, Lomonosov had been keenly interested in electricity. At a public meeting of the Academy in November 1753, he had delivered an address in which he speculated at length on its role in various meteorological phenomena.[116] Three years later he had begun but failed to complete a memoir in which he outlined a theory ascribing electricity to the motion of particles of the extremely subtle aether that, he supposed, filled the pores of all bodies.[117] It must have been extremely galling to him to find that Aepinus had no sympathy whatever for the approach he had adopted, but was content instead to base his account on unexplained forces acting between particles at a distance. Not once was either Lomonosov's name or the kind of explanation he had advanced for electrical and magnetic phenomena mentioned in the whole three hundred and ninety pages of Aepinus's book: worse, when Aepinus tendered his manuscript to the Academy, Lomonosov was not even chosen as one of those who were to advise on whether the book merited publication.[118] Granted that no evidence exists of any ill-feeling over this sequence of events, it would be astonishing if a man of Lomonosov's fiery temperament resisted feeling considerable bitterness at what had come to pass, or giving vent to his bitterness from time to time.

A further source of friction that developed at this time—and one that is thoroughly documented—arose from Lomonosov's efforts to resuscitate the by then moribund university attached to the Academy. At the very time when Aepinus was preparing his report on Shuvalov's so-called night-vision telescope, Lomonosov challenged him publicly at a session of the Academy as to whether he was willing to give lectures in the university. The matter was evidently a sensitive one, since, rather than answering immediately, Aepinus submitted a carefully considered reply in writing.[119] He there took some pains to point out

[116] M. V. Lomonosov, *Oratio de meteoris vi electrica ortis*, St. Petersburg, 1753; reprinted in his *Polnoe sobranie sochinenii*, Moscow/Leningrad, 1950–59, III, 15–99.

[117] Lomonosov, "Theoria electricitatis methodo mathematica concinnata," *Polnoe sobranie sochinenii*, III, 265–313.

[118] *Protokoly*, II, 428.

[119] Ibid., pp. 430–31. As early as September 1757, Aepinus had been ordered to

that he was not required under the terms of his contract to deliver such lectures; specific provision had been made in the regulations of the Academy for the appointment of a second professor of natural philosophy who would be charged with that task. Furthermore, no provision had been made in his salary for such lecturing duties. Nevertheless, he said, he was prepared to cooperate, but in the circumstances he felt entitled to lay down the conditions under which he would undertake the work. Lomonosov later described these conditions as "impossible," and he regarded Aepinus's insistence upon them as tantamount to a refusal to undertake lecturing duties.[120] Six months later, when his proposals for the reorganization of the university were accepted, Lomonosov used the new powers he thus acquired to overrule all Aepinus's carefully formulated conditions and ordered him peremptorily, along with a number of other academicians, to begin lecturing at once.[121]

Lomonosov's opinion notwithstanding, the conditions that Aepinus had set out appear, on the surface, to have been minimal. He wanted it made clear that in agreeing to lecture, he was not committing himself to doing so in perpetuity. He also wanted it spelled out that his duties in the Academy would remain his primary responsibility, and that he would be expected to lecture only so long as his health—still somewhat uncertain—permitted. He demanded freedom to fix the times at which his lectures would be given, and he insisted that the students attend them at his house. This latter cannot have been much of an obstacle for potential students, however, since had Aepinus not set such a condition, he himself would apparently have been expected to walk to the Academy whenever bad weather prevented the use of a carriage.[122] If Aepinus's conditions really were impossible, as Lomonosov charged, then what made them so must have been the remaining one: "No students must be given to me to instruct other than those of whom it is known for certain, that my trouble with them is not taken in vain." [123] Here, surely, is the rub. Throughout its entire history, the basic problem confronting the Academic University had been its inability to attract suitable students, and it may well have

---

give lectures on experimental physics (Lomonosov, *Polnoe sobranie sochineniĭ*, IX, 471), but apparently he had simply ignored this instruction. It is therefore easy to understand why he adopted a very cautious approach when the matter was brought up again.

[120] Bilyarskiĭ, *Materialy*, p. 470.
[121] Pekarskiĭ, *Istoriya*, II, 680.
[122] *Protokoly*, II, 431.
[123] Ibid.

been the case that no students could have been found who could satisfy Aepinus's requirements regarding their competence. If so, his conditions were certainly not as ingenuous as they appear at first.

From this point onward, relations between Aepinus and Lomonosov went from bad to worse. In October 1760, for example, a dispute arose over a notable discovery that had been made by the St. Petersburg academician J. A. Braun during a particularly cold spell of weather in December of the previous year. Braun had found that, under the excessively cold conditions then prevailing, if a thermometer was immersed in a freezing mixture of spirit of niter and powdered snow, the mercury in it would solidify. Traditionally, mercury had been regarded as an essentially liquid substance; for many years, indeed, a pure and unalloyed mercury had been held to be the very principle of fluidity in things. Hence Braun's discovery that it could be solidified caused a flurry of excitement, and within a few days a number of other St. Petersburg scientists, including both Aepinus and Lomonosov, hastened to confirm the report. Aepinus seems to have taken the lead in most of the subsequent experimenting, collaborating with Braun, the academician J. E. Zeiher, the physician K. F. Kruse, and the "first apothecary" J. G. Model, and developing what quickly became the standard procedure used within the group for obtaining solid mercury. Lomonosov, meanwhile, pursued his own independent series of investigations.[124] Afterward, Aepinus was delegated to prepare a report on the various experiments for the periodical press, while Braun, as discoverer of the phenomenon, composed a more formal paper for presentation to the Academy and subsequent publication by that body.[125]

At a meeting of the Academy on 6 October 1760, the academicians were confronted with an angry memorandum from the chancellery concerning Aepinus's report. This had clearly been inspired by Lomonosov. It asserted that in St. Petersburg the publication of Aepinus's report had been suppressed "on account of several inaccuracies," and complained that, despite this, the report had now been published in the Leipzig *Nova acta eruditorum*. The chancellery demanded to know who had sent a copy to Germany, and on what authority.[126]

Without waiting for anyone to answer the chancellery's question,

---

[124] "Account of artificial cold produced at Petersburg. By Dr. Himsel, in a letter to Dr. De Castro . . . ," *Phil. Trans.*, LI, 1759–60, 670–76.

[125] Braun, *De admirando frigore artificiali quo mercurius est congelatus dissertatio*, St. Petersburg, 1760.

[126] *Protokoly*, II, 456.

Aepinus demanded to know what inaccuracies his report was supposed to contain. Lomonosov replied that the report falsely attributed the initial idea of the congelation experiment to Zeiher rather than to Braun, and it also rejected—unjustifiably, in his opinion—his own claim to have achieved a much greater degree of cold than anyone else. Müller, the secretary of the Academic Conference, interjected that he, not Aepinus, had been responsible for the passage referring to Zeiher, but that in any case this had been deleted from the report before copies had been sent abroad. He pointed out that there was no trace of the offending passage in the printed reports in either the *Nova acta eruditorum* or the *Journal des sçavans*.[127] Furthermore, Müller showed that, in both Aepinus's original document and the versions of this published in the two foreign journals, Lomonosov's experiment was reported exactly as he had described it. The only report he had sent to Leipzig, Müller insisted, was the one drawn up by Aepinus; and that this should be published had been affirmed at the time by the whole Academic Conference.

Despite Müller's assurances, Lomonosov, and Braun too, had good reason for being suspicious. A report containing the offending passages had in fact been sent abroad, but to the Academy of Sciences in Paris rather than to Leipzig. It had been sent by a Frenchman who had taken part in some of the St. Petersburg experiments, Pierre-Isaac Poissonnier, and had promptly been published, with commentary, in the September 1760 issue of the *Journal étranger*.[128] Evidently a somewhat garbled account of this publication had come to Lomonosov's ears. Convinced for some reason that the offending report had been published in Leipzig, he and Braun now charged that it had been published in another leading Leipzig journal, the *Commentarii de rebus in scientia naturali et medicina gestis*. Since in fact no report at all had appeared in this particular journal, their complaint immediately lost all semblance of credibility. Aepinus did not bother to reply to the new attack, and Müller simply protested that he knew nothing about the matter. Müller then began to read a paper to the meeting, but because of repeated interruptions by Lomonosov, the session had to be abandoned.

[127] Cf. *Nova acta eruditorum*, February 1760, pp. 74–78, and also *Journal des sçavans, combiné avec les mémoires de Trévoux* (Amsterdam), June 1760, pp. 259–67 (the article in question did not appear in the Paris edition of the *Journal des sçavans*).

[128] *Journal étranger*, September 1760, pp. 203–24. Poissonnier's report subsequently also served as the basis of the account of the St. Petersburg experiments given in *HAS*, 1760 (publ. 1766), Hist. pp. 27–30.

A month later, Lomonosov returned once more to the attack, this time submitting to the president of the Academy, Count Kiril Razumovskiĭ, a formal statement of his dissatisfaction with Aepinus's conduct and, in particular, with his management of the physics apparatus. Lomonosov charged that the collection of apparatus had been "lying thrown about in corners mouldering and rusting without being used at all," to the point where hardly any trace remained of its former excellent condition. Aepinus, he said, had disregarded his duty in this respect "since his very entry into the service of the Academy." [129] These were serious accusations, particularly because Lomonosov was a high official in the Academy's chancellery when he made them. Their truth has been taken for granted by most historians. But the fact that Razumovskiĭ ignored them completely suggests that he for one regarded them as baseless, and Aepinus's publication record makes it clear that they must have been, for this record reveals that during the period in question Aepinus was pursuing an extremely active program of high-quality experimental investigations in a number of different fields. Lomonosov's attack, in other words, was completely unwarranted. Aepinus may not have found a use in his experiments for all the equipment that had been acquired for the Academy by his predecessors. But even granting this does not alter the picture very much. Lomonosov, it seems, simply let his personal animosity toward Aepinus get the better of his judgment on this occasion.

It was the long-awaited transit of Venus across the face of the sun, due in June 1761, that led to the worst disputes of all. Following the death in mid-1760 of Grischow, professor of astronomy and director of the Academy's observatory, Aepinus was temporarily placed in charge of the Academy's astronomical activities while he trained Stepan Rumovskiĭ, his friend from his days in Berlin and now an adjunct in the St. Petersburg Academy, to the point where Rumovskiĭ could take over.[130] This process took time, however, and in the meantime, within a few months of assuming his new responsibilities, Aepinus contributed an essay on the forthcoming transit to the journal *Works and Translations of Benefit and Amusement*. In this he explained why it was that the scientific world was awaiting the transit with such eagerness, and outlined the considerations that astronomers

---

[129] Pekarskiĭ, *Istoriya*, II, 701.

[130] Lomonosov, *Polnoe sobranie sochineniĭ*, IX, 341. There is a suggestion in a letter from Müller to Euler (June 1760, Yushkevich et al., *Die Berliner und die Petersburger Akademie*, I, 151) that Aepinus was first offered the position on a permanent basis but turned it down.

had to take into account in their calculations concerning this event.[131]

Aepinus's essay provoked a furious outburst from Lomonosov,[132] who found its approach far too condescending. Aepinus's decision to avoid using technical terms such as "ecliptic" and "horizon" and to rely on more common phrases was, Lomonosov insisted, an insult to those for whom the piece had been written. Furthermore, he proclaimed, Aepinus's diagram illustrating the path of the planet across the sun's disc was seriously in error, because it took no account of the fact that the orbit of Venus was inclined at a small angle to the ecliptic. Lomonosov began spreading reports around the city that Aepinus's article was worthless—at least he made no attempt to deny having done so when Aepinus accused him of it. Aepinus responded by preparing a defense of what he had written, with the intention of having it included in the next month's issue of the journal. A week after he had lodged this with the Academy, Lomonosov read a spirited rejoinder at a session from which Aepinus, in order (so he said) to avoid further disputes, had purposely absented himself.[133] In the end, neither Aepinus's defense nor Lomonosov's rejoinder was published, but Aepinus's original article was re-issued separately as a pamphlet by the Academy Press.[134]

In this particular case, Lomonosov's dispute with Aepinus clearly originated in their differing conceptions of the function of the Academy's monthly journal. Aepinus took the title of the journal literally. He wrote his article simply for the amusement and edification of his readers. As he made clear at the beginning of the article, it was meant merely to give some understanding of what he and his colleagues were doing, and why they were doing it: "My intention is solely to give society an adequate idea of this phenomenon, and to show the real and incomparable benefit that the whole human race as well as the sciences can expect from a precise observation of it." [135] Aepinus conceded at once when challenged by Lomonosov that he had ignored the inclination of the orbit of Venus toward the ecliptic in preparing his diagram. But he clearly regarded this as a relatively unimportant factor whose incorporation into his discussion might well

---

[131] Aepinus, "Izvestiye o nastupayushchem prokhozhdenii Venery mezhdu Solntsem i zemleyu," *Sochineniya i perevody*, October 1760, pp. 359–71.

[132] *Protokoly*, II, 459–60; reprinted in Lomonosov's *Polnoe sobranie sochineniĭ*, IV, 325–31.

[133] *Protokoly*, II, 459–60. Cf. AAN, raz. v, э–7, no. 28.

[134] See Appendix below, item **12A**. Unfortunately, no indication is given as to who authorized this publication.

[135] Aepinus, *Sochineniya i perovody*, October 1760, p. 360.

have rendered the whole thing too complicated for the majority of his readers. Lomonosov, however, had wanted much more than a simplified and "popularized" account. He envisaged the article's being used by untrained observers in the remote parts of Siberia as a detailed set of instructions as to what they might expect to see.[136] More generally, he assumed that the article would be used by people who themselves would want to make precise observations of the transit, whereas Aepinus had assumed that his readers would want nothing more than to be given a general idea of what was going on. For Lomonosov, that is, Everyman still had an active and useful role to play even in this most abstruse kind of scientific research. For Aepinus, science had become more specialized; it was for him the prerogative of the scientists alone, theirs to practice and then to explain and justify to the general public.

In 1760, Aepinus's attitude was not yet widely held within the scientific community. Yet there is certainly no need to see in it, as Lomonosov apparently did, a disdain for Russian society in particular. More likely, it simply reflected Aepinus's background: he was after all, trained primarily as a mathematician, and mathematics more than any of the other sciences had become by then a subject open only to the initiated. On the other hand, the continuing intellectual isolation of the St. Petersburg Academy within Russian society could only have encouraged the development of such ideas: the differences in outlook between the scientists and the society that supported them were more marked in St. Petersburg than they were in most other European centers at this time.

As the time of the transit itself approached, Aepinus found himself charged with the primary responsibility within the Academy for all the observations that were to be made throughout the Russian Empire. His duties in this connection were twofold. First, he was to outfit the expeditions the Academy was sending to make observations from two separate locations in Siberia, one under the command of the academician Nikita Popov, the other under Rumovskiĭ, whom once again Aepinus was to instruct in the art of making the necessary observations. Second, he was expected to make observations himself in St. Petersburg.

Unfortunately, Aepinus's preparations for the latter task created further friction with Lomonosov. In order, he said, to avoid being disturbed as he made the observations during the transit, Aepinus refused to share the Observatory with two of Lomonosov's protégés,

[136] Bilyarskiĭ, *Materialy*, p. 479.

Andreĭ Krasil'nikov and Nikolaĭ Kurganov; and he persisted with his refusal even in the face of a direct instruction from Lomonosov (using his authority as a councillor of the chancellery) to give way. An appeal was promptly lodged with the Senate, which ruled against Aepinus. When the latter presented a memoir stating again his reasons for wanting to observe the transit alone, Lomonosov drew up an extraordinarily vituperative point-by-point reply.[137] He insinuated that it was not interruptions that Aepinus feared, but the presence of witnesses to his clumsiness and ignorance as an observer. And what did it matter anyway, Lomonosov wanted to know, if those witnesses by their shuffling around should drown out the ticking of the clock? Any competent observer should be able to count off the seconds just as well in his head! He insisted that, despite the aspersions Aepinus had cast upon the abilities of his two nominees, they were in fact widely acknowledged (by whom, it is not clear) to be able astronomers. In another document from the same period, Lomonosov denigrated Aepinus's own prior experience in astronomy, asserting that in Berlin "there is not a single good instrument, but perhaps a rusty quadrant."[138]

It is hard to tell at this distance how much Aepinus's objections to Lomonosov's protégés were based on genuinely held grounds, and how much on his by then lively antipathy toward Lomonosov and all his works. At one point, the latter alleged that Aepinus, despite his professed need for isolation if he was to make the observations properly, had in the meantime invited along a number of his nonscientific cronies to keep him company.[139] If this was the case, Aepinus was certainly less than candid in his protestations. But Aepinus's earlier career as a scientist makes it clear that Lomonosov's response to his argument (a response which has, unhappily, been accepted at face value by far too many historians) was entirely without foundation. Lomonosov's insinuations that Aepinus was incompetent are totally refuted by the contrary and extremely high opinion that such men as Euler and Tobias Mayer held of his talents as an observational astronomer. Lomonosov's disparaging comment about the standard of the equipment that had been available to Aepinus in Berlin was also completely false. His memoranda were simply drawn-out screams of rage against a stubborn opponent; they are worthless as objective assessments of Aepinus's work.

The most unfortunate aspect of this whole sorry story was its out-

[137] Pekarskiĭ, *Istoriya*, II, 730–33.
[138] Bilyarskiĭ, *Materialy*, p. 493.
[139] Pekarskiĭ, *Istoriya*, II, 731.

come. When the Senate's decision in Lomonosov's favor was upheld, Aepinus washed his hands of the transit altogether, and left Krasil'nikov and Kurganov to observe it unaided. As he had predicted, their efforts turned out to be unsatisfactory.[140] Meanwhile Aepinus himself, undoubtedly the best qualified person in all Russia at that time to make the critical observations, sat idle at home deprived of the instruments he needed for the task.

Never again did the dispute between the two men flare up so violently; indeed, from this point on, though considerable ill-feeling evidently remained, things quieted down considerably. This did not mean, however, that Aepinus was again free to pursue his research without interruption, because other factors now intervened that soon afterward were to disrupt his scientific work even more completely. At the very time that the dispute with Lomonosov was building up, Aepinus also found himself launched on a new career as a courtier. His success in this direction, indeed, only exacerbated the situation with Lomonosov, because the latter passionately resented his rival's rapid social advancement; and in this regard it is easy to feel considerable sympathy for Lomonosov, since Aepinus's success came remarkably quickly, and his fortunes prospered exceedingly—at the expense of his scientific pursuits—just when some of Lomonosov's own favorite projects were suffering for want of support.

Aepinus's rise to imperial favor may have begun with his public lecture before the Empress Elisabeth and her court on 7 September 1758. But whether it was then or on some subsequent occasion that he first attracted attention, what is certain is that only a few months later, in early 1759, he was commissioned to prepare an exposition of the "system of the world" for the edification and amusement of the then Imperial Grand Duchess, soon to become the Empress Catherine the Great. His essay was extremely well received—so well, indeed, that he was appointed there and then by Catherine as her personal tutor in natural philosophy. And from this time onward, we are told, he was always treated by Catherine with marks of "the most exceptional favor." [141]

[140] Ibid., p. 734.
[141] Koppe, *Jetztlebendes gelehrtes Mecklenburg*, I, 12–13; also Rumovskiĭ to J. A. Euler, 22 November/2 December 1764 (AAN, fond 1, op. 3, no. 46, l. 274–75), and Aepinus's own dedication to his essay, which was published a few years later under the title *Beschreibung des Welt-Gebäudes*, St. Petersburg, 1770. The entry on Aepinus in the *Russkiĭ biograficheskiĭ slovar'*, St. Petersburg, 1896–1918, XXIV, 266–67, errs in dating his winning of Catherine's favor to 1763, and to the public lecture he gave in July of that year. On Aepinus's essay, see R. W. Home, "The scientific education of Catherine the Great," *Melbourne Slavonic Studies*, no. 11, 1976, 18–22.

Presumably as a result of his success with Catherine, Aepinus was appointed early in 1761 as director of studies at the Imperial Corps of Noble Cadets. This evidently carried considerable prestige, since his new position led to his being advanced one degree in the imperial table of ranks, to the level of collegial counsellor (*kollezhskiĭ sovetnik*).[142] It also doubled his income: when Aepinus had first been appointed to the St. Petersburg Academy, his salary was 860 roubles per annum;[143] now, for his work at the Corps of Cadets, an extra 900 roubles per annum were added to this.[144] Furthermore, this brought Aepinus into contact with an entirely different social group from the one in which he had been working at the Academy, because the Corps of Cadets was the institution at which the children of the upper levels of the Russian nobility received their education. In all, therefore, the effect of Aepinus's new appointment was to set him somewhat apart from the other members of the Academy. He was now in a much higher income bracket than most of his colleagues, and his new rank placed him, socially, above all but the most senior of them. (Lomonosov, for example, had reached the rank that Aepinus now enjoyed only after many years' service, when he was appointed a councillor in the Academic chancellery, and he was enraged by the younger man's sudden promotion to the same rank; he was even more upset when his efforts to win himself a new promotion proved unavailing.)[145] Aepinus's new appointment also considerably reduced the amount of time he was able to devote to his science, for it was certainly no sinecure. Lomonosov soon accused him of spending more time on frivolous social pursuits than on his academic duties, while Euler, foreseeing what would happen, disapproved of the whole thing[146] (though a few years later, while negotiating the reform of the Academy, he found it useful to cite Aepinus as an example of the way academicians could play a role in the wider life of the nation).[147]

Nevertheless, Aepinus was not yet removed entirely from his science or from the Academy. Throughout 1760 and 1761 he continued to read papers regularly at academic meetings, and in July 1762 another major work on electricity, his *Recueil de différents mémoires sur la*

[142] Müller to Euler, 24 April/5 May 1761, Yushkevich et al., *Die Berliner und die Petersburger Akademie*, I, 173.
[143] AAN, fond 21, op. 3, no. 3, 1. 6.
[144] Lomonosov, *Polnoe sobranie sochineniĭ*, x, 690.
[145] Pekarskiĭ, *Istoriya*, II, 720.
[146] Ibid., p. 698. Euler to Müller, 31 July/11 August 1761, Yushkevich et al., *Die Berliner und die Petersburger Akademie*, I, 177. Cf. also Müller's letter to Euler, 13/24 July 1761, ibid., p. 176.
[147] Euler to Müller, 9/20 August 1766, ibid., p. 263.

tourmaline, was issued by the Academy Press.[148] In this, Aepinus republished the historic paper on the electricity of the tourmaline that he had read to the Berlin Academy some years earlier, and that had since appeared in the Academy's *Mémoires.*[149] But now he backed up his argument with a detailed account of the experiments upon which the paper had been based. He also included in the book two papers that had since been published challenging some of his conclusions, and he set out his replies to these in detail. In all, the book constituted as complete a statement as could be given of the current state of knowledge in this area.

While still a young man in Rostock, Aepinus had brought his work to the attention of the Swedish Royal Academy of Sciences; then, the Academy had not even deigned to reply to his letter. How completely the situation had now been transformed! In July 1760, almost certainly at the urging of Wilcke (who had gone home to Stockholm a few months after Aepinus left Berlin for Russia), but apparently without any prompting from Aepinus himself, the Swedish academicians were pleased to elect him a foreign member of their organization.[150]

Aepinus evidently continued to be highly regarded by his fellow academicians in St. Petersburg too. Twice more, in 1761 and again in 1763, he read one of the papers at the Academy's annual public meeting before the reigning empress (Elisabeth in 1761, Catherine in 1763) and the imperial court. On the first of these occasions, he presented some extremely interesting and typically lucid speculations concerning surface temperature distributions around the Earth, basing several of his arguments on data collected during the great ten-year expedition to Siberia that the Academy had organized some years earlier.[151] Aepinus clearly recognized the importance of the world's oceans in moderating climatic conditions, and he made some perceptive suggestions as to why they were able to function in this way. Discounting on the basis of thermometric evidence the suggestion that salt water was less susceptible than other things to cold, he argued that it was simply the internal mobility of the oceans, and the consequent need to vary the temperature of a whole body of water and not just its surface, that was the crucial factor involved. In contrast, he

---

[148] See Appendix, item **14A.**

[149] *Mém. Acad. Berlin,* XII, 1756 (publ. 1758), 105–21.

[150] E. W. Dahlgren, *Kungl. Svenska Vetenskapsakademien Personförteckningar 1739–1915,* Stockholm, 1915, p. 121.

[151] Aepinus, *Cogitationes de distributione caloris per tellurem,* St. Petersburg, 1761.

pointed out, on land, seasonal variations of temperature affected only the surface layers of ground; and here he cited in support of his position reports that had been made to the Academy concerning permafrost conditions in Siberia, and the superficial nature of the summer thaw there.

Aepinus's 1763 paper [152] was a much less substantial piece of work. Interestingly, however, it too was concerned with large-scale meteorological questions, and it ended with a broad hint to his patron, Catherine, that only in a far-flung realm such as hers could a systematic series of meteorological observations be undertaken with any hope that they would lead eventually to an understanding of the factors affecting climate.

Aepinus, indeed, seems to have developed a general interest in geographical matters at this period. This was very likely inspired by his new responsibilities in the Academy's observatory following Grischow's death in 1760. One of his first duties in this connection was to present a formal opinion on the Academy's plans for the mapping of European Russia,[153] and soon afterward he became involved in the planning of the Rumovskiĭ and Popov expeditions to Siberia to observe the transit of Venus. In addition to the two geographically oriented addresses Aepinus delivered at public meetings of the Academy, he was responsible for new editions in both German and Russian of Georg Wolfgang Krafft's standard textbook on geography, his *Kurze Einleitung zur mathematischen und natürlichen Geographie . . . zum Nutzen der russischen studirenden Jugend*, brought out in 1764.[154] Presumably these were published for his students at the Corps of Cadets. But Krafft's book had first been published in 1738. Aepinus therefore included in the new edition an extensive series of footnotes drawing attention to new discoveries made during the intervening years, and these reveal his impressive command of the general literature of the subject. In fact, by this time he had clearly become a recognized authority on geographical affairs, and, naturally, the authorities sought his advice on a number of occasions. In addition to becoming further involved in the Academy's mapping projects, he was also consulted by the Admiralty in 1765 in connection with the naval expedition (the second in as many years) that was to be sent out under Captain Chichagov in search of a northwest passage into the Pacific

---

[152] Aepinus, *Abhandlung von den Luft-Erscheinungen*, St. Petersburg, 1763.

[153] V. F. Gnucheva, *Geograficheskiĭ Departament Akademii Nauk XVIII veka*, Moscow/Leningrad, 1946, p. 70.

[154] See Appendix, items **16A** and **16B**.

Ocean.¹⁵⁵ In 1768, when the Academy wanted to reorganize its Geography Department, Aepinus was coopted onto the commission that was set up to inquire into the matter.¹⁵⁶

Aepinus's scientific productiveness had dried up some years before this, however. Although articles by him continued to appear in the Academy's *Novi commentarii* throughout most of the 1760s, a study of the *procès-verbaux* of the Academy's weekly meetings reveals that these were all read to the Academy in the early years of the decade.¹⁵⁷ Despite Lomonosov's insinuations, it was not a taste for frivolity but the increasing burden of his other duties that lay behind the falling-off. Granted that it does not seem to have taken Aepinus long to have had annulled Lomonosov's decree concerning his lecturing duties in the Academic University—there is some doubt whether it was ever enforced ¹⁵⁸—his new pedagogical activities at the Corps of Cadets severely curtailed the amount of time he was able to devote to his science. Furthermore, what time he had for this tended to be taken up more and more by administrative matters rather than by research. The administration of the observatory turned out to be particularly demanding, the more so because it was under Aepinus's direction that this was eventually rebuilt after having been badly damaged by fire some years earlier.¹⁵⁹

In addition, Aepinus, like many of the other academicians, was regularly called upon to prepare statements of opinion on a wide variety of matters. A few examples illustrate the range of topics dealt with. In July 1761 a number of people were asked to comment on a scheme of Lomonosov's to have samples of various rocks, sands, and clays collected and sent to the Academy for examination for possible traces of useful minerals. Several of those asked commended the scheme, but others expressed doubts about the practicability of the method proposed for gathering the samples (about which Lomonosov himself, it seems, had doubts). Aepinus and Zeiher condemned the method outright as impracticable and unreliable.¹⁶⁰ On another occasion, in November 1761, Aepinus was asked by the Academic Chancellery for comments on the calendar for 1762 that had been prepared

---

¹⁵⁵ Gnucheva, *Geograficheskiĭ Departament*, p. 77. V. A. Perevalov, *Lomonosov i Arktika*, Moscow/Leningrad, 1949, pp. 404–405. The admiralty had previously depended upon Lomonosov's advice on such matters, but he died in early 1765.

¹⁵⁶ Gnucheva, *Geograficheskiĭ Departament*, p. 87.

¹⁵⁷ *Protokoly*, II, passim.

¹⁵⁸ Lomonosov, *Polnoe sobranie sochineniĭ*, x, 235.

¹⁵⁹ *Protokoly*, II, 502; cf. Lomonosov, *Polnoe sobranie sochineniĭ*, IX, 341.

¹⁶⁰ Bilyarskiĭ, *Materialy*, p. 536.

by Lomonosov's protégés Krasil'nikov and Kurganov. Aepinus reported "large and inexplicable" errors in the calculations, which led the Academy to transfer responsibility for the preparation of future calendars to Aepinus's friend and pupil Stepan Rumovskiĭ.[161] In September 1763, an Academic Commission was set up, with Aepinus as one of its members, to discuss ways and means of implementing a recently promulgated imperial decree establishing an agriculture section within the Academy.[162] During 1764, Aepinus, together with the apothecary Model and the mineralogist Lehmann (who now was also an academician in St. Petersburg), was instructed to prepare a reply to a series of questions from the governor of the Orenburg province concerning the "geographic, historical and physical description" of a salt deposit recently discovered in that district.[163] In the same year, Aepinus was active in obtaining full membership of the Academy for the young historian August Ludwig Schlözer, then living in St. Petersburg and later to acquire lasting renown as a professor at Göttingen.[164] Some months later, when the Academy was considering what should be done to round off the education of one of its students, Aepinus submitted a very sensible set of proposals that was eventually adopted by his colleagues.[165]

During the early 1760s, Aepinus was clearly one of the most respected St. Petersburg academicians, even though he was spending only part of his time on academic affairs. His outstanding intellectual abilities were widely recognized—Schlözer subsequently described him as "indisputably the most learned person in the whole Academy at that time" [166]—and so too was his judgment in administrative matters. He was regularly appointed to important academic committees; and in 1767, when an election was held within the Academy to choose a representative to the Empress Catherine's famous Commission on Laws, only G. F. Müller and Jakob von Stählin, the two most senior professors, received more votes than he did.[167] His life seemed to have

---

[161] Ibid., p. 549.

[162] Lomonosov, *Polnoe sobranie sochineniĭ*, VI, 601.

[163] N. M. Raskin, *Khimicheskaya laboratoriya M.V. Lomonosova*, Leningrad, 1962, p. 206.

[164] Bilyarskiĭ, *Materialy*, p. 701.

[165] *Protokoly*, II, 540; for Aepinus's actual proposals, see AAN, fond 3, op. 1, no. 287, ll. 27–28.

[166] *August Ludwig Schlözers öffentliches und Privatleben, von ihm selbst beschrieben. Erstes Fragment*, Göttingen, 1802, p. 198. The Abbé Chappe d'Auteroche also expressed a high opinion of Aepinus's talents at about this time, in his *Voyage en Sibérie fait par ordre du Roi en 1761*, Paris, 1768, I, 210.

[167] *Protokoly*, II, 592.

settled into a very .comfortable pattern with part of his time spent teaching at the Corps of Cadets and the rest spent in administrative duties, plus a little research, at the Academy. But not for long! Early in 1765, Aepinus's circumstances were transformed yet again: the Empress, as a further mark of her favor, appointed him one of the tutors to her son, the Tsarevich Paul Petrovich.

It was this new appointment that finally brought to an end Aepinus's career as a professional scientist. It required him to wait upon his royal pupil each day, and this made it impossible for him to attend sessions of the Academic Conference. Catherine, recognizing this, issued a special decree whereby Aepinus was to retain all the rights and benefits of his academic position (including a salary that had by then increased to 1000 roubles per annum) but was at the same time relieved of all his academic duties. The Academy was authorized simultaneously to appoint a new professor of physics to take over those duties.[168]

As a result of these new arrangements, Aepinus now found himself drawing three salaries simultaneously, each of 1000 roubles: one from the Academy, one from the Corps of Cadets, and one from the grand duke's allowance to cover the latest appointment.[169] It appears that he relinquished his position at the Corps of Cadets within a few months, but it is unlikely that he would have suffered a reduction in salary thereby, since at about the same time he began to be seconded with increasing frequency to official governmental duties in the College of Foreign Affairs.[170]

By this time Aepinus was clearly well launched on a career as courtier and civil servant, and he retained only tenuous links with the Academy. Occasionally his former colleagues called upon him for advice—most notably in connection with the reform of the Geography Department in 1768 and the preparations for observing the second transit of Venus in 1769, but, such incidents apart, he had left his science behind.[171] For a number of years most of his time must have been devoted to the bringing up of his royal pupil, but as the young Tsarevich grew to manhood, his duties in this regard would have be-

[168] AAN, fond 3, op. 1, no. 2240, l. 26.
[169] Rumovskiĭ to J. A. Euler, 22 November/2 December 1764; AAN, fond 1, op. 3, no. 46, ll. 274–75.
[170] Aepinus was listed among the staff of the corps in the official calendar for 1765 (*Adres-kalendar' rossiĭskiĭ na leto ot Rozhdestva Khristove 1765*, St. Petersburg, 1765, p. 47) but not in the volume for 1766. S. A. Poroshin, *Zapiski, sluzhashchiya k istorii Ego Imperatorskago Vysochestva . . . Pavla Petrovicha*, 2nd. ed., St. Petersburg, 1881, col. 279.
[171] See above, p. 51. *Protokoly*, II, 641, 689.

come less onerous. Simultaneously, however, the College of Foreign Affairs seems to have begun calling upon his services with increasing frequency, and it was almost certainly in recognition of this that he was promoted a further step up the imperial table of ranks some time during the first half of 1769, to the level of councillor of state (*statskiĭ sovetnik*).[172] It was probably at this time, indeed, that he came to be looked on as part of the regular establishment of the College; certainly he was so regarded by the beginning of 1771 at the latest.[173]

Aepinus's activities at the College of Foreign Affairs illustrate one fascinating yet generally unrecognized way in which mathematical talents were put to use in the eighteenth century, since the post he occupied within the College was head of the Cipher Department.[174] Nor was he the first academician to be appointed to this sensitive office: from 1742 until his death in December 1764, Christian Goldbach, one of the foundation professors who had been brought to St. Petersburg in 1725, had performed the same duties.[175] It was clearly no coincidence that within a couple of months of Goldbach's death Aepinus was first called upon for help. The episode provides a striking illustration of the way in which the eighteenth-century Academy was able to serve the Russian government as a source of expertise that it would otherwise have found extremely difficult to procure. Aepinus evidently carried out his new tasks to the satisfaction of his employers, since in 1773 or 1774 he was promoted again, to the rank of active councillor of state (*deĭstviteĺnyĭ statskiĭ sovetnik*), and by 1780 only two officials in the college were ranked ahead of him, namely Panin, the minister, and Osterman, the vice-chancellor.[176]

Very little evidence has survived from this period of Aepinus's life. We know that he joined the important and influential St. Petersburg Free Economic Society in 1766, soon after its foundation, and that in

[172] Cf. ibid., p. 689.

[173] *Adres-kalendar'*, 1771, p. 69. I have not been able to consult a copy of the volume for 1770, but Aepinus's name was not included in the College entry for 1769 (which would, of course, have been printed quite early in that year).

[174] P. van Wonzel, *Etat présent de la Russie*, St. Petersburg/Leipzig, 1783, p. 85. In the annual volumes of the *Adres-kalendar'*, Aepinus's position was consistently characterized as "pri osoblivykh dolzhnostyakh."

[175] A. P. Yushkevich and E. Winter, eds., *Leonhard Euler und Christian Goldbach: Briefwechsel 1729–1764*, Berlin, 1965, p. 5.

[176] *Adres-kalendar'*, 1774, p. 66. I have not seen the volume for 1773. Ibid., 1780, p. 66. A sample of Aepinus's work as a cryptographer survives among his papers in Leningrad (AAN, rav. v, op. 9–7, no. 32). His technique is discussed in Albert C. Leighton, "Some examples of historical cryptanalysis," *Historia mathematica*, IV, 1977, 319–37.

1770 he actually served a term as its president.[177] In the latter year he published an article in the society's journal on the use of lightning conductors for the protection of buildings and saw through the press both German and Russian editions of the essay on the structure of the world that he had drawn up for Catherine more than a decade before.[178] The article on lightning conductors indicates that, even though Aepinus was no longer a practicing scientist by this time, he still kept up with at least some of the latest scientific literature. The article contains several extended quotations from the most recent edition of Franklin's *Experiments and Observations on Electricity*, published in London only a year earlier. A few years after this, Johann Bernoulli III insinuated after a visit to St. Petersburg that Aepinus by then preferred to forget his scientific past, but to judge by the indignant reaction of those who knew him, the charge was quite ill-founded.[179] What is more, we know that at about this time, in 1778 in fact, Aepinus took the trouble to have one of the newly developed achromatic telescopes sent to him from England,[180] and that, far from wishing to play down his scientific background, as Bernoulli alleged, he was inspired by the arrival of this instrument to bring out several new and quite important scientific publications during the succeeding few years.

Aepinus's new flurry of scientific activity was concerned with two very different subjects. On the one hand, his observation of the moon through his new telescope prompted him to put forward some very interesting suggestions as to the likely origin of the multitude of craters he could see so clearly. He was struck by the resemblance between the lunar surface and that famous volcanic region around Naples, the Campi Phlegraei, which had been so splendidly depicted in Sir William Hamilton's recently published work, *Campi Phlegraei* (Naples, 1776). This led Aepinus to speculate that the lunar surface, too, had been largely formed by volcanic action. Though prevented

---

[177] H. Mohrmann, *Studien über russisch-deutsche Begegnungen in der Wirtschaftswissenschaft (1750–1825)*, Berlin, 1959, p. 111.

[178] Aepinus, "Razsuzhdenie o sredstvakh sluzhashchikh k otvrashcheniyu povrezhdeniya proizvodimago gromovym udarom i molnieyu domam i drugim stroeniyam," *Trudy Vol'nago Ekonomicheskago Obshchestva*, chast' XVI, 1770, 257–85. Appendix, items **17A, 17B(i)**.

[179] Johann Bernoulli, *Reisen durch Brandenburg, Pommern, Preußen, Curland, Rußland und Pohlen, in den Jahren 1777 und 1778*, Leipzig, 1779–80, IV, 19–20. A. Lauch, *Wissenschaft und kulturelle Beziehungen in der russischen Aufklärung zum Wirken H.L.Ch. Bacmeisters*, Berlin, 1969, p. 360.

[180] Aepinus, "Ueber den Bau der Mondfläche . . . ," *Schriften der Berlinischen Gesellschaft Naturforschender Freunde*, II, 1781, 1.

from writing up his ideas for a time by the pressure of his official duties, he eventually did so in his usual careful, cogently argued manner and sent a copy of the paper to Hamilton, and another to Berlin, where it was published in the *Schriften* of the recently formed Berlinische Gesellschaft Naturforschender Freunde.[181] It is not clear why Aepinus chose to have the paper published in Germany rather than in St. Petersburg—the covering letter he sent with it sheds no light on this—but his action prompted the Berlin Society to elect him an honorary member forthwith.[182]

The other line of research prompted by his acquisition of the achromatic telescope earned Aepinus membership in yet another scientific organization. For some months at the end of 1783 and the start of 1784, his health broke down and he was forbidden by his doctor to occupy himself with anything "except what could afford the soul peace and contentment" [183] (a harbinger here, perhaps, of the mental collapse that was to bring his career to a close in the 1790s). Under these conditions, Aepinus turned once more to his science and took up again a long-held ambition to produce a better microscope than those then generally in use. The lenses from his new telescope were ready to hand, and he used them successfully to construct the first achromatic microscope. This was in fact nothing more than a simple adaptation of the achromatic telescope, modified according to well-known principles of optics for microscopic work. But it worked; and Aepinus reported in rapturous terms on the clarity and beauty of the images it produced. So marked was the improvement in performance, he felt, that he confidently expected that microscopes fashioned according to his new design would everywhere quickly supplant those commonly in use. However, he did not seek to profit from this invention personally. Rather, he proclaimed his belief that this belonged not to himself alone but to mankind in general, and he hastened to draw up a pamphlet setting out full details of its construction and the advantages it afforded in use.[184]

In Göttingen, at least, the invention was much appreciated.

---

[181] Ibid. On the significance of this paper, see R. W. Home, "The origin of the lunar craters: an eighteenth-century view," *Journal for the History of Astronomy*, III, 1972, 1–10. Concerning Hamilton's receiving a copy, see P. I. Bartenev, ed., *Arkhiv Knyazya Vorontsova*, Moscow, 1870–95, XXIX, 222.

[182] Archives of the Berlinische Gesellschaft Naturforschender Freunde, Museum für Naturkunde an der Humboldt-Universität zu Berlin: Aepinus to J. G. Gleditsch, 27 October/7 November 1780, and the Society's minutes, 23 January 1781.

[183] [Aepinus], *Description des nouveaux microscopes, inventés par Mr. Aepinus*, St. Petersburg, 1784, p. v.

[184] Ibid.

Aepinus's friend, the physician Georg von Asch, regularly corresponded with the Society of Sciences there, and he quickly sent the society a copy of Aepinus's initial report. Von Asch had actually looked through Aepinus's microscope, and could testify personally as to its merits. An account of the invention was promptly published in the *Göttingische Anzeigen von gelehrten Sachen*,[185] and Johann Friedrich Blumenbach, the famous Göttingen anthropologist, constructed one of the instruments in accordance with the instructions Aepinus had given. His report on its performance was highly favorable, and the members of the Göttingen Society were sufficiently impressed to elect Aepinus a corresponding member of their organization.[186] Elsewhere, however, Aepinus's hopes for his invention were not realized, and people continued to use microscopes of traditional design until well into the nineteenth century.[187]

Productive though it was, this new burst of scientific activity was no more than a temporary diversion for Aepinus from the heavy burden of his regular official duties. Indeed, his work at the College of Foreign Affairs must have been especially demanding at this particular period because of the hectic round of diplomatic activity surrounding Catherine's sponsorship of the controversial Treaty of Armed Neutrality of 1780. One source even credits Aepinus with the actual authorship of the treaty document, but Catherine herself tells us it was the work of someone else.[188] On at least one occasion, an attempt seems to have been made to use Aepinus's earlier career in science to open channels of diplomatic communication. In early 1783, news reached St. Petersburg that Franklin and the other American representatives in Paris had signed a preliminary peace treaty with the British and thereby brought the American War of Independence to a successful conclusion. Aepinus promptly wrote to Franklin, carefully drawing attention beneath his signature to his own involvement in diplomatic affairs as a servant of the Russian government. He recorded his gratitude for the way in which the American had formerly

---

[185] *Göttingische Anzeigen von gelehrten Sachen*, 5 August 1784; and also, with greater detail, 18 October 1784.

[186] A. Buchholz, *Die Göttinger Rußlandsammlungen Georgs von Asch*, Giessen, 1961, pp. 106–108.

[187] For a general discussion of Aepinus's microscope, see Sobol', *Istoriya mikroskopa*, pp. 331–39.

[188] J. Castéra, *History of Catharine II. Empress of Russia*, London, 1800, p. 433. Catherine is quoted in I. de Madariaga, *Britain, Russia and the Armed Neutrality of 1780*, London, 1962, p. 174. This work describes in detail the circumstances surrounding Catherine's sponsorship of the treaty.

approved his scientific endeavors, and congratulated him on having secured his country's liberty in the recent negotiations.[189]

Shortly before this, Aepinus had had yet another set of responsibilities thrust upon him. In May 1780, Catherine learned from the Emperor Joseph II of Austria of the success of a new school system that had recently been established throughout the Habsburg dominions. Her interest aroused, she sought the advice of a number of people as to what might be done along these lines in Russia. Aepinus was one of those whom she consulted, and he proceeded to draw up a set of detailed recommendations for her consideration.[190]

Broadly speaking, Aepinus proposed the introduction into the Russian Empire of the Austrian school system in its entirety. This was a three-tiered system comprising elementary schools (*Trivial-* or *Landschulen*) in the parishes, secondary schools (*Haupt-* or *Stadtschulen*) in the towns, and higher schools (*Normalschulen*) in the various provincial capitals. These latter were intended chiefly to supply trained teachers to staff the two lower grades of school, and each of them was to incorporate its own *Trivialschule* and *Hauptschule*. The *Normalschulen* should, Aepinus said, be established first, and then the rest of the scheme should be gradually implemented as more and more teachers completed their training. Aepinus was convinced that the supply of teachers was the central problem to be overcome. In Russia, he argued, there were very few suitably educated people who could be recruited into the scheme at the outset, and so the system would for the most part have to generate its own supply. But this meant that expansion at the lower levels in the system would of necessity be quite slow for a number of years. To get the scheme off the ground, Aepinus suggested that as many teachers as possible who were already trained in the system should be recruited from the Orthodox and Slavic-speaking parts of the Austrian dominions. Their first students should be pupils brought together from the existing Russian schools and seminaries, who should all be destined to become teachers in due course. Austrian textbooks, amended where necessary to take account of local conditions, should be translated into Russian and printed for use throughout the system. To supervise the whole, a two- or three-man commission should be set up, with a high-ranking official at its head.

---

[189] Aepinus's letter has been published by N. N. Bolkhovitinov, "1783 god. Peterburg-Filadel'fiya," *Nauka i zhizn'*, 1976 (no. 7), pp. 65–67.

[190] *A. L. Schlözer's Stats-Anzeigen*, Bd. 3, Heft xi, 1783, 260–78. This document was apparently composed some time during 1781, since Aepinus remarked at one point in it that he had by then been in Russia for some twenty-four years.

Catherine found Aepinus's proposals entirely satisfactory, and resolved to implement the scheme at once. The Emperor Joseph sent her copies of the Austrian textbooks, and also approved the transfer to St. Petersburg of one of his principal Serbian school administrators, Teodor Janković, to act as adviser to the Russian officials. Catherine, meanwhile, announced the formation of a three-man Education Commission under the chairmanship of her one-time favorite P. V. Zavadovskiĭ, with Aepinus and the Empress's private secretary P. I. Pastukhov as its other members. Planning proceeded rapidly, with the commission meeting twice weekly, and Janković busily preparing Russian translations of the textbooks. In August 1783, Aepinus and the Privy Councillor and Senator Dietrich Osterwald reported favorably to the empress on the possibility that the many privately operated foreign or "German" schools within her borders should adopt the general principles of the scheme. To encourage them to do so, they proposed that the Lutheran school attached to the Church of St. Peter in the capital be developed, with the help of a government grant, into a model for the remainder. During the same year, the first teacher-training institution opened its doors in St. Petersburg with Janković as its director, and a number of elementary and secondary schools were soon founded in the capital as well. Thereafter the number of schools increased steadily, with a major expansion into the provincial towns in 1786, and a more modest rate of growth in the years after that. By the time the Education Commission was replaced by a full-scale Ministry of Education in 1802, it could look back with some pride on its record. Certainly it had not achieved all that its founders had hoped for it, and it was still the case that only a tiny fraction of Russian children was being educated. Nevertheless, there had been a substantial increase in the numbers of teachers, schools, and children going to school: a solid foundation had been laid for future development.[191]

Yet for Aepinus himself, the commission seems to have turned sour soon after its establishment. As schools began to be established, it became evident that his recommendations were being only partially put into effect. In particular, by 1786, when a new schools statute

[191] D. A. Tolstoĭ, "Gorodskiya uchilishcha v tsarstvovanie imperatritsy Ekateriny II," *Zapiski Imperatorskoĭ Akademii Nauk*, LIV, 1887, Supplement I; S. V. Rozhdestvenskiĭ, *Istoricheskiĭ obzor deyatel'nosti Ministerstva Narodnago Prosveshcheniya 1802–1902*, St. Petersburg, 1902; M. I. Demkov, *Istoriya russkoĭ pedagogii, Chast' II: Novaya russkaya pedagogiya (XVIII-ĭ vek)*, 2nd. ed., Moscow, 1910; N. Hans, *History of Russian Educational Policy 1701–1917*, New York, 1964. For the report on the "German" schools, see Tsentral'nyĭ gosudarstvennyĭ arkhiv drevnikh aktov, Moscow, fond 17, op. 1, ed. khr. 77.

was promulgated, it was evident that, contrary to Aepinus's advice, the system that was emerging differed from the Austrian model in a number of respects. Most notably, instead of the three-tiered Austrian structure, the Russian system involved only two classes of school, the so-called "major" and "minor" schools. The teacher training that Aepinus had recommended as a major part of the curriculum at the third stage, and a crucial part of the system as a whole, was to a large extent eliminated. The single pedagogical seminary that had been opened was maintained in St. Petersburg, but this could not supply the whole empire. As a result, the acute shortage of adequately trained teachers was never overcome.

Undoubtedly a combination of lack of funds and a desire to show quick results determined the kind of school system that Catherine's Education Commission developed. But since it is clear from Aepinus's initial set of proposals that he saw the training of sufficient teachers as the central problem facing the commission, and one that should determine the rate of expansion of the system as a whole, it is certain that he would have bitterly opposed the kinds of changes introduced into his scheme. Some evidence has survived of disagreement between him and Janković, who was almost certainly responsible for the changes.[192] When Aepinus was forced to take leave from his official duties during his illness in 1784, his opponents no doubt profited, and by early 1785 he had clearly lost the fight: an anonymous letter printed in Göttingen in June of that year reported how "the excellent ideas of this perceptive man, and his way of regarding this business, will have no further influence on it." "He himself," the report added, "is already looking on it as something that is no longer any concern of his."[193] A year later, when, as a member of the Education Commission, he had to sign the new statute that had been drawn up, he composed a memorandum to the empress acknowledging the defeat of his own proposals and abdicating responsibility for the form the statute had acquired.[194] Although he continued for many years thereafter to be listed in the official annual calendar as a member of the commission, his active involvement in its work seems to have ceased at this point.

Another affair in which the empress sought Aepinus's advice had a happier outcome. For some years in the 1770s, the director of the Academy of Sciences was S. G. Domashnev, an overbearing and abrasive individual whose conduct led to many complaints from the

---

[192] Castéra, *History of Catharine II*, p. 480.
[193] *Schlözer's Stats-Anzeigen*, Bd. 7, Heft XXVI, 1785, 521–22.
[194] Tolstsoĭ, "Gorodskiya uchilishcha . . . ," p. 213.

academicians. By 1782 these had become so frequent that Catherine appointed a high-level commission of inquiry, of which Aepinus was a member, to look into the matter. The commission's report led to Domashnev's dismissal from office in January 1783 and his replacement by the Princess Dashkova.[195]

By this period of his life, Aepinus had clearly advanced to a very high level in the service of the Russian government. That his standing at court also remained high is signified by the conferral on him in 1783 of a knighthood, first class, in the Order of St. Anne. His friends included the future ambassador to England, Count S. R. Vorontsov, and two leading functionaries at the court of the Grand Duke Paul, F. H. La Fermière and A. L. (later Baron) Nikolaï.[196] He was also evidently on close terms with General F. W. von Bauer, a talented military engineer recruited by Catherine to overhaul the administration of the Russian army; when von Bauer died in 1783, Aepinus acted as executor of his estate.[197]

Another friend was Dr. Matthew Guthrie, a Scottish physician who resided prosperously in St. Petersburg for many years. Both Aepinus and the English traveler William Coxe were interested observers at Guthrie's experiments during the winter of 1784–1785 on the freezing of mercury.[198] In 1789, at Guthrie's instigation, Aepinus drew up some comments in his usual clear-headed style on the relationship between atmospheric temperature, humidity, and the degree to which everyday objects might become electrified by casual friction. Guthrie sent these to his friends in Edinburgh, where they were published soon afterwards in the *Transactions of the Royal Society of Edinburgh*.[199]

At some time prior to 1793, however, and possibly even as early as 1786, Aepinus withdrew almost entirely from polite society and, apparently with Catherine's blessing, from his duties at the College of Foreign Affairs.[200] By the late 1790s, his mental faculties had become

---

[195] K. V. Ostrovityanov, ed., *Istoriya Akademii Nauk S.S.S.R.*, Moscow/Leningrad, 1958–, I, 321.

[196] Aepinus is mentioned from time to time as a mutual friend in letters exchanged between Vorontsov, La Fermière, and Nikolaï over nearly twenty years commencing in the mid-1770s, published in P. I. Bartenev, ed., *Arkhiv Knyazya Vorontsova*, Moscow, 1870–95. For detailed references, see the index volume, *Rospis' soroka knigam arkhiva Knyazya Vorontsova s azbuchnym ukazatelem lichnykh imen*, Moscow, 1897.

[197] AAN, raz. v, op. э–7, nos. 30, 31.

[198] Coxe, *Travels into Poland, Russia, Sweden, and Denmark*, London, 1784–90, III, 249.

[199] Aepinus, "Two letters on electrical and other phenomena," *Trans. Roy. Soc. Edinburgh*, II, 1790, Pt. II, Sect. I, 234–44.

[200] La Fermière, who retired to a country estate early in 1793, reports that Aepinus

seriously impaired.[201] In 1798 his erstwhile pupil, now Tsar Paul, rewarded him for his services to his adopted country with the customary final promotion (in his case to the rank of privy counsellor [*tainyi sovetnik*]), and formally released him from all his duties. He retired to Dorpat in Estonia, and there he died on 10 August 1802. The funeral oration was delivered by the professor of physics at the newly re-opened Dorpat University, G. F. Parrot, with whom Aepinus had become friendly during his last years.[202]

What kind of a man was he, this Aepinus who made such a notable contribution both to eighteenth-century science and to the development of his adopted country? Unfortunately, few of his personal papers have survived, nor have any portraits, so it is hard to arrive at a clear picture of him as an individual. There is an unmistakable earnestness about his early published writings and in the few surviving letters from the early period such as his correspondence with the Royal Swedish Academy. And there can be no question about the diligence with which he pursued his studies, even after his graduation. Yet he was no mere drudge. In virtually the only personal reminiscence that he recorded later in life,[203] he remarked that he had spent many an hour during his youth in the tower of St. Jacob's Church in Rostock admiring the view—and a splendid view it must have been, too, across the harbor to the open country beyond. Talented and ambitious, he yet seems to have made friends easily and—with the notable exception of Lomonosov—to have made few enemies. Although similar characteristics are not necessarily shared by all the members of a family, of course, it is worth recalling how popular was Aepinus's father; and when Thomas Nugent visited Mecklenburg, he found Angelius

---

had already withdrawn from society by then (La Fermière to S. R. Vorontsov, 15 November 1795, *Arkhiv Knyazya Vorontsova*, XXIX, 293). That Aepinus's effective retirement from the College of Foreign Affairs may date from as early as 1786 is suggested by the College's seeking a suitable mathematician then to take up ciphering duties similar to those he had previously performed (S. R. Vorontsov to I. A. Osterman, 6/17 November 1786, ibid., IX, 408–409).

[201] J. M. Wehnert, *Mecklenburgische Gemeinnützige Blätter*, Bd. VII, Parchim, 1803, p. 99; also J. C. Eschenbach, *Annalen der Rostockschen Academie*, Bd. XII, Rostock, 1805, p. 128.

[202] Parrot to the St. Petersburg Academy of Sciences, 14 March 1827, Sobol', *Istoriya mikroskopa*, pp. 342–43. Some years after Aepinus's death, Parrot took the trouble to have two exemplars of the final improved version of Aepinus's telescopic microscope constructed by the Leipzig instrument maker J. H. Tiedemann. One of these survives in Moscow.

[203] *Trudy Vol'nago Ekonomicheskago Obshchestva*, chast' XVI, 1770, 270.

Aepinus a most agreeable host.[204] It seems likely that Franz Ulrich was of a similarly sociable disposition. Certainly he was quickly accepted into the group gathered around Euler in Berlin, and once in St. Petersburg he seems to have soon established friendly relationships with most of the other academicians there. Furthermore, his preferment by Catherine was clearly based on his personal qualities, since there can have been no conceivable political reasons for it. Catherine must have found him personable enough, or he would never have been admitted into her circle of intimates. Yet strangely, he never married, so far as is known (he was certainly still a bachelor in 1764),[205] and we are told that, late in life at least, he had a distinct aversion to writing letters, even to his friends.[206]

While little is known about Aepinus's personality, we know even less about his religious convictions. Given what we know of his father, we may be sure that he would have been brought up along orthodox Lutheran lines. Wolffian rationalism would probably have had a considerable impact on his thinking during his undergraduate days. However, the only writings we have that bear on these matters come from much later in his career, and consist merely of scattered remarks in some of his scientific works about God's wisdom and skill as a creator. There is about these perhaps a slight flavor of that characteristically optimistic form of deism so prevalent in eighteenth-century thought; and Voltaire's aphorism, "The catechist teaches God to children; Newton teaches him to the wise," is actually quoted in German on the title page of Aepinus's *Beschreibung des Welt-Gebäudes* (1770).[207] But beyond this, the evidence is too sparse to suggest any definite conclusions about his beliefs.

Where his work was concerned, Aepinus was self-assured without being arrogant. He wrote confidently as the equal of the famous men in his field such as Franklin and Musschenbroek, and Franklin's highly favorable assessment makes it clear that he had the respect of his contemporaries. He also wrote in a lucid style, which undoubtedly reflects one other important trait of his about which we may speak with com-

[204] Nugent, *Travels through Germany*, I, 192.

[205] Rumovskiĭ to J.A. Euler, 22 November/2 December 1764, AAN, fond 1, op. 3, no. 46, ll. 274–75. An index entry to a supposed wife in A. P. Yushkevich et al., eds., *Leonhardi Euleri Opera omnia, Series Quarta A : Commercium epistolicum*, I, Basel, 1975, 611, is incorrect: the wife mentioned in the letter cited is Euler's, not Aepinus's.

[206] La Fermière to S. R. Vorontsov, 15 November 1795, *Arkhiv Knyazya Vorontsova*, XXIX, 293.

[207] Cf. Voltaire, *Dictionnaire philosophique, portatif*, "London," 1764, p. 43: "un catechiste annonce Dieu aux enfants, et Newton le démontre aux sages."

plete confidence, namely the unfailing clarity and incisiveness of his thinking. From Aepinus's first publication in 1747 to his last in 1790, there is no mistaking his ability to go directly to the heart of whichever problem occupied his attention, and to subject the issues involved to penetrating and detailed scrutiny. This quality is apparent not only in his creative scientific writing but also in the more pedestrian report-writing he was required to do as a St. Petersburg academician, and later in the meticulous drafting of his plan for a school system for the Russian Empire. It was this clarity of mind above all else, rather than any transient spark of genius, that enabled him to make a number of important contributions to the science of his day. And nowhere is it more in evidence than in the work that principally concerns us here, his work on the theories of electricity and magnetism.

CHAPTER TWO

The Electrical Background

The eighteenth century was the golden age of electrical investigation. Restricted to a single obscure and little-known phenomenon at the beginning of the century, by the end electricity had become a major field of scientific endeavor, with ramifications spreading into a number of other disciplines. The basic laws had been established, so far as the phenomena of static electricity were concerned, and a well-articulated general theory had come to be almost universally accepted.

Aepinus played a central role in the development of this theory. When he took up the study of electricity, the principal phenomena had already been identified, and Benjamin Franklin had enunciated the main outlines of a successful theory to account for them. Aepinus's contribution was to identify and resolve, in a way that was vital for the further evolution of the subject, a series of structural difficulties in what Franklin had proposed. Aepinus set Franklinian electrical theory on a sound, precisely formulated and rigorously consistent footing that brought it to the point where it could be rendered completely mathematical once the law of force between electric charges had been determined, something Charles Augustin Coulomb was to do conclusively in the next generation. At the same time, Aepinus devised a number of ingenious experiments to clarify the phenomena with which he was principally concerned. Here, too, his labors were highly beneficial, for he showed that some of the best-known of all electrical experiments, if observed closely and accurately enough, agreed with the amended theory he had put forward rather than with the versions that had been proposed by Franklin and his supporters. Aepinus not only strengthened the Franklinian theory immeasurably, therefore, in a way that Franklin himself could never have done, he also provided his strengthened version with an impressive new level of empirical confirmation.

At the time Aepinus first took up the study of electricity in earnest, during his two-year sojourn in Berlin in the mid-1750s, electrical theory was in a state of flux. Franklin's revolutionary ideas had been published just a few years earlier, and the debates these had provoked were far from settled. The Abbé Nollet in particular, whose theories had dominated electrical thinking prior to Franklin's appearance on the scene, had vigorously defended his opinions against Franklin's challenge, and indeed had done so to such effect that, in France at least, his views continued to hold sway until well into the 1770s.[1] Meanwhile, in eastern Europe especially, a mechanistic, Cartesian-style aether continued to be invoked by such people as the Eulers and Lomonosov in order to account for the phenomena in a quite different manner again.

It would be of considerable interest to discover what made Aepinus such a convinced upholder of Franklin's cause. Nothing in his background, so far as we know, would have particularly pointed him toward Franklin's mode of thinking. If anything, the reverse is the case. Aepinus would almost certainly have been introduced to the study of electricity by the few pages on the subject in the textbook by G. C. Hamberger that he used in his classes, and here it was approached from a very orthodox and rather old-fashioned point of view. When a body is rubbed, Hamberger suggested, certain relatively coarse particles that it contains are agitated and ejected into the surrounding air. Here they dissolve and form a kind of denser atmosphere around the body, and this in turn gives rise in some unspecified way to the observed effects. If air is not present, Hamberger felt, the atmosphere will not form, and this is why the usual effects are not observed in this case. But Hamberger evidently had doubts about the adequacy of his proposal, because he ended by lamely confessing: "I readily admit, however, that not all the phenomena of electricity can be explained in this way." [2]

In the first (1727) edition of his book, the only electrical phenomena Hamberger discussed were the traditional attractions of nearby light objects to an electrified body, together with some ingenious experiments devised by Francis Hauksbee in the first years of the eighteenth century using linen threads in the vicinity of a rubbed glass globe.[3] He was aware that Hauksbee and others had reported that objects attracted to an electrified body were afterward repelled from it, but he said he had not so far been able to observe this for himself. Subse-

---

[1] On this, see R. W. Home, "Electricity in France in the post-Franklin era," *Actes du XIVe Congrès International d'Histoire des Sciences, Tokyo-Kyoto, 1974*, Tokyo, 1974–75, II, 269–72.

[2] Hamberger, *Elementa physices*, Jena, 1727, pp. 287–90; 3rd ed., Jena, 1741, pp. 451–56.

[3] Francis Hauksbee the elder, *Physico-Mechanical Experiments on Various Subjects*, 2nd ed., London, 1719, pp. 68–78. Hamberger took his account of the experiments from 'sGravesande's well-known text, *Physices elementa mathematica, experimentis confirmata*, Leyden, 1720.

quently he must have succeeded in doing so, because in the third (1741) edition, the one Aepinus probably used, he modified his account to include repulsion along with attraction as a fully attested electrical phenomenon. However, the only other alteration of any significance that he made in this edition was the inclusion of a brief paragraph concerning the electricity reputedly associated with agitated barometers; he failed altogether to mention the important discoveries made by Stephen Gray and C. F. Du Fay during the 1730s on electrical conduction and insulation, or Du Fay's discovery that there are two distinct modes of electrification possible in nature. As a result, Hamberger's discussion, which in 1727 had been reasonably adequate, had become much less so by the time Aepinus used his book in the 1740s.

During the 1740s and early 1750s, a number of much more advanced works on electricity were published in Germany by such authors as Johann Heinrich Winkler, Andreas Gordon, and Georg Matthias Bose,[4] as well as a German edition of Du Fay's electrical papers,[5] an interesting elaboration of the great French investigator's hints about electrical "vortices" by the Halle student C. G. Kratzenstein (whom Aepinus may have met when he visited Halle in the mid-1740s), and an excellent history of the whole subject by Daniel Gralath.[6] Also, the Berlin Academy of Sciences held a prize competition in 1745 on the theme, "the cause of electricity," and four of the entries, including that of the winner, J. S. Waitz, were subsequently published as a collection.[7] Unfortunately it is not known whether the young Aepinus read any of these works. Nor do we know how well he was acquainted early in his career with Nollet's voluminous writings from the same period. Later in life he claimed that he had once been an adherent of Nollet's

---

[4] Winkler, *Gedanken von den Eigenschaften, Wirkungen und Ursachen der Electricität*, Leipzig, 1744; *Die Eigenschaften der electrischen Materie und des electrischen Feuers*, Leipzig, 1745: *Die Stärke der electrischen Kraft, des Wassers in gläsernen Gefässen, welche durch den Musschenbroekischen Versuch bekannt geworden*, Leipzig, 1746. Gordon, *Versuch einer Erklärung der Electricität*, Erfurt, 1745; 2nd ed. enlarged, Erfurt, 1746. Bose, *Recherches sur la cause et sur la veritable téorie de l'électricité*, Wittenberg, 1745; *Tentamina electrica*, Wittenberg, 1744-47.

[5] Du Fay, *Versuche und Abhandlungen von der Electricität derer Körper*, Erfurt, 1745.

[6] Kratzenstein, *Theoria electricitatis more geometrico explicata*, Halle, 1746. Gralath, "Geschichte der Electricität," *Versuche und Abhandlungen der Naturforschenden Gesellschaft in Dantzig*, I, 1747, 175-304; II, 1754, 355-460; III, 1756, 492-556.

[7] Waitz, *Abhandlung von der Electricität und deren Ursachen*, plus three anonymous essays on the same subject, Berlin, 1745. Most bibliographies wrongly show Waitz as the author of the entire collection.

views, but had been driven by the empirical evidence to abandon them in favor of Franklin's ideas.[8] Yet the way he presented this story gives the impression that it was a literary embellishment rather than a strictly accurate historical report.

Apart from Hamberger's rather inadequate discussion, indeed, there is only one set of ideas about electricity that we can be confident Aepinus knew well at an early date. These are the ideas of the Eulers, father and son, with which Aepinus would have become thoroughly familiar soon after he arrived in Berlin in March 1755. The Eulers' thinking, so far as electricity was concerned, lay squarely within the Cartesian mechanical-aether tradition. Hence, like Hamberger's, their views were generally far removed from the kind of theory that Aepinus was soon to espouse. Nevertheless, it seems likely that he drew his initial inspiration from their work. Later on, as his own ideas developed, the older Euler's example would have continued to help him in other and more general ways. In particular, Euler's inclusion of unexplained Newtonian-style forces in his astronomical computations must have helped Aepinus to reconcile himself to doing likewise, though rather more freely, in his own work, even in areas where Euler had not been willing to have recourse to such things.

For the most part it is impossible to separate the thinking of the two Eulers on the subject of electricity. The dissertation that Johann Albrecht Euler entered late in 1754 for the St. Petersburg Academy's prize competition was based on certain of his father's ideas, but we do not know which ideas among those set out in the essay belonged to the older man.[9] For the purpose of this discussion, no attempt will be made to distinguish the contributions of the two men, and for ease of reference I shall write as though the entire scheme was the creation of the son alone.

In his 1754 essay, young Euler sought to explain all the phenomena of electricity in terms of particular modifications of a subtle aether that filled the vast reaches of the heavens and also pervaded the pores of all ordinary matter. Like many other eighteenth-century scientists, he took the existence of such an aether for granted, and, although he did not say so in his essay, his principal authority for doing so was his

---

[8] Aepinus, *Recueil de différents mémoires sur la tourmaline*, p. 134.

[9] J. A. Euler, "Disquisitio de causa physica electricitatis," in *Dissertationes selectae Jo. Alberti Euleri, Paulli Frisii et Laurentii Beraud quae ad Imperialem Scientiarum Petropolitanam Academiam An. 1755 missae sunt* . . . , Petropoli/Lucae, 1757. Euler's paper was also included in a similar collection published in St. Petersburg in the previous year, *J. A. Euleri Disquisitio de causa physica electricitatis . . . una cum aliis duabus dissertationibus de eodem argumento*. I have not seen this.

father, who for many years had steadfastly maintained that such an aether existed and acted as, among other things, both the mechanical cause of gravity and the carrier of the undulations that in his view constituted light. Young Euler argued that, so far as the cause of electricity was concerned, recourse to some kind of subtle matter was essential; simply to invoke attractive and repulsive forces of some kind was not good enough, first because such forces were in themselves mere occult qualities, and second because the phenomena in question displayed such manifest signs of the presence of a subtle matter that its involvement could not be doubted. Yet if some subtle matter had to be invoked, Euler continued, it was necessary at the same time to guard against that unphilosophical presumptuousness that invented a new subtle matter for each new set of phenomena. Having established the existence and chief properties of one particular subtle matter, the aether, he suggested, we must ask ourselves whether its known properties suffice to account for the phenomena that now confront us. If they do, "it would be quite ridiculous to introduce a peculiar electrical matter over and beyond the aether."

In common with other eighteenth-century aether theorists, young Euler attributed to his aether both an extreme rarity and an extremely high degree of elasticity. These properties were made plain, he argued, by the behavior of light, which, following his father's example, he took to be a wave phenomenon and not a stream of fast-moving corpuscles as was maintained by large numbers of his contemporaries. He then set out to demonstrate that, given this combination of properties, all the usual electrical appearances could be adequately explained. In every case, the key to Euler's explanation was the assumption—and perhaps this was the idea for which his father would later claim credit—that inequalities could arise and be maintained for a time in the distribution of aether within the pores of ordinary matter. That they could be maintained despite the aether's extreme subtlety and elasticity was not a totally unreasonable suggestion, he insisted, because no matter how subtle the aether, it was still possible that certain classes of body might have pores so cramped and narrow that it was denied free passage through them. He argued that from the phenomenon of the "lucent barometer"—the appearance of light in the evacuated region above the mercury in a barometer when this is shaken—it could be concluded that glass was a substance of this kind. So too, he went on, must air be, since the experiment does not succeed if there is air in the region above the mercury.

Generally, Euler proposed, bodies can be divided into those whose pores offer free passage to the aether and those whose pores do not. The former are to be identified with electrical conductors, the latter with electrical insulators (though it should be noted that Euler himself did not use these terms). Electrification, Euler suggested, consists in the expulsion of some of the normal quantity of aether from the pores of a body. If the body in question is completely surrounded by substances such as air that firmly retain the aether in their pores and only let it go with great difficulty, it will be some time before the deficiency is made up, and the body will remain electrified until then. If, on the other hand, it is touched at some point by a body that readily permits aether to leave its pores, the deficiency will immediately be made up, and the electricity will be destroyed. If such a body is only brought near to the electrified object, without actually touching it, the aether will burst forth and irrupt into the electrified object to restore the equilibrium. This will give rise to a violent agitation in the aether that will manifest itself as light, and, if the degree of agitation be sufficient, as fire. Hence the appearance of sparks is explained. If the object brought near is some part of a person's body, for example a finger, the rapid discharge of aether from it will be felt as a kind of concussion.

Euler next considered why insulators, or to use the standard eighteenth-century term, "electrics *per se*," can be electrified by rubbing, whereas conductors cannot. His explanation followed from his principles. When glass, for example, is rubbed, its pores are compressed and some aether is squeezed out; but when the rubber passes on and the pores by their elastic power regain their original state, they are, given the nature of glass, such that the expelled aether cannot get back in, and so the glass is left deficient. On the other hand, if it is a conductor that is being rubbed, even if some aether is squeezed out during the rubbing, this will be able to re-occupy its former place immediately afterward, so that no electrification will result. Bodies of this latter kind can be electrified only derivatively, through contact with electrics *per se* that have themselves already been electrified by rubbing.

Traditionally, the attraction and subsequent repulsion of nearby light objects had been regarded as *the* principal electrical phenomenon in need of explanation. Euler attempted to account for this by invoking a stream of aether passing, though not so violently as to produce a spark, from the nearby objects to the glass tube or other electrified body. According to the laws of hydrodynamics, he said, this current of aether will cause a reduced air pressure in the region through which it is passing, and the pressure of the air on the further side of the objects involved will then propel them forward into the low-pressure region. They will adhere to the tube only long enough, however, to give up some of their aether to it and thereby "restore the equilibrium" in that most elastic substance. Thereafter, Euler suggested, the objects themselves will begin receiving aether gradually from the surrounding air, chiefly—though his argument that this should be the case is unconvincing—from the region furthest from the tube. If this happens, it will in the same way as before lead to a correspondingly lower air pressure in that region, as a result of which the little objects will be driven in that direction; that is, they will move away from the tube.

The explanation of various other electrical phenomena, such as, for example, the appearance of a vivid light inside an evacuated globe when it is rubbed, followed fairly easily from Euler's principles. But his attempt to account for that other remarkable electrical phenomenon, the Leyden experiment, was quite inadequate. He referred to the experiment only in its initial form, as it had been discovered by Musschenbroek in Leyden late in 1745 (and somewhat earlier, Euler maintained, by Kleist in Pomerania). That is to say, he considered only the case where a glass bottle partly filled with water was suspended from the prime conductor of an electrical generating machine by a metal hook whose lower end dipped into the water in the bottle (Fig. 1). It had been found that, when the person holding the bottle touched the hook, he experienced a severe shock. Euler attempted to explain this by suggesting that water has an especially large number of very open pores; it is of such a nature, in other words, that it can yield up much larger quantities of aether than other bodies, and can correspondingly become much more strongly electrified provided it is stored in an insulating container such as the glass bottle in Musschenbroek's experiment.

But to dismiss such a spectacular phenomenon so casually was most unsatisfactory. Euler's explanation ignored a number of well-known but puzzling features of this crucial experiment. Why, for example, was it necessary to provide the outer surface of the bottle with a conducting layer, such as the hand of the person conducting the experiment, connected to ground? Why was it that the bottle delivered a more powerful shock, the *thinner* the glass of which it was made? Finally, to attribute the success of the experiment to some special property of water was to overlook entirely the fact that the experiment succeeded equally well if the water was replaced by lead shot, or indeed if the bottle itself was replaced by a flat sheet of glass with separate metal coatings on its two sides. To pretend that Musschenbroek's discovery warranted only the few lines Euler devoted to it was a truly remarkable piece of dissimulation.

Despite such shortcomings, Euler had made a brave attempt at ex-

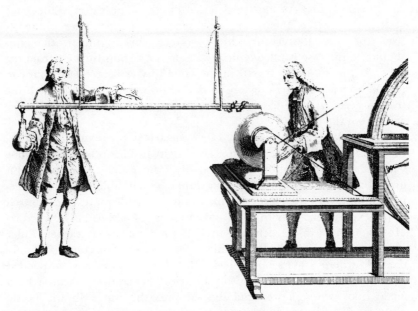

Figure 1. The Leyden experiment

plaining a particularly recalcitrant set of phenomena, and Aepinus would have found very helpful his ideas and the discussions these must have engendered among the group that met at the Eulers' house. Since it was almost certainly the Eulers who first encouraged Aepinus to take up the study of electricity, he may well for a while have adopted their views, or something like them. At any rate, some months after moving to Berlin he made it clear in an article published in the Rostock *Gelehrte Nachrichten* that he accepted in general terms the older Euler's position concerning the existence and nature of the aether, citing with approval the master's calculations purporting to show that the resistance of the aether would have observable effects on the motions of comets.[10] Two years later he evidently still felt the same way on this particular subject, since in early August 1757 he jotted down in his notebook the question: "If the comet of 1682 were to approach the sun in a straight line, could it be retarded a whole year by reason of the resistance of the aether?"[11] By this time, how-

[10] [Rostock] *Gelehrte Nachrichten*, 1755, p. 527. Aepinus's authorship of this article has not been established with absolute certainty, but can scarcely be doubted.

[11] AAN, raz. v, op. э-7, no. 1, l. 22: "an cometa A. 1682, si in recta ad solem accederet, ex resistentia aetheris toto anno retardari queat?"

ever, Aepinus had become a convinced adherent of Franklin's views on electricity, and in his *Essay* of 1759 the aether was relegated, so far as electrical matters were concerned, to an implied role as the mechanical cause of the forces of attraction and repulsion upon which the theory Aepinus was setting forth was founded; that it might be the immediate cause of the phenomena themselves was explicitly denied.[12] In his *Beschreibung des Welt-Gebäudes*, written in the same year, Aepinus still wrote confidently of "die unbegreiflich dünne Luft, so die Himmel füllet,"[13] but a few years later, as, perhaps, he became more reconciled to the probability that the forces he had invoked would remain unexplained, he was prepared to cast doubt on even this.[14]

Euler's essay was sent to St. Petersburg in late 1754, and was not actually published for another two years. As a result, it was a latecomer to the electrical scene; the two leading theories of the day, those of Nollet and Franklin, had by then been in print for several years. Of these, Nollet's had been the first to appear.

Nollet began his scientific career as an assistant to the leading French electrical investigator of the 1730s, Charles-François de Cisternai Du Fay, and then to the noted naturalist Réaumur.[15] During the late 1730s he acquired a considerable reputation through his courses of public lectures in Paris on experimental physics, and in 1739, the year Du Fay died, he was elected *adjoint mécanicien* in the Paris Academy of Sciences. In 1742 he was promoted within the Academy to *associé*, and in 1757 he was appointed *pensionnaire mécanicien* in succession to Réaumur. For almost thirty years up until his death in 1770, Nollet reigned supreme over French experimental physics, and throughout that time he devoted himself to the study of electricity above all else. His first major publication on the subject was his paper, "Conjectures sur les causes de l'électricité des corps," which was read to the Paris Academy in April 1745 and subsequently published in the *Mémoires* of the Academy for that year.[16] Here he announced all the basic premises of the theory to which he was to remain committed for the rest of his life.

Nollet, like the other leading electrical theorists of the period, regarded "electricity" not as a cause but as an effect, namely the power that certain bodies acquired when rubbed of attracting and then

[12] See below, pp. 240, 243.

[13] Aepinus, *Beschreibung*, p. 45.

[14] Aepinus, editor's note in *Hrn. Georg Wolfgang Kraffts Kurze Einleitung zur mathematischen und natürlichen Geographie*, 1764, pp. 49–50.

[15] J. Torlais, *L'abbé Nollet 1700–1770: un physicien au siècle des lumières*, Paris, 1954.

[16] *HAS*, 1745, Mém. pp. 107–51.

repelling nearby light objects.[17] He recognized, of course, that there were a number of other phenomena regularly associated with electrified bodies, for example the yielding of sparks and the sensation of the "electric wind," that peculiar cobweb-like feeling experienced when an electrified object is held near a sensitive part of the skin such as the face. But these were, in his opinion, of secondary importance; his central concern was, as it was for all electrical theorists before Franklin, to account for the attractions and repulsions. To do so, Nollet turned to the subtle, all-pervading elastic fluid that in his opinion filled the spaces between the particles of ordinary matter. Electricity, he concluded, arose from a disturbance in the equilibrium of this fluid brought about by the rubbing. The fluid Nollet had in mind, however, was not the universal aether that was subsequently to be invoked by young Johann Albrecht Euler. Rather, he had recourse to the all-pervading material Fire whose existence had been powerfully argued a few years earlier by the renowned Leyden professor Herman Boerhaave.[18]

When a body is rubbed, Nollet said, a stream of particles of Fire is squeezed out from its surface. However, these particles are not pure and unadulterated Fire, but are surrounded by an envelope of sulphureous and oily matter deriving from the body being rubbed. As a result of the outflow of subtle matter, spaces void of it will be left in the body, and in order to fill these, the subtle matter in all the surrounding bodies will by virtue of its great elasticity immediately rush in. An electrified body is thus surrounded, according to Nollet, by a double stream of fiery matter, one moving out from the body, the other moving in. These streams can, he argued, be plainly seen if the electrification is carried out in a darkened room, in the form of luminous rays darting out from the surface of the body and, if the electricity is powerful enough, similar rays darting in toward the body from the projecting parts of neighboring objects. They can be heard as the characteristic crackling noise associated with strongly electrified objects, and smelled, and also felt as in the phenomenon of the "electric wind."

It had been observed that the luminous rays emitted by an electrified body were not distributed uniformly over the surface of the body. Rather, divergent clusters of rays emerged from a limited number of points on the surface. Since according to Nollet these rays were

[17] Heathcote has made a similar point concerning the usage of seventeenth-century writers such as Gilbert; N. H. de V. Heathcote, "The early meaning of *Electricity:* some *pseudodoxia epidemia,*" Annals of Science, XXIII, 1967, 261–75.

[18] Boerhaave, *Elementa chemiae,* Leyden, 1732.

simply emergent streams of Fire rendered visible, he concluded that an electrified body emits the fiery matter from only a relatively small number of its pores. The incoming streams, on the other hand, are distributed more uniformly over the entire surface of the body.

With this settled, Nollet was able to account at once for the electrical attractions and repulsions. Any body near an electrified object will, he said, be acted upon by the two streams of fiery matter, since these fill the whole space surrounding the object and are not subtle enough, because of the sulphureous sheaths on the particles composing them, to pass through a body without affecting it. The motion of the body will, then, be a direct response to the stronger of the two streams. But the rays of the outward-moving streams are highly divergent. Hence if the neighboring body is small, it will be more likely to find itself in a region of inward-moving matter than in one where this is moving outward, except when it is very close to the electrified object, that is, in a region where the outward-moving beams have not yet had a chance to spread out. Thus in general the body will move toward the electrified object until it is very close to it. Unfortunately, there will also be occasions when the body chances to be placed initially directly in the path of an outward-going ray, and it should then be repelled without being first attracted. Nollet simply asserted that this was occasionally observed to happen. To account for the well-known fact that after a body has come very close to, or touched, an electrified object, it will move away again, Nollet took up Du Fay's suggestion that the attracted body itself becomes electrified, and adapted the idea for his own ends. In the process of becoming electrified, he argued, the previously attracted body acquires its own "atmosphere" of effluent matter; since this serves effectively to increase its volume, it will now be subjected to the action of a greater number of the outgoing rays from the object that first attracted it, and these will drive it away until it reaches a position where the impulses from the two streams acting upon it are equal.

Nollet recognized, of course, that some substances act as electrical insulators. They do so, he said, because their pores permit the streams of electrical matter to enter or leave only with great difficulty. To account for the generation of sparks and flame when a finger, for example, is brought near an electrified body, Nollet argued that because the finger is a conductor, and hence allows the streaming matter to enter and leave its pores with greater ease than does the surrounding air, there will be a concentration of both inward and outward moving streams in the direction of the finger. If the rays composing these streams become sufficiently forceful, the particles of the two streams will collide head-on with enough force to break open their envelopes of sulphureous matter and release the active Fire within. Then, just as a glowing cinder can kindle a much larger blaze, these free particles of Fire will activate surrounding inactive Fire particles, and light and heat will be generated. A conductor becomes unelectrified when a spark is drawn from it, Nollet suggested, because in these circumstances, with two conductors involved, the electrical streams will be even more condensed and will be traveling even more rapidly than usual. The impact between the two streams will then be so great that an actual rebounding will occur, and this, he said, will stop the effluence of matter at once.

In January 1746, only a few months after Nollet first announced this theory, Musschenbroek's letter announcing the discovery of what came to be known as "the Leyden experiment" reached the Paris Academy of Sciences, and Nollet was forced to consider its implications for his theory.[19] He soon concluded that the new phenomenon could be readily explained without amending any of the principles he had already enunciated; all that was required was the addition of a further postulate ascribing a peculiar electrical character to glass (and also to porcelain, with which Nollet had found the experiment to succeed as well). Glass, he suggested, shares with porcelain an ability to be electrified by communication combined with an ability to retain the electricity thus acquired, even when touched by an unelectrified body. Sulphur, wax, and so on can be electrified by friction, just as glass can be; but in the circumstances of Musschenbroek's experiment it was a question of being electrified by communication, something to which these substances are hardly susceptible at all. The metals, on the other hand, can readily be electrified by communication, but they also lose that virtue as soon as they are touched. Glass and porcelain alone satisfy both conditions.

The reason the shock delivered by the Leyden jar is so much more severe than that from a simple electrical spark is, Nollet said, that instead of an impact between a stream from an electrified body and one from an unelectrified one, there is in this case a much more powerful impact between streams emanating from two electrified bodies, the bottle and the prime conductor of the electrical machine. It might be wondered, he went on, why one does not experience the same shock if the prime conductor is simply grasped in two places, for the conditions specified would then seem to be fulfilled equally well. But in fact they would not be, since as soon as one grasped the bar with one

[19] Nollet, "Observations sur quelques nouveaux phénomènes d'électricité," *HAS*, 1746, Mém. pp. 1–23.

hand, it would lose the whole of its electrification, whereas, on account of the peculiar nature of glass, this does not happen when the bottle is used.

Nollet's attempted explanation of the Leyden experiment did not accord at all well with certain aspects of the theory he had put forward previously. There was no disputing his statement that the bottle retained its electrification despite its resting on a grounded conductor; it was soon found that the bottle could actually be detached from the prime conductor and carried around in the hand for several hours and yet retain its electrification. But this fact was nevertheless at odds with the account he had previously given, and to which he continued to adhere, of the way an electrical conductor functioned. In that account, the loss of electrification when a body was touched by a conductor did not depend in any way on the nature of the material emitting the electrical effluvia; it depended only on the fact that, once emitted, the effluvia found a readier passage through the conductor than through the surrounding air. To ascribe what happened in the Leyden experiment to a special property of glass was unsatisfactory because, on his own terms, the properties of the glass were not in any way relevant to the problem.

Notwithstanding difficulties such as this, and indeed not even acknowledging their existence, Nollet was content to rest on his laurels so far as electrical theory was concerned. In the following twenty years and more during which he continued to dominate French electrical science, he published many works defending his views from the onslaughts of Franklin and his supporters in particular, but at no stage did he consider it necessary to modify or abandon any of the principles he had announced. Instead, he contented himself with adducing further "proofs" from time to time of what he had put forward initially, and with showing how various newly discovered phenomena could be reconciled with his doctrines.

Franklin, meanwhile, suddenly emerged in the early 1750s as a serious rival to Nollet. Working largely in isolation in colonial Philadelphia, and seemingly in ignorance of some of the experimental evidence that had sorely troubled his European contemporaries, Franklin addressed himself particularly to the problems surrounding the charging and discharging of bodies, including the Leyden jar, rather than to the traditional questions concerning the electrical attractions and repulsions.[20] The theory he devised, like those of Nollet and Euler,

[20] I. B. Cohen points out that Franklin refers in his *Account of the New-invented Pennsylvanian Fireplaces*, Philadelphia, 1744, to an experiment described by J. T. Desaguliers in the second volume of his *Course of Experimental Philosophy* pub-

attributed the phenomena of electricity to disturbances in the equilibrium of an all-pervading subtle elastic fluid. For Franklin, however, this was a specifically electric fluid, different from both Boerhaave's all-pervading Fire and the universal aether invoked by Euler. Also, the disturbances he envisaged did not involve the excitations of masses of the fluid, as they did for Nollet, but alterations in the amounts of it present in different parts of a system of bodies.

All ordinary matter, Franklin supposed, is saturated with some "natural" quantity of electric fluid. When an electric *per se* is rubbed, a transfer of fluid takes place either from the rubber to the body being rubbed or vice versa. The body thus comes to have either a surplus or a deficiency of fluid; in Franklin's terminology, it comes to be electrified "positively" or "plus," or "negatively" or "minus." When a spark is drawn from, say, a positively electrified body, the surplus fluid rushes from the body and as a result this no longer shows any signs of electrification. In Franklin's theory, in other words, there was a straightforward and obvious link between the drawing of a spark and the loss of electrification that invariably accompanies this, something which was certainly not the case in the system put forward by Nollet. More than this, in Franklin's theory we meet for the first time a clear and consistently applied notion of an electric "charge." In an excitation theory such as Nollet's, there was simply no place for such a concept.

Franklin made his ideas known to the "electricians" of Europe by means of a series of letters to friends, especially the botanist Peter Collinson, at the Royal Society of London. The first letter of any significance was dated 25 May 1747; William Watson, the leading English electrical investigator of the period, included a long extract

---

lished in London earlier in the same year (Cohen, *Franklin and Newton: An Inquiry into Speculative Newtonian Experimental Science and Franklin's Work in Electricity as an Example Thereof*, Philadelphia, 1956, p. 245). Since Desaguliers in the same volume reprinted in its entirety his prize-winning *Dissertation concerning Electricity* that had been honored by the Bordeaux Academy of Sciences a couple of years earlier, Franklin could clearly have made himself familiar thereby with current work on this subject. However, his subsequent failure to attend to the problems posed by the attractions and repulsions suggests that he did not. The emphasis on the phenomena of charging and discharging faithfully reflects instead the orientation of Franklin's earliest attested reading on the subject of electricity, the brief account published in the April 1745 number of the *Gentleman's Magazine* of some of the latest German investigations (*Gent. Mag., xv,* 1745), 193–97; cf. Cohen, ed., *Benjamin Franklin's Experiments,* Cambridge, Mass., 1941, p. 19, and J. L. Heilbron, "Franklin, Haller, and Franklinist history," *Isis,* LXVIII, 1977, 539–49, where the author of the article in question is identified as none other than the famous Göttingen professor Albrecht von Haller).

THE ELECTRICAL BACKGROUND

from it in a paper published in the Society's *Philosophical Transactions* for that year. Succeeding letters were read and discussed at meetings of the Society, but they did not appear in print until they were gathered together by Collinson and published in 1751 as a slim volume entitled *Experiments and Observations on Electricity, made at Philadelphia in America*. A year later, a French translation was published in Paris at the instigation of the great naturalist Buffon. In 1753 and 1754, further letters were published as the second and third parts of the work. These, however, were in the main devoted merely to extending and elaborating the ideas that had already been put forward. Franklin's book went through four more English and two more French editions before the end of the century and, in addition, appeared in a German translation by Aepinus's friend and student Wilcke (1758), and also in Italian (1774).

In part, the book's popularity was due to Franklin's successful identification of lightning with the electrical discharge. Others before him had pointed out the similarity between the two, but Franklin went further and suggested an experiment that would test the hypothesis directly. The quick success achieved when the experiment was actually tried near Paris in 1752 created a sensation, and people from all walks of life clamored to read his suggestions as to how man could at last bring the terrors of the lightning bolt under control. These developments are of only peripheral interest to us here, however, and then only insofar as they converted the study of electricity from a fascinating parlor game into a serious study of something that played an obviously important role in the workings of the world. Far more important for our purposes are Franklin's contributions to electrical theory.

Franklin's letter of May 1747 [21] began with a discussion of "the wonderful effect of pointed bodies, both in drawing off and throwing off the electrical fire," and went on to describe the circumstances in which a man could receive a charge and then pass it on to another. It was to account for the latter that he set out his theory of electricity for the first time. "We had for some time been of opinion," he said, "that the electrical fire was not created by friction, but collected, being really an element diffused among, and attracted by other matter, particularly by water and metals. . . . To electrise *plus* or *minus*, no more needs to be known than this, that the parts of the

---

[21] *Benjamin Franklin's Experiments,* pp. 171–78. This work is a modern edition of Franklin's *Experiments and Observations on Electricity*. In it, the letter in question is dated 11 July 1747, as it is in the definitive 1769 edition of the book. Cohen shows, however, that the true date is 25 May 1747.

tube or sphere that are rubbed, do, in the instant of the friction, attract the electrical fire, and therefore take it from the thing rubbing: the same parts immediately, as the friction upon them ceases, are disposed to give the fire they have received, to any body that has less."

Notably, in his first formal communication on electrical matters, Franklin paid no attention whatever to what had always been the most important question confronting electrical theorists, the cause of the electrical attractions and repulsions. Nor did he do so in any of his other early writings. Attractions and repulsions were mentioned infrequently and then but incidentally as well-known experimental facts. Nor would these facts have obtruded upon Franklin's consciousness as frequently as they would have done with his predecessors; whereas it had once been customary to test a body for electricity by testing its ability to attract light objects, Franklin, coming to the subject in an age when much stronger degrees of electrification were the norm, almost always tested bodies by seeing whether he could draw a spark from them. Perhaps more important, the number of phenomena confronting the would-be theorist had expanded enormously, in both range and dramatic impact, during the preceding few years, with the result that the attractions and repulsions were no longer the obvious starting point they had been in the past.

It was in his second letter, dated 28 July 1747 (1 September 1747 in the printed version) that Franklin turned his attention to the most spectacular of all the new phenomena, the Leyden experiment.[22] The key to his analysis was his recognition that bodies can be electrified either "plus" or "minus." In a charged Leyden jar, he concluded, one surface of the glass acquires a positive electricity, the other a negative. To explain why the two electricities do not immediately neutralize each other within the body of the glass, he supposed that glass is absolutely impermeable to the electric fluid: if this is the case, a surplus of fluid on one surface will not be able to pass through the glass to overcome the deficiency on the other side, but will only be able to overcome it if an external conducting link is provided from one surface to the other. What is more, Franklin supposed the positive and negative charges on the two surfaces to be equal in magnitude: "whatever quantity of electrical fire is thrown in at the top," he said, "an equal quantity goes out of the bottom." And unless this egress is possible, the experiment will not succeed. The violent shock that had made the apparatus famous is experienced if the experimenter's body forms part of the external circuit linking the two surfaces: it "is oc-

[22] *Benjamin Franklin's Experiments*, pp. 179–86.

casioned by the sudden passing of the fire through the body in its way from the top to the bottom of the bottle."

Not all of these ideas were fully developed in Franklin's letter of 28 July 1747. In particular, Franklin did not realize at the time that it was the glass and not the water inside it that played the crucial role in the experiment. It was only in his next letter, written almost two years later, in April 1749, that he set this situation right. What happened in this instance is, in fact, symptomatic of developments more generally within Franklin's thinking. The vital core of Franklin's contribution to electrical theory is set out in his first two letters on the subject (Letters II and III in the collected edition). Here we find the crucial notion that electrification involves an accumulation of electric matter rather than its excitation. Here, too, we find the doctrine of "plus" and "minus," which was seen by Franklin's contemporaries as the central novelty of the theory, and also the application of this to the explanation of the Leyden experiment. These are the fruits of Franklin's genius in its first flush of creative originality. Subsequently, he became much more aware than he had been at the outset of the baffling complexities posed by the phenomena at hand. Simultaneously, he became better acquainted with the thinking of his European contemporaries concerning them. Initially he seems to have had a brief report in the April 1745 number of the *Gentleman's Magazine* of recent German developments as virtually his only guide, but beginning in early 1747 with some of William Watson's first writings on the subject, he gradually became more and more well-versed in the electrical literature.[23] Sometimes, as in the case of the Leyden jar, Franklin's re-thinking of a problem led him to correct something he had put forward earlier. But more often we find him in his later letters trying to extend the theory to take previously unperceived problems into account, and simultaneously trying to account in more detail for what he had already proposed. In one way or another, the quick flash of genius was replaced by the more humdrum labor of elaborating upon his first ideas so as to establish these upon a firmer foundation.

Questions obviously needed to be answered concerning the peculiar "electrical fire" that lay at the heart of the whole theory. What were

[23] See n. 20 above, and also J. A. Leo Lemay, *Ebenezer Kinnersley: Franklin's Friend*, Philadelphia, 1964, pp. 48–59; but cf. Cohen, *Franklin and Newton*, p. 390, and Bernard S. Finn, "An appraisal of the origins of Franklin's electrical theory," *Isis*, LX, 1969, 367–68, both of whom regard Watson as Franklin's "master and guide" in the study of electricity. Cohen also invokes the possibility that Collinson may himself have drawn up an account of the new discoveries for Franklin's benefit.

its properties, and how did they give rise to the phenomena with which Franklin was concerned? Toward the end of Letter V (dated, like Letter IV, 29 April 1749), Franklin considered the relationship between electrical fire and what he called "common fire"; and while he refrained from asserting positively that they were different things, his account strongly implied that they were. Another year and more passed, however, before he eventually went more deeply into the nature of the electrical fire in his paper, "Opinions and Conjectures concerning the Properties and Effects of the electrical Matter . . .," the last major paper included in his original published pamphlet of 1751.[24]

"The electrical matter," Franklin wrote on this occasion, "consists of particles extremely subtile, since it can permeate common matter, even the densest metals, with such ease and freedom as not to receive any perceptible resistance. . . . Electrical matter differs from common matter in this, that the parts of the latter mutually attract, those of the former mutually repel each other. . . . But though the particles of electrical matter do repel each other, they are strongly attracted by all other matter." The electric matter was, in other words, in eighteenth-century terms a typical subtle elastic fluid: in fact it was but one of several such fluids invoked by the theoreticians of the age to account for various classes of phenomena. In this form, with its elasticity attributed to a mutually repulsive force between the particles of which it was composed, it was modeled directly on Newton's picture of the structure of air, for at one point in his *Principia*,[25] Newton showed that if air is composed of particles repelling their neighbors with a force inversely proportional to the distance between them, it will obey Boyle's law. Since air manifestly does obey Boyle's law, it was immediately though illicitly concluded by Newton's followers that it is indeed of this nature. Thereafter, it became a commonplace among Newtonians to explain all other cases of elasticity in similar terms. In particular, the extremely rare but highly elastic particulate aether that Newton invoked in the 1717 edition of his *Opticks,* and which Franklin took as a further model for his electric matter, was assumed to have a like structure.

It has often been assumed that Franklin introduced the forces mentioned above into his theory of electricity in order to explain the observed macroscopic attractions and repulsions. This, however,

---

[24] *Benjamin Franklin's Experiments,* pp. 213–36. The letter that accompanied this paper was dated 29 July 1750.

[25] Newton, *Mathematical Principles of Natural Philosophy,* trans. Andrew Motte, rev. Florian Cajori, Berkeley, 1934, p. 300.

is incorrect. He introduced them to account for the phenomena with which he had been chiefly concerned throughout his investigations, namely those involving the sharing of charges between conductors brought into contact.[26] Through the action of these forces, he said, "common matter is a kind of spunge to the electrical fluid," and this is why an equilibrium in the distribution of the fluid will immediately establish itself in any connected system.

In one of his early letters, Franklin described an experiment in which a suspended cork ball was first attracted and then repelled by the wire of a charged Leyden jar; he found that, while the ball was in this latter condition, it would be attracted to the outer coating of the jar. Given his explanation of the charged jar, he was able to conclude immediately that the cork ball, when charged "plus," would be repelled by a positively charged body, but attracted by one that was neutral or negatively charged.[27] Shortly afterward, he also recognized the awkward fact that two negatively charged bodies repelled each other just as two positively charged ones did.[28] At the time he framed these various generalizations, however, they were merely summaries of certain of his empirical results. He made no attempt at that stage to reconcile them with the general principles he had already set out concerning the nature of electrification. Not until he wrote his paper "Opinions and Conjectures" did he tackle this problem. Even then he did little more than hint at what he had in mind, yet he did say enough to make it clear that he did not see the macroscopic attractions and repulsions as immediate consequences of the forces that he supposed acted between particles of electric fluid and those of ordinary matter. For reasons difficult for us to grasp today, he invoked as a vital intermediate link in the explanatory chain the notion of an electrical "atmosphere" surrounding a positively charged body: attractions and repulsions between physical objects at the macroscopic level were to be explained in terms of interactions of some kind between an electrical atmosphere and an uncharged (or negatively charged) object or between two atmospheres, as the case might be, and these interactions in turn were to be explained by attractions at the microscopic level between the various particles involved.[29]

What were these atmospheres? How did their interactions with matter or with other atmospheres explain the macroscopic attractions

---

[26] *Benjamin Franklin's Experiments,* p. 213.
[27] Ibid., p. 185.
[28] Ibid., p. 199.
[29] R. W. Home, "Franklin's electrical atmospheres," *British Journal for the History of Science,* vi, 1972, 131–51.

and repulsions? How were these interactions to be explained in terms of the forces Franklin postulated between the various particles? About the first of these questions at least, he was quite specific, and what he said makes it clear that he envisaged the atmospheres rather differently from his predecessors such as Hamberger and Nollet: "In common matter there is (generally) as much of the electrical as it will contain within its substance. If more is added, it lies without upon the surface, and forms what we call an electrical atmosphere; and then the body is said to be electrified." [30] That is, Franklin envisaged an electrical atmosphere as simply a static layer of electric fluid surrounding a charged body; a close analogy could be drawn between such an atmosphere and the common atmosphere surrounding the earth. A little later he added that "the form of the electrical atmosphere is that of the body it surrounds," but concerning the other questions he was much less definite.[31]

In one of his early letters Franklin spoke of a "sphere of attraction" around an electrified ball, but did not link this explicitly with his notion of an electrical atmosphere surrounding the ball. From his discussion of the well-known experiment in which a feather enclosed in a sealed glass vessel was affected by a nearby electrified tube, however, one gains a distinct impression that what was supposed to bring about the attraction was an actual contact between a mass of electric fluid and the feather: "When an additional quantity of the electrical fluid is applied to the side of the vessel by the atmosphere of the tube, a quantity is repelled and driven out of the inner surface of that side into the vessel, and there affects the feather, returning again into its pores, when the tube with its atmosphere is withdrawn." [32] In the case of repulsions between two positively charged bodies (forces due to negatively charged bodies were not considered at all), only once did Franklin give any indication of his thinking. He described an experiment in which two balls, A and B, were suspended from insulating threads, with A being positively charged; "bring A into contact with B," he wrote, "and half the electrical fluid is communicated, so that each has now an electrical atmosphere, and therefore they repel each other." [33]

Such a sketchy account was clearly not enough to satisfy either friend or foe among the electricians of Europe. Franklin's supporters were forced to elaborate considerably upon what he had said, and,

[30] *Benjamin Franklin's Experiments*, p. 214.
[31] Ibid., p. 216.
[32] Ibid., p. 233.
[33] Ibid., p. 216.

not surprisingly, they often deviated from what he seems to have had in mind. Nor was this the only point in the theory where conceptual difficulties arose. The function of the glass in the Leyden experiment proved equally perplexing.

Franklin's explanation of Musschenbroek's famous bottle depended crucially, we have seen, upon his postulate that even quite thin glass was absolutely impermeable to the electric fluid. But to any traditionally minded electrical theorist, this could only have appeared as nonsense, since a large number of well-attested phenomena, including an experiment devised by Newton himself, seemed to show beyond all doubt that glass, so long as it was not too thick, was on the contrary completely transparent to the electric matter. In Newton's experiment, for example, a piece of glass set in a brass ring was laid on a table so that a small volume of air was enclosed by the ring between glass and table. In this air space, Newton put some tiny scraps of paper. He found that, if he then rubbed the upper surface of the glass with a coarse cloth, the pieces of paper "began to be attracted and move nimbly to & fro." [34] Hauksbee a few years later modified the experiment slightly in a way that made its implications to the electrical theorist even plainer; instead of rubbing the glass, he merely held a piece of rubbed sealing wax above it, and yet the scraps of material underneath were once again set in motion. To an eighteenth-century electrician, nothing could be more obvious in this latter experiment than that the effluvia emitted by the sealing wax were passing through the glass and then having their usual effect on the little bodies on the other side.

Franklin had to offer an alternative explanation for phenomena such as these, and the passage quoted earlier concerning the motion of a feather in a sealed glass container reveals his answer. No electric matter actually passes through the glass, he insisted; rather, an excess of fluid on one side exerts a repulsive influence on the fluid in the other side (glass normally containing within its pores large quantities of the electric matter) and drives it out into the adjacent air. Once there, it acts like any other electrical atmosphere on light objects with which it comes in contact.

Plausible though this may sound in isolation, it yet bristles with unresolved problems. If the electric matter can pass readily into or out of the surface of a piece of glass, as Franklin's account clearly demands, why can it not pass all the way through? Surely the struc-

[34] Newton to Henry Oldenburg, 7 December 1675, *The Correspondence of Isaac Newton*, ed. H. W. Turnbull et al., I, 364.

ture of the internal parts of a sheet of glass is not likely to be so very different from that nearer the surface? And yet Franklin's theory requires that there be an impenetrable barrier down the middle. Also, granting for a moment the existence of this barrier, how are we to understand the ability of the glass nevertheless to transmit a repulsive *influence*, so that electric matter accumulated on one side can force fluid out from the other? Whittaker regarded Franklin's introduction of this latter notion as the germ of the action-at-a-distance theory of electricity,[35] but Franklin himself, had he thought about things in these terms, would probably have resisted any such imputation. Certainly many of his supporters would have wanted to do so, and one of the most influential, the Italian Giambatista Beccaria, went to considerable lengths to provide a mechanistic account that avoided any suggestion of electrical forces acting at a distance. Let us suppose, he said, that the basic structural unit of glass is a hollow rectangular-based pyramid with flexible sides. And let us imagine that a thin sheet of glass is built out of these units by forming two layers fitted together so that the points of the pyramids forming one layer occupy the spaces between the points of the pyramids forming the other. Then the two surfaces of the glass will be mosaics formed by the open bases of the pyramids. Now, Beccaria went on, fill the pyramids with electric fluid, and we have glass in its normal state. If more fluid is forced into the cells whose open bases form one surface, their walls will flex outward; this will compress the adjoining cells that face the other way, and fluid will be forced out from them.[36]

Even as ingenious a model as this, however, could not resolve all the conceptual difficulties involved. If anything, it made it less clear than ever why the Leyden experiment succeeded better, all other things being equal, the thinner the glass that was used. And surely neither it nor any other attempted explanation that relied on the internal structure of the glass could withstand an experiment Franklin performed to test his own hypothesis that the impenetrability was due to the pores getting progressively narrower toward the middle of any sample: "I ground away five-sixths of the thickness of the glass, from the side of one of my phials, expecting that the supposed denser part being so removed, the electric fluid might come through the remainder

[35] E. T. Whittaker, *A History of the Theories of Aether and Electricity*, London, 1910, p. 46.

[36] Beccaria, *Dell'elettricismo artificiale, e naturale*, Turin, 1753, pp. 101–102. To be fair, it should be remarked that Beccaria did not take this model seriously as a representation of the actual state of things. He put it forward merely to demonstrate the "mechanical possibility" of Franklin's doctrines.

of the glass, which I had imagined more open; but I found myself mistaken. The bottle charged as well after the grinding as before." [37] Confronted by such a baffling result, Franklin simply gave up trying to understand how glass could act as it did. But of its absolute impermeability to electric matter he remained convinced.

Not long after Franklin's first collection of letters appeared in print, his friend and erstwhile collaborator in electrical investigating, Ebenezer Kinnersley, reported to him some surprising results he had recently obtained.[38] If a piece of rubbed glass was brought near a cork ball suspended from a silk thread, the ball was, as usual, at first attracted to the glass and then, after making contact, repelled. In Franklin's terms, as we have seen, the ball was supposed to acquire a positive charge (that is, an electrical atmosphere) during its contact with the electrified glass, and the repulsion followed because this atmosphere interacted with that of the glass. What startled Kinnersley was his discovery that when, while the ball was still being repelled, he brought near to it a piece of rubbed amber, sealing wax, or sulphur, this substance did not repel it too but, on the contrary, attracted it. Assuming with Franklin that the ball had acquired a positive charge from the glass, he could only conclude that the things that subsequently attracted it were charged not "plus" but "minus."

Though both Kinnersley and Franklin were unaware of it at the time, Kinnersley's discovery (but not, of course, the Franklinian interpretation he placed upon it) had in fact been anticipated almost twenty years earlier by Du Fay. Du Fay had concluded from a similar series of experiments that two different modes of electrification were possible in nature. These he had designated "vitreous" and "resinous" to indicate the chief classes of bodies that gave rise to them when rubbed. And he had concluded from his experiments that two similarly electrified bodies would repel each other, while two differently electrified bodies would attract.[39]

Du Fay's discovery had achieved a considerable notoriety when it was first announced, but by the late 1740s it had become discredited and almost forgotten. This was due chiefly to Nollet's energetic dis-

---

[37] *Benjamin Franklin's Experiments,* p. 333. Franklin performed this experiment soon after sending his paper "Opinions and Conjectures" to London in July 1750, but he did not publish his results until many years afterward. He mentioned them in a letter to Dr. John Lining of Charlestown, South Carolina, written in March 1755 but not published until 1769.

[38] Kinnersley to Franklin, 3 February 1752, *Benjamin Franklin's Experiments,* pp. 250-52.

[39] Du Fay, *HAS,* 1733, Mém. pp. 457-76.

counting of it in the meantime, notwithstanding his having accepted it completely in his first published work, the only one published during his mentor Du Fay's lifetime.[40] Thereafter, Nollet argued repeatedly that there was no real difference in kind between the electricities involved in experiments like Du Fay's, but rather a mere difference in degree. Substances such as sulphur and resin, he insisted, do not become electrified in a wholly different way from glass, it is just that when rubbed they do not acquire as strong an electricity as glass does. He pointed with much justice to the small scale and extreme variability of the effects concerned; this, he said, could easily deceive even such a careful observer as Du Fay.

Nollet's position was not altogether unreasonable. The phenomena involved were quite ephemeral and difficult to determine precisely. Nevertheless, his stand must also have been conditioned to some extent by the fact that it is exceedingly difficult to conceive how his general theory of electricity could be modified to permit two different modes of electrification, whereas it could readily accommodate differences of degree within the single mode he had postulated. Nollet, one suspects, had become enmeshed at this point by "l'esprit de système," that too great fondness for hypotheses of which Bacon had warned and against which he himself continued to inveigh regularly.

Franklin had already insisted that a body could be electrified in two different ways, "plus" and "minus," well before Kinnersley made his discovery. But he was initially led to this notion by theoretical considerations rather than by experimental results. From his conception of electrification as an accumulation of electric fluid rather than the traditional excitation of it, it necessarily followed that whenever one body became electrified, another must have been deprived of some of its normal quantity: the only alternative was that the additional fluid had actually been created by the friction, and this was unthinkable. It was only in the later stages of Franklin's argument that experiment played a role: experiments with insulated electrical machines, in which fluid was presumably transferred from the machine itself to the prime conductor, showed that both the machine and the conductor gave all the usual signs of being electrified. Hence it followed that, whichever one was lacking in fluid, it might just as properly be said to be electrified as the other. As direct confirmation of his idea, Franklin was able to offer only a single observation—if two people were insulated and then electrified, one by touching the

[40] Nollet, *Programme ou idée générale d'un cours de physique expérimentale*, Paris, 1738, p. 103.

machine and the other by touching the conductor, a stronger spark would pass between them than would pass between either of them and a third man standing on the floor, and afterward neither of them would show any signs of being electrified.[41]

Within this context, Kinnersley's vindication of Du Fay's work and his re-interpretation of it in Franklinian terms provided welcome if wholly unexpected support for his friend's views. It also came at an opportune time, when the battle between the Franklinists and Nollet and his supporters was about to be joined in earnest. Furthermore, since there was no room in Nollet's theory for a second possible mode of electrification, it bore directly on one of the crucial distinguishing features at issue in the dispute: so central was the question, indeed, that Franklin's theory soon came to be popularly known as the theory of "plus" and "minus" rather than by any other name. And of course Nollet was able to launch his counterattack long before news of Kinnersley's work reached Europe, since he was concerned with a matter by no means restricted in its application to the phenomena that Kinnersley dealt with; the distinction was, for example, equally bound up with the explanation Franklin had given of such things as the Leyden experiment. In France a fierce debate broke out within the Royal Academy of Sciences between Nollet and the Franklinist Jean Baptiste Le Roy. In Italy, Beccaria leapt to Franklin's defense. And the question of the "contrary electricities" became the main focus of the first investigations of Aepinus and Wilcke when they took up the study of electricity in Berlin.

As indicated earlier, very little has come down to us concerning Aepinus's initiation into electrical research, and what little information there is has mostly to be inferred from incidental remarks in his writings and those of his friends. The richest source of information is undoubtedly Wilcke's doctoral thesis, *De electricitatibus contrariis*, which he composed in Berlin and presented to the Philosophical Faculty at Rostock only a few months after Aepinus's departure for St. Petersburg; it seems reasonable to suppose that Aepinus would have been thoroughly familiar with most of what Wilcke set out in that work.

The precise nature of the relationship between Wilcke's electrical investigations and those of Aepinus at this period is, however, difficult to untangle. Historians have generally assumed that Wilcke provided the chief inspiration and driving force, and that Aepinus became involved only when the peculiar properties of the crystal called tour-

---

[41] *Benjamin Franklin's Experiments,* p. 175.

maline were brought to his attention.[42] Yet there are grounds for questioning this. For one thing, we know that Aepinus did not begin his tourmaline research until late 1756 or even early 1757, since both Aepinus and Wilcke tell us that this is when Aepinus discovered the electrical character of the effect,[43] and elsewhere Aepinus says that he proved that electricity was involved at the very outset of his investigation of the stone.[44] But if this marked the beginning of his electrical studies in general, and not just his work on the tourmaline, it seems beyond belief that he could have reached such a remarkable degree of sophistication in both experiment and theory as he displayed in his report to the Berlin Academy in the period of less than three months that intervened before this was presented. Second, when Johann Albrecht Euler amended his aether-based theory to take into account the fact that two different modes of electrification and not just one were possible, it was Aepinus and not Wilcke whom he credited with demonstrating the distinction to him—and this notwithstanding that it was Wilcke who presented a thesis on the subject.[45] This suggests that, despite first appearances to the contrary, Aepinus was involved with Wilcke throughout most of his investigations and indeed that he was the dominant partner all along, and not just toward the end when the tourmaline arrived on the scene. In fact Wilcke himself almost conceded as much, freely acknowledging his debt to Aepinus, "to have had whom sometimes as a partner in this work, has more than once been to my advantage." [46]

Taking this evidence into account, it was most likely at the younger Euler's instigation rather than Wilcke's that Aepinus began his electrical investigations. Also, he almost certainly took up this study well before the tourmaline was brought to his attention, and perhaps not long after his arrival in Berlin in March 1755. Whether Wilcke's interest was also aroused by Euler or whether he was already interested in electricity when he arrived in Berlin a few months after Aepinus is not certain, but it may have been he who first directed Aepinus's attention to the problem of the two electricities. Certainly Euler was at that time quite unaware of the significance of this phenomenon,

[42] For an authoritative statement of this point of view, see John L. Heilbron's article on Aepinus in the *Dictionary of Scientific Biography*, New York, 1970–78, I, 66–68.

[43] See below, p. 237 and Wilcke, *Abhandl. Königl. Schwed. Akad. Wiss.*, 1766, p. 99.

[44] Aepinus, *Mém. Acad. Berlin*, XII, 1756, 107.

[45] J. A. Euler, *Mém. Acad. Berlin*, XIII, 1757, 125.

[46] Wilcke, *De electricitatibus contrariis*, p. 93: "quem in hisce laboribus socium quandoque habuisse, plus semel profuit."

whereas Wilcke some years earlier had actually observed at first hand some experiments by the Uppsala professor Samuel Klingenstierna which clearly confirmed Du Fay's distinction.[47] In any event, Aepinus and Wilcke then seem to have done most of their researching together, with Aepinus (if we have understood the implications of Euler's statement correctly) generally taking the lead.

Their investigations convinced both men of the reality and also the importance of Du Fay's distinction. If they had ever seriously entertained Nollet's theory, this must have caused them to abandon it. Euler's theory was more resilient in this respect, as Euler himself showed in the paper he read to the Berlin Academy late in 1757.[48] To accommodate the idea of the two electricities, Euler said, all he had to do was to allow that a body could become electrified by either condensing or rarefying the aether in it, instead of only by rarefying it as he had supposed earlier. Having done this, however, Euler was forced to concede that it became much more difficult to give a coherent explanation of the electrical attractions and repulsions; he now appeared to be driven to the highly implausible conclusion that the directions of the electrical motions were independent of the direction of flow of the aether! Euler struggled desperately to circumvent this new difficulty, but without any real success.

Franklin was in fact no more successful than Euler in explaining the attractions and repulsions exerted by bodies charged "plus" and "minus." But it now appeared that he was no less successful either; all attempts to explain the motions in purely mechanical terms had broken down. Moreover, Franklin's theory did have the considerable advantage that it could account much more coherently than any of its rivals for the Leyden experiment. Also, the notions of "plus" and "minus" entered into his explanation of this phenomenon in a way that seemed to be directly confirmed by experiment. Both Wilcke and Aepinus became convinced Franklinists, so that when the latter took up the study of the tourmaline he automatically analyzed its behavior in Franklinian terms. Wilcke meanwhile not only wrote his Franklin-oriented doctoral thesis, he also prepared a German translation of Franklin's papers for publication.[49] This, like the thesis, appears to have been close to completion, if not actually completed, by the time Aepinus left for St. Petersburg.[50]

[47] Ibid., p. 7.
[48] J. A. Euler, *Mém. Acad. Berlin*, XIII, 1757, 125–59.
[49] *Des Herrn Benjamin Franklins Esq. Briefe von der Elektricität* . . . , Leipzig, 1758.
[50] A.J.D. Aepinus, introduction to Wilcke, *De electricitatibus contrariis*, p. 11.

It is impossible to determine from Aepinus's own writings his familiarity with the writings of the other electricians of his day, since he was always sparing in referring to the works of others. Fortunately, Wilcke was much more free with his citations. Since the two men worked together so closely, and even shared the same house throughout this period,[51] we can be confident that Aepinus would have become acquainted during his time in Berlin with nearly all the works mentioned by his friend in either his thesis or his edition of Franklin's letters, even if he had not read them earlier.

Among the authors Wilcke cited, Franklin naturally took pride of place. All the classic writers on electricity are mentioned—Gilbert, Guericke, the members of the Florentine Accademia del Cimento, Boyle, Hauksbee, the "very celebrated" Gray and Du Fay—while Waitz's prize-winning essay, William Watson's *Sequel*, and John Ellicott's remarkable paper in the 1748 volume of the *Philosophical Transactions* are referred to with much respect.[52] In connection with the specific question of the contrary electricities, Wilcke cited a large number of authors who had confirmed Du Fay's distinction. Perhaps the most notable of these were his own teachers in Uppsala, Klingenstierna and Strömer, together with Kinnersley, Gralath, and Richmann.[53] Wilcke also described Canton's discovery that unpolished glass acquires, when rubbed, a charge opposite to that acquired by ordinary glass and the same investigator's very important experiments on what is now called "electrostatic induction." [54] Various of Nollet's electrical writings, too, are referred to with evident familiarity. Clearly Wilcke and, by implication, Aepinus, had by early 1757 at the latest become thoroughly familiar with the work of most of the leading electrical authors of the age.

Aepinus's friend Lehmann first drew his attention to the strange behavior of the then little-known crystal called tourmaline in late 1756. Samples of this stone had begun to be imported into Europe from Ceylon in the early years of the eighteenth century for use as jewelery. The Dutch and German jewelers who for many years en-

[51] Ibid., p. 10.

[52] Watson, *A Sequel to the Experiments and Observations tending to illustrate the Nature and Properties of Electricity*, London, 1746. Ellicott, "Several essays towards discovering the laws of electricity . . . ," *Phil. Trans.*, XLV, 1748, 195–224.

[53] Klingenstierna and Strömer, *Handlingar Kungl. Svenska Vetenskapsakademien*, 1747, p. 141. Kinnersley, Letter VIII in *Benjamin Franklin's Experiments*. Gralath, *Versuche und Abhandlungen der Naturforschenden Gesellschaft in Dantzig*, I, 1747, 508ff. Richmann, *Commentarii Academiae Scientiarum Imperialis Petropolitanae*, XIV, 1744–46, 296.

[54] *Phil Trans.*, XLVIII, 1753–54, 780–85, 350–58.

joyed a virtual monopoly over the tourmaline trade soon discovered in working them that their stones possessed the remarkable property of attracting a coating of ash to themselves as they were heated in the fire. A reasonably reliable description of the phenomenon was published in 1717 by Louis Lémery,[55] but it was then more or less forgotten by the learned world until Lehmann brought it to Aepinus's notice.[56]

Aepinus's proof that the effect was an electrical one was interesting enough in itself, for in no other case had simply heating a body been found sufficient to electrify it. His successful analysis of the behavior of the stone in Franklinian terms of "plus" and "minus" was even more significant. Aepinus announced his results at a time when electrical theory was still in a very unsettled state, when any new discovery was bound to be scrutinized closely by all the parties involved to see what it might contribute to their arguments. His strong support for Franklin's position was therefore particularly timely.

In England, the first published report of Aepinus's work appeared in the supplement to the 1758 volume of the *Gentleman's Magazine*.[57] This was a summary based directly, it seems, on the original paper published in the Berlin *Mémoires* for 1756, which had issued from the printers only a short time before. Franklin, Canton, and Benjamin Wilson all quickly verified the most important of Aepinus's results, and strongly supported his interpretation of them in terms of the theory of "plus" and "minus."[58] In France, news of Aepinus's discovery appeared in print earlier, in the May 1757 number of F.-V. Toussaint's *Observations périodiques sur la physique, l'histoire naturelle et les arts*,[59] the report being based on a précis of Aepinus's paper that the latter had drawn up and sent to an unidentified friend in Paris just prior to departing for St. Petersburg.[60] The French response was, as might have been expected, very different from what happened in England. An essay was promptly published that at-

---

[55] *HAS*, 1717, Hist. pp. 7–8.

[56] Linnaeus did give a brief description of the stone in the preface to his *Flora Zeylanica*, Stockholm, 1747, p. 8, and even coined for it the name "lapis electricus." However, this seems to have been nothing more than an inspired guess.

[57] *Gentleman's Magazine*, XXVIII, 1758, 617–19.

[58] Franklin to William Heberden, 7 June 1759, in his *Experiments and Observations on Electricity*, 4th ed., London, 1769, pp. 375–78: Canton, *Gentleman's Magazine*, XXIX, 1759, 424–25 (this report being published over the *nom de plume* "Noncathoni," an obvious anagram for "Iohn Canton"): Wilson, *Phil. Trans.*, LI (Pt. I), 1759, 308–39.

[59] Pages 341–45.

[60] Aepinus, *Recueil*, p. 130.

tempted to understand the behavior of the tourmaline in terms of Nollet's general theory of electricity. This purported to be the work of a Neapolitan nobleman, the Duc de Noya Carafa, but in fact the noted naturalist and friend of Nollet, Michel Adanson, was largely responsible for it.[61] For many years thereafter, this essay was regularly cited by French authors as the standard authority on the subject, but in fact, compared with Aepinus's article in the Berlin *Mémoires,* not to mention his *Recueil de différents mémoires sur la tourmaline* published a few years later, it was a much inferior piece of work. Its author was far less successful than Aepinus in reconciling the phenomena with his own general theoretical position. Outside France, therefore, Aepinus's work on the tourmaline considerably strengthened the position of the Franklinists in the general debate over electrical theory, while the position of Nollet and his effluvialist supporters was correspondingly weakened.

Aepinus's work on the tourmaline catapulted him into the front rank of electrical investigators of the period. What is more, it brought him into direct contact with the leading electricians in England, for it led to a correspondence and a lively controversy between him and Benjamin Wilson, F.R.S., who was at this period perhaps the most active of all the English electrical workers.

Aepinus's analysis of the behavior of the tourmaline particularly excited Wilson's attention. He quickly established himself as the foremost British authority on the subject, and his efforts were crowned soon afterward when the Royal Society of London bestowed on him its Copley Medal for 1760, for, in the president's words, "the very many curious Experiments, relating to Electricity, that he has made upon the Tourmaline and other Bodies; and which he has laid before

---

[61] *Lettre du Duc de Noya Carafa sur la tourmaline, à Monsieur de Buffon,* Paris, 1759. In the catalogue of the Bibliothèque Nationale, Paris, it is assumed that "Duc de Noya Carafa" was a pseudonym used by Adanson. Yet there really was such a person as the duke, and what is more he took an active interest in philosophical matters, to the point of being elected a Fellow of the Royal Society of London in March 1759. Whether he actually wrote much of the book, however, is debatable. The *Lettre* begins with an account of how the duke first became acquainted with the tourmaline, and he frankly concedes (p. 7) that most of the research described was done by Adanson. The latter certainly regarded the book as his own, for in the catalogue of his library he listed it as "Duc Noya Adanson Lettre sur la Tourmaline" (George H. M. Lawrence, ed., *Adanson: The Bicentennial of Michel Adanson's "Familles des plantes",* Pittsburgh, 1963–64, I, 273). Yet this claim does not necessarily conflict with what was said in the *Lettre;* probably what happened was simply that the duke wrote the introductory section, but then left Adanson to do the rest. For the latter's friendship with Nollet, see Lawrence, *Adanson,* I, 31.

this Society since our last Anniversary Meeting." [62] Aepinus, it will be recalled, had announced in his original paper on the tourmaline that the stone acquired opposite charges on two opposite faces when it was heated. Wilson, and Canton and Franklin too, quickly confirmed this once they gained access to some crystals to experiment with. Aepinus had likewise discovered that the stone behaved differently according to whether it was heated evenly or unevenly, and this too was confirmed by the three British workers. However, Wilson found in addition to this that, when the particularly large stone he had at his disposal was heated unevenly, it sometimes acquired the same charge on both faces, something Aepinus had stated categorically did not happen.[63]

Aepinus was perplexed by this result. He initially supposed that the trouble stemmed from slight differences in the experimental arrangements in the two cases, but even when he repeated Wilson's experiments exactly as the Englishman had described them, he still could not reproduce the latter's results. Obviously embarrassed about seeming to contradict "un aussi habile observateur" as Wilson, and yet confident that he had not made a mistake himself, he freely confessed his bewilderment: "I find it extremely difficult to imagine that Mr. Wilson should have fallen into error here, yet I am convinced that on my side there is assuredly no mistake. Thus I do not know what to make of it." [64] Wilson in turn reiterated his faith in the results he had obtained, and in doing so adopted a rather patronizing tone toward a colleague so far from the scientific centers of western Europe. Aepinus, he unjustly remarked, had not really described his exact procedures sufficiently clearly, but "I am inclined to believe some material circumstance has been omitted in the method; or that our apparatus's are essentially different; (though you seem to have had a regard to some of the necessary requisites for making the experiment properly:) otherwise, I cannot apprehend why you were not able to succeed: because the experiment always answers with me." [65]

When Joseph Priestley came to write his *History of Electricity* a few years later, he found the whole affair most confusing, but in the end decided that Wilson's apparatus was more accurate than Aepinus's, so that his results could be better depended upon.[66] However,

---

[62] Royal Society, Journal Book, entry for 1 December 1760.
[63] Wilson, *Phil Trans.*, LI, 1759, 318ff.
[64] Aepinus, *Recueil*, p. 192.
[65] Wilson, *A Letter . . . to Mr. Æpinus . . .* , London, 1764, p. 15; also *Phil. Trans.*, LIII, 1763, 446.
[66] Robert E. Schofield, ed., *A Scientific Autobiography of Joseph Priestley (1733-*

Priestley's judgment cannot be regarded as definitive, since he reached it without reading Aepinus's *Recueil,* the only place where Aepinus described in any detail the apparatus he had used in his experiments.[67] In Sweden, Wilcke and young Torbern Bergman proceeded more circumspectly; they sought to reconcile the conflicting results by arguing that, though in normal circumstances the two opposite faces of a tourmaline acquired opposite charges, in reality they acted quite independently of each other, so that under appropriate conditions both Aepinus's and Wilson's results were possible.[68] Nevertheless, uncertainties remained, and the debate was never satisfactorily resolved.

The behavior of the tourmaline is, unquestionably, extremely complex. In this instance, the conflicting observations almost certainly arose from the fact that when a crystal of the substance is heated unevenly, gross and unpredictable piezoelectric effects become superimposed on the primary pyroelectric-cum-piezoelectric phenomenon, and these depend critically on, among other things, crystal size. The particularly large size of Wilson's crystal could explain the result he obtained. In addition, Wilcke and Bergman may well have been correct; Wilson's stone may have been large enough so that the face furthest from the source of heat actually may have been in its cooling phase while the opposite face was still being heated. This would explain his results equally well. In any event, both Aepinus's and Wilson's results could have been quite valid. There simply was no definitive answer in this case.

When Aepinus presented his original memoir on the tourmaline to the Berlin Academy of Sciences, he added to it an account of another equally interesting discovery, to which he had probably been led by his work on the tourmaline.[69] Aepinus argued in opposition to Franklin that the Leyden experiment ought to succeed equally well using other insulating materials besides glass. To prove it, he described how he had succeeded in producing the characteristic shock delivered by the Leyden jar using not glass but a layer of air as his dielectric! [70] Aepinus's discovery of the so-called "air condenser" was fully reported in the article in Toussaint's *Observations périodiques,* and also, though somewhat less adequately, in the summary of Aepinus's paper

---

1804), Cambridge, Mass., 1966, pp. 18, 28. Priestley, *The History and Present State of Electricity,* 2nd ed., London, 1769, p. 301.

[67] Ibid. p. [714].

[68] Wilcke, *Abhandl. Königl. Schwed. Akad. Wiss.,* 1766, p. 112.

[69] See below, p. 108.

[70] Aepinus, *Mém. Acad. Berlin,* XII, 1756, 119–21.

published in the *Gentleman's Magazine*. Only a few years later, Priestley felt able to describe it as "one of the greatest discoveries in the science of electricity since those of Dr. Franklin." [71] When it was first announced, however, Aepinus's invention attracted no comment, in either London or Paris, from those who took up with such interest his work on the tourmaline. Nor did Nollet see any need to take account of it, or of the tourmaline experiments either for that matter, when re-stating his views on electrical theory shortly afterward.[72]

Yet Priestley's judgment of the importance of Aepinus's discovery was completely justified. To begin with, the experiment provided dramatic confirmation of Aepinus's novel view of the role of the dielectric in the Leyden phenomenon. It made it clear that this depended not, as Franklin had thought, on some special property peculiar to glass, but simply on the fact that glass was an electrical insulator. In addition, Aepinus's discovery cast serious doubts on the very existence of electrical atmospheres of the kind Franklin had envisaged around positively charged objects. In the circumstances of Aepinus's experiment, any such atmosphere on the surface of the positively charged plate should certainly have extended as far as the second, earthed plate nearby, and this made it extremely difficult to understand why the fluid of which the atmosphere was composed did not immediately flow into this plate and away to ground. In other words, given the existence of Franklin-style atmospheres, it should have been quite impossible for Aepinus's experiment to succeed. Finally, and following on from this, Aepinus's invention bore directly upon the more extended question, of crucial importance for the whole basis of Franklinian electrical theory, of the inductive effects of electrostatic charges. This topic had been opened up only a short time before by John Canton, in a paper published in the *Philosophical Transactions* for 1753.[73] Its importance had been recognized immediately by Franklin himself, who confirmed and extended Canton's results in what turned out to be his last major piece of scientific research.[74] In due course, it was to become the focal point for the reformulation of Franklin's theory that Aepinus carried through in his *Essay* of 1759.

The apparent lack of interest in Aepinus's "air condenser" is therefore astonishing, especially on the part of Franklin's supporters. The

---

[71] Priestley, *History*, p. 243.

[72] Nollet, *Lettres sur l'électricité*, II, Paris, 1760; also *Leçons de physique expérimentale*, VI, Paris, 1764.

[73] Canton, *Phil Trans.*, XLVIII, 1753–54, 350–58; reprinted in *Benjamin Franklin's Experiments*, pp. 293–99.

[74] *Benjamin Franklin's Experiments*, pp. 302–306.

most likely explanation is that no one knew what to do with it. On the one hand, it made an absolute mockery of Nollet's purported explanation of the Leyden phenomenon; and on the other, the upholders of Franklin's views, imbued as they were with the idea of electrical atmospheres surrounding charged bodies, were just as unable to accommodate it within their general conceptual scheme. Only Aepinus himself saw how to cut the Gordian knot by abandoning the notion of atmospheres altogether and reducing everything to forces acting at a distance between the various particles involved.

We have seen that Franklin gave only a very sketchy indication of his views about atmospheres in his earliest published papers on electricity. Probably he thought they were so obvious and straightforward that no further elaboration was needed, but in reality this was far from being the case. Either because they did not understand what he had in mind or, if they did, because they felt this to be unsatisfactory, the majority of his supporters adopted a rather different approach. Eventually, in 1755, confronted now with the need to cope with Canton's induction experiments as well, Franklin tried to rectify the situation. By then, however, it was too late, and in any event his attempt amounted to little more than a restatement of some of the problems that beset his theory. Many of his most important followers continued to prefer the alternative accounts they had developed for themselves in the meantime.

Let us consider once more the basis of Franklin's theory: electrically neutral matter contains within it as much electric fluid as is required to saturate it, and the surplus it acquires when it becomes positively charged gathers around it in the form of an atmosphere. A question immediately arises: Why does the atmosphere remain around the body? Franklin gave two different answers to this question within the space of a few pages in his original pamphlet of 1751. The first was that it was held in place by the attraction between the particles of the body and those of the electric fluid constituting the atmosphere, just as the air is held in place around the Earth by gravitational attraction: "The form of the electrical atmosphere is that of the body it surrounds. . . . And this form it takes, because it is attracted by all parts of the surface of the body. . . . Without this attraction, it would not remain round the body, but dissipate in the air." [75] A few pages further on, in discussing the rubbing of an evacuated glass tube, Franklin invoked the insulating properties of the surrounding air: "*In vacuo* the electrical fire will fly freely from the inner surface . . .

[75] Ibid., p. 216.

but air resists its motion; for being itself an electric *per se* [i.e. an insulator], it does not attract it, having already its quantity." [76]

These two explanations are completely contradictory; if the atmosphere is held to the body by an attractive force, it should not leak off freely *in vacuo*. Furthermore, in the former explanation it is asserted unequivocally that, were it not for the attraction, the atmosphere would "dissipate in the air"; but in the latter we are told that the air resists such dissipation because of its insulating characteristics. Other problems arise as well because, according to the theory, any ordinary body contains a large amount of electric fluid and this will necessarily exert a repulsive force on any nearby fluid. Unless, therefore, the attraction between particles of ordinary matter and particles of the fluid is consistently greater than the mutual repulsion between particles of the fluid, there will be no resultant attraction to retain an atmosphere in its place; nor will there be any resultant attraction of the kind Franklin wanted to invoke in order to explain the sharing of charge between a charged conductor and an uncharged one when they come in contact, or the attraction between an uncharged body and a positively charged one. Franklin's characterization of the force between particles of ordinary matter and those of the fluid as "strong," [77] a term he did not use when discussing the repulsion between particles of fluid, may possibly imply a recognition of this point; if so, he certainly did not stress it. What is more, if he did recognize the point, how are we to understand his assertion, when discussing the evacuated tube, that the atmosphere will "fly freely" from the body *in vacuo*? This surely implies that there is no resultant attractive force retaining the atmosphere in place, and hence that the attractive force due to the ordinary matter of the body is exactly balanced by the repulsive force due to the fluid within it. But, as has just been pointed out, Franklin's explanation of other effects besides the retention of atmospheres on bodies depended on these forces being unequal. Franklin offered no escape from this dilemma. His readers in England and on the Continent were left to do the best they could to understand what he meant, and it is hardly surprising that they failed to follow him completely.

The explanation Franklin's theory gave for the observed repulsion between two positively charged bodies also raised problems. In his first group of published papers, Franklin had done no more than hint that when the atmosphere surrounding two bodies came into contact,

[76] Ibid., p. 233.
[77] Ibid., p. 213.

they repelled each other. But by analogy with air, it was much more reasonable to suppose that the two atmospheres would simply merge into a single atmosphere, and symmetry would suggest that then the two lumps of ordinary matter would if anything migrate to the center of the common atmosphere; that is, they would appear to attract each other. This was one of the problems that Franklin tried to solve in 1755, but the best he could do even then was to assert baldly that mixing did not occur: "Electric atmospheres, that flow round non-electric bodies, being brought near each other, do not readily mix and unite into one atmosphere, but remain separate, and repel each other." [78] Similarly, to explain the facts of electrostatic induction, he had to assume against all analogy that, when an atmosphere came into contact with the fluid within a neutral body, the two bodies of fluid remained distinct and did not merge into a single mass: "An electric atmosphere not only repels another electric atmosphere, but will also repel the electric matter contained in the substance of a body approaching it; and, without joining or mixing with it, force it to other parts of the body that contained it." [79] It was this kind of inconsistency between the behavior that had to be attributed to the electric fluid in Franklin's theory, and the normal laws of fluid dynamics, that Franklin's supporters sought particularly to avoid in their own elaborations of the theory.

Finally, nowhere did Franklin attempt to explain the observed repulsion between two negatively charged bodies; even in 1755, all he could do was state the facts of the matter. Some years later Kinnersley suggested to him that this and all other cases of repulsion could be reduced to differential attractions between the bodies concerned and the surrounding air,[80] but Franklin did not regard this as a satisfactory solution to the problem: "You know I have always looked upon and mentioned the equal repulsion in cases of positive and of negative electricity, as a phaenomenon difficult to be explained. I have sometimes, too, been inclined with you, to resolve all into attraction; but besides that attraction seems in itself as unintelligible as repulsion, there are some appearances of repulsion that I cannot so easily explain by attraction." [81]

[78] Ibid., p. 302.
[79] Ibid.
[80] Ibid., p. 349.

[81] Ibid., p. 365. Does Franklin's admission that attraction was "unintelligible" mean that he was by then (1762) aware of the problems his theory confronted even in accounting for attractions, or is it simply a recognition that all the forces he had invoked lacked a "mechanical" explanation of the kind demanded by the Cartesians?

Despite uncertainties of this kind, it is evident that Franklin saw electrical attractions and repulsions as consequences of static interactions between electrical atmospheres, or between atmospheres and lumps of ordinary matter. By contrast, in the variant accounts developed by his followers, the interactions were envisaged as dynamic ones, involving an actual transfer of fluid as the observed motions occurred. These accounts therefore bore a certain resemblance to those provided by the more traditional theorists of the day such as Nollet and J. A. Euler. The resemblance was, however, no more than superficial. In the traditional accounts, the motions were supposed to be a direct mechanical consequence of the impact of streams of fluid on the bodies moved; indeed, the reduction of the motions to such straightforwardly mechanical terms had always been seen as the chief aim of the electrical theorist. But in the theories of Franklin's followers, this was no longer the case; despite the fact that a flow of fluid was supposed somehow to give rise to the electrical motions, it was recognized that it did not do so in any simple mechanical way. In Franklin's theory, and also in the variants adopted by his followers, the traditional aim of explaining the attractions and repulsions mechanically was abandoned as the price that had to be paid to secure a satisfactory explanation of that more recently discovered and equally perplexing, but much more spectacular phenomenon, the Leyden experiment.

The dynamical approach adopted by Franklin's followers developed naturally out of the ideas that Franklin himself had set out in his early writings on electricity. All ordinary matter, Franklin had suggested, was under normal conditions saturated with its "natural" quantity of electric fluid; electrification amounted to an alteration in the distribution of fluid among the various samples of matter involved. Since the fluid was also highly elastic, such a state of affairs would always be extremely unstable. What was more natural, then, than that Franklin's supporters should have understood him to be saying that all electrical phenomena, including attractions and repulsions, occurred as the equilibrium of the fluid was restored after its initial disruption during the process of electrification? In this regard, the widespread preference among his supporters for the verbs "condense" and "rarefy," rather than Franklin's own "accumulate" and "subtract," when explaining what was involved in charging bodies "plus" and "minus," is most revealing: Franklin's readers simply took it for granted that what counted in electricity, as in all other systems of fluid mechanics, were not so much the absolute quantities of fluid involved as the hydrodynamically more relevant variables, their densities.

Similarly, and much to Franklin's chagrin, his followers tended to modify the basic law of electrical attractions to read that any two unequally electrified bodies, rather than any two bodies carrying unlike charges, would attract each other. That this could happen is a further reminder of how difficult it was to obtain reliable experimental results in this area, where the phenomena were so small-scale and so subject to distortion through changing external circumstances. Nollet, too, had been able to exploit this difficulty for his own ends. In the present case, what mattered to Franklin's supporters was that electrical repulsions then became a very special case indeed, occurring, so it was said, only between two equally charged bodies; the attractions that were supposed to happen in all other cases could meanwhile be happily subsumed under the single standard "hydrodynamical" rubric of "restoring the equilibrium."

As Franklin's European contemporaries struggled to make sense of his doctrine of atmospheres, they could not help but be influenced by the very different notions about electricity that were already familiar to them from sources closer to home. Hence they tended to re-interpret his doctrines, especially his doctrine of "plus" and "minus," in semi-effluvial terms. Those who followed Franklin maintained with him that a body could become electrified in two contrary ways, by either gaining or losing electric fluid. But because the electric fluid was extremely elastic, the surplus fluid on any positively charged body would tend to flow away from it, while fluid would tend to flow toward any negatively charged body from bodies in its vicinity. Nollet's double stream of electric matter was replaced by a single stream, outward from a body charged "plus," inward toward one charged "minus"; as John Canton expressed it, "according to Mr. *Franklin,* excited glass *emits* the electrical fluid, but excited wax *receives* it."[82]

Within this general framework, two different approaches can be discerned. According to some, the streams of electric matter began flowing as soon as the bodies became electrified, and continued as long as they remained in this condition. On this view, the "atmosphere" surrounding an electrified body was simply the region occupied by the streams of electric matter flowing inward or outward as the case may be. Others believed with Franklin that when bodies became electrified, they acquired essentially static charges that would remain in place so long as there were no conductors nearby to carry the charge away. These people differed from Franklin, however, in assuming that the electrical attractions and repulsions, like all other electrical phe-

[82] Ibid., p. 294.

nomena, were brought about by a flow of electric fluid to "restore the equilibrium."

Nollet took Franklin to be upholding the first of these alternatives.[83] What is more, he took it for granted that Franklin meant to account for electrical attraction or repulsion in the traditional way in terms of the impact of an appropriately directed stream of electric fluid on the body in question. Time and again he professed himself completely at a loss to understand how Franklin intended to explain the fact that a single electrified body could both attract and then repel a nearby light object; to him, this simple observation was enough to prove that in the region surrounding the body there were two oppositely directed streams, not just the one he thought Franklin was advocating.

Canton was the most notable of Franklin's supporters who seems actually to have adopted our first alternative. Nowhere, however, did he spell out his views in a connected fashion, and they have to be reconstructed from scattered hints [84] in his published papers and in Priestley's *History and Present State of Electricity* (Priestley having been in close contact with Canton during the preparation of this work). For Canton, electricity seems at this time to have been a state in which a body was surrounded by an atmosphere of streaming electrical effluvia, either emitting or absorbing fluid according to whether it had previously acquired a surplus, if it was electrified "plus," or a deficiency, if electrified "minus." Sometimes, Canton indicated, this streaming fluid could be observed directly, as when a polished glass tube was rubbed in the dark; the fluid being emitted could then be easily seen as "a great number of diverging pencils of electric fire" thrown out by the tube after each stroke of the cloth.[85] Unfortunately he never set out in any detail his views within this general framework on the causes of the observed attractions and repulsions. From the few hints he gave, however, it is at least clear that, like Franklin, he had abandoned the attempt to provide a direct mechanical explanation for them.[86]

The second of our two alternatives was strenuously upheld by perhaps the most vociferous of all Franklin's early supporters, the Italian Giambatista Beccaria, and through his writings it obtained a wide

[83] Nollet, *Lettres sur l'électricité*, Paris, 1753, p. 63, and repeatedly in his later writings.
[84] The evidence is brought together in R. W. Home, "Franklin's electrical atmospheres," *British Journal for the History of Science*, VI, 1972, 142–45.
[85] Canton, *Phil Trans.*, XLVIII, 1753–54, 780–82.
[86] *Benjamin Franklin's Experiments*, pp. 294–95.

currency during the 1750s and 1760s. Electrification, according to Beccaria, was essentially a static condition in which a body contained either more or less than its usual complement of electric fluid. In his view, however, all the signs of electricity, including the attraction and repulsion of nearby light objects, were due to "the vapor expanding itself from a body in which it is in greater quantity into one in which there is less." [87] Yet a flow of fluid outward from a body electrified "plus" clearly could not provide a direct mechanical explanation of the attraction of neighboring bodies toward the electrified body; the cause of the motion remained mysterious and unknown. Beccaria ascribed it to "the force of the electric vapor which expands itself from the body in which there is more into the one in which there is less until equilibrium is reached," [88] but when faced with the necessity of explaining how such a force could give rise to the observed motions, he declined to do so, at least for the time being. In support of his stand, he cited Newton's well-known positivistic views on methodology.

To account for the repulsions that were also observed in the vicinity of electrified bodies, Beccaria sought to reduce them to differential attractions. An equilibrium in the electric fluid could be restored, he said, by a slow flow of fluid and hence, according to the principles already enunciated, an attraction, between each body and the external air. The separation of the bodies followed, he argued, because only the external air was involved; the air between the bodies was replete with the fluid constituting the atmospheres of the two bodies.[89]

Beccaria's *Dell'elettricismo artificiale e naturale* was only the first of a number of important works he published on electricity. In 1758 his next major production appeared, and in it he expounded the same general principles he had put forward five years earlier.[90] Once again his chief concern was to establish a workable system of electric fluid dynamics that would explain the various appearances. Hence he argued at the outset that, through its "natural diffusive force," the electric "vapor" would make continual efforts to expand. If the quantity of vapor in one of the bodies in a material system was altered in a different proportion from that of the other, the diffusive force would strive to restore the equilibrium; it was the resulting motion of fluid across the intervening air from one body to another that gave rise to all those effects included under the general rubric "elec-

---

[87] Beccaria, *Dell'elettricismo artificiale e naturale*, p. 17.

[88] Ibid., p. 40.

[89] Ibid., pp. 26–28.

[90] Beccaria, *Dell'elettricismo. Lettere di Giambattista Beccaria . . . dirette al chiarissimo sig. Giacomo Bartolomeo Beccari . . .* , Bologna, 1758.

tricity."[91] The air offered a considerable resistance to this passage of fluid, however, with the result that an excess of fluid could form temporarily in and around a suitably insulated body and thus form the usual Franklinian electrical atmosphere.[92]

Just as in 1753, Beccaria again included the attraction of light objects among the phenomena brought about by the flow of electric fluid. Now, however, he no longer retreated into Newtonian-style agnosticism when it came to explaining how such a flow could cause two bodies to approach each other. Instead he described a series of experiments that demonstrated, he said, that the air, too, played a crucial role. More specifically, he suggested that what happened was that, in flowing from one body to the other, the electric fluid drove out the intervening air, upon which the pressure of the external air drove the bodies toward each other into the resulting central low-pressure region: "If the vapor diffusing itself from A, which has more of it, into B, which has less, displaces the air between A and B, the air which presses upon A and B from the outside will prevail; whence A and B, impelled by this excess of pressure, will come together with an equal motion proportional to the quantity of vapor which passes between the bodies."[93]

Apart from Aepinus and Wilcke, Canton and Beccaria were probably the most important Franklinist electricians in the first twenty

[91] Ibid., pp. 4–5.
[92] Ibid., p. 12. Some years afterward both Priestley (*History*, p. 245) and Beccaria himself (*Elettricismo artificiale*, Turin, 1772, p. 173) claimed priority for the latter, over Aepinus, as the first person to abandon belief in Franklinian-style atmospheres—though Priestley did insert the word "probably" in his account. In support, both men referred to "Elettricismo artificiale, p. 54," which can only refer to Book I of Beccaria's *Dell'elettricismo artificiale e naturale* of 1753. The relevant passage is somewhat obscure but does not really support the claim; and neither does the explanation (noted above) of the mutual repulsion of two equally charged bodies, which was given only a few pages earlier. Quite apart from this, however, Beccaria's description of what happens when the prime conductor of an electrical machine draws its charge from the globe being rubbed proves that even in 1758 he was still expounding very orthodox views on the subject of atmospheres; the charge developed on the globe, he assured his readers, was constantly discharged into the chain (i.e. the prime conductor), "and thus both in and around this an excess of vapor will be accumulated, which through its diffusive force will expand itself somewhat into the air, but from the resistance and inaction of this it will be retained around the chain, and, as it were, thrown back against this."
[93] Ibid., pp. 42–43. Beccaria at first presented his views somewhat hesitantly, but in a letter to Franklin soon afterward he spoke with rather more confidence (*Phil. Trans.*, LI, 1759–60, 514–26; English translation in Leonard W. Labaree, ed., *The Papers of Benjamin Franklin*, New Haven, 1959–, VII, 300–14). His ideas have been described more fully in Mario Gliozzi, "Beccaria nella storia dell'elettricità," *Archeion*, XVII, 1935, 15–47.

years of the theory's existence. The fact that they both adopted views rather different from what Franklin had intended concerning the proposed atmospheres and their role in the phenomena of electricity reinforces what has been said already; though Franklin's theory was much more successful than any of its competitors in accounting for a number of crucial electrical phenomena, and especially for the otherwise quite baffling behavior of the Leyden jar, it too had its weaknesses. For the most part, these centered on Franklin's doctrine of atmospheres and on the related difficulties in accounting in a coherent fashion for the observed electrical motions. As we shall see, it was on precisely these points that Aepinus focused his attention when he set out in his *Essay* of 1759 to establish Franklinian electrical theory on a more secure foundation. It is to that work itself that we must now turn.

CHAPTER THREE

Electricity in the *Essay*

Although Aepinus's *Essay* of 1759 is one of the most important eighteenth-century works dealing with electricity, Aepinus himself regarded its contributions to electrical science as peripheral to its contributions to the study of magnetism. At the very beginning of his book, in the dedication to Count Kiril Razumovskiĭ, the president of the St. Petersburg Academy of Sciences, he states that the investigation of magnetism is the chief task he has set himself in what is to follow, and similar statements appear in the body of the work. This attitude had a considerable effect on the scope of Aepinus's undertaking. His approach to the phenomena of magnetism was inspired, as we have noted, by the analogy with electricity that his tourmaline experiments had revealed to him. He makes no attempt in his book to treat systematically all the phenomena of electricity, but deals only with those for which he can perceive an analogue in the magnetic realm: "It would be completely foreign to the scope of this essay," he explains, "to bring in here the other phenomena of electricity, since my intention is to consolidate the theory of magnetism rather than that of electricity." An understandable pride in what he was, in passing, contributing to the science of electricity does occasionally manifest itself. Yet for the most part Aepinus failed to emphasize sufficiently the novelty of what he was doing. This contributed directly to the relative neglect of the work when it was first published. Few if any of Aepinus's contemporaries recognized the importance of what he was saying, and only in the next generation did his ideas bear fruit.

Aepinus at once makes plain his indebtedness to Franklin when he presents four "universal propositions" that are his starting point and that, so far as he is concerned, embody the whole of "Franklin's very elegant theory of electricity, which agrees astonishingly with the phenomena." Following Franklin, Aepinus ascribes all electrical phenomena to a certain extremely subtle elastic fluid whose particles mutually repel each other. These particles are, he says, attracted by particles of ordinary matter. He asserts a radical difference between insulators ("electrics *per se*") and conductors so far as the electric fluid is concerned, namely that the fluid can move with the greatest of ease within the pores of conducting bodies, whereas it moves only

107

with great difficulty within the pores of bodies electric *per se*. And, finally, he maintains that while some electrical phenomena are brought about by a transit of electric fluid from one body to another, others, and the traditional attractions and repulsions in particular, occur without an actual movement of fluid.

Notwithding Aepinus's professed dependence on Franklin in framing these propositions, they represent an advance beyond the ideas of the great American in one respect. Immediately after stating his four propositions, Aepinus reiterates what he had said earlier, in his paper on the tourmaline, when discussing his invention of the "air condenser." He insists that there is no need to ascribe a special property to glass, namely an absolute impermeability to the electric fluid, in order to account for the Leyden experiment. Rather, it is sufficient to invoke the insulating power it shares with all other bodies electric *per se*: "I am astonished," he says, "that the sagacious man did not observe that this impermeability . . . constitutes the essence of bodies electric *per se*, and is not proper to glass, but is common to all bodies electric *per se*."

Aepinus, however, almost certainly did not make this observation directly from the first principles of Franklin's theory. During his initial investigation of the electrical properties of the tourmaline, Aepinus found that when he rubbed the stone very gently, being careful not to heat it in doing so, it, like glass, normally acquired a positive charge over its whole surface. If one side of the crystal was rubbed while the other was grounded, however, the latter became charged "minus," while the side being rubbed became charged "plus" as usual. After pondering the matter for a time, Aepinus realized that he was dealing here with an induction effect made possible by the insulating character of his crystal, and he went on to prove that the same thing happened when any other electric *per se* was treated in the same way.[1] While Aepinus nowhere made the point explicitly, the conclusion is almost irresistible that it was his recognition of the principle involved in these experiments that led to his improved understanding of the Leyden jar. In any event, it follows from Aepinus's argument in the *Essay* that, if sufficient care is taken in setting things up, it should be possible to replicate the Leyden experiment using as the dielectric other electrics *per se* besides glass. Aepinus himself, of course, had already demonstrated two years earlier, in the case of air, the correctness of this conclusion, and since then Wilcke had succeeded with a variety of other materials.[2] Aepinus's argument was therefore no

[1] Aepinus, *Recueil de différents mémoires sur la tourmaline*, pp. 49–51.
[2] Wilcke, *Handlingar Kungl. Svenska Vetenskapsakademien*, 1758, pp. 250–82.

longer as novel in 1759 as it once had been. Nevertheless, it seems to have been this later presentation that finally caught the attention of others. Franklin, for one, was sufficiently persuaded to abandon his previous position.[3]

Except on this point, Aepinus was much closer to Franklin's thinking than were most others who had taken up the cudgels for the American. Most of Franklin's supporters, we have seen, wanted to associate all electrical phenomena, including the attractions and repulsions, with a flow of electric fluid from one body to another. Aepinus realized that, despite Franklin's having invoked a kind of fluid to account for the various electrical phenomena, the analogy with ordinary fluids could not be pressed too far. Granted that in Franklin's formulation the electric fluid was like other fluids in that it tended always to flow back whenever its equilibrium was disturbed, it was also a veritable *power*, which by its mere presence could exert an influence on ordinary bodies. In particular, it was the mere *presence* of fluid, not its motion, that gave rise to the attractions and repulsions.

Aepinus, however, saw the implications of this latter point much more fully than had Franklin. In doing so, he took the step that, more than any other, set his work apart from all that had gone before. If the electric fluid really does act on other matter by mere influence, the notion of an electrical atmosphere becomes superfluous, and the way is open to the development of a fully fledged action-at-a-distance theory of electricity. Aepinus makes explicit, in his introductory remarks, his intention of relying throughout the work on unexplained forces acting at a distance. In the first of his four "universal propositions," he asserts unequivocally that the forces exerted by particles of the electric fluid act "even over rather large distances." As the work proceeds, he derives a multitude of theorems that are all based on the presumption that forces of this kind act in both the electrical and magnetic realm to bring about the observed phenomena; nowhere does Aepinus have recourse to electrical atmospheres, or anything like them, in his explanations. Finally, at the beginning of the fourth and last chapter, he explicitly denies that these atmospheres exist. "It is manifest," he says, "that I never consider magnetic or electric matter as clinging outside the body or ambient to it, so it is clear that I am not using the words vortex or atmosphere in their proper sense. Whenever these words occur in what follows, nothing else must be thought to be denoted by them but what at other times is usually called

---

[3] Franklin to John Winthrop, 10 July 1764, Labaree, *The Papers of Benjamin Franklin,* XI, 254.

the sphere of activity. For me these words designate nothing but the space to which the attraction or repulsion, electric or magnetic, sensibly extends itself in any direction around a given body."

Aepinus was fully aware that this position was contrary to that of most of his contemporaries. Though Isaac Newton's *Principia* had been published almost three quarters of a century earlier, and though others such as Euler had subsequently made free use of an unexplained gravitational force in their astronomical calculations, there had been little tendency to introduce into science additional long-range forces acting at a distance. Such other forces as had been invoked were either exclusively short-range, such as the one used to account for cohesion, or acted only between particles at the microscopic level, such as the interparticulate forces of attraction and repulsion that eighteenth-century chemists relied upon to explain the various reactions with which they were concerned. So far as electricity is concerned, it is easy to understand why recourse to long-range macroscopic forces had not been seriously considered. Until almost the middle of the eighteenth century, the attraction and subsequent repulsion of nearby light objects had been the chief, and for a long time virtually the only known electrical phenomenon. In such circumstances, the electrical theorist had always taken his central task to be the explanation of these effects in terms of other and more general physical principles. To invoke a new set of unexplained forces on their account was simply to say that electricity acted because it acted; that is, it was to explain nothing at all.

During the 1740s and 1750s, however, the discovery of several spectacular new electrical phenomena created an entirely different situation, which is reflected in the writings of men such as Watson, Ellicott, and, above all, Franklin. As the number and complexity of known phenomena increased, people became concerned simply to discover a coherent pattern that would make sense of these; as a result they worried less and less about reducing electricity as a whole to traditional mechanical terms. Electrical theory, in other words, began to be studied in its own right, and not merely as an adjunct to some wider mechanical theory of nature. Not until this had happened was it possible for an electrical theorist to argue, as Aepinus does in his book, that the use of unexplained forces of attraction and repulsion was justified if one thereby succeeded in reducing the phenomena to order (Intro. § 6). Even then, this was not an easy step to take. Franklin had not been able to take it, nor had any of his other followers, not even one such as Beccaria who quite frankly gave up trying to explain the macroscopic motions mechanically. Aepinus was very much a pioneer in this respect.

Similar considerations apply in the magnetic realm. Here, eighteenth-century investigators inherited from the past a much wider range of phenomena than was the case with electricity, but here, too, they seldom invoked forces acting at a distance to account for them. Once again, Aepinus was an innovator in introducing such forces into his account. In the magnetic case, however, he justifies his procedure not simply on general grounds of the kind just indicated but also through the success of the analogy he is drawing between magnetism and the field where he has already introduced unexplained forces to great effect, electricity. He also draws a very striking parallel between what he is doing when he reduces a set of complex phenomena to these unexplained forces, and the then standard mathematical practice of resting content once it could be shown how a given geometrical problem could be reduced to an assumed quadrature of the circle, even though it was well known that the quadrature itself had not been achieved.

In Aepinus's day, a formal proof that the quadrature was impossible had not yet been given. Nevertheless, at about this time the Paris Academy of Sciences passed an official resolution that it would no longer waste its time examining so-called "solutions" to the problem, and it was generally accepted among mathematicians that the quadrature could not be done.[4] Aepinus evidently felt the same way about the problem of explaining the various electrical and magnetic forces. "I leave that," he remarks, "to the further discussion of those who are happy to spend time in investigations of that kind." He himself had better things to do.

Aepinus's attitude to his forces is very similar to Newton's attitude to gravity, and so he does not hesitate to call upon the Englishman's authority in support of his stand. Aepinus, like Newton, is convinced that his forces act, but he is unable to discover what causes them to do so. Furthermore, just as Newton always insisted that even though he did not know the cause of gravity, there must certainly be one,[5] so Aepinus unequivocally declares that he takes the proposition "a body cannot act where it is not" to be "an indubitable axiom." And he criticizes certain "incautious disciples of the great Newton" for upholding the contrary, and also for asserting that forces of attraction and repulsion are innate in matter. (Here he doubtless has particularly in mind Roger Cotes, the author of the controversial preface to the second [1713] edition of Newton's *Principia*.) In only one respect,

[4] E. W. Hobson, *"Squaring the Circle": A History of the Problem*, Cambridge, 1913, chs. I, III.

[5] E.g. Newton to Richard Bentley, 25 February 1692/3, I. B. Cohen, ed., *Isaac Newton's Papers and Letters on Natural Philosophy*, Cambridge, 1958, pp. 302–303.

in fact, does Aepinus's attitude differ significantly from Newton's. For much of his life Newton seems to have believed that the cause of gravity was non-material; certainly he was at least prepared to entertain seriously the possibility of this.[6] Aepinus, however, espouses a position much more akin to that of Euler. The only alternative he can see to a material cause is some incorporeal spirit, and this he finds completely unacceptable: "I cannot be induced to believe," he says, "that this happens in the world." Presumably, therefore, the forces must be brought about by various actions in that universal medium, the aether; but Aepinus does not even bother to state this explicitly, and he shows no interest at all in pursuing the matter further.

Aepinus, in fact, displays throughout all his work the same highly sophisticated form of empiricism that Newton had developed earlier, and his reluctance to concern himself with hypothetical aetherial mechanisms is symptomatic of this. For Aepinus, the facts of experience are absolutely paramount. The scientist must take these as his starting point, and his proper task is to reduce them to order, to "their proximate causes and primitive forces," not to invent mechanisms lacking all empirical support in order to "explain" them. Yet, unlike many of Newton's less perceptive followers, Aepinus is prepared to allow hypotheses a place in his science; indeed, he insists on describing his own crucial theoretical insight, the analogy between electricity and magnetism, as an "hypothesis." At the same time, he acknowledges the difficulties, so familiar to present-day philosophers of science, inherent in confirming his or any other hypotheses empirically, namely that while empirical evidence may show a particular hypothesis to be false, it can never prove conclusively that it is true. "I am fully aware," Aepinus says, "that it cannot be certainly concluded from the agreement of an hypothesis with the phenomena that we have reached the true cause." Hence he puts his magnetic theory forward "as probable rather than as certain." He claims, however, that it is strongly supported not only by its success in handling the magnetic phenomena but also by its similarity in form to what has been found to apply in other parts of physics: "It is quite likely," he argues, "that nature produces analogous phenomena by analogous means." Even Newton, he says, could claim no other kinds of support than these for his position; hence the theory he is now putting forward should be regarded as probable to the same degree as Newton's.

[6] Ibid. Also A. Koyré, *From the Closed World to the Infinite Universe*, Baltimore, 1957, p. 209; A. R. and M. B. Hall, eds., *Unpublished Scientific Papers of Isaac Newton*, Cambridge, 1962, pp. 194–97; J. E. McGuire, "Force, active principles and Newton's invisible realm," *Ambix*, xv, 1968, 154–208.

Another way Aepinus's approach resembles Newton's is his consistent use of mathematics in framing his arguments. This alone gives the work an entirely different appearance from anything published on the subject previously, because until this time electrical theory had always been wholly qualitative in nature. Indeed, Aepinus's mathematics may have prevented most electrical investigators of his day from appreciating the importance of his analyses. Despite the confidence Aepinus expressed in his preface that any electrician worthy of the name would have no trouble in coping with his calculations, this was manifestly not true. On the other hand, most of the mathematics Aepinus uses is elementary, and so from the purely mathematical point of view his work was of little interest. It therefore fell between two camps. Too mathematical for the experimentalists, it was yet too experimentally oriented and insufficiently abstract for the mathematicians. The work was, in fact, one of the earliest examples we have of that genuinely mathematical physics to which we have since become accustomed, in which mathematical analysis is united to experiment in a highly fruitful alliance. It was as such that its merits eventually came to be appreciated, but they were certainly not recognized at first.[7]

Unfortunately, Aepinus's mathematical arguments were rendered more difficult than necessary for his non-mathematically inclined readers by the large number of typographical errors in the mathematical formulae presented in the book. These were no doubt partly a result of the haste with which the work was printed. In addition, part of the blame almost certainly rests with Aepinus's new career at court, which had just begun when his manuscript was sent to the printery; his new duties must have greatly reduced the time he was able to devote to correcting the proofs. Some correcting was clearly

---

[7] Dorfman has argued (*Teoriya*, p. 537) that, insofar as Aepinus's book marked the beginning of quantitative calculation in the theory of electricity, it was an "historically inevitable step" after the work of Richmann and Lomonosov; Dorfman has stressed in particular the importance of Richmann's invention of an electrometer in the early 1740s and his subsequent use of it in his electrical experiments. However, nothing in Aepinus's book suggests that his quantification of electrical theory was related to the work of his Russian predecessors. Furthermore, the "tourmaline" memoir that Aepinus presented to the Berlin Academy just prior to his departure for St. Petersburg, and the accounts that he and Wilcke subsequently gave of the events leading up to the discovery of the air condenser described in that paper, make it clear that his semi-mathematized theory of electricity was fully worked out before he ever set foot in the Russian capital (see below, § 72, and Wilcke, *De electricitatibus contrariis*, pp. 93ff.). A much more likely influence so far as Aepinus's use of mathematics in his theory is concerned is Leonhard Euler.

done, since the Latin text is almost free from error. Perhaps Aepinus arranged for someone else to do the job for him, someone who knew Latin but not mathematics. There is a hint of an apology for the typographical inadequacies of the finished product in the preface, which Aepinus composed in October 1759, shortly before printing was completed, but that can have come as small consolation to readers who would have been struggling even in the best of circumstances to cope with such unfamiliar material.

One advantage of mathematical reasoning is that it allows one to carry through longer and more complicated arguments than one could otherwise handle; in Aepinus's phrase, it enables one "to escape . . . the prolixity of the usual language." This in turn explains one of the principal strengths of mathematical physics in general, one which Aepinus understood perfectly and calls upon explicitly in support of what he is doing. With the aid of mathematics, he argues, one can arrive at consequences that are complex and quite remote from the premises from which one began, and that therefore, to use the modern terminology, may be said to have a very low prior probability. If these consequences are then found to agree with the experimental phenomena, the premises from which they have been derived acquire a correspondingly high degree of probability. Any errors in the premises will become obvious, Aepinus says, because they will be amplified by the length of the chain of deductions. Also, there can be no suspicion that the theory has been manufactured in a *post hoc* fashion specifically to account for the phenomena in question, and this greatly increases its credibility.

Yet in certain respects, Aepinus's brand of mathematical physics is far removed from that which has developed subsequently. In particular, none of Aepinus's calculations, as they stand, is either based upon or leads to quantitative experimental determinations, and his theory as a whole therefore never becomes more than semi-quantitative at best. All of Aepinus's mathematics rests squarely on the notion of static electrical charges, "plus" or "minus" as the case may be, that he took from Franklin and further refined for himself by abandoning the Franklinian doctrine of atmospheres. In Aepinus's theory, masses of electric fluid and lumps of ordinary matter exert forces on each other at a distance, and Aepinus's various calculations amount to determining the effects of particular distributions of fluid and ordinary matter on each other. He assumes that the individual forces that enter into his computations will depend directly on the quantities of fluid or ordinary matter involved, and he assumes, too, that the forces will decrease as the distance between the interacting masses increases. Beyond this, however, he cannot go. In particular, as he repeatedly confesses throughout his book, he simply does not know the precise manner in which the forces vary with the distance. But only if this law was known could his analyses lead to precise quantitative results. Until the law could be determined, therefore, the best Aepinus could hope for was to arrive at increasingly detailed and specific qualitative predictions; numerical data, such as those that his predecessor Richmann had hoped to supply, remained entirely beyond his reach. Yet despite such limitations, Aepinus's book is an impressive piece of work, and in it he reaches some very important conclusions.

Aepinus's constant reiteration of his ignorance of the precise form of the force law is striking. This makes it all the more remarkable that when Joseph Priestley published his famous *History and Present State of Electricity* a few years later, he felt able to dismiss Aepinus's calculations out of hand, precisely on the grounds that they assumed a form of the law that was known to be incorrect: "He that reads the first chapter, as well as many other parts of [Aepinus's] elaborate treatise," Priestley wrote, "may save a good deal of time and trouble by considering, that the result of many of his reasonings and mathematical calculations cannot be depended upon; because he supposes the repulsion or elasticity of the electric fluid to be in proportion to its condensation; which is not true, unless the particles repel one another in the simple reciprocal ratio of their distances, as Sir Isaac Newton has demonstrated, in the second book of his Principia." [8]

Though Priestley's readers must have been considerably influenced by his sweeping condemnation of Aepinus's work, most authors on electricity have either ignored it or at best treated it with great caution. Dorfman, however, in his recent Russian edition of Aepinus's book, has accepted and even amplified Priestley's criticism.[9] He sees as the source of the difficulty Aepinus's habit of assuming that whatever electric fluid is contained in each of the interacting samples of ordinary matter entering into his calculations is distributed uniformly throughout the sample. This implies, says Dorfman, that for Aepinus the repulsive action of the fluid is in each case proportional to its volume density, which is what Priestley asserted. Hence, Dorfman argues, Priestley's appeal to Newton's "Boyle's Law" theorem is legitimate; Aepinus's assumption of a uniform distribution of fluid in each case does commit him to the manifestly incorrect idea that electrical and magnetic forces are in the simple inverse ratio of the distances rather than any higher powers of these.

[8] Priestley, *History*, p. 436.
[9] *Teoriya*, pp. 504, 514, 542.

Dorfman's argument is ingenious, but can be questioned on several grounds. To be sure, Aepinus usually assumes uniform distributions of fluid in each of the interacting units in the various situations he discusses. Yet there is no real evidence that it was this that Priestley had in mind when he dismissed Aepinus's work so cavalierly. It is equally possible that Priestley's eye may simply have been caught by Aepinus's assertion, repeated several times (§§ 50, 67, 73), that the violence of the shock delivered by the Leyden jar depended on the high degree of condensation of the electric fluid that this device permitted. In this case, however, Priestley's criticism would have been entirely without foundation, as it appeared to Aepinus himself, and as it also appeared, for example, to the eminent Edinburgh professor John Robison, who referred scornfully to Priestley's "very slight and almost unintelligible account" of Aepinus's work.[10] A careful reading of what Aepinus says makes it clear that the only reason an increased condensation of fluid increases the violence of the commotion is that it leads to a greater quantity of fluid flowing across in the brief period of the discharge, so that "degree of condensation" has no other connotation here than "accumulation" or "charge."

But neither can Dorfman's alternative explanation be maintained. Even though Aepinus usually assumes uniform distributions of fluid, it is simply not the case that this effectively commits him to assuming that the repulsive actions of the masses of fluid are proportional to their volume densities. On the contrary, whenever Aepinus sets out to calculate the forces exerted by the various bodies in a given situation, it is always and unequivocally the actual quantities of fluid present that enter into his calculations. Neither the volumes of the bodies present nor, consequently, the densities of fluid within them are ever mentioned, nor do they need to be.

Furthermore, on most of the occasions on which Aepinus assumes uniform fluid distributions, he does so purely as a mathematical device, in order to simplify his calculations. Insofar as these cases are concerned, he carefully warns that his assumption is "a gratuitous hypothesis" that "cannot occur in nature." He also insists, however,

[10] Aepinus's reaction to Priestley's account is revealed by his inscription in the presentation copy of the *Essay* (now preserved at the Biblioteca Nazionale Braidense, Milan) which he sent to the Italian mathematician Paolo Frisi. The book was a gift, he indicated, "Pour M^r. Frisius de la part de l'Auteur, qui prie ce Celebre Savant, de ne vouloir pas juger de ses travaux, qui ont rapport à l'Electricité, que sur ses propres ouvrages, qu'il a l'honneur de lui présenter, et nullement sur les extraits, infideles, inexacts et incomplets, que Mr Pristley [*sic!*] en a donné dans son Histoire de l'Electricité." For Robison's opinion, see his article, "Electricity," *Encyclopaedia Britannica*, 3rd ed., *Supplement*, Edinburgh, 1803, I, 538.

that his procedure in making it is quite legitimate: "It can be tolerated as long as the conclusions deduced from it are such as are little changed even if the fluid is considered unequally distributed, as is really the case." He goes on to analyze a particular situation in which fluid density will not be uniform but will vary continuously throughout the sample (§ 193; cf. §§ 25, 182). In only one case, to be discussed later, does Aepinus actually commit himself to a uniform distribution of fluid as a physical hypothesis, and though in fact we now know that the distribution he here assumes is inconsistent with an inverse-square law of force (since this predicts that the charge on a body will be confined to a very thin surface layer of the body), it was clearly not this particular instance that led Priestley to make his comment about the basis of Aepinus's calculations.

It is possible to go even further than this in rebutting Priestley's criticism. Not only does Aepinus not imply that the electrical force law is one of simple inverse proportionality (and he was certainly a good enough mathematician to see the implications of Newton's "Boyle's Law" theorem for his own work), he explicitly denies that the law takes this form. He does so on one of the rare occasions where he wishes to illustrate his argument by means of a worked example, and so needs to assume temporarily a particular form for the law. In this instance, the example comes from the magnetic realm. Aepinus, having assumed for the sake of simplicity a law of simple inverse proportionality, firmly assures his readers (§ 306) that "whatever the function according to which the magnetic actions are exercised, it does not coincide with the law assumed but differs from it"! On another occasion, and for a similar purpose, he assumes that the law is inverse-square in character (§§ 334ff.), but he does not wish to commit himself to this either. He does concede elsewhere, however, that this is perhaps the most likely form the law might take: "the analogy of nature seems to militate on behalf of this law" (§ 30).

Aepinus's theory of electricity is founded, then, on an explicit notion of masses of ordinary matter and electric fluid exerting unexplained forces on each other at a distance; these forces are supposed to vary with distance according to some law that remains to be determined, but which most likely involves an inverse-square relationship. His theory of magnetism, which will be discussed in the next chapter, is constructed on an identical basis.

Aepinus begins his analysis by refining Franklin's discussion of the general relationship between ordinary matter and the electric fluid. He does so by considering the pair of forces, one attractive and one repulsive, exerted on a neighboring particle of electric fluid by a piece of ordinary matter containing within it a certain quantity of fluid. In general, of course, the attractive force exerted by the ordinary matter will be either greater or less than the repulsive force exerted by the mass of fluid, according to the relative quantities of ordinary matter and fluid present. Hence there will be a flow either toward or away from the body as the case may be, and this will continue until a situation is reached where the two forces exactly balance each other. Aepinus identifies the quantity of fluid that needs to be present in a body to produce this equilibrium situation with what Franklin had earlier called the "natural quantity" for the body in question. But whereas for Franklin this was always a somewhat imprecise notion, with Aepinus it has acquired a strict, mathematically specified meaning; a body contains its natural quantity of fluid when it "contains neither a lesser nor a greater quantity of fluid than is exactly sufficient to produce an equilibrium between the attracting and repelling forces."

The advantage of Aepinus's procedure is immediately apparent when one recalls the inadequacy of Franklin's discussion of what keeps an electrical charge on a body. We have seen that within the space of a few pages, Franklin gave two contradictory answers to this question, namely that the charge is held in place by the attractive force of the matter of the body, and that it is prevented from escaping by the insulating power of the surrounding air. Aepinus's analysis disposes of the first alternative at once. Whenever a body contains any other quantity of fluid than the natural, the situation is inherently unstable. Hence, whether a body be charged positively or negatively, it ought to be able to retain its charge only if it is entirely surrounded by insulating materials. However, Aepinus also recognizes that the nature of the body itself ought to affect its capacity to retain a charge. He argues that, since the electric fluid moves with more difficulty within the pores of bodies electric *per se*, such bodies when grounded ought to lose their charge more slowly than conducting materials do. To conclude the discussion, he points out that "experiment agrees excellently with this reasoning" in every respect.

Perhaps Aepinus's most famous theoretical innovation was his introduction into electrical (and magnetic) theory of a force between the various particles additional to the forces postulated by Franklin. He introduces this in the context of an analysis of the various forces acting when two uncharged bodies are placed near each other (§ 28). Calling the bodies A and B, and remembering that since each is uncharged it must contain its natural quantity of fluid, Aepinus designates the force exerted by A on the fluid in B by $a-r$, where a is the attraction between the matter of body A and the fluid in B, and r is the repulsion exerted on the latter by the fluid in A. Furthermore, the fluid in A will attract the matter of B; this force Aepinus represents by A. Finally, in case the matter of A interacts with that of B, he includes a fourth term x in the account to represent this. All in all, therefore, the two bodies will attract each other with a force $a-r+A+x$, where it remains to determine the value of x. To do this, Aepinus argues first that $a-r$, being the force exerted by the body A on fluid outside it, has to be zero, since A is uncharged. As well, it is known that the total force $a-r+A+x$ is zero, since experience reveals that uncharged bodies do not exert a force on each other. Putting these two together, Aepinus obtains $A+x=0$, or $x=-A$. With a little more analysis, introducing only the very reasonable additional assumption that the force exerted by masses of either ordinary matter or electric fluid are proportional to the quantities of material involved, Aepinus is able to show that $A=a$, or, in all, $a=r=A=-x$.

Two conclusions follow from this. First, since it is known from experience that the non-interaction of the two uncharged bodies does not depend on the distance between them, all four forces must always be equal at equal distances, so they must all follow the same law of variation with distance. The element of uncertainty arising from Franklin's having characterized the attraction between particles of fluid and particles of ordinary matter, but not the mutual repulsion between particles of electric fluid, as "strong," is thus completely removed. There can no longer be any question of these two forces following different laws.

Second, and much more surprisingly, Aepinus is driven to the conclusion that the force x acting between particles of ordinary matter is a repulsive one! He confesses that when he first obtained this result, he was "somewhat horrified at it," chiefly because it seemed to be diametrically opposed to the celebrated law of gravitational attraction that Newton had enunciated not so many years earlier. What is more, there is no denying that this strang new force has since acquired a considerable notoriety among historians of electricity. Very often, however, those who have written on the subject have failed to appreciate the argument that lay behind Aepinus's innovation, and have treated it simply as an ad hoc modification of Franklin's theory to enable it to account for the observed mutual repulsion between two negatively charged bodies. Yet this is seriously to underestimate the significance of what Aepinus is doing at this point. Granted that once his proposal is accepted, the previously baffling interaction between two negatively charged bodies acquired a ready explanation (§ 35), it is clear that Aepinus's reasons for putting the idea forward went much deeper than this. With his skilled mathematician's eye, Aepinus perceived that if the principles Franklin had laid down were taken literally and their consequences rigorously traced out, they led to absurd conclusions. In particular, they led to the ludicrous proposition that two unelectrified bodies exerted electrical forces on each other. Hence Aepinus's additional postulate, or something formally equivalent to it, far from being a merely incidental modification of Franklin's theory, was absolutely essential if the foundations of the theory were to be rendered coherent. Only if this were done could the theory be reformulated in mathematical terms. In other words, Aepinus's innovation was an important step in the subjection of this particular branch of physical theory to mathematical analysis.

Necessary though his new force may have been, one can nevertheless understand Aepinus's initial feelings of horror when he found himself forced to invoke it, since to challenge Newton's law of gravity was unthinkable. Aepinus soon saw, however, that his work and Newton's were not in conflict after all. His force manifested itself only when the bodies concerned contained more or less than their natural quantities of electric or magnetic fluid. In the case of interactions between uncharged or unmagnetized bodies, which were the only ones Newton had considered, it need never be taken into account.[11] Since, so far as Aepinus is concerned, neither the force of gravity nor his new repelling force is innate to matter, but both are caused, in ways as yet unknown, by extrinsic material agencies, there is no contradiction in asserting that the two of them act simultaneously: "It is no contradiction to say that the same body at the same time and at the same moment is enticed by two extrinsic and opposed forces."

Having established the nature of the forces acting in the electrical and magnetic realms, in the remainder of his treatise Aepinus is able to analyze a wide range of phenomena in terms of them. Sometimes this leads to surprising results, as, for instance, when he considers the simple case of the action of a charged body on a nearby uncharged one. Traditionally, the attraction and subsequent repulsion that occur in this case had been the best known of all electrical phenomena. Yet now (§ 33), when Aepinus adds up the forces that are acting, including

---

[11] Whittaker incorrectly states in *History of the Theories of Aether and Electricity*, p. 49, that Aepinus suggested that gravity might be a residual force arising from a slight lack of equality between the force due to the ordinary matter and the force due to the fluid in neutral bodies. Aepinus made no such suggestion; the earliest mention of it that I have found is in Thomas Young's *Course of Lectures on Natural Philosophy and the Mechanical Arts,* London, 1807, I, 660.

his new force of repulsion between the particles of ordinary matter, he concludes that in these circumstances there ought to be no interaction whatever between the two bodies! Far from being dismayed by such a turn of events, however, Aepinus actually exults in it. "Although these assertions of mine can seem to be discordant with and opposed to everyday experience," he says, "they must be considered completely true. For not only do they flow from the fundamental principles of Franklin's theory so necessarily that if they are not true the whole of Franklin's theory must collapse, but also they agree so well with experience that physicists should have arrived at these laws through this alone if they had attended sufficiently to all the circumstances."

The latter part of Aepinus's comment reveals the source of his satisfaction on this occasion, and the episode well illustrates the power of his method more generally. Both here and in many other instances throughout his book, we find a remarkable combination of rigorous arguments leading to precisely articulated conclusions and skillful experiments that confirm them. In the present case, although Aepinus's analysis yields the unexpected result that any completely uncharged body should be neither attracted nor repelled by a nearby charged one, it also indicates that for all but the most perfect of insulating materials, in the circumstances of the experiment at least one pair of equal and opposite charges ought always to be induced on the body in question. That is, no ordinary body brought near a charged one ought to remain uncharged. The attraction that had always been observed in such cases becomes, for Aepinus, the algebraic sum of the forces acting on the set of induced charges. Moreover—and this is the kind of thing that makes his work so impressive—by a series of ingenious experiments Aepinus manages to confirm the production of these induced charges in practice (§ 124). It is, of course, their production that he is referring to when he implies that his predecessors had not "attended sufficiently to all the circumstances" of the situation. And yet it was surely no accident that it was he rather than anybody else who actually sorted things out and proved that induction played an important role in even this simplest of all electrical phenomena. The design of the requisite experiments is always theory-laden, particularly in cases like this, and at the time, Aepinus was the only one who had the necessary degree of theoretical insight. In other words, only he knew what to look for.

Much of Aepinus's analysis of the various phenomena of electricity follows this pattern. Relying as he does on forces acting at a distance, and being thus unencumbered by the complexities and confusions inherent in the theory of electrical atmospheres, he achieves a much clearer view than earlier workers such as Franklin and Canton of the nature of electrostatic induction in general. This in turn enables him to see that induction, far from being confined to the particular classes of phenomena investigated by these men, necessarily plays a role in every electrical phenomenon. He realizes, in other words, that the mere presence of a charged body will always lead to a rearrangement of the electric fluid in all other bodies in its vicinity. What is more, his analytical methods enable him to determine, albeit only in a semi-quantitative way, the effects of such rearrangements in a wide variety of cases, so that he is able to carry his study far beyond the point reached by the earlier Franklinists. He is, of course, able to apply the whole of his analysis to the magnetic case equally well.

To illustrate the power of Aepinus's approach, let us look at his discussion of that most famous of all electrical devices, the Leyden jar. Here, as so often in his work, Aepinus assumes as his starting point the basic correctness of Franklin's analysis—so much so, indeed, that to an unperceiving reader it may well have seemed that he was adding very little to what the master had said. In reality, this is far from being the case. Though the general structure of the explanation remains as it was with Franklin, in the degree of exactness that his treatment of detail achieves, Aepinus improves considerably on the work of the American, and casts much light on the role of induction more generally in electrical phenomena.

For the purposes of his analysis, Aepinus considers not the traditional glass bottle but, what was known to deliver the shock equally well, a plate of glass coated on both sides with sheets of metal (§ 45). He assumes that one of these metal sheets is connected by means of conducting wires to the earth, and the other to an electrical generating machine. If the latter receives a positive charge from the machine, it follows immediately from Aepinus's principles that electric fluid will be driven out of the opposite sheet and away to ground. On the other hand, if the machine delivers a negative charge, the flow will be in the other direction.

Thus far, Aepinus agrees entirely with Franklin. But now he argues that if a given charge is transferred to the first metal sheet, fluid will flow out from or into the other until this has acquired a charge opposite in kind to the first and of such a magnitude that the net force on a particle of fluid in the conducting wire connecting to ground is reduced to zero. This enables him to compute in his usual semi-quantitative way the magnitude of the induced charge, and at once he is led to conclude that, contrary to Franklin's assertion, this is not quite equal to the inducing charge on the other sheet. Rather, because of the finite thickness of the glass, it ought to be slightly less.

Aepinus is, of course, well aware of Franklin's views on this matter, and a little later in his book (§ 78) he discusses one of the experiments upon which Franklin based his conclusion.[12] In this experiment, a man standing upon an insulating support holds an electrified Leyden jar in one hand and discharges it through his body by touching the hook with his other hand. He experiences the shock as usual, yet afterward neither he nor the jar shows any sign of being electrified either "plus" or "minus." Aepinus argues that this experiment is not as conclusive as Franklin took it to be. He shows that so long as the difference between the charges on the two opposite coatings of the jar is small in comparison with the total quantity of fluid natural to the two coatings and the connecting link between them (in this case the man on the insulating stand), any residual charge will be unobservable. As always, he offers some interesting experimental support for his position. He points out that, in his "air condenser," the metal coatings are much further apart than they are when glass is the dielectric. Hence, according to his principles, the difference between the two charges should be much greater in this case, and perhaps even large enough to be observable. Sure enough, when he tries the experiment, his body acquires a charge of the appropriate sign, positive if the leading plate is electrified positively, and negative if this is electrified negatively.

Rarely in his discussion of the Leyden experiment does Aepinus oppose Franklin's opinion in the way he has done here. For the most part, his aim is to sharpen and extend the Philadelphian's account rather than to challenge it. There is, however, one other point on which he actively disagrees with the American, and this concerns Franklin's explanation of the well-known fact that there is a limit to the amount of charge that can be stored in any given bottle. According to Franklin, this limit is reached when enough fluid is added to one side of the glass to drive out the last of the fluid from the opposite side: "no more electrical fire can be thrown into the top of the bottle," Franklin proclaimed, "when all is driven out of the bottom."[13] Aepinus, by contrast, concludes that the limiting factor is simply the insulating power of the surrounding air. This must, he says, eventually be exceeded if the charge on the jar is allowed to build up indefinitely.

During the course of his discussion of this point, Aepinus obtains a mathematical expression for the magnitude of the force tending to

[12] Cf. *Benjamin Franklin's Experiments,* p. 185.
[13] Ibid., p. 181.

dissipate the charge on the plate. Now, armed with this, he is able to resolve many of the other puzzling features associated with the phenomenon. His calculations make it plain, for example, that as a result of the proximity of the induced charge of opposite sign on the further side of the dielectric, the force tending to dissipate the initial charge is very much less than it would be for an unaccompanied charge of the same magnitude. Hence it at once becomes clear why much greater charges can be stored in Leyden bottles than on single conductors; a much greater charge can be imposed before the dissipative force becomes sufficiently great to overcome the insulating power of the surrounding air. It also becomes clear why the hook of a charged Leyden jar gives only weak signs of electricity, even when it carries a very large charge: once again, the presence of the neighboring induced charge greatly reduces the dissipative force on which these effects depend.

An essential feature of Franklin's explanation of the Leyden phenomenon was that, during the charging process, the outer coating of a bottle necessarily acquires a charge opposite in sign to that acquired by the hook. Yet after the bottle is charged, its outer coating gives none of the usual signs of being electrified; it neither draws a spark from an unelectrified object brought near it nor attracts nearby light objects to it. Aepinus sees this as potentially a serious embarrassment for Franklin's theory.[14] However, he immediately resolves the difficulty by pointing out that, according to his analysis, the coating ought always to be in a state where it neither attracts nor repels nearby particles of fluid. In other words, from the point of view of an external object, the effect of the induced charge ought always to be exactly counterbalanced by that of the inducing charge; hence it is that the coating behaves toward such objects just like an uncharged body.

If one coating of a charged Leyden jar gives no signs of being electrified, and the other gives only very weak signs, how is it that, when the coatings are connected, the jar nevertheless delivers such a powerful shock? Aepinus recognizes the validity of this question, but, given the inductive connection between the two charges involved, he has no difficulty in answering it in terms of his general principles. In doing so, he is even led to the remarkable conclusion (§§ 66–67), not

[14] Aepinus here assumes that no contact is made with the hook during these experiments. In the experiments described by Franklin, on the other hand, this condition was always violated, and electrical effects were observed as usual. However, Franklin's remark that the coating will not "receive in, unless the [hook] can at the same instant give out" (*Benjamin Franklin's Experiments,* p. 190) suggests that he was aware of what happened, or failed to happen, at other times.

confirmed experimentally for another hundred years, that the discharge ought to be oscillatory in nature!

On several occasions in his discussion of the Leyden jar, Aepinus comes very close to the modern notion of capacitance, and he fully understands the dependence of this on both the effective area of the conducting sheets and the distance between them. This is particularly evident in his account of the experiments leading up to the invention of the air condenser (§§ 72–75). Air is not as good an insulator as glass; hence in order to prevent a spark passing between the metal sheets, these had to be held further apart in Aepinus's experiments than was normally the case when glass was used. Aepinus's growing recognition of the need to compensate for this by greatly increasing the surface areas of his conducting sheets is clearly displayed here. Nevertheless, it would be wrong to credit Aepinus with actually arriving at the idea of capacitance in the modern sense, for essential to this is the notion of electrostatic potential, and, as Maxwell noted long ago,[15] this is lacking in Aepinus's account.

How close Aepinus came to the ideas of potential and capacitance is, however, a matter of contention. In his discussion of the air condenser, Aepinus states that it is the "violence of the commotion" that depends on the area of the conducting sheets and the distance between them. Elsewhere, he says this depends particularly on "the degree of condensation of the electric fluid" (§ 50), and Dorfman, for one, has seen the latter notion as closely related to the idea of potential.[16] The way in which Aepinus uses the concept suggests, however, that Dorfman has somewhat overstated the case. Aepinus's further discussion of the operation of the Leyden jar indicates that the only reason an increased condensation of fluid increases the violence of the commotion is that it leads to a greater quantity of fluid flowing across in the brief period of discharge; that is, "degree of condensation" seems here to imply "charge" rather than "potential."

Later, in discussing the sharing of charge between two conductors, Aepinus assumes the final equilibrium condition to be one in which the fluid has been reduced to a uniform density throughout (§§ 195, 197). We alluded earlier to this as the only occasion in the book where he assumes a uniform distribution of fluid as a physical hypothesis and not merely a mathematical device. Aepinus's rule is incorrect, and indeed incompatible with an inverse-square law of force. Much more satisfactory is the rule stated by Henry Cavendish twelve years after

[15] *The Electrical Researches of the Honourable Henry Cavendish, F.R.S.*, ed. J. Clerk Maxwell, Cambridge, 1879, p. 382 (editorial notes).
[16] *Teoriya,* pp. 505, 544.

Aepinus's book was published, namely that in equilibrium a uniform *pressure* is established throughout the fluid in a connected system.[17] Cavendish thereby effectively arrived at the notion of potential, as Aepinus did not, and his use throughout his analysis of the mathematical construct of slender canals of fluid linking conductors at a given "pressure" confirms his complete mastery of the idea.

Aepinus's discussion of the Leyden jar provides one example of the power of his new approach to the theory of electricity. As stated earlier, the discussion makes it apparent that the key to his success was his clear understanding of the importance of induction, or "communication" as he called it, and his ability to handle this in his computations.

Aepinus's detailed analysis of what happens when two charged bodies are brought near each other further illustrates the same point. Establishing the general rules that apply in this situation was perhaps the greatest of Du Fay's achievements in his classic series of researches during the 1730s. When the bodies concerned carry the same kind of electricity, Du Fay reported, they repel each other; but when one is electrified vitreously and the other resinously (or one positively and the other negatively, to use the terminology later developed by Franklin), they attract. Paradoxically, Aepinus's calculations lead him to challenge the validity of Du Fay's rules, or at least the first of them. He is led to do so by his recognition that the two bodies involved must in fact always exert mutual inductive effects on each other; that is, the distribution of electric fluid within each of the bodies will be altered by the presence of the other body nearby.

Applying his usual methods of computation, Aepinus is able to analyze, in qualitative terms at least, the effect of this redistribution of fluid (§§ 128ff.). He concludes that while Du Fay's rule concerning the force acting between two oppositely charged bodies remains valid, the rule dealing with the force between bodies carrying charges of like sign requires modification. Specifically, he shows that if the two bodies are brought sufficiently close together, or if one of the two charges involved is very much weaker than the other, the inductive effects may be sufficient to change the normal repulsive force into an attraction. Hence Aepinus is led to reformulate Du Fay's rule in this case. If two charged bodies repel each other, they must, he agrees, be carrying charges of like sign. But he insists that the converse of this does not always hold; bodies carrying charges of like sign do not always repel each other. And, as we have by now come to expect, he then verifies this conclusion by an elegant series of experiments.

[17] Cavendish, *Phil. Trans.*, LXI, 1771, 594; reprinted in his *Electrical Researches,* p. 9.

Aepinus's mastery of the basic principle of electrostatic induction is evident throughout his *Essay*. The examples of his approach that have been discussed so far reveal the way his understanding of the principle enabled him to establish electrical theory on a much more coherent and consistent footing than had ever been done before. To appreciate fully the magnitude of Aepinus's achievement, however, it is necessary to look not at the *Essay* itself, but at the first of the two appendices that he added to his book at the last moment, in September 1759, some time after the remainder of the work had gone to press. Here, Aepinus successfully analyzes in terms of his general principles an interesting experiment that had been devised by his predecessor as professor of physics in the St. Petersburg Academy of Sciences, G. W. Richmann, and that had been seen by Richmann as a refutation of the Franklinian explanation of the Leyden phenomenon in terms of the impermeability of glass to the electric fluid.[18] Not only is Aepinus's discussion once again admirably precise and consistent in its details, on this occasion it also relies on a quite demanding and much more sophisticated piece of mathematical analysis than is used elsewhere in the book. The effect of this is to highlight even further the contrast between Aepinus's work and the wholly non-mathematical approach of his predecessors, and, as a consequence, the extent to which his *Essay* marks the beginning of an entirely new and mathematical style of electrical theorizing that in fact did not come fully into its own for another half-century or more.

As described by Aepinus, Richmann's apparatus consists of a pane of glass fitted with metal coatings on its opposite faces, just as in one common version of the Leyden experiment. Now, however, each metal coating has a thread attached to it, which serves as an electroscope. As usual, to charge the device, one coating is connected to an electrifying machine and the other to ground. Once this has been done, the thread attached to the electrified coating stands out at an angle to the coating, while the other hangs limply. If the connection between the second coating and ground is then broken, the first thread sinks until it stands out only about half as far as before, and simultaneously the other rises until it stands out at about the same angle as the first. Thereafter, both threads will very slowly sink until eventually they both hang vertically once more. However, if, before this happens, either coating is earthed, its thread will immediately collapse, while the other springs out to about twice the distance it stood at previously.

---

[18] Cf. Richmann, "De indice electricitatis et ejus usu in definiendis artificialis et naturalis electricitatis phaenomenis, dissertatio," *Novi commentarii Academiae Scientiarum Imperialis Petropolitanae*, IV, 1752–53, 324. The experiment Aepinus discusses is a slightly simplified version of the one actually described by Richmann.

And this can be repeated many times before the electricity is exhausted.

Before turning to Aepinus's explanation of this sequence of events, we need to be clear about what it is that a deflection of a thread measures in such an experiment. Aepinus nowhere explicitly discusses the question, but his opinion is nevertheless quite plain: the deflection of a thread measures "the force of the plate" to which it is connected, that is, it measures the repelling force that the plate exerts on an adjacent particle of electric fluid. Since it is precisely upon his calculations of such forces that most of Aepinus's analyses throughout his book depend, he can now simply transfer many of his earlier results to the present case.

Aepinus's explanation of Richmann's experiment hinges on the fact that the air surrounding the apparatus is a less than perfect insulator. This means, of course, that there must be a gradual leakage of charge from the electrified coating. Aepinus recognizes that this will of necessity have important consequences for any induced charges on the second coating. He develops and solves a differential equation linking the rate of change of the induced charge with that of the charge inducing it, and this enables him to obtain expressions for the forces of the two coatings at each stage of the experiment, and to show that these are in perfect agreement with the experimental results obtained by Richmann. Because the first coating is steadily losing charge, its force will, of course, constantly decrease throughout the experiment. So long as the second coating is grounded, its force necessarily remains zero. Once this coating is disconnected from earth, however, its force will, according to Aepinus's analysis, rise to a maximum and then decrease again. Aepinus shows that only a very small decrease in the charge on the first coating is required to bring the second to its maximum force, so that this point ought to be reached very quickly. He also shows that this maximum is nearly but not quite half the initial force of the first coating. What is more, he finds that, at the same time as the force of the second coating rises to this value, that of the first decreases to precisely half its initial value; but he also shows that once the forces of the two coatings thereby become approximately equal, the effects produced by further leakages of charge are much less dramatic, and indeed amount to no more than both forces decreasing very slowly until eventually they become imperceptible.

Toward the end of Aepinus's discussion, he points out that his equations suggest that quite different results ought to be obtained if a thicker insulating layer is used than the glass employed by Richmann. Typically, having first illustrated the point by means of a worked numerical example, he then verifies it experimentally. Finally, he considers what ought to happen if one or the other coating is earthed following the removal of the original ground connection. Once again, he provides a worked numerical example, and once again he shows that Richmann's observations follow directly from his equations. In all, Aepinus's analysis is extraordinarily successful and provides a fitting climax to his discussion of the theory of induction as set out in the earlier sections of his book.

In his discussion of the Leyden jar, Aepinus considered the question of how a charged jar could be as highly electrified as it obviously was and yet attract and repel nearby objects so feebly. His answer, it will be recalled, was that the effects of the two nearly equal charges of opposite sign close together in the jar very nearly canceled each other out. In discussing this point, Aepinus described a number of other situations in which two opposite charges were for a time in close proximity to each other and so initially scarcely betrayed their presence. One of the experiments he described (§ 59) has since attracted the attention of historians because of its close connections with the invention a few years later by Alessandro Volta of that remarkable device, the electrophorus.

The experiment seems to have been one on which Aepinus and his friend Wilcke had collaborated in Berlin, and Wilcke described the apparatus, though not the particular use to which Aepinus put it, in his thesis *De electricitatibus contrariis*,[19] well before Aepinus's *Essay* appeared in print. The apparatus consists of a hemispherical metal dish on an insulating stand. This is filled with molten sulphur, which is allowed to cool. Once the sulphur has solidified, an insulating handle is attached to it so that it can be removed from the dish or replaced at will. So long as the sulphur segment remains in the dish, the apparatus gives no signs of being electrified, but if the sulphur is withdrawn, both it and the dish are found to be strongly electrified, one positively and the other negatively. If the sulphur is returned to the dish, all signs of electricity disappear, but whenever the two are separated they reappear once more. If the dish is touched while the two are separated, it loses its charge, and if the sulphur is then returned to its place, the combination exhibits a charge of the same kind as the sulphur previously showed when separated.

This apparatus is similar to that subsequently described by Volta,[20] for there, too, the essential components were a cake of non-conducting material (resin in Volta's experiments) and an insulated sheet of metal,

[19] Pages 44–45.
[20] Volta to Joseph Priestley, 10 June 1775, *Le opere di Alessandro Volta*, Milan, 1918, III, 95–108.

and these were alternately brought into contact and separated. Yet Volta was thoroughly justified in claiming that his invention was a new one, and in persisting with this claim even after he became aware of what Aepinus and Wilcke had done.[21] In Aepinus's discussion of his experiments, he shows how his results, and especially the complete absence of electrical effects whenever the two electrified surfaces are in contact, can be accounted for by the theory he has already enunciated. Elsewhere, in his *Recueil de différents mémoires sur la tourmaline,* he merely used his observations to support his conclusion that two opposite charges were always generated in any electrificatory process. Volta, however, by adopting an experimental procedure that differed in one crucial respect from Aepinus's, could legitimately make much more dramatic claims for the apparatus. Volta made a point of briefly earthing his metal plate while it was in contact with the sheet of electrified resin. It thereby became electrified by induction, and did so without causing any reduction in the charge on the resin; and this could be repeated as often as one wished. Hence Volta did indeed have at his disposal, as he claimed, a virtually perpetual source of electricity; and it was on this basis, and not as Aepinus had left it, that the apparatus found its niche in history.

An experiment closely related to these was reported to the St. Petersburg Academy of Sciences as early as 1755 by the Jesuit missionaries who were then working in Peking. A thin glass plate was electrified by friction and placed on the glass lid of a magnetic compass. The needle of the compass then immediately rose from its pivot and adhered to the underside of the glass for some hours before returning suddenly to its normal position. If the upper glass plate was then removed, the needle would again rise and remain stuck to the lower plate for a considerable time, but as soon as the other plate was brought back, it would fall away again. This could be repeated over and over again.

Aepinus described this, along with a number of other curious experiments on electrical induction, in a paper he read to the St. Petersburg Academy in 1758,[22] but, strangely, he made no mention of it in

---

[21] Volta to Joseph Klinkosch, – May 1776; ibid., 135–45. Volta appears not to have known any of Aepinus's work at first hand at the time he made his initial discovery of the electrophorus. Soon afterward, however, he read and was impressed by Aepinus's *Recueil de différents mémoires sur la tourmaline,* and was inspired thereby to look for a copy of his *Essay* as well (Volta to Giovanni Francesco Fromond, 3 August 1775, ibid., pp. 112–13). A reading of the *Recueil* would have been sufficient, however, to bring the sulphur-and-dish experiment to Volta's attention, since an account of this was included there (p. 69).

[22] Aepinus, "Descriptio ac explicatio novorum quorundam experimentorum electricorum," *Novi commentarii,* VII, 1759, 277–302.

his *Essay*. In the paper he read to the Academy, however, he also gave a complete explanation of the phenomenon in terms of the charge induced on the glass lid of the compass, its necessarily slow transfer under the circumstances of the experiment into the compass needle in contact with it, and the equally slow return of the charge to the glass once the inducing charge was removed. Yet his explanation did not satisfy everyone. Beccaria, in particular, attempted an alternative account in terms of a peculiar notion that he called "avenging electricity" (*elettricità vindice*).[23] Ironically, it was Volta's dissatisfaction with Beccaria's notion that in turn led him to his invention of the electrophorus a few years later.[24]

Aepinus's publication of the *Essay* in 1759 brought about a dramatic advance in man's understanding of electricity. He provided both a sophisticated, thorough theoretical analysis and an impressive array of experimental support for his position. Many of his experiments were entirely new, others were old experiments newly rendered precise and definitive. Yet the book makes no effort to provide an exhaustive treatment of the subject; rather, Aepinus restricts his attention to those phenomena for which he can see an analogue in the magnetic realm. This leads him to consider at length the forces electrified bodies exert on other bodies and on each other in a variety of situations. In particular, as we have seen, he concerns himself with the phenomena of electrostatic induction. But he completely ignores several whole classes of electrical phenomena, including, for example, those concerned with the production of light and sound in electrical discharges. Yet it is clear that he does so by choice and not out of any inability to handle the subject. When an occasion later presented itself, he was able to place before the St. Petersburg Academy a masterly analysis of one of the trickiest of all electrical discharge phenomena, the so-called "mercurial phosphorus" or "lucent barometer."[25]

Further evidence of Aepinus's interest in the full spectrum of electrical phenomena is found in his notebook that has survived from the late 1750s, the period of his greatest scientific activity.[26] Here, a number of jottings make plain Aepinus's concern to relate the phenomena of electricity to wider considerations about the structure of matter in general, and to connect the electrical forces he was investigating to the

---

[23] Beccaria, *Experimenta atque observationes quibus electricitas vindex late constituitur, atque explicatur*, Turin, 1769, pp. 44ff.

[24] Volta, *Opere*, III, 138.

[25] Aepinus, "De electricitate barometrorum disquisitio," *Novi commentarii*, XII, 1766–67, 303–24.

[26] AAN, raz. v, op. э–7, no. 1.

various other kinds of forces that seemed to act in the world. Unfortunately, his comments amount at best to disconnected sentences, and often to rather less than this. Hence they are only scattered clues to his thinking; it is impossible to deduce a coherent pattern from them.

Some of Aepinus's notes are extensions of the work of others. A query as to whether sulphur whose surface has been made rough by grinding will charge positively, for example, is clearly related to Canton's discovery that glass so treated acquires a negative charge when rubbed.[27] So too are questions about how roughly ground glass will behave as a "shock-glass" (*ErschütterungsGlaß*) in the Leyden experiment, and the effect of similar treatment on the electrical properties of the tourmaline.[28] Most of his speculations seem, however, to be entirely original.

At one point, for example, Aepinus asks himself whether salt crystals or chemical growths such as the *arbor dianae* take up different shapes if electricity is applied during the formative process.[29] Likewise, he wonders what happens when electrified water is frozen, and whether the "tears" formed when molten glass is dripped into cold water differ in any way if either the glass or the water is electrified.[30] He queries whether an electrified window-pane can become iced over,[31] and how dew forms on bodies that are electrified.[32] He ponders the effect of electrification on the rate at which water evaporates,[33] and conversely, he wonders whether boiling or freezing reduces the electricity of the water.[34] He asks whether, when metal is dissolved in electrified acid, this affects the electricity of the acid in any way,[35] and also whether glass glowing red hot can be charged at all.[36] In by far the longest connected passage dealing with electricity in the entire notebook, he considers the origin of the electricity that is generated in the tourmaline and in other cases: "The tourmaline has the electric matter in itself, as is evident from the experiment where it becomes

[27] Ibid., l. 18. Cf. Canton, *Phil. Trans.*, LXVIII, 1753–54, 780–85.
[28] AAN, raz. v, op. э–7, no. 1, l. 18; ibid., l. 28^{ob}.
[29] Ibid., l. 17^{ob}.
[30] Ibid.
[31] Ibid.
[32] Ibid., l. 33^{ob}. Here, Aepinus refers to Du Fay, "Mémoire sur la rosée," *HAS*, 1736, Mém. pp. 352–74.
[33] AAN, raz. v, op. э–7, no. 1, l. 17^{ob}. Aepinus does not mention Nollet's earlier work on this subject (*HAS*, 1747, Mém. pp. 207–42).
[34] AAN, raz. v, op. э–7, no. 1, l. 19^{ob}.
[35] Ibid.
[36] Ibid., l. 17^{ob}.

warm from the solar rays. Other bodies only take from or give the electric matter into another. Fused-sulphur bodies do not belong to the class of the tourmaline. Fused-sulphur bodies, if they touch air alone do not become electric, hence the contraction of these bodies accomplishes nothing here, but they do give [electric] matter into other bodies. Glass and metals become positively electric from fused-sulphur bodies applied. Very close application is needed so that they cohere." [37]

Some cryptic remarks in which Aepinus speculates on the relationship between electricity and light are also worth mentioning. Here, he asks himself "whether electric[ity] is propagated by means of rays of light reflected from a metal mirror" and then "whether electr[icity] has an influence on the refract[ion], reflect[ion] and inflexion of rays." [38]

A number of Aepinus's jottings, such as the pregnant remark that "electric attraction and repulsion must be examined more closely," [39] deal with the traditional concerns of electrical investigators. Of special interest is evidence that Aepinus carried out other research on attraction and repulsion in addition to that he eventually reported in his *Essay*. In particular, he states that two similarly electrified bodies repel each other even in a vacuum, though he also warns that the experiment is very difficult on account of the rapid loss of charge in these circumstances.[40] Also, at one point he comments on possible forms which the law of force might take: "No law of attraction in nature," he says, "can have more than two real, equal roots, one positive and the other negative. Hence I doubt whether the ratio $1/x^3$ occurs." [41] Elsewhere, he comments favorably on the electrometer sub-

---

[37] Ibid., l. 28^{ob}: "Der Tourmalin hat die electrische Materie in sich selbst, uti patet ex experimento, ubi radiis solaribus calefit. Andere Körper nehmen oder geben erst, die electrische Materie von oder in einen andern. Die corpora sulphurea fusa, gehören nicht zu des tourmalins Claße. Corpora sulphurea fusa si non contingant nisi aërem non fiunt electrica. hinc contractio horum corporum nihil hic efficit. sed dant materiam in alia corpora. Vitrum et metalla fiunt ab adplicatis corp. sulphureis fusis positive electrica.

"Necessaria est intima adplicatio ut cohaereant."

[38] Ibid., l. 17^{ob}: "an per radios lucis a speculo metall. reflexas, propagetur electric.? an electr. habeat influxum in refract. reflex. et inflexionem radiorum." Cf. also ibid., l. 35^{ob}: "an in diffractione luminis, aliquid valeat electricitas."

[39] Ibid., l. 17^{ob}: "Attractio et repulsio electrica accuratius examinanda."

[40] Ibid., l. 33^{ob}: "duo corpora aut posit. aut negat. electrica, etiam in vacuo se repellunt, sed exper. difficillimum videtur, quia electrica materia ex his corporibus statim emanabit." Whether electrical forces persisted in vacuo had long been a matter of great contention among electricians.

[41] Ibid., l. 13^{ob}: "Lex attract. nulla in natura potê [i.e. potest] habere plures

mitted to the Paris Academy of Sciences a few years earlier by the refugee Irishman and recently appointed member of the Academy, Patrick D'Arcy.[42]

Some of Aepinus's notes reappear virtually unchanged in the *Essay*, such as the blunt assertion that a succession of charges of opposite sign—so-called "consequent points"—can be induced on a glass rod.[43] Others we recognize immediately as important steps toward the mature positions of the *Essay*. In this category, for example, is the recognition of the important point that when wax is electrified by being poured while molten into a metal dish and allowed to cool, the dish, if it is suitably insulated, also acquires a charge, but of opposite sign.[44]

Much more tantalizing are some remarks that are clearly related to some of the leading novelties of the *Essay*, but that are insufficient to reveal the actual evolution of Aepinus's ideas. The following questions, which appear together in the notes, obviously reflect a stage through which Aepinus passed on the way to rejecting altogether the doctrine of electrical atmospheres, yet without a context in which to set them, they remain unilluminating (indeed, even the meaning of one of them remains uncertain):

> In a vacuum is there also a wind going out from an electrized point?
>
> [By means of such?] methods can there be found a method of investigating whether air extends to bodies which are positively or negatively electric?
>
> Can there be a negatively electric body endowed with a positive atmosphere? Investigate
> 1) which sort of electricity such a body communicates
> 2) which sort of electricity is acquired by a body immersed in this atmosphere and touched by a non-electric body. [45]

---

quam 2 radices reales, =les unam posit. alteram negat. hinc dubito an ratio $1/x^3$ locum habeat."

[42] Ibid., l. 18. Cf. D'Arcy, "Mémoire sur l'électricité, contenant la description d'un électromètre . . . ," *HAS*, 1749, Mém. pp. 63–74.

[43] Ibid., l. 28^{ob}: "An einer cylindrischen Glaß Stange, die ex una parte electrisirt wird, können puncta consequentia entstehen." Cf. below, § 202.

[44] Ibid., l. 21^{ob}. Cf. below, § 59.

[45] Ibid., l. 21:

"Ob in vacuo aus einer electrisirten Spitze auch ein Wind gehet?

"ob ŏt [?] Methoden gefunden werden kann methodus, investigandi an aër *p*tineat [i.e. pertineat] ad corpora positive, an negative electrica.

"Dari potê [i.e. potest] corpus negat. electricum, athmosph. positiva præditum. Qv.

Figure 2. Aepinus's notebook diagram illustrating the continuity of the force law.

Likewise, the assertion that "two bodies electrified with the same kind attract each other, in conformity with the law of continuity," accompanied by the diagram in Figure 2,[46] reveals a new factor that Aepinus may have had in mind when he denied Du Fay's "law" that similarly electrified bodies repelled each other. Once again, however, that is all it does: the lack of context again makes it impossible to take the matter any further.

In all, therefore, Aepinus's notebook tell us disappointingly little about his evolution as an electrical investigator, and the fact that he is extremely sparing of autobiographical and other historical remarks in his more formal writings makes the disappointment even more acute. Unless new evidence comes to light—and the possibility of this happening seems very slight—we must reconcile ourselves to remaining largely ignorant of the chains of thought that led to the doctrines set out in his book.

Yet the importance of those doctrines remains undiminished. The ideas about electricity that Aepinus inherited from Franklin, while vastly more successful than rival systems in accounting for such spectacular new phenomena as the Leyden experiment, were nevertheless incompletely worked out, and in particular, their relevance to the electrician's traditional problem of explaining the electrical attractions and repulsions remained unclear. Franklin's doctrine of electrical atmospheres proved of little help in this regard, and the variants developed by followers such as Canton and Beccaria were no more satisfactory. Also, Franklin's discussion was entirely qualitative, and, as Aepinus showed, its premises, when taken literally and in conjunction with each other, yielded conclusions that were in direct conflict with the most elementary and best known of all electrical phenomena.

Aepinus transformed Franklin's ideas into a rigorous, precisely articulated and internally consistent account that was in excellent detailed agreement with experiment over a broad range of electrical

---

1) qualem tale corpus communicet electricitatem.

2) qua[le]m acquirat electric. corpus huic athmosph. immersum, et tactum a corpore non electrico."

[46] Ibid., l. 13^{ob}: "duo corpora electr. ejusd. gen se attrahunt, ex lege continuitatis."

phenomena. By abandoning altogether the notion that a charged body was surrounded by an atmosphere of electric matter, and by relying on forces acting at a distance, he was able to develop a powerful new theory of electrostatic induction that not only accounted satisfactorily for a number of previously baffling observations but also eventually opened the way to a number of important new discoveries. Perhaps most important of all, he transformed what had previously been a purely experimental branch of physical inquiry into an at least partly mathematized discipline. In Aepinus's hands, the science of electricity acquired an entirely new dress, one that was not to be fully exploited for another generation and more. But in the meantime, his work was sufficient clearly to point the way for those equipped adequately to pursue such investigations.

CHAPTER FOUR
Magnetism

Eighteenth-century physics faced no more baffling task than to explain satisfactorily the mysterious powers associated with the magnet, or lodestone. Almost to a man, those writing on the subject began by bemoaning the complexity and seeming capriciousness of the phenomena at hand. When it came to accounting for these, they found it exceedingly difficult to agree among themselves on more than a few fundamentals. Indeed, so far as the theory of magnetism was concerned, the situation grew worse, if anything, as the century progressed. The general agreement on fundamentals that had existed at the outset gradually eroded as the years went by, and it was only toward the end of the period that a new consensus was reached, one which took Aepinus's work as its starting point and inspiration. We shall see that Aepinus's work constituted a veritable epoch in the history of magnetism, and was held to do so by those who came after him.

The eighteenth century inherited from the past much more information about magnetism than it did about electricity. From the theoretical point of view, the magnet presented a fascinating challenge to anyone who sought, according to the dictates of the new science, to reduce all natural phenomena to mechanical interactions among particles of matter. In addition, the subject was of great practical interest on account of its relevance to the then pressing problem of long-distance navigation at sea. For all these reasons, magnetism regularly attracted attention in scientific circles, notwithstanding the frustrations to which it almost always gave rise.

The starting point for all subsequent writers on magnetism was William Gilbert's famous treatise *De magnete,* published in 1600.[1] Gilbert, having systematically tested and rejected all manner of fantastic legends that had grown up over the centuries concerning the powers of the lodestone, established its basic properties on a secure experimental footing. These included the specificity of magnetism

[1] William Gilbert, *De magnete, magneticisque corporibus, et de magno magnete tellure; physiologia nova, plurimis et argumentis et experimentis demonstrata,* London, 1600; English translation by Silvanus P. Thompson, London, 1900, reprinted New York, 1958.

to lodestone and to iron and certain of its ores, and the distinct difference between magnetism and the attractive virtue acquired by amber and other "electricks" when they were rubbed. Gilbert found, too, that every lodestone had two distinct "poles" at opposite points of its surface, and that, whenever it was free to do so, it turned until the axis joining these points lay approximately north-south. Hence he distinguished the two poles as "boreal" and "austral." He found that, while a lodestone would attract iron with any part of its surface, it would attract another lodestone only if the two were aligned with unlike poles facing each other. If, on the contrary, two like poles faced each other, repulsion and not attraction would occur. In addition, these attractions or repulsions were not affected by the interposition of even quite thick pieces of other matter, provided it was not iron. The power of a lodestone could, Gilbert found, be considerably increased by "arming" it, that is, by fitting iron caps on its poles. Convinced that the Earth as a whole was a giant magnet, Gilbert realized that the north-seeking poles of his lodestones or magnetized needles (for iron, too, could acquire the magnetic virtue if "touched" by a lodestone) must be opposite in kind to the north or boreal pole of the Earth; hence he designated them austral poles, and similarly the south-seeking poles he called boreal. Gilbert was aware that in most parts of the world, the compass needle did not point due north and south, but deviated somewhat to east or west. He was not, however, aware that this declination (or "variation," as he called it) itself varied with time; indeed, he insisted that it did not. Finally, he knew that a magnetic needle that was free to move in a vertical rather than the usual horizontal plane would, if left to itself, take up a position somewhat inclined to the horizon, and would incline the more, the higher the terrestrial latitude in which it was located.

Gilbert, far from resting content with establishing the principal magnetic phenomena, had wished to found a whole new cosmology upon them. By the eighteenth century, such hopes had long perished; magnetism was now seen merely as one minor, though still very interesting, branch of physical inquiry. In large measure, the phenomena were still accepted to be as Gilbert had described them, though a closer acquaintance had considerably refined men's knowledge of most of them. The variability of the declination, for example, had become well known, to the point where Edmond Halley in 1692 could make a serious and very influential attempt to account for it systematically on a world-wide basis.[2] Also, the induction of magnetism in iron and

---

[2] Halley, *Phil. Trans.*, No. 195, Oct. 1692, pp. 563–78.

the general relationship of iron to the lodestone, though still not well understood, were at least better known than they had been. The design and construction of the mariner's compass had continued to improve, with a corresponding gain in its precision and reliability as a scientific instrument.

In explaining the properties of the magnet, Gilbert had relied entirely upon traditional notions such as the specific "form" of the lodestone and the "mutual love" for each other that led two stones to "take delight in their mutual proximity." However, it was precisely this kind of explanation that was rejected so vehemently by later seventeenth-century scientists. The revolution that occurred in science during this period brought with it the demand that all natural phenomena, magnetism included, be explained in purely mechanistic terms. So far as magnetism was concerned, the mechanism proposed by Descartes [3] set the pattern for most of those that followed. Indeed, Cartesian-style mechanisms predominated not merely for the remainder of the seventeenth but for much of the eighteenth century as well, until such time as Aepinus's entirely different approach won acceptance in its stead.

Descartes' theory assumed that streams of subtle aethereal matter passed along the axis of any magnet from one pole to the other along channels specially adapted to receiving them, and then returned again to the first pole through the encompassing medium to establish a continuous circulation. The special channels through which the subtle matter flowed were, Descartes said, peculiar to lodestone and iron. Accepting Gilbert's view that the Earth itself was a magnet, Descartes assumed that it, too, was surrounded by circulating subtle matter—a different circulation, however, from the vortex he had earlier invoked to account for gravity. The new circulation enabled him to explain in a very straightforward manner the directive property of lodestones and compass needles, and also their tendency to incline their axes toward the Earth. Both effects were due, he said, to a pressure exerted by the streaming magnetic matter that surrounded the Earth on the inner walls of the channels in any magnetic material it chanced to meet in its way. This forced the channels to orient themselves with the direction of flow. The declination he attributed to unequal deposits of lodestone and iron in different regions disturbing the symmetry of the circulation surrounding the Earth. Variations in the declination (which, unlike Gilbert, he accepted as a

[3] Descartes, *Principia philosophiae* (1644), Pt. IV, §§ 133–83; in his *Oeuvres*, ed. C. Adam and P. Tannery, Paris, 1897–1913, VIII, 275–311. I have also used the French edition (1647), published in *Oeuvres,* IX.

genuine phenomenon), he ascribed in part to man's working these deposits and hence changing the global distribution of iron. More important, he assumed that natural processes would continually corrupt existing ores and produce new accumulations elsewhere.

To explain the attraction one lodestone exerted on another was only a little more difficult. Descartes argued that, when two stones were brought together with unlike poles facing each other, the streaming matter leaving the pole of one would, on finding in its path pores in the other suitably disposed to receive it, continue through these rather than diverging to one side or the other. In so doing, it would drive out the air that was normally between the two stones. This, receding behind them, would drive them toward each other. If, on the other hand, the stones were brought together with like poles facing each other, they would separate because their streams of matter, on finding themselves confronted in this case by channels they were unable to enter, "must have some space between the two lodestones through which they can pass."

But how was it that the channels provided a ready passage for the subtle matter in one direction but not in the other? Descartes replied that they had, projecting into them, many little points, or filaments. These were supposed to act like the valves in the veins; they would lie flat and allow matter to pass over them in one direction, but would ruffle up and prevent matter from flowing the other way.

Ordinary unmagnetized iron, Descartes suggested, was furnished with channels just like those in a magnet, but now the filaments lay higgledy-piggledy in every direction. When a lodestone was brought near, the streams of subtle matter associated with the latter would enter the channels in the iron, and, by their momentum, force the filaments to arrange themselves according to the direction of flow. In other words, the iron became magnetized. This happened very quickly, Descartes said, after which the iron would be attracted as if it were an ordinary lodestone. Under normal circumstances, the induced magnetism would last only so long as the lodestone remained nearby, but if it remained there for a long time, the filaments thus lying constantly in the one direction would gradually lose their flexibility, and the magnetism that had been induced would become more lasting. If steel were used rather than iron, the induced magnetism would last still longer, because in steel the filaments are less flexible than they are in iron, and "cannot so readily be reversed."

Descartes' theory of magnetism was part of a vast explanatory scheme that sought to embrace all the phenomena of the natural world. In his mature physics, at the time when he propounded his

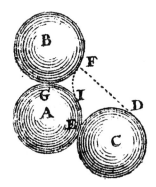

Figure 3. Descartes' diagram showing how grooved particles are produced.

theory of magnetism, Descartes distinguished, as he had not done earlier,[4] between two different kinds of subtle matter. These he called simply the "first element" and "second element." He supposed that most of the universe was filled with the second element, whose spherical particles acted as the transmitter of light and, swirling in vast cosmical vortices, as the cause of gravity. Magnetism, however, he attributed to the first element, the extremely mobile "cosmic dust" that, he supposed, filled the spaces between the spherical particles. More specifically, he called upon the largest particles of the first element, that is, those which, in cross-section, just filled the space left when three spheres came in contact (the space FGI in Descartes' diagram, Fig. 3). These particles were little fibers, triangular in cross-section, with concave sides. In passing rapidly, as was their wont, axially through the vortices of the second element, they became twisted, like little screws. In this form, Descartes referred to them as "grooved particles" (*particulae striatae; parties canelées*), and it was their threaded surface that he invoked to account for the various magnetic phenomena.

In lodestone and in iron, Descartes said, and in no other kinds of ordinary matter, there are channels whose inner surfaces are rifled in such a way that the little grooved particles find a ready passage through them. Elsewhere, the particles move with much more diffi-

[4] Rosaleen Love, "Revisions of Descartes's matter theory in *Le Monde*," *British Journal for the History of Science*, VIII, 1975, 127–37.

culty, since they have to work their way through pores that are much less well fitted to receive them. Some of the grooved particles will have been twisted one way, some the other, and in magnetic bodies there will be channels to fit each kind, some with right-hand and others with left-hand rifling. We have already mentioned the little hairlike points projecting into these channels, allowing the grooved particles to pass through the channels in one direction but not in the other. However, Descartes saw no reason to suppose that all the particles flow in the one direction around a magnet, so that, for example, they all enter at the north pole, or all leave from it. Rather, he supposed that there are some channels in which the particles pass from north to south, and others in which they pass from south to north. In other words, he maintained that there are two streams of grooved particles flowing in opposite directions through and around any magnetized body.

Descartes' theory of magnetism was enormously influential. Yet those who took his work as their starting point often did not follow him in every detail of his account. In particular, the complexities of Descartes' discussion of the grooved particles were frequently passed over in silence, if they were not rejected outright, and likewise many authors opted for one rather than two streams of subtle matter. The speculations of Christiaan Huygens concerning magnetism well illustrate these tendencies. In a paper read in two parts to the Paris Academy of Sciences on 25 May and 1 June 1680,[5] Huygens, like Descartes, ascribed the phenomena of magnetism to a circulation of subtle matter through any magnetized body along channels lined with hairlike projections, and back through the surrounding air. Huygens, however, explicitly rejected both the notion of grooved particles and suitably rifled channels to receive them, and the doctrine that there were two streams of subtle matter flowing in opposite directions around a magnet. He held that the mere fact that the channels lined with bristles provided an exceptionally free passage for the subtle matter in one particular direction would be sufficient to give rise to and maintain a distinctive magnetic circulation, and he dismissed the idea of two opposing streams of subtle matter as "contre toute apparence de raison." Additionally, Huygens attributed the magnetization of a piece of iron, not to an action on the bristles lining channels that were already formed, but to a re-arrangement of the particles of iron so as to form the channels themselves. At first, he also advocated a rather different explanation of magnetic attraction,

---

[5] Huygens, *Oeuvres complètes,* The Hague, 1888–1950, xix, 574–81.

attributing it not to the driving out of the intervening air by an expanded circulation, but to the tendency of this circulation to contract into as small a volume as possible, and hence to make itself as nearly spherical as possible. In his later writings, however, he reverted to something more like the Cartesian account, except that he supposed it to be the grosser aether rather than the intervening air that was displaced, and the resulting aether pressure rather than air pressure to be what drove the bodies toward each other.[6]

Cartesian ideas continued to dominate the proceedings of the Paris Academy of Sciences in the early years of the eighteenth century. In no field is this more apparent than in that of magnetism. In papers on magnetic topics contributed by Philippe de la Hire, his son Gabriel-Philippe, Louis Lémery, Réaumur and Du Fay [7] during the first third of the century, it was taken for granted that, in its basic outline at least, Descartes' theory was correct. Fontenelle, in his summaries of the papers in the annual "Histoire" section of the Academy's journal, more than once seized the opportunity to stress the point.[8]

Not that all these authors accepted Descartes' ideas uncritically. To a man, they passed over the "grooved particles" in silence; only Du Fay, and Fontenelle in summarizing his paper, mentioned the supposed double stream of subtle matter, and even they did so but to reject it explicitly in favor of a flow in one direction only. Both Réaumur and Du Fay were concerned particularly with what happened to iron when it became magnetized, and Réaumur produced an interesting anticipation of the theory of "molecular magnets" later espoused by Coulomb and others. He supposed that an unmagnetized piece of iron already contained a large number of tiny magnetic whorls, but that only when these were properly aligned and put "in a state where they combined to produce the same effect" would the iron become magnetized and display an overall polarity.

Du Fay followed Descartes more closely at this point. Like Descartes, he ascribed the magnetization of a piece of iron to the aligning of the filaments in already existing channels so that they all lay the same way. To this he added the hypothesis that the filaments, each having one end attached to the inner wall of a channel and the other free to move, were gross enough that they would tend to fall under

[6] Ibid., pp. 586, 600–601.

[7] Philippe de la Hire, *HAS*, 1717, Mém. pp. 275–84. Gabriel-Philippe de la Hire, ibid., 1705, Mém. pp. 97–109. Lémery, ibid., 1706, Mém. pp. 119–35. Réaumur, ibid., 1723, Mém. pp. 81–105. Du Fay, ibid., 1728, Mém. pp. 355–69; 1730, Mém. pp. 142–57; 1731, Mém. pp. 417–32.

[8] Ibid., 1705, Hist. p. 6; 1723, Hist. p. 3.

their own weight whenever they were free to do so. By this means he was able to provide a novel explanation for the fact, famous since observations were first made on the iron crosses on the steeples at Aix-en-Provence and Chartres, that iron bars became magnetized, with their north-seeking pole downward, if held in a vertical position for a time. The magnetization was due simply to the filaments all inclining downward under their own weight! If the bar was inverted, the magnetism, too, ought to reverse itself, as it had in fact been found to do. This would be expected to happen the more rapidly, the freer the filaments were to move, and hence would happen much more rapidly with soft iron than with steel. A further interesting consequence followed from Du Fay's hypothesis, one which tended to give the theory as a whole a more settled appearance than it had had previously. Granting Du Fay's premise, one could determine the direction of flow of the magnetic circulation! It followed from the theory that the flow in a vertical iron bar would be from top to bottom, since this was the direction favored by the downward inclination under their own weight of the filaments in the channels. But it was known that the upper end was a south-seeking pole. Hence such a pole was one at which the stream of subtle matter entered a magnet, whereas a north-seeking pole was one from which the stream emerged! So pleased was Du Fay with this deduction that he thereupon sought to remove the long-standing confusion in nomenclature between north and north-seeking, and south and south-seeking poles, by re-naming them "exit pole" (*pole de sortie*) and "entry pole" (*pole d'entrée*), respectively.

During this period, many other authors besides those publishing in the Paris Academy's *Mémoires* adopted the theory of circulating subtle matter in order to account for the phenomena of magnetism.[9] Perhaps the highest point in the theory's development was reached in 1746, when the Paris Academy of Sciences announced the results of the prize competition it had set some years before on the subject, "The explanation of the attraction of the magnet and iron, the pointing of the magnetized needle toward the north, its declination and inclination."

The Academy had originally set this topic for its 1742 prize competition, but then and again in 1744 the judges had decided that none of the entries merited the prize. In 1746, they decided to distribute the prize, by then swollen to three times its normal value, between the authors of the three best entries received over the years. When the

---

[9] The ideas of a number of them have been described in Jean Daujat, *Origines et formation de la théorie des phénomènes électriques et magnétiques*, Paris, 1945.

seals were broken, one of these proved to be none other than Leonhard Euler. The Auvergnat scholar Etienne-François Du Tour was another, while the third work was composed jointly by Daniel Bernoulli and his father, Jean (i.e. Johann I) Bernoulli.[10]

All three of the prize-winning essays were based squarely upon the tradition of circulating subtle matter deriving from Descartes, though at the same time their authors were as one in rejecting many of the mechanical details of the Cartesian system, above all the notion of grooved particles and special rifled channels to receive them. The Bernoullis captured perfectly the spirit of all three contributions when they wrote that they "had begun [their] researches on the magnet by reading the system of Descartes." "It is not to be doubted," they went on, "but that if this founder of the true philosophy had had a sufficient knowledge of mechanics, and the general laws of motion, he would have pushed the system of the world much farther, and he would perhaps have left nothing to be desired in the theory of the magnet; but unfortunately this great philosopher was almost entirely devoid of such knowledge." Given that the science of mechanics had made a series of dramatic advances since Descartes' day, their task was clear: "It is thus by combining the assistance of mechanics with that of Descartes' principles, that we are going to try to clarify our question."[11] Similar ambitions fired the authors of the other two essays. Like the two Bernoullis, both Euler and Du Tour were convinced of the basic correctness of Descartes' explanation of magnetism.[12] They, too, saw that the mechanics of the scheme had not yet been satisfactorily worked out. And they, too, took their chief task to be the rectification of this situation.

Euler's theory involved a unidirectional circulation of subtle matter through and around a magnet. Ostensibly, this subtle matter was the same all-pervading universal aether to which Euler had had recourse in explaining various other physical phenomena. In fact, however, in order to develop his account, Euler was driven, as Descartes had been,

---

[10] *HAS,* 1744, Hist. p. 64; ibid., 1746, Hist. p. 122. The Academy subsequently published the three prize-winning essays in the fifth volume of its *Recueil des pièces qui ont remporté les prix de l'Académie,* Paris, 1752. The details of this twice-postponed prize have often been misreported by historians, probably led astray by the fact that when Euler's entry was published it was dated 1744, the year it was submitted, rather than 1746, when it was actually awarded the prize.

[11] Daniel and Jean Bernoulli, "Nouveaux principes de méchanique et de physique, tendans à expliquer la nature & les propriétés de l'aiman," *Recueil des pièces,* v, 118.

[12] Euler, "Dissertatio de magnete," ibid., pp. 3–47. Euler's essay was also published in his *Opuscula varii argumenti,* Berlin, 1746–51, III, 1–53. Du Tour, "Discours sur l'aiman," *Recueil des pièces,* v, 51–114.

to invoke what amounted in practice to a new subtle fluid. He did so by singling out the "subtlest part" of his aether and attributing the phenomena of magnetism to its behaving differently from the rest.

All matter, Euler argued, is endowed with many pores that offer an unimpeded passage to the aether. Bodies such as iron, however, which are susceptible to magnetization, are, he suggested, also endowed with certain pores that are so narrow that only the subtlest part of the aether can penetrate them. The process of magnetization involves a rearrangement of the particles of such bodies so that these narrow pores form continuous channels from one side of the body to the other. In addition, he supposed that these channels would be fitted with valves like those in the veins. These he, like many an earlier writer, took to be very probably composed of fine threads projecting into the channels; as usual, their function was to permit the particles of aether to pass through the channels in one direction but not in the other.

For the most part, the structure just outlined is very similar to that proposed by earlier writers such as Huygens. However, the increasing mechanical sophistication of the eighteenth century becomes evident as Euler's account proceeds. The aether, Euler said, was highly elastic.[13] Its normal state of equilibrium would be disturbed in the vicinity of a body endowed with the special magnetic channels he had described, because, as it pressed by virtue of its elasticity on the openings of the channels, only its subtler parts would be able to enter. These would therefore become separated from the general mass of aether, and would afterward, he argued, find some difficulty in insinuating themselves again into such small interstices in the ambient mass as those they had vacated. Hence when they emerged from the far ends of the channels, they would be unable to mingle at once with the ordinary aether they encountered there, and would instead be deflected back in the way other authors had supposed. Some of them would, he conceded, be absorbed into the surrounding aether before they reached the open ends of the channels again, but these would be replaced by new secretions of the subtler matter into the channels, and thus a perpetual circulation would be established of the kind usually imagined by magnetic theorists.

[13] Mechanist that he was, Euler was not for a moment willing, as some others were, to admit elasticity as a primitive term in his theory. Rather, he believed that it, too, ultimately had to be explained in mechanical terms. On this occasion, he actually suggested (pp. 14–15) a possible explanation, namely that the aether was itself composed of myriads of tiny vortices in which "very subtle matter is driven very rapidly in a circle," with the spaces between these being filled with other, even smaller, vortices.

Almost incidentally, Euler went on to suggest that the mere presence of a magnetic circulation around the Earth might explain the action of gravity. Since the circulation "is generated by the elastic force of the aether," he argued, "it is necessary that this elastic force is perceptibly diminished around the Earth." He demonstrated that, if this diminution were inversely proportional to distance from the center of the Earth, bodies swimming in the aether around the Earth would be subjected to a central force inversely proportional to the square of that distance. With this proved, he argued conversely that since, like the Earth, the sun and the other planets also exert inverse-square gravitational forces on bodies in their vicinity, they too must be surrounded by magnetic circulations of the kind he had described!

For Euler, as for his predecessors, accounting for the supposed continuous circulation of subtle fluid was only a preliminary. The real tasks were to show why magnets aligned themselves as they did in the presence of another, or in relation to the Earth as a whole, and to explain the magnetic attractions and repulsions. Here, Euler proved a traditionalist. Like so many others, he attributed the orientation of a magnet to its aligning itself with the direcon of fluid flow wherever it happened to be; the only real novelty in his presentation was that he spelled out the dynamics of the situation in a more acceptable form than had usually been done. Likewise, he relied upon inequalities generated in the pressure of the surrounding aethereal medium to explain the magnetic attractions. Only in accounting for the repulsion of two like poles does his account differ very much from that of, say, Huygens. Most previous authors had seen this as a straightforward result of the head-on collision of two opposing streams of subtle fluid. Euler envisaged a more complicated pattern of interacting streams, one that enabled him to deal reasonably successfully with Musschenbroek's report that, though two like poles often repelled each other, they might also attract each other if the circumstances of the experiment were varied only slightly and in seemingly insignificant ways.

One other point upon which Euler adopted a novel approach concerned the number and location of the poles of a magnet. In his theory, the poles were merely regions where entrances to or exits from the channels through a magnetic body were concentrated. The theory required that the channels be continuous through the body, but in Euler's opinion there was no need for their ends to lie diametrically opposite each other, or even close to this. On the contrary, the cluster of channels in a particular magnet might pass through it well off-center, or the channels might diverge in the form of a "V." In the latter case, he said, the magnet ought to exhibit three poles, two of one kind and one of the other. There might even be two separate clusters of channels passing through different regions of a magnet, in which case it would manifest no fewer than four poles. Some years after the Paris prize competition, Euler read a paper to the Berlin Academy of Sciences in which he took advantage of the extra flexibility his theory provided and attempted a systematic account of the magnetic declination on the assumption that the Earth had an off-center magnetic axis.[14]

Euler's magnetic theory was a product of the early 1740s. His opinions on the subject changed little with the years, however, and in the section on magnetism in his widely read *Letters to a German Princess*—composed in the early 1760s, published ten years later in 1772, and subsequently reprinted many times in several different languages—he remained a powerful advocate of the circulation theory.[15] Only in two relatively minor respects did he deviate from what he had set out earlier. First, he now frankly acknowledged that the matter comprising the circulating streams was a special kind of subtle matter, and not merely the "subtlest part" of a single all-pervading aether. Second, and testifying to the success with which Muschenbroek's views about the irreducibly errant nature of many magnetic phenomena had been combatted in the intervening period, he abandoned his somewhat complicated account of what happened when two like poles approached each other, in favor of the simpler and more traditional view that a collision between the two opposing streams caused a mutual repulsion.

The Bernoullis' prize-winning essay proposed a very different mechanism to account for the formation and maintenance of the magnetic circulation, but in other respects their theory, too, was along traditional lines. They rejected, as being contrary to the general economy of nature, the idea that magnetism involved the action of some special kind of subtle matter. Rather, they supposed that the circulating streams were composed of the same aether that in their opinion manifested itself in so many other natural phenomena. Like everybody else, they also supposed the aether to be highly elastic. In accounting for this, however, they introduced some quite new ideas.

[14] Euler, "Recherches sur la déclinaison de l'aiguille aimantée," *Mém. Acad. Berlin*, XIII, 1757, 175–251. Some years later, Euler published a modified version of his theory in which he also attempted to take the magnetic dip into account (*Mém. Acad. Berlin*, XXII, 1766, 213–64).

[15] I have used the English translation by Henry Hunter, London, 1802. The section on magnetism is in vol. II, pp. 209–86; it appeared in the third and final volume of the original edition, St. Petersburg, 1768–72.

Some years earlier, Daniel Bernoulli had sought to explain the elastic properties of air by assuming it to be composed of a multitude of particles moving rapidly in all directions.[16] These particles, he supposed, continually collided with each other and hence changed the directions of their motions. His theory was, in fact, a remarkable anticipation of the nineteenth-century kinetic theory, but it won singularly little support at the time. Now, however, he and his father adopted a similar explanation for the elasticity of the aether, and developed the idea into a kinetic theory of liquids and solids as well. A magnet, they suggested, is composed of tense, elastic, roughly parallel fibers that are vibrating very rapidly. As the fibers vibrate, they alternately squeeze and relax the subtle matter in the hollows or channels between them. When a particular "cell" in a channel is squeezed, the aether will be forced into the adjacent one. To prevent it from moving back when the first cell relaxes, the Bernoullis invoked the usual device of certain filaments or valves lining the hollows and permitting a flow in one direction only. The presence of these would result in a general drift of fluid along the channels and not a mere ebb-and-flow. The intestinal motions that normally gave rise to the elasticity of the aether would thus be converted, in part at least, to a progressive motion along the channels, and the elasticity inside these would be correspondingly reduced. The tendency of the flowing matter to conserve the motion it had acquired, even after it emerged from the channels, together with the tendency of external aether to be sucked constantly into the open ends of the channels, would, the Bernoullis argued, lead to a continuous circulation of the kind usually imagined by magnetic theorists. To complete the picture, the Bernoullis held that some channels would permit aether to flow in one direction and some in the other, so that in their theory, as in that of Descartes, a magnet was supposed to be surrounded by twin streams of matter moving in opposite directions through and around it.

With the circulation of subtle matter established, the Bernoullis turned to explaining the various phenomena of magnetism. In general, their explanations followed traditional lines. Like Euler, however, they agreed that the poles of a magnet need not be diametrically opposite, and also that the channels need not be strictly parallel, and indeed might even separate into branches, so that a magnet might have a multiplicity of poles. Also, their belief in a double stream of subtle

[16] D. Bernoulli, *Hydrodynamica, sive de viribus et motibus fluidorum commentarii*, Strasbourg, 1738, ch. x (English trans., New York, 1968, pp. 226–74).

matter forced them to modify somewhat the usual explanation of the repulsion of two like poles. According to the general principles of their theory, it would seem that, when two like poles were brought near each other, the streams emerging from each would find some channels in the other that would offer them ready access. Hence it would appear that the two circulations, rather than pressing against each other, ought to combine into one as they did when two unlike poles came together, and so an attraction ought to result in this case as it did in the other. To avoid this possibility, the Bernoullis sought a mechanism that would prevent the streams from one magnet from entering the channels of the other, even though the filaments in the latter were inclined the right way to let them pass. Their solution was to return to the oscillatory nature of the mechanism generating the streams of matter. The streams would be effectively blocked from passing from one magnet to the other, they said, because the oscillations of the fibers in the two magnets would not be "in harmony" (*dans une agitation harmonieuse*).

The third prize-winning essay, that by Du Tour, was a less substantial piece of work than the other two. Du Tour took the ideas of Du Fay as his starting point, but sought to overcome what he saw as a major difficulty in them, namely Du Fay's failure to specify a cause that might maintain the perpetual circulation of magnetic matter around a magnet. Du Tour's solution was to suppose that the channels passing through the magnet alternately expanded and contracted. This would, he said, lead to the magnetic fluid's being squeezed along them. Like the Bernoullis, he invoked the usual hairlike projections in the channels to explain why the motion of the fluid was continuous and not an ebb-and-flow phenomenon. However, he presented the entire mechanism as simply an ad hoc hypothesis. Unlike the Bernoullis, he made no effort to incorporate the idea of alternate expansions and contractions into a more general kinetic theory of nature.

Du Tour proposed a slightly different mechanism to account for magnetic attractions. Instead of invoking the pressure of the surrounding aether, he argued that what drove two magnets together when their separate circulations joined into one was the pressure of the larger circulation surrounding the Earth. One other novelty worth mentioning is Du Tour's explanation of the magnetic declination and its irregular pattern across the surface of the globe. The Earth, he suggested, might contain several large magnets clustered around its axis with their south poles grouped near its south pole and their north poles near its north pole; but their axes might not be parallel, and some might be stronger magnets than others. Deposits of iron might also, he conceded, play a role.

The three winning entries in the 1746 prize competition, published together by the Paris Academy in 1752, evidently reflected a scientific opinion that was still well entrenched on the Continent at mid-century. Magnetism was caused by subtle matter circulating through and around magnetic bodies. But what kind of matter it was that circulated, what the mechanism was that kept it in motion, and how the interactions of various circulations with each other or with other matter brought about the various magnetic phenomena—all these questions were still to be resolved.

But what of Britain? Newton's anti-Cartesian views came to dominate science there much sooner than they did elsewhere. What of Holland, where Newton's ideas first made substantial inroads into Continental modes of thought, as early as the second decade of the eighteenth century? Was the outcome of the Paris Academy's prize competition merely a very late manifestation of the well-known reluctance of European and particularly French scientists to adopt the Newtonian approach?

Unfortunately, most British scientists of the period who wrote on the subject confined themselves to the practicalities of magnetic research, and were extremely reticent concerning the causes of the phenomena they were dealing with. It is therefore difficult to give an unequivocal answer to this question. Yet the very silence of British physicists, when compared with their outspoken condemnation of other aspects of the Cartesian tradition, suggests that the theory of magnetic circulations retained at least a measure of credibility with them. So too, and even more so, does the fact that a version of the theory was able to get published as late as 1747 in the standard-bearer of Newtonian science, the *Philosophical Transactions* of the Royal Society of London.

Newton's own published writings do not make clear his opinion on the subject. At times his choice of phrase seems to imply that magnetic forces acted, like gravity, in some unexplained fashion at a distance, but elsewhere he ascribes them to the mechanical action of certain subtle material effluvia. Nowhere, however, does he provide a systematic exposition of his views.

For the most part, when Newton refers to magnetism, he mentions it merely in passing, as a force that is already known to act in the world. Very early in his *Principia,* for example, he uses well-known facts about the lodestone to illustrate his definitions concerning forces in general. The fact that "the magnetic force is greater in one loadstone and less in another, according to their sizes and strength of intensity" is used to illustrate the proportionality of any central force to "the efficacy of the cause that propagates it from the centre, through the spaces round about." Likewise, in order to explain what he means by the "accelerative quantity" of a central force, Newton calls upon the fact that "the force of the same loadstone is greater at a less distance, and less at a greater." [17] A little further on, he illustrates the equality of action and reaction with an experiment using lodestone and iron.[18] When he needs an example of a force whose action extends only to neighboring particles, he cites the fact that the attractive power of a magnet is interrupted by a nearby iron plate.[19] Later, magnetic and electric attractions are brought forward as examples showing that "all attraction towards the whole arises from the attractions towards the several parts." [20] In the *Opticks,* too, in Query 29, magnetism provides a useful analogy, this time for a point Newton wishes to make in connection with the anomalous refraction of light by Iceland crystal.[21] On other occasions in Newton's published writings, however, he mentions magnetism only to distinguish between the magnetic force and his new force of gravity. In opposition to a large number of seventeenth-century authors, he insists that the two are quite different, even to the extent that they seem to vary according to different laws.[22]

None of these remarks in fact commits Newton to a particular theory of magnetic action, but his discussing the magnetic force so frequently in parallel with the force of gravity might well have given his readers the impression that he accorded them both the same causal status. For much of his life, Newton was convinced that the cause of gravity was non-mechanical, and his followers were well aware of this. The suspicion that he held a similar view about the cause of magnetism was therefore perhaps inevitable. Furthermore, the famous opening lines of the thirty-first query in the *Opticks* would have done nothing to dispel this idea. Here, Newton presented gravity, magnetism and electricity as "Instances shew[ing] the Tenor and Course of Nature," to support his suggestion that "the small Particles of Bodies [have] certain Powers, Virtues, or Forces, by which they act at a distance . . . upon one another for producing a great Part of the Phænomena of Nature." [23] Admittedly, Newton immediately qualified what he had said by adding, "What I call Attraction may be per-

[17] Newton, *Mathematical Principles of Natural Philosophy,* p. 4.

[18] Ibid., pp. 25–26.

[19] Ibid., p. 302.

[20] Ibid., p. 415.

[21] Newton, *Opticks,* pp. 373–74.

[22] E.g. Newton, *Mathematical Principles,* p. 414. Also *The Correspondence of Isaac Newton,* II, 341–42, 360–62.

[23] Newton, *Opticks,* pp. 375–76.

form'd by impulse, or by some other means unknown to me. I use that Word here to signify only in general any Force by which Bodies tend towards one another, whatsoever be the Cause." Yet later in the same query, having mentioned gravity, fermentation and cohesion as examples of "certain active Principles," "general Laws of Nature, by which the Things themselves are form'd," Newton added "magnetick and electrick Attractions" to the list of "manifest Qualities" whose causes yet remained occult and unknown, but were very possibly, he seemed to imply, non-mechanical.[24]

Yet it would be a mistake to accept any such implication so far as magnetism is concerned. Though the evidence is sparse, it is sufficient to show that at almost all periods of his life Newton actually held, like all but the Aristotelians among his contemporaries, that magnetic phenomena were caused by mechanical processes, namely circulating streams of subtle material effluvia. Even if we do not do so in other cases, with respect to magnetism we ought to take literally Newton's caveat that what he called an attraction might be caused by mechanical impulse.

Newton's early belief in magnetic circulations rather like those described by Descartes is revealed by an as yet unpublished manuscript dating from 1666 or 1667, now preserved in the University Library, Cambridge. Here, Newton actually sketches the likely patterns of flow in various circumstances![25] Similar views are set forth in a document dated by its editors to a period a few years later, 1673–1675, and published by them under the title "De Aere et Aethere." At one point in this paper, while arguing for the existence of various kinds of matter subtler than air, Newton remarks without further elaboration: "I believe everyone who sees iron filings arranged into curved lines like meridians by effluvia circulating from pole to pole of the [lode-]stone will acknowledge that these magnetic effluvia are of this kind."[26] The same doctrines appear again in other writings from this period, namely the "Hypothesis explaining the Properties of Light" that Newton sent to the Royal Society of London in late 1675, and another still unpublished manuscript in the Cambridge University Library. In the former of these, Newton twice refers casually to "Magnetic effluvia" in a way that indicates that he takes their existence

---

[24] Ibid., p. 401.

[25] Cambridge University Library, Add. MS 3974, f. 1–3.

[26] A. R. and M. B. Hall, *Unpublished Scientific Papers of Isaac Newton*, p. 220 (English trans., p. 228). R. S. Westfall argues (*Force in Newton's Physics*, London, 1971, pp. 409–10) that this document ought to be given a somewhat later dating, namely 1679, than that provided by the Halls.

for granted.²⁷ However, this document, like "De Aere et Aethere," was not published at the time; not until it was included by Thomas Birch in his *History of the Royal Society of London* in 1757 did its contents enter the public domain. Those of the second document have still not done so. In it, Newton discusses magnetism at greater length than anywhere else. Throughout, he assumes the existence in the vicinity of a magnet of peculiar magnetic "streams"; indeed, he specifies (though in terms somewhat different from those used by Descartes) two separate and "unsociable" streams entering a magnet at its two opposite poles and passing through it in opposite directions.²⁸

Testimony that Newton held similarly mechanistic views about magnetism during the 1690s comes from two sources. First, in a draft addition to the *Principia* written during the early years of the decade, Newton argues that the particles of bodies must have "a certain wonderful and exceedingly ingenious (*artificiosam*) texture" in order that the material be sufficiently porous to "allow magnetic effluvia and rays of light to pass through them in all directions and offer them a very free passage."²⁹ Second, the mathematician David Gregory, while visiting Newton in 1694, recorded it as his host's view that the magnetic virtue "seems to be produced by mechanical means."³⁰ Another casual allusion to "magnetick Effluvia" in the first edition of the *Opticks,* published ten years later, tells the same story.³¹ Finally, in one of the famous "aether" queries that Newton added to the second English edition (1717) of this work, he uses the existence and activity of the magnetic effluvia, which he again takes as beyond dispute, to justify certain assumptions he is making about the aether: "If any one would ask how a Medium can be so rare," he suggests, "let him tell me . . . how the Effluvia of a Magnet can be so rare and subtile, as to pass through a Plate of Glass without any Resistance or Diminution of this Force, and yet so potent as to turn a Magnetick Needle beyond the Glass." ³²

---

²⁷ I. B. Cohen, ed., *Isaac Newton's Papers and Letters on Natural Philosophy,* Cambridge, 1958, pp. 180, 183.

²⁸ Cambridge University Library, Add. MS 3970.3, f. 473–74. Westfall summarizes Newton's account in his *Force in Newton's Physics,* p. 332.

²⁹ Newton, *Unpublished Scientific Papers,* p. 314 (English trans., p. 316).

³⁰ Newton, *Correspondence,* III, 335 (English trans., p. 338). Gregory himself seems to have been less confident than his host: cf. his "Notae in Newtoni Principia," Royal Society of London, MS Gregory 210, f. 79 (insert).

³¹ Newton, *Opticks* (1st ed., 1704), Book II, p. 69. Cf. reprint ed., New York, 1952, p. 267.

³² Newton, *Opticks* (reprint ed.), p. 353.

A number of scholars have recently argued that during his lifetime Newton's views changed concerning the possibility of finding mechanical explanations for the various forces he had introduced into his science. More precisely, they have seen him as much more sanguine about the aether as an explanatory mechanism in the early and late periods of his life than he was in between.[33] It by no means follows, however, that Newton's views about the cause of magnetism varied in the same sort of way, since in his thinking he seldom conflated the "magnetic effluvia" with his more general explanatory mechanisms. (Magnetism is, for example, a conspicuous absentee from the list set out in the "Scholium Generale" added to the 1713 edition of the *Principia* of phenomena that might be explained in terms of the "certain most subtle spirit" introduced at that point.)[34] If Newton's faith that magnetism was caused mechanically ever wavered, it apparently did so only once, and that very briefly. In a draft "Conclusion" to his *Opticks*, which he drew up during the early 1690s, Newton seems to put the "attractive vertue" of the particles of a magnet in the same category as those other seemingly unexplained forces, gravity and electricity: "The particles of bodies have certain spheres of activity," he here asserts, "wthin wch they attract or shun one another. For ye attractive vertue of the whole magnet is composed of ye attractive vertues of all its particles & the like is to be understood of the attractive vertues of electrical & gravitating bodies." More importantly, there is the evidence just cited which shows that even during the 1690s, the period during which Newton seems to have abandoned all attempts to explain certain classes of forces mechanically, his conviction that the attractive power of the magnet was brought about by mechanical means was as strong as ever. In the light of his subsequent allusions to "magnetic effluvia" in the various published versions of the *Opticks*, there seems no reason to doubt that thereafter he always continued in this belief. It is less certain that he continued to believe in the same circulatory mechanism as before, but what we know of the views of his closest disciples during this period suggests that he did.

Confirmation that we have correctly understood Newton's scattered remarks about magnetism comes from the writings of one of his earliest

[33] Henry Guerlac, "Newton's optical aether: his draft of a proposed addition to his *Opticks*," *Notes and Records Roy. Soc. London*, XXII, 1967, 45–57; McGuire, "Force, active principles, and Newton's invisible realm," *Ambix*, xv, 1968, 154–208; Westfall, *Force in Newton's Physics*, ch. VI, esp. pp. 363–77, 391–95.

[34] Newton, *Mathematical Principles*, p. 547. Magnetism is included in a similar list drawn up by Newton on another occasion (Cambridge University Library, Add. MS 3970, f. 241; quoted by Joan L. Hawes, "Newton's two electricities," *Annals of Science*, XXVII, 1971, 95–103, p. 97).

followers, Dr. Samuel Clarke. Clarke produced one of the best known of all expositions of Newtonian natural philosophy, in the unusual guise of notes appended to his series of Latin editions of Jacques Rohault's *Traité de physique,* a work that was itself thoroughly Cartesian in outlook. He was, we know, in close contact with Newton. He it was who prepared the Latin edition of the *Opticks* and saw it through the press on behalf of the master. Later, he acted as spokesman for the Newtonian position in a famous exchange of letters with Leibniz, and it has been established that he did so in close consultation with Sir Isaac himself. He would therefore have been well able to ensure that in his editions of Rohault's book, as elsewhere, the positions he adopted faithfully reflected the master's views.[35]

In one of his notes, responding to a typical attack by Rohault on the admissibility of the very notion of attraction in science, Clarke quoted extensively from the thirty-first query of the *Opticks,* and included both the opening passage, in which magnetism is cited as an example of a force acting at a distance, and the passage, noted above, in which Newton mentions magnetism in close conjunction with his idea of "active principles" underlying various classes of natural phenomena.[36] Much more significant than this for our purposes, however, is Clarke's failure to respond at all to the whole chapter that Rohault devoted elsewhere in his book to an exposition of the Cartesian theory of magnetism. Far from attempting to rebut Descartes' theory, Clarke added but three notes to the entire chapter, and these merely unimportant glosses.[37] Given that elsewhere in the book Clarke seized every possible opportunity to advocate the Newtonian viewpoint, his failure to comment here is most conspicuous, and suggests that he accepted at least the general outline of the Cartesian account. This in turn suggests that Newton did likewise.

Comments by some other early Newtonians lead to the same conclusion. John Keill was lecturer in natural philosophy and later Savilian Professor of Astronomy at the University of Oxford. Like

[35] Rohault's *Traité* was first published in 1671. The first edition of Clarke's Latin translation appeared in 1697, and his notes were steadily expanded in succeeding editions. I have used the English translation of 1723, *Rohault's System of Natural Philosophy, illustrated with Dr. Samuel Clarke's Notes taken mostly out of Sir Isaac Newton's Philosophy,* London, 1723; reprinted New York, 1969. On Clarke's notes, see M. A. Hoskin, " 'Mining all within': Clarke's notes to Rohault's *Traité de physique," The Thomist,* xxiv, 1961, 353-63. For Clarke's connections with Newton, see A. Koyré and I. B. Cohen, "Newton and the Leibniz-Clarke correspondence," *Archives internationales d'histoire des sciences,* xv, 1962, 63-126.

[36] *Rohault's System,* I, 54-55.

[37] Ibid., II, 163-87.

Clarke, he was in close personal contact with Newton, being deeply involved, in particular, in defending Newton's rights in his famous dispute with Leibniz over the invention of the calculus. In his *Introduction to Natural Philosophy* (London, 1720)—though not in the original Latin version of this—Keill revealed his acceptance of traditional ideas about magnetism by alluding in a manner very similar to Newton's to "magnetick Effluvia" that freely pervaded the pores of dense matter.[38] John Harris edited the popular *Lexicon Technicum*, which appeared in two bulky volumes in 1704 (the same year as the first edition of Newton's *Opticks*) and 1710. The chief influence on the 1704 volume was Robert Boyle, but by the time the second volume appeared, six years later, Harris had become a dyed-in-the-wool Newtonian. So far as magnetism was concerned, however, Harris made no attempt in 1710 to retract his earlier opinion that "the Operation of the *Magnet* depends on the Flux of some fine Particles which go out at one Pole, then round about and in again at the other."[39] Apparently he saw no need to do so. Finally, the astronomer Edmond Halley, who worked closely with Newton for many years, also accepted the circulation theory, and at one point attempted to explain the aurora borealis in terms of magnetic effluvia emerging from the north pole of the Earth and either shining faintly by themselves or causing nearby matter to glow as a result of friction.[40]

Yet not all British scientists after Newton were prepared to commit themselves so openly to the traditional theory. Most authors of articles in the *Philosophical Transactions* dealing with magnetism—men such as William Derham, the elder Hauksbee, Brook Taylor, and George Graham—avoided the question of what might be causing the phenomena they were studying. Rather, they used theoretically neutral terms such as "magnetic virtue," "polarity," and "attractive virtue," and did not elaborate further. Servington Savery did at least refer in passing to the cause of the "invisible Force" by which one lodestone

---

[38] Keill, *Introduction*, p. 67. I have used the 1745 edition of this work, but Robert E. Schofield (*Mechanism and Materialism: British Natural Philosophy in an Age of Reason*, Princeton, 1970, p. 29) indicates that a similar passage appears in the earlier edition. The original Latin version was published in 1702.

[39] Harris, *Lexicon Technicum: or, an Universal English Dictionary of Arts and Sciences*, London, 1704, art. "Magnet"; and vol. II, London, 1710, art. "Magnetism." For Harris's conversion to Newtonianism, see Geoffrey Bowles, "John Harris and the powers of matter," *Ambix*, XXII, 1975, 21–38.

[40] Halley, *Phil Trans.*, XXIX, 1716, 406–28. See also J. Morton Briggs, Jr., "Aurora and Enlightenment: eighteenth-century explanations of the aurora borealis," *Isis*, LVIII, 1967, 491–503, p. 492.

acted on another, but only to describe it as "unknown." [41] Desaguliers alluded at one point to Du Fay's idea that magnetization involved "Threads or Beards fix'd at one End . . . laid all one way" in the iron, but he failed to indicate how these were supposed to account for the phenomena.[42] And neither in the printed summary of his immensely popular course of lectures on experimental philosophy nor in the treatise that grew out of the lectures did he consider the theory of magnetism at all.[43] In the absence of a published lead from the master (and it must be remembered that some of the most important evidence described above concerning Newton's real views could not be cited by his followers), Desaguliers, like most other British writers of the period, preferred not to commit himself. Yet the fact that he deliberately drew attention to Du Fay's ideas, and that while doing so he stressed the conformity of his experiments with those ideas, suggests that he thought Du Fay was on the right track. Furthermore, some independent evidence survives concerning the underlying beliefs that he and others accepted at this period. "The opinion that principally prevails among the moderns," we read in the popular *Cyclopaedia* published by Ephraim Chambers, "is that of Des Cartes." [44]

One author who was prepared to commit himself in public was the instrument maker and itinerant scientific lecturer Benjamin Martin. In a brief discussion of magnetism in his popular work, *Philosophia Britannica,* Martin defined magnetic poles as "points . . . which emit the Magnetic Virtue," and described the traditional experiment with iron filings patterns as "a very curious Method of rendering visible the Directions which the Magnetic Effluvia take in going out of the Stone." [45]

Another exception was Dr. Gowin Knight (1713–1772), and since he was undoubtedly Britain's most important magnetic researcher in the first half of the eighteenth century, his views on the subject are significant. In the early 1740s, Knight developed a much-improved method of making artificial magnets. His achievement was reported to the Royal Society, but for commercial reasons he kept the method

[41] Savery, *Phil Trans.*, XXXVI, 1729–30, 295–340.

[42] Desaguliers, ibid., XL, 1737–38, 385–87.

[43] Desaguliers, *Physico-Mechanical Lectures: or, an Account of what is Explain'd and Demonstrated in the Course of Mechanical and Experimental Philosophy,* London, 1717; and also his *Course of Experimental Philosophy,* 2 vols., London, 1734–44.

[44] Chambers, *Cyclopaedia: or, an Universal Dictionary of Arts and Sciences,* 5th ed., London, 1741–43, art. "Magnetism."

[45] Martin, *Philosophia Britannica: or, a New and Comprehensive System of the Newtonian Philosophy* . . . , Reading, 1747; 2nd ed., London, 1759, I, 42–44.

itself secret.[46] This prompted others—Canton and Michell in England, Duhamel, Le Maire, and Antheaulme in France—to tackle the problem for themselves, and eventually they had the same success. Their work will be discussed in some detail below. Knight, meanwhile, turned his attention to the design of the mariner's compass, a field in which he introduced some truly revolutionary improvements.[47]

In 1747, just one year after the Paris Academy of Sciences announced the result of its prize competition on the theory of magnetism, Knight read a paper to the Royal Society in which he set out his views on the same subject. These views, it turned out, were very similar to those that had been crowned in Paris.[48] "The magnetic Matter of a Loadstone," he maintained, "moves in a Stream from one Pole to the other internally, and is then carried back in curve Lines externally, till it arrives again at the Pole where it first entered, to be again admitted." As proof, he, like so many other authors of the period, cited the patterns formed by iron filings scattered on a sheet of paper held above a magnet. As to the direction of the stream, he remarked at one point that "we have some Reason to think it enters at the North Pole." Here, however, nomenclature problems arise: by "north pole" Knight, like most of his contemporaries, meant the south-seeking pole of the magnet. Knight did not say why he thought the magnetic stream entered at that end, but he was almost certainly following Du Fay on this point. Magnetic attractions occurred, Knight said, as a result of "the Flux of one and the same Stream of magnetical Matter" through the two bodies involved. On the other hand, two unlike poles repelled each other because of "the Conflux and Accumulation of the magnetic Matter" in the space between them.

A few months after this, Knight further elaborated his ideas and incorporated them into an all-inclusive matter theory, in his *Attempt to Demonstrate that all the Phaenomena in Nature may be explained by Two Simple Active Principles, Attraction and Repulsion . . .* (London, 1748; 2nd ed., 1754). As Boerhaave had done earlier in his *Elementa chemiae* (Leyden, 1732), Knight argued that "there are in Nature two Kinds of Matter, one attracting, the other repelling." Unlike Boerhaave, however, who had supposed his repelling matter, or Fire, to be weightless, Knight assumed it to be attracted to other matter: "Those Particles of Matter, that mutually repel each other, seem in respect of other Matter also subject to the general Law of Attraction." Upon this basis, he developed an elaborate account of a

---

[46] *Phil. Trans.*, XLII, 1744–45, 161–66, 361–63; XLIV, 1746–47, 656–64.
[47] W. E. May, *A History of Marine Navigation*, Henley-on-Thames, 1973, p. 69.
[48] Knight, *Phil. Trans.*, XLIV, 1746–47, 665–72.

variety of natural phenomena, in most cases taking as his starting point the notion that each particle of attracting matter would be surrounded by an atmosphere of repelling matter. In the case of magnetism, Knight assumed that it was his "repelling matter" that circulated through and around a magnet. He supposed that iron, and iron alone, had pores within it so constructed that "more repelling Matter may be contained in its Pores than what would belong to the Space occupied by them." In such a situation, he argued, if this matter were once put in motion, its elastic force would lead to a perpetual circulation. As he put it, "those Pores from whence the Stream moves will be supplied by a Conflux of the same Matter from without; and the Stream of repellent Matter, coming out of the Body at the opposite Side, will be carried round to the Side where it first began, to supply the Place of what enters the Body there." Once established, "circulations" of this kind acted in the way he had described in his earlier paper to bring about the various magnetic effects.

As Britain's leading magnetic researcher, Knight cannot simply be dismissed as an isolated unreconstructed Cartesian in an otherwise hostile Newtonian environment. The Royal Society's interest in his work was not confined to his experimental results. His theoretical ideas were also given an attentive hearing, and the fact that they were subsequently published in the *Philosophical Transactions,* as well as in his book, emphasizes their respectability. We are forced to conclude, therefore, that such views retained a wide measure of credibility in England, as elsewhere, even in the late 1740s. As with Newton himself, then, so with his followers. The triumph of Newtonianism did not carry with it the automatic rejection of everything Cartesian. Magnetic circulations had certainly not yet been abandoned.

On the other hand, a few questioning voices had begun to be heard. One of the earliest was that of Newton's successor at Cambridge, William Whiston. Like his contemporaries, Whiston ascribed the directive power of a magnet to "magnetic effluvia" circulating continually around it. He was unable, however, to imagine a mechanism to explain the magnetic attraction, and argued on this basis that its cause must be non-mechanical, like that of gravity.[49] Servington Savery's insistence that the cause of magnetism was unknown implied, among other things, a healthy skepticism so far as the circulation theory was concerned. Not long after Savery's paper was published, doubts were also expressed in no less august a place than the *Histoire*

[49] Whiston, *The Longitude and Latitude found by the Inclinatory or Dipping Needle . . . ,* London, 1721, p. 11.

of the Paris Academy of Sciences.⁵⁰ On this occasion, Fontenelle described some experiments by Pierre Le Monnier that seemed to prove that iron, far from giving an easy passage to the magnetic matter, was opaque to it. He also reported Le Monnier's conclusion, based largely on the details of various iron filings patterns, that the magnetic matter did not circulate through and around a magnet as was commonly supposed. He failed to make clear, however, what alternative Le Monnier was advocating. He ended his account in typically dramatic fashion: "What point of physics will be fixed," he cried, "if the magnetic vortex is not?"

Far more influential than the doubts of Savery or Le Monnier, or even those of Whiston, was the outspoken criticism of the traditional theory by Pieter van Musschenbroek. It was easy to ignore the others; it was impossible to ignore Musschenbroek, unquestionably the most important magnetic researcher of his generation, whose work was cited repeatedly by subsequent eighteenth-century writers on the subject. Beginning with a letter to Desaguliers published in the *Philosophical Transactions* of 1725,⁵¹ Musschenbroek seized every opportunity to stress the speculative and experimentally unsupported character of the circulating-fluid theory. More than this, he argued over and over again that his experiments showed that magnets do *not* act by means of corporeal effluvia. He insisted that the cause of magnetism remained unknown, and, following the example set by Newton in other cases, he was willing to allow that it might even be non-corporeal.

Musschenbroek's most authoritative work concerning magnetism was his *Dissertatio physica experimentalis de magnete,* published in Leyden in 1729. He also included a discussion of the subject in his extremely popular textbook of experimental physics, which over the years appeared in a number of editions in several European languages.⁵² Both books included extensive attacks on the circulation theory. "I would say that I have never observed an Hypothesis more opposed to Experiments and truth than this one," he proclaimed in the preface to the *Dissertatio,* "I am of the opinion that there does not

⁵⁰ *HAS,* 1733, Hist. pp. 13–17.
⁵¹ Musschenbroek, *Phil. Trans.,* xxxiii, 1724–25, 370–78.
⁵² Musschenbroek, *Epitome elementorum physico-mathematicorum in usus academicos,* Leyden, 1726; I have used John Colson's English translation of the second edition, *The Elements of Natural Philosophy,* London, 1744, and also Sigaud de la Fond's French rendering of the greatly-expanded final (1762) version of the Latin text, *Cours de physique expérimentale,* Paris, 1769. Musschenbroek took care to bring his text up to date from one edition to the next, but his opinions concerning magnetism seem not to have changed very much with the years.

exist a magnetic Fluid save in the mind of one imagining it." [53] Early in his career he seems not to have understood that those he was attacking had supposed a *circulation* of magnetic fluid. His arguments were directed, rather, against the notion of effluvia flowing out from a magnet (or perhaps in toward it) from all directions at once.[54] Later, he became better informed, but not thereby persuaded to accept the traditional opinion. On the contrary, he dismissed the idea of a fluid flowing through channels lined with filaments as a mere "fiction." People ought to be more frank, he said, and admit they did not know the cause of magnetism.[55]

Musschenbroek's objections to the traditional theory derived principally from his discovery that the interposition of a barrier between a pair of magnets affected the force they exerted on each other only if the barrier were made of iron. This was not a new discovery, of course—the fact had been known to Gilbert—but Musschenbroek added a new refinement by actually measuring the forces involved. By analogy with the way a sheet of glass reduced the intensity of a beam of light without stopping it altogether, he expected that any non-ferreous barrier would weaken the magnetic force to some extent. His discovery that it did not persuaded him that the entire analogy was misconceived. Musschenbroek's measurements also convinced him that the repulsive force between two like poles was rather weaker than the attractive force between two opposite poles of the same pair of magnets. He argued that if the effluvial theory were correct, it must follow from his result that less fluid left a magnet than entered it, a proposition that led to consequences so absurd that the theory had to be rejected. He also showed that neither a blast of air nor a current of heat rising from a stove and passing between a magnet and a compass needle had any effect on the magnetic force. These results, too, he regarded as incompatible with the idea of effluvia.

Unfortunately, Musschenbroek was unable to devise a better explanation for the phenomena of magnetism than the one he rejected. Convinced that the traditional theory was false, he could do nothing better by way of an alternative than advocate a wait-and-see attitude, a suspension of belief in all theory until more information could be obtained. Psychologically, such a position lacks persuasive power. In the absence of firm evidence, speculation has an insidious appeal. Furthermore, Musschenbroek's arguments were by no means as conclusive as he said they were. His experiments were open to a variety of interpretations, and upholders of the circulation theory felt they

[53] Musschenbroek, *Dissertatio,* pp. 4, 5.
[54] Ibid., pp. 63–67.
[55] Musschenbroek, *Cours de physique expérimentale,* I, 469.

had an explanation for them. Only when confronted by a viable alternative, namely the one Aepinus set out in his *Essay*, did the old ideas really begin to lose sway, and this is why Aepinus's successors saw his work rather than Musschenbroek's as marking the watershed in the history of the subject. Nevertheless, Musschenbroek's work was highly esteemed in its day, and the stand he adopted must have raised doubts in the minds of at least some of his readers.

One instance where Musschenbroek's influence is clearly documented occurs in the famous *Encyclopédie* of Diderot and d'Alembert.[56] In the article "Magnétisme," d'Alembert mentions the three winning essays in the 1746 prize competition as the most plausible attempts to date to discover the cause of magnetism. He says that, having studied the phenomena, it is difficult to reject the commonly accepted theory. But he also mentions that objections have been raised to this and refers particularly to Musschenbroek's *Essai de physique*, and also to the 1733 volume of the Paris Academy's *Histoire* containing Le Monnier's views. In his "Preliminary Discourse" to the *Encyclopédie,* d'Alembert sides, if anything, a little more strongly with Musschenbroek's opinion, and asserts that "the necessary enlightenment concerning the physical cause of the properties of the magnet [is] lacking."[57] On this occasion, however, he mentions no names. In line with the preliminary discourse, a neutral position is also generally maintained with regard to theory in the article "Aimant," written by Le Monnier's son, Louis-Guillaume Le Monnier. Yet at one point a hint is given of this author's views; these, in contrast to those of his father, appear to have been traditional rather than skeptical, since he speaks of iron filings obeying "les écoulemens magnétiques."

One other mid-century author who was clearly skeptical about the circulting-fluid theory was the Englishman John Michell. Since his book on magnetism, like Musschenbroek's, was frequently cited by contemporaries, his views may also have carried some weight. Only an attentive reader, however, would have become aware of them, since for the most part Michell eschewed theoretical speculation. His unfavorable opinion of the circulation theory emerges only in a single footnote, and then only by implication.[58]

During the 1740s and early 1750s, then, the period in which

---

[56] Diderot and d'Alembert, eds., *Encyclopédie, ou dictionnaire raisonné des sciences, des arts et des métiers*, Paris, 1751-65. I have used the edition published in Livorno, 1770-75.

[57] D'Alembert, *Preliminary Discourse to the Encyclopedia of Diderot*, trans. Richard N. Schwab, Indianapolis, 1963, p. 23.

[58] Michell, *A Treatise of Artificial Magnets*, Cambridge, 1750, pp. 17-18.

Aepinus was growing to scientific maturity, magnetic science continued to be dominated by one form or another of the theory of circulating subtle matter. Only a few voices had as yet been raised against the theory.

We know very little about Aepinus's own introduction to the study of magnetism. There is no doubt, however, that he learned early in his career that not everyone accepted the traditional theory without question. There are only two figures, Georg Erhard Hamberger and Leonhard Euler, whose work we can be sure Aepinus knew before he became immersed in his own magnetic researches around the beginning of 1757. Euler's views have already been discussed. Hamberger's are those that are relevant here. He, like Euler, was convinced that magnetism must have a material cause, but he was much less sanguine than Euler that the basic mechanism involved had been discovered. Like Euler, he took it as certain that the "primary efficient cause" was some kind of fluid that surrounded any magnet, but he insisted that "what this fluid is like, whether it is at rest or moving, and if it moves what its motion is like, and what is the disposition of the pores, and how the phenomena arise from these things, can be explained only hypothetically and not adequately at that." [59] In line with this, he himself refrained from setting out any elaborate hypotheses in his textbook. The most he was prepared to do in this regard was to refer to the writings of those other well-known textbook authors, Johann Christoph Sturm and Nicolas Hartsoeker, where different versions of the circulation theory were set out in some detail.[60]

It is impossible to say whether Hamberger's skepticism inspired Aepinus to doubt the circulation theory even as a youth. We do know, however, that once he took up the study of magnetism in earnest, he rejected that theory altogether; and what prompted him to do so was the analogy he perceived between a magnet and an electrified tourmaline. In place of the traditional ideas, he proposed a theory of magnetism completely analogous to Franklin's theory of electricity in the form to which he himself had newly elaborated it.

This was, as Aepinus admitted, an unusual angle from which to approach his subject. Most writers had accepted Gilbert's sharp distinction between electricity and magnetism, and had sought to account for the two classes of phenomena in quite different ways. Musschenbroek was typical in this respect. He raised the question

[59] Hamberger, *Elementa physices* (1741), p. 309.

[60] Ibid., p. 299. Hamberger's references are to Sturm, *Physicae electivae sive hypotheticae,* II, Nürnberg, 1722, and Hartsoeker, *Conjectures physiques,* Amsterdam, 1706.

briefly at one stage, but only to emphasize the differences involved. His strongest arguments were exactly the same as Gilbert's of over a century before. He pointed to the fact that magnetism was a property of iron and its compounds alone. If its cause were similar to that of electricity, he asked, "would not the magnetick effluvia attract and repel all kinds of bodies they meet with, as those of electrical bodies certainly do?" He also relied heavily on the fact that magnetic effects were not interrupted by matter other than iron placed between a magnet and the iron it was attracting, whereas electricity was disturbed by any intervening substance.[61]

Despite the general acceptance of such arguments, the fact that both electricity and magnetism apparently depended on invisible causes acting between separated bodies led to their being regularly coupled in the writings of the period. In 1748 the Académie Royale des Sciences, Belles Lettres, et Arts of Bordeaux even went so far as to conduct a prize competition on the theme, "Connections between the magnet and electricity." The winner was the Jesuit scholar Laurent Béraud, whose entry was published in the same year.[62] Unlike Musschenbroek, Béraud concluded that magnetism and electricity were related. "It is the same matter," he said, "but differently modified, that produces the magnetic and electrical phenomena." This matter was, he asserted, nothing other than the aether itself. So far as electricity was concerned, Béraud espoused a theory that was, in its details, very similar to Nollet's. Only in his choice of active matter did he differ significantly from the Paris academician. In the case of magnetism, he rejected grand Cartesian-style circulations in favor of static distributions of molecular or submolecular whorls in the aether. In a magnet, he said, there are large numbers of tiny aethereal whorls aligned with their axes pointing north-south in a series of parallel channels adapted to receiving them. There is, however, no continued flow of matter along the channels. At first glance, Béraud's scheme is reminiscent of the famous aether mechanism proposed by James Clerk Maxwell a hundred years later. The resemblance is, however, no more than superficial, and the mechanics of Béraud's scheme, so far as they are explored in any detail, are far from satisfactory.

Béraud was unusual in proclaiming so vigorously the identity of the causes of magnetism and electricity. However, his position was,

[61] Musschenbroek, *The Elements of Natural Philosophy*, I, 210–11.

[62] Beraut [sic], *Dissertation sur le rapport qui se trouve entre la cause des effets de l'Aiman, et celle des phenomenes de l'Electricité*, Bordeaux, 1748. See also Pierre Barrière, *L'Académie de Bordeaux, centre de culture internationale au XVIIIe siècle*, Bordeaux, 1951, p. 130.

at bottom, little different from that of any other aether theorist. Such people always tried to reduce all natural phenomena, including electricity and magnetism, to different modifications of the one underlying substance. The relationship proposed, whatever it amounted to in detail, was very different from that Aepinus put forward in his *Essay*. Far from wishing to identify the underlying cause of magnetism with that of electricity, Aepinus insists that the two are entirely different. What he is concerned to establish is not an identity but an analogy, and he goes out of his way to emphasize the point, remarking (§ 4) that "I in no way consider the magnetic and electric fluids as one and the same thing, as do those who toil to derive the phenomena of both electricity and magnetism, and many other things, from one single extremely subtle fluid, namely the aether. It is my supposition that these fluids are endowed with very different properties that are not compatible with a single underlying material." [63]

Aepinus recognizes that his stand might lead to Occamist-inspired accusations that he is multiplying entities without necessity, but he firmly rejects any such imputation, insisting that his position is based upon the "contemplation of nature itself." Yet he seems not to recognize that he is, by his choice of phrase, apparently negating what he said earlier, and ruling out the possibility of ever finding a mechanism to explain the various interparticulate forces upon which his analysis depends. It was no doubt Euler's recognition of this that led him to describe Aepinus's forces as "arbitrary," and other mechanists, too, must have shared his reservations. Orthodox Newtonians, however, would have found the form of Aepinus's argument familiar, and would have had little difficulty in accepting his conclusion.

Aepinus regarded the exposition of his revolutionary theory of magnetism as his primary task in the *Essay*. He places much less emphasis in the book on his contributions to electrical science, and, indeed, he frequently treats his electrical investigations as mere preliminaries to his discussion of corresponding situations in the magnetic realm. This he can do because in his hands the structures of the two theories are identical; the same theorems can be applied in both sets of circumstances. Magnetic phenomena, he supposes, arise from the action of a subtle "magnetic" fluid, analogous to the electric fluid in Franklin's theory. The particles of this fluid mutually repel each other, and can permeate the pores of ordinary matter. Most natural bodies, he says, exert no action at all on the magnetic fluid, but it is

[63] M. I. Radovskiĭ ("Issledovaniya Epinusa v oblasti elektromagnetizma," *Elektrichestvo*, 1940, 67–70) is quite wrong in seeing in Aepinus's work an anticipation of Oersted's famous discovery of electromagnetic action.

attracted by particles of iron and other ferreous materials. Such materials are, so far as the magnetic fluid is concerned, analogous to electrics *per se*, in that the magnetic fluid can pass through their pores only with considerable difficulty. The softer the iron, the less difficulty magnetic fluid has in moving through its pores. There are, however, no substances known, not even the softest of irons, that are truly analogous to electrical conductors; that is, there are no substances known whose particles attract the magnetic fluid while their pores give open passage to it.

The contrast between Aepinus's theory and the traditional explanations of the phenomena of magnetism is immediately apparent. Not for a moment does Aepinus need to have recourse to circulating streams of subtle matter. Instead, he reduces all magnetic phenomena to the net effect of forces, themselves unexplained, exerted by static masses of magnetic fluid on each other or on masses of ferreous matter. Just as in the electrical case, Aepinus's analyses amount merely to adding up the forces acting in any given set of circumstances. Once again his approach proves to be extraordinarily successful.

The notion of a "natural" quantity of fluid arises in Aepinus's magnetic theory in the same way as it does in his theory of electricity. So, too, does the idea that iron can become magnetized in two different ways, "plus" and "minus," according to whether it contains more or less than its natural quantity of fluid. One part of an ordinary magnet is, according to Aepinus, always magnetized "plus," while the other is magnetized "minus" (he recognizes that, though "unipolar" magnets are in principle possible in his theory, they do not normally occur in nature). In Aepinus's theory, a magnet is analogous in many ways to a Leyden jar, and many of the theorems he has developed for this can be applied immediately to the magnetic case.

In the electrical case, Aepinus concluded that it was the insulating character of the surrounding air that allowed a charge to remain on a body. In the magnetic case, however, he cannot avail himself of this device, since neither air nor any other surrounding medium except iron is supposed to offer any barrier to the magnetic fluid. To account for the continued existence of magnetic "charges," Aepinus invokes the difficulty with which magnetic fluid moves through the pores of ferreous bodies. He argues that once a magnet has been formed, the excess of fluid at one end will have great difficulty in moving through the iron to the other end to make up the deficiency there, and the resistance it experiences will be the greater, the more hardened the iron that is used. Another possibility is that fluid might flow out from the end magnetized "plus" into the surrounding medium, while other fluid flows from the surroundings into the end magnetized "minus." To counter this, Aepinus points out that the presence of an opposite pole nearby, namely at the other end of the bar, greatly reduces the force giving rise to such a flow. The dipole arrangement will therefore last for some time—and longer, of course, the harder the iron that is used—even though a magnetic monopole, were such a thing to be formed, would decay very rapidly.

On the other hand, even a dipole will decay in time, and indeed, if the initial degree of magnetization is so high that the dissipative force is greater than the resistance offered to the magnetic fluid by the ferreous matter concerned, there will be a rapid decrease in magnetization as soon as the magnetizing force is removed, until the point is reached where the dissipative force again becomes less than the resisting one. Reasoning in this way, Aepinus is led to the notion of magnetic *saturation,* which he defines as "the degree of magnetic force which can subsist for a reasonable length of time without sensible diminution in a magnetized body left to itself, and beyond which, if the magnetic force is increased, the excess is quickly destroyed" (§§ 86, 204). Obviously, the degree of magnetization required for saturation, so defined, will vary from substance to substance, being least in soft iron and greatest in hardened steel.

As in the electrical case, Aepinus's analyses with regard to magnetism are mathematical in form. Once again, however, he is unable to advance beyond the semi-quantitative. Because of his ignorance (which he again freely confesses) of the law according to which the forces involved vary with distance, he again cannot provide precise quantitative predictions, but only increasingly detailed and specific qualitative ones. Yet here, unlike the situation with respect to electricity, Aepinus had behind him a long tradition, stretching back at least to the great Newton himself, of attempts to discover the form of the force law. Nevertheless, Aepinus's assessment was correct. The efforts of his predecessors had proved unavailing. Their results were vitiated by a failure to take the dipole character of magnetism into account, and, in addition, inductive effects often gave a random appearance to the results obtained. Only with the work of Coulomb, in the generation after Aepinus, was the law finally established on a satisfactory basis.

Newton's attempted determination of the law of magnetic force is recorded in a single sentence in the *Principia,* at a point where he wishes to draw a distinction between the power of gravity and that of the magnet. The latter, he says, "in receding from the magnet decreases not as the square but almost as the cube of the distance, as nearly as I could judge from some rude observations." Newton gives no indication of the experimental procedure he adopted, but it is clear from the wording he uses that he was not trying to measure the force between two magnetic poles, as Coulomb subsequently did, but rather the attractive force exerted by a given magnet as a whole on a piece of iron placed at various distances from it. It is little wonder, therefore, that he failed to obtain Coulomb's inverse-square relationship.[64]

Newton's report was not originally included in his book, but was added only in the second (1713) edition of the work.[65] By the time the report appeared in print, the Royal Society had heard other reports, from both the elder Hauksbee and the young mathematician Brook Taylor, of efforts to discover the law. Hauksbee and Taylor initially experimented together with the great lodestone belonging to the Society, but for some reason they later presented separate reports, including data obtained with two different magnets. Initially, only Hauksbee's paper was published in the *Philosophical Transactions*. Following Hauksbee's death a few months later, however, Taylor, evidently somewhat aggrieved by what had happened and believing that his data were superior, submitted a new report, which was published in the *Philosophical Transactions* soon afterward.[66]

In fact, neither Hauksbee's paper nor Taylor's as originally published was very helpful. Both papers merely presented sets of raw data, without any attempt to extract a functional relationship from the numbers. However, Taylor did take up the question elsewhere, both in his initial report to the Royal Society and again, in very similar terms, in a letter to Sir Hans Sloane dated 25 June 1714 that was

[64] Newton, *Mathematical principles*, p. 414. The force exerted by a magnetic dipole on a single magnetic pole varies approximately as $1/r^3$ at distances large with respect to the length of the dipole. However, it was not even this that Newton was observing, but something much more complicated, in effect the resultant force acting on a *pair* of poles induced in the piece of iron, poles whose strengths themselves vary with distance, and depend as well on the hardness and geometry of the iron used, and on the positioning of the iron with respect to the axis of the dipole. Newton's and several other eighteenth-century attempts to discover the law of magnetic force are discussed in L. Tilling, "The Interpretation of Observational Errors in the Eighteenth and Early Nineteenth Centuries" (unpublished Ph.D. thesis, University of London, 1973), pp. 28–57.

[65] Alexandre Koyré and I. Bernard Cohen, eds., *Isaac Newton's Philosophiae naturalis principia mathematica: the Third Edition (1726) with Variant Readings*, Cambridge, 1972, II, 576.

[66] Hauksbee, *Phil. Trans.*, XXVII, 1710–12, 506–11; Taylor, ibid., XXIX, 1714–16, 294–95. Taylor's original report is in the Royal Society's *Letter Book original*, XIV, 394.

eventually published in the *Philosophical Transactions* in 1721.[67] What he had to say is most revealing of the conceptual difficulties involved in the subject.

Taylor, like Newton (and like Hauksbee, too, for that matter) was concerned with the "power" of a magnet and the way it varied with distance. "Power" again meant simply the general attractive power exerted by the magnet on a piece of iron, not the much more refined notion of the force exerted by one isolated magnetic pole on another. There is a clear echo here of Gilbert's well-known distinction between the attractive power of a magnet and its directional properties. The poles are seen not so much as centers of magnetic force, but simply as the ends of the axis according to which the magnet aligns itself with respect to the Earth or another magnet. Taylor complains that "if it were known what point within the Stone, and what point in the Needle [it attracts] are the Centers of the Magnetical power, it would be easy to find the true powers of the Magnet at all the distances observed," but for want of such knowledge he has had to make arbitrary assumptions about the locations of these points. When he computes the forces from the center of the needle and "the Extremity of the Loadstone," he cannot find a regular law, and so he tries to work backward and find points that he might use as the centers of power, such that "the Law, by which the Magnetism alters, could be reduced at all distances to any one certain power of those distances." But he finds that for this to be the case, the center of the lodestone's attractive power must be assumed to lie outside the stone! Hence he is driven to the dissatisfying conclusion that "the power of Magnetism does not alter according to any particular power of the distances, but decreases much faster in the greater distances, than it does in the near ones." Whiston, on the other hand, concluded that the difficulty arose from insufficient care having been taken to maintain a constant orientation of the needle with respect to the magnet during the experiment. His own results showed, he said, that there was a definite law, and indeed, that it was sesquiduplicate in form.[68]

Problems similar to those that had troubled Taylor plagued Musschenbroek's attempts to determine the law, and for similar reasons.[69] Nevertheless, the Dutchman's experiments were for many years regarded as definitive. Musschenbroek's apparatus consisted of a balance, from one arm of which a small magnet was suspended directly

---

[67] Taylor, *Phil. Trans.*, XXXI, 1720-21, 204-208.
[68] Whiston, *The Longitude and Latitude*, pp. 13-15.
[69] Musschenbroek, *Phil. Trans.*, XXXIII, 1724-25, 370-78; also *Dissertatio . . . de magnete*, pp. 13-23, and *Elements of Natural Philosophy*, I, 205-208.

above a larger fixed magnet. Weights could then be added to the pan on the other arm to counterbalance the force between the two magnets. To enable this force to be measured at different separations of the two magnets, the entire balance could be raised or lowered by a pulley (see Fig. 4).

Because Musschenbroek's experiments involved two magnets, rather than a single magnet and a piece of iron, we would expect him to take more account of the polarities of his magnets than his English predecessors had done. Sure enough, he arranged his two magnets in such a way that their axes were aligned, normally with a pair of unlike poles facing each other. It is perhaps significant, however, that in his initial description of his experimental arrangement, he did not mention this; the point only becomes clear later in his paper, when he describes some variations on the original arrangement.[70] In any event, Musschenbroek's apparatus was obviously not designed with any thought of isolating and measuring the forces acting between individual poles. It simply measured the total force exerted by one magnet on another. If unlike poles were facing each other, this force was an attractive one, if like poles were adjacent, it was repulsive, and the apparatus served equally well to measure either. Not surprisingly, Musschenbroek's initial results gave no more hint of a regular relationship between force and distance than Taylor's had done: "I can only conclude," he said, "that there is no proportion between forces and distances."[71] Later, however, Musschenbroek claimed to have discovered definite laws of magnetic action, but a different law in each different set of circumstances. If he used a cylindrical magnet and an identical unmagnetized piece of iron, the force was, he claimed, inversely as the space between the bases of the cylinders. If he used a spherical magnet in the end-on position to a cylindrical one of the same diameter, the force was inversely as the sesquiplicate of the empty space enclosed in a cylinder bounded by the base of the cylindrical magnet and a mid-plane through the spherical one (i.e. if this space was v, the force was as $v^{-3/2}$). A spherical magnet attracted an iron ball of the same size inversely as the fourth power of the space between them.[72]

[70] Musschenbroek, *Phil. Trans.*, xxxiii, 1724–25, 371–72, 375.

[71] Ibid., p. 374.

[72] Musschenbroek, *The Elements of Natural Philosophy*, I, 205–208. Benjamin Martin obtained the sesquiplicate ratio using a lodestone (presumably spherical) and an iron bar of square cross-section (Martin, *Philosophia Britannica*, 2nd ed., p. 47). The Dublin professor Richard Helsham, using slightly different apparatus, found the force between a lodestone and a flat sheet of iron to be inverse-square in

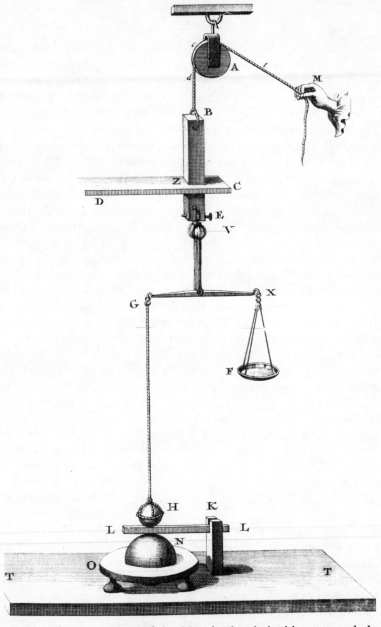

Figure 4. The apparatus used by Musschenbroek in his attempted determination of the law of magnetic action (Tab. I, Fig. 2 in his *Dissertatio physica experimentalis de magnete* [1929]). The rather complicated suspension shown linking the balance arm to the rope over the pulley was designed to prevent the apparatus twisting to and from above the large magnet NO.

A rather different experimental arrangement was described in a footnote in the copiously annotated edition of Newton's *Principia* brought out by Thomas Le Seur and François Jacquier.[73] Here, the quantity measured was the deflection of a compass needle from the magnetic meridian produced by a magnet placed in the plane of the needle, at some distance from it on a line through its center and perpendicular to the meridian. The couple on the needle due to the presence of the magnet was therby determined, and was found to vary inversely as the third power of the distance.

Unlike all these early investigators, John Michell made it clear that he recognized the need to treat the two poles of a magnet as separate centers of force.[74] His work thus constitutes a landmark in the history of the subject. "Each Pole," Michell asserted, "attracts or repels exactly equally, at equal distances, in every direction." It was, he said, the failure of Taylor, Musschenbroek, and others who had experimented on the subject to recognize this fact that had prevented them from finding the true law of magnetic action. He, however, had succeeded in discovering it. "The Attraction and Repulsion of Magnets decrease," he announced, "as the Squares of the distances from the respective Poles increase"; at least, he thought this relationship "very probable," even though he did "not pretend to lay it down as certain, not having made experiments enough yet, to determine it with sufficient exactness." Unfortunately, Michell gave no indication of what experiments he *had* performed, and so his readers were left with no more than his unsubstantiated assertion to go on. Hence his discovery of the law—and it undoubtedly was a genuine discovery—was never regarded as definitive. Aepinus, for one, was obviously unimpressed by it. Though he knew Michell's work well and refers to it frequently, not once does he mention this aspect of it, despite the fact that so far as his own theory is concerned Michell had at least been trying to measure the appropriate quantity, whereas no one else had been.

---

character (Helsham, *A Course of Lectures in Natural Philosophy*, London, 1739; 2nd ed., London, 1743, pp. 19–20). In St. Petersburg, G. W. Krafft confirmed Musschenbroek's result for two equal spheres, but his efforts to extend the rule to spheres of differing dimensions proved unavailing (Krafft, *Commentarii Academiae Imperialis Scientiarum Petropolitanae*, XII, 1740, 276–87).

[73] *Philosophiae naturalis principia mathematica*, auctore Isaaco Newtono . . . perpetuis commentariis illustrata communi studio PP. Thomae Le Seur et Francisci Jacquier, Geneva, 1739–42, III, 40–42. An English translation of the relevant footnote is included in Robert Palter, "Early measurements of magnetic force," *Isis*, LXIII, 1972, 544–58, where it is also suggested that the author of the note was the Genevan mathematics professor J. L. Calandrini (1703–1758).

[74] Michell, *A Treatise of Artificial Magnets*, pp. 17–20.

Strictly speaking, when Aepinus refers to the law of magnetic force, he means the force acting between particles of magnetic fluid or between particles of fluid and particles of ferreous matter. Given the structure of his theory, however, this can be determined directly by measuring the force between a pair of magnetic poles. As indicated already, Aepinus is not prepared to commit himself to any particular form of the law, Michell's work notwithstanding. As in the electrical case, the furthest he is prepared to go is to concede that "the analogy of nature militates on behalf of" the inverse-square form.

The forces that appear in Aepinus's theory of magnetism are identical in form with those he invoked in the electrical case. Hence it follows automatically that he must introduce into his magnetic theory a repulsive force between the particles of ferreous matter, in just the same way as he had to conclude, when developing his electrical theory, that the particles of ordinary matter repel each other. Euler's charge that Aepinus's forces are "arbitrary" perhaps finds its justification here more than anywhere else in the work. Ferreous bodies are, of course, subject to electrical as well as magnetic effects. Aepinus is therefore committed to the notion that the particles of such bodies exert two separate forces of repulsion on each other, while at the same time they also continue to attract each other with the usual gravitational force. Operationally, the two repulsive forces *must* be carefully distinguished. One of them can be neutralized by the presence of electric fluid, the other by the presence of magnetic fluid. But the presence or absence of magnetic fluid cannot affect the electrical behavior of a piece of iron, nor can the presence or absence of electric fluid affect its magnetic properties.

As remarked earlier, Aepinus is able to transfer many of the theorems he developed in his discussion of the Leyden jar to the magnetic case. But whereas in the electrical case the opposite charges on the two sides of the device could be and indeed always were unequal in magnitude, in a magnet they are, he says, generally equal. This is so, he argues, because magnetization almost always arises from a redistribution of fluid within a piece of iron or other ferreous matter, rather than an interchange of fluid with external bodies. "On account of the very great difficulty with which the fluid moves in the pores of ferreous bodies," he says, "it is scarcely possible for any magnetic matter to enter a body of this kind from outside by any means, or for it to be evacuated from it." Yet it must be confessed that it is difficult to see how fluid can redistribute itself within the body of the iron, if its motion is resisted as powerfully as this. Aepinus appears to be unaware of the problem, but it is nevertheless one that emerges at several points in the development of his theory. It was eventually dissolved by Coulomb, with his assumption that magnetism is essentially a molecular phenomenon in which magnetic fluid is confined within each molecule of the iron. Fluid can, Coulomb supposed, redistribute itself within each molecule, as it does when the iron becomes magnetized, but it cannot flow from one molecule to the next. Whatever the cause preventing this flow—and Coulomb forbore from speculating on the subject—it is clearly no longer the same cause as that which hinders the flow of fluid within each molecule.[75]

Difficulties emerge for Aepinus when he considers what happens when a magnet is broken in two. In Aepinus's day it was already well known that when this was done new poles always appeared at the point of separation, in such a way that each piece of the original magnet acquired its own pair of opposite poles. On the other hand, from Aepinus's theory it appears that very often the piece broken off ought, rather, to have a single polarity: "plus" if it is broken off the end originally magnetized "plus," "minus" if broken off the other end. Benjamin Franklin, one of the first recorded converts to Aepinus's theory of magnetism, saw this as a grave weakness in his account, and Coulomb, too, drew attention to it and in fact used it as his justification for supposing that magnetization was essentially a molecular rather than a macroscopic phenomenon.

Yet Aepinus himself is not unaware of the problem, and he discusses it at some length in his book (§§ 96ff.). His theory is, he concludes, well able to cope with the situation. However, Aepinus's explanation depends on an assumed ability of magnetic fluid to flow from the piece cut off into the surrounding medium (or in from the surroundings if the piece is magnetized "minus"). He is therefore forced to posit this kind of situation as an exception to his previously stated rule that fluid is generally confined within each individual sample of iron. Aepinus argues that when a positively magnetized piece is separated from a magnet, the force tending to dissipate fluid from it into the surroundings becomes much greater than it was before, since the negatively magnetized further end of the original magnet is no longer there to counter the dissipative tendencies of the fluid itself. Hence it is possible that fluid will flow into the surroundings, and it may even do so to the extent that part of the piece of iron becomes magnetized negatively. Whether it will actually do so cannot, he says, be predicted from the theory, but can only be determined by experiment. A similar condition applies if the piece is cut from the negatively

[75] Coulomb, *HAS*, 1789, Mém. pp. 488–90.

magnetized iron; it is possible that fluid will flow in from the surroundings, but only experiment can tell whether it will actually do so, and to what extent.

It comes as no surprise that, when Aepinus carried out the test, he found that a flow does indeed take place, nor that he found that it does so to such an extent that part of the excised piece acquired an opposite magnetization to that it had before. What is interesting is the way he managed to devise experiments that add considerably to the plausibility of his analysis, even in an instance such as this where his theory is notoriously at its weakest. In each of his experiments, Aepinus carefully determined the position of what he calls the "magnetic center" of the excised piece of iron, that is, the neutral point separating its positively and negatively magnetized zones. He does not, in fact, say in his *Essay* how he located this point, but we learn elsewhere that he used iron filings patterns, locating the magnetic center at the center of symmetry of the curves traced out by the powder.[76] In a typical experiment, immediately after separation the magnetic center of the excised piece is some distance from its geometrical center. Aepinus found, however, that in time it migrates closer toward the geometrical center, and he interpreted this as indicating just what he had expected, a continued slow flow of fluid from the iron. He also found that the magnetic center can be made to migrate further toward the geometrical center by various processes that, according to his theory, open up the pores of the iron somewhat. This, he says, enables the magnetic fluid to flow more freely, and hence more of it to escape from the iron. The processes he employed were to hurl the iron repeatedly against a stone pavement and to heat it. The migration he observed is, he argues, exactly what should happen, according to the principles he has enunciated.

Later in his book, Aepinus uses similar experiments to support the astonishing claim that under certain circumstances a short piece of iron can be given the same kind of magnetism along its entire length! He does so while examining the way a magnet induces magnetism in nearby pieces of iron. He suggests that if the magnet is sufficiently powerful, and the piece of iron sufficiently short, the magnetic fluid that would normally move to one end of the iron might be driven entirely out of the iron. If this should happen, he points out, the iron would be left with the one kind of magnetism throughout. He then reports that when he put the matter to the test, this is exactly what happened; more specifically, as the powerful magnet he was using

---

[76] Aepinus, *Acta Acad. elect. Mogunt.*, II, 258–59.

was brought closer to a short piece of iron wire, the magnetic center in the latter moved away toward the further end of the wire, until eventually he was no longer able to find a magnetic center anywhere in the wire (§ 189).

The magnetic properties of a piece of iron are often exceedingly complex, different domains within the iron being able to behave relatively independently of each other. The writings of eighteenth-century magnetic investigators clearly reflect the complexities involved, and nowhere more so than in their frequent references to what they called "consequent points." Here, as in so many other aspects of eighteenth-century magnetic research, Musschenbroek's work served as the starting point for most later investigations, though he was not, in fact, the first to record the phenomenon.[77] He reported, and others confirmed, that when magnetized bars of iron, especially longer bars, were tested at various points along their surface with a small compass needle, they often exhibited not just the normal pair of opposite poles at their two ends, but a succession of alternate poles along their entire length.

Aepinus was one of those who confirmed Musschenbroek's report, and in his book he attempts to account for the phenomenon in terms of his general theory (§§ 200–201). His proposal is not very plausible, however, since it involves some ad hoc assumptions. He considers what should happen if fluid is subjected to some external force tending to drive it along a bar. He supposes that a mass of fluid from one end will travel a little way along the bar, thereby leaving a negative region behind it and creating a positive region in its new location. But, he says, "the fluid will have to stop here for a moderate period of time at least, since because of the difficulty with which it moves through the pores of the body it cannot flow swiftly through the remaining parts of the body." Thus stopped, it in turn will act like an ordinary positive body and repel fluid in the next section of the bar, fluid that will then gather a little further along to create another positive region. And so on. The ad hoc character of Aepinus's assumption is evident, but so too are the reasons for it. The phenomena in question are simply too difficult for any eighteenth-century or even any nineteenth-century theorist to handle. Only in relatively recent times has any significant progress been made.

Of central importance in Aepinus's magnetic theory, as in his theory of electricity, is the notion of induction. Just as in the electrical case,

[77] Musschenbroek, *Dissertatio . . . de magnete*, pp. 143ff. Brook Taylor had described the same phenomenon somewhat earlier (*Phil. Trans.*, XXXI, 1720–21, 206–207).

any inequality in the distribution of magnetic fluid in one body necessarily gives rise to inequalities in neighboring bodies, and hence the mere presence of a magnet is always accompanied by inductive effects. Naturally, many of Aepinus's theorems about electrical induction developed earlier carry over directly into the magnetic realm. Once again Aepinus proves remarkably skillful at devising experiments to test these theorems in practice.

It follows from the theory, for example, that a magnet will neither attract nor repel a piece of iron so long as it remains completely unmagnetized, but also that any ordinary piece of iron in the vicinity of a magnet will have a pair of poles induced in it, and will therefore be attracted on their account. Aepinus easily shows by experiment that, when iron is attracted by a magnet, it has magnetism induced in it. But this is not sufficient; he also needs to show that the induced magnetism is essential for the attraction to occur. To this end, he designs further simple but effective experiments in which the normal induction is inhibited, and he finds that then the attraction no longer occurs (§§ 159ff.).

Aepinus's theory also leads him to some important conclusions about the action of two magnets on each other. He recognizes that each magnet will have inductive effects on the other, so that neither will remain in its original condition. In addition, these effects will be greater, the closer together the magnets. They will be greater, too, the more freely magnetic fluid can move within each magnet, that is, the softer the iron of which each is made.

Taking the last of these points, for example, Aepinus is led to the interesting and somewhat counterintuitive conclusion that, with unlike poles facing each other in each case, a given magnet will sometimes attract a weaker magnet more powerfully than it will a stronger one (§§ 173–74). This is possible, he argues, provided that the weaker magnet is made of softer iron than the stronger one, for then the additional magnetism induced in the weaker one may be sufficient to overcome its previous deficiency. Once again, Aepinus asserts, experiment bears him out completely. He is obviously particularly pleased with his deduction on this occasion: "So it has been my lot again," he boasts, "to give with the aid of our theory a simple explanation for a phenomenon which those who knew of it through experience alone could have found nothing but extremely difficult." And well he might boast, since the ability to generate unlikely conclusions that turn out to be true is the mark of a highly successful theory.

When Aepinus turns to the case where like poles of two magnets are presented to each other, he again obtains an interesting result. Common wisdom had it, of course, that two like poles repelled each other. However, Musschenbroek had found that, in some of his experiments, an attraction occurred.[78] Aepinus makes no mention of this, but his computations provide him with a ready explanation for what the Dutchman had observed. They also enable him to specify, at least in general terms, the circumstances in which an attraction ought to occur. In effect, what is needed is for the inductive power of one of the magnets to be so great as to force a reversal of the poles of the other. Aepinus points out that bringing the two magnets closer together should increase the likelihood of this happening, as should using an especially powerful magnetic force, and making the magnets of softer iron. Using a compass needle and a strong magnet, he is again able to provide experimental confirmation for his analysis (§ 181).

Aepinus by no means restricts his attention to phenomena such as these. Indeed, the main thrust of his discussion is toward a much more practical problem, namely to improve the methods of fabricating artificial magnets. As we remarked earlier, these methods had been revolutionized a few years before by the work of Dr. Gowin Knight. Knight kept his technique secret, but reports of his successes inspired others to take up the subject for themselves, and ushered in a period of rapid progress.

An initial report triggered by Knight's work was read to the Paris Academy of Sciences in February 1745 by the academician Henri-Louis Duhamel du Monceau.[79] In this paper, Duhamel set out the results of research he had undertaken with one Le Maire, "Ingénieur pour les instrumens de Mathématique." The traditional method of magnetizing a piece of iron had been simply to rub it with one of the poles of a lodestone. Duhamel and Le Maire employed the simple stratagem, suggested by Le Maire, of holding the piece of iron on a larger magnetized strip of iron while it was rubbed. By this means, Duhamel reported, they were able to produce artificial magnets as strong for their size as the two samples of Knight's wares they had available to them.

Duhamel returned to the subject a few years later, prompted by news that Knight had succeeded in producing still stronger magnets.[80] On

---

[78] Musschenbroek, *Dissertatio . . . de magnete*, pp. 33–34. Musschenbroek reports that the same phenomenon had been observed earlier by Pierre Polinière, and refers to this author's *Expériences de physique*, p. 279.

[79] Duhamel, *HAS*, 1745, Mém. pp. 181–93.

[80] Duhamel, ibid., 1750, Mém. pp. 154–65.

this occasion, the Frenchman was working with Antheaulme as well as Le Maire, and he was able to report the production of magnets even stronger than those newly received from England. The procedure now followed by Duhamel and his companions was rather complicated. For our purposes, however, the central part of the operation was to place an already magnetized bar at each end of the bar to be magnetized and in line with it, in such a way that a north pole was adjacent to one end and a south pole to the other, and then to stroke the middle bar with two other magnetized bars, the stroking being done with two opposite poles moving simultaneously from the middle of the bar to its two ends, each pole moving toward the pole of like name already placed there.

Meanwhile, in England, an even more effective method was discovered by John Michell in Cambridge and, independently, by John Canton in London.[81] Michell gave the method the name "the double touch." Like Duhamel's method, it consisted of stroking a bar of iron with a pair of opposite magnetic poles. But now, instead of drawing the poles apart at each stroke, they were held a small fixed distance apart and moved together back and forth several times along the iron. For convenience, the upper ends of the stroking magnets were bound together, while a small piece of wood or something similar was clamped, in order to keep them slightly apart, between the pair of poles that actually did the stroking. Michell's treatise emphasized the need to employ high-quality hardened steel if the best results were to be achieved, and the advantages of placing other bars of iron (which he called "supporters") at each end of the bar while it was being stroked.

Canton's report was very brief and did not indicate whether theoretical considerations had guided his research at any point. In Michell's case they clearly had, particularly in connection with the use of "supporters." Michell saw that the effects of the intense magnetizing force in the region between the stroking poles would be somewhat negated by a force in the opposite direction in the regions outside the two poles, and he felt that the "supporters" would counter this to some extent.[82] As indicated earlier, although Michell said enough to reveal that he rejected the traditional circulating-fluid theory altogether, he scarcely more than hinted at his theoretical

---

[81] Michell, *A Treatise of Artificial Magnets,* Cambridge, 1750; Canton, *Phil. Trans.,* XLVII, 1751–52, 31–38. Michell in fact accused Canton of appropriating his ideas without acknowledgment, but the charge was almost certainly without foundation, and was so seen by people at the time (Clyde L. Hardin, "The scientific work, of the Reverend John Michell," *Annals of Science,* XXII, 1966, 24–47).

[82] Michell, *Treatise,* p. 33. Cf. below, § 241.

views anywhere in his *Treatise*. By contrast, Duhamel evidently had no doubts at all about the circulation theory, and he used its vocabulary as a matter of course throughout his discussion. So too did Antheaulme in the essay that won the St. Petersburg Academy's prize in 1759. Indeed, Antheaulme even hinted that it was this theory that had led him to a newly improved version of the double touch method described in his paper.[83]

It is to Antheaulme's work that Aepinus graciously defers in the preface to his *Essay*. Both men discovered quite independently a way of obtaining even better results than those achieved by Michell and Canton. Their modification was a simple one. They recommended that instead of holding the pair of stroking magnets together and approximately vertical while magnetizing by the double touch method, these should be tilted away from each other until they were almost horizontal, and thus almost lying on the bar being magnetized. With the magnets held in this position, however, the ends actually doing the stroking should still, they said, be held a constant distance apart throughout the entire stroking process.[84]

Antheaulme's claims notwithstanding, theory and practice were not nearly so closely integrated in his account as they are in Aepinus's *Essay*. In the latter, they are intimately connected at every stage. For example, Aepinus is able to account with ease for the common but counterintuitive observation that a magnet is able to communicate its power to a nearby piece of iron without losing any of its own strength (§ 208). In Aepinus's terms, "communication" of magnetism is merely induction; that is, the mere presence of the magnet causes a redistribution of fluid in the iron, and so imparts magnetic power to it. But in this, "there is no cause which could lessen the force of the magnet." Indeed, as Aepinus points out, the whole process ought, if anything, to *increase* the power of the magnet, since the induced magnetism in the iron ought to tend in its turn to increase the existing disequilibrium in the distribution of fluid in the magnet. Once again, Aepinus is able to point to experimental evidence to support his position, namely the observation that "the strength of a magnet is increased slowly but significantly by iron placed close to it." Elsewhere, a similar line of argument enables him to account for Réaumur's remarkable observation that a magnet can lift a greater weight of iron if this is itself placed on something made of iron, for example an anvil (§ 167).[85]

[83] Antheaulme, *Mémoire sur les aimants artificiels,* Paris, 1760, pp. 25–26.
[84] Ibid. See also below, §§ 252ff.
[85] Cf. Réaumur, *HAS,* 1723, Mém. p. 126.

Still more impressive, however, is Aepinus's theoretical analysis of the existing methods of magnetizing iron, including the double touch (§§ 218ff.). One by one he identifies the strengths and limitations of these methods. He shows why the method of the double touch gave such dramatically improved results over those that had been achieved earlier, and why the "supporters" advocated by Michell still further improved things. He describes methods for minimizing the production of consequent points and explains why they are effective. Above all, he shows why, in terms of his general principles, the modifications he is suggesting—including the crucial one also devised by Antheaulme—lead to an improved result. Throughout, theory and experiment progress hand in hand in a thoroughly controlled manner.

Aepinus is well aware of the practical benefits of the innovations he is proposing, and in particular he refers on several occasions to the relevance of his work to the improvement of magnetic compass needles. He does not pursue this question at any length in the *Essay*, instead promising his readers a separate book devoted to the subject. The promised work never appeared, but Aepinus did take up the question again, though only briefly, in three less formal papers published at about the same period.[86] In these, he particularly stressed the need to magnetize compass needles in such a way that their magnetic centers coincided with their geometrical centers. An elementary argument was enough to show that, if these two points did not coincide, the directive power of the needle would be seriously impaired, no matter how strongly it was magnetized. Testing a number of commercially produced needles, he found that, as he had more than half-expected would be the case, in most of them the two centers far from coincided. Fortunately, the remedy was to hand. The method of the double touch, Aepinus pointed out, whether in its original form as devised by Michell and Canton or in the improved version that he was advocating, automatically insured, if carried out correctly, that the magnetic center was in the center of the needle. Nevertheless, he was not optimistic that his recommendations would be heeded. "Perhaps half a century will pass," he sadly remarked, "before [my studies] are used to some advantage. It is not an easy task to draw people's attention toward the common good of humanity." [87]

The final chapter of Aepinus's *Essay* is chiefly devoted to the behavior of small magnets in the vicinity of large ones, particularly with reference to the Earth as a whole. Like most other writers on the subject in the century and a half since Gilbert, Aepinus accepts that

[86] See Appendix below, items **31A(i), 31B, 33.**
[87] Aepinus, *Sochineniya i perevody*, November 1758, p. 506.

the Earth is a large magnet; more precisely, he takes it that "there is included within the globe of the earth quite a large core made of matter which is ferreous in character, and imbued with magnetic force." However, he takes seriously the objection that, though the Earth clearly exerts a directive influence on any freely suspended magnet, it does not attract this, or repel it if its poles are aligned the other way, nor does it attract a piece of unmagnetized iron with any other force than simple gravity. Aepinus first describes some experiments that confirm that there is no attraction. With this established, he undertakes a theoretical analysis of the situation, and shows that the objection is nevertheless not a valid one. "Nothing should be feared from it," he concludes; "rather, it . . . is in our power to show that everything happens just as it ought under the hypothesis of a magnetic core."

The analysis is a lengthy one, which Aepinus illustrates with a numerical example in order to bring out his point more clearly. For simplicity, he also assumes an inverse-square force law throughout. However, the argument is not, in terms of Aepinus's principles, a difficult one. He simply demonstrates that if two similar and equal magnetic needles are placed in relation to two magnets, one large and the other small, at distances such that the directive influences they experience are equal, the larger magnet attracts its needle with a much smaller force than that with which the smaller magnet attracts the needle on which it is acting. In other words, "one cannot draw any conclusions about the strength of the attractive force exerted by a magnet from observations of the strength of its directive force." Nor, he goes on to show, can one draw any conclusions about the strength of the attractive force from the strength of the magnetism induced in a non-magnetic iron rod. He proves that a rod may well have a sensible magnetism induced in it by a distant large magnet, yet not be sensibly attracted by it. More generally, then, one may distinguish the "sphere of activity" of a magnet from its "sphere of attraction," as "some rather diligent observers of the magnet" had already done.[88]

Though Aepinus accepts that the Earth has a magnetic core, he, unlike those working in the Cartesian tradition, offers no explanation. Rather, he regards it as one of those phenomena "for which we can recognize no efficient mechanical cause, and which must be derived from the immediate action of the creator of the world" (§ 289). Here, much more explicitly than earlier in the *Essay,* we see how Aepinus has finally abandoned the hope of explaining all electrical and mag-

[88] Cf. Gilbert, *On the Magnet,* pp. 103–104.

netic phenomena in wholly mechanical terms. Like Newton, Aepinus accepts that many irreducibly non-mechanical phenomena occur that can only be explained in terms of God's immediate activity in the world. Indeed, he implies that to maintain otherwise—to insist that every natural event must have a mechanical cause—verges dangerously on the atheistic. "It cannot be denied that there are innumerable such phenomena," he says, "unless one dares to affirm that the whole world has been formed according to mechanical laws alone and without any concurrence of some intelligent being, or to take the position either that God does not exist at all, or that he has taken no part in the production of the world except as an inert spectator." But he is not prepared to accept these alternatives: "not only *a priori* but also *a posteriori* we are certain that no cause of the phenomena can be recognized other than the action of the very wise creator of the world." On the other hand, Aepinus freely confesses his ignorance concerning God's purpose in imbuing the Earth with magnetic force. He explicitly rejects the simplistic utilitarian explanation that God made the Earth magnetic to help men find their way over the world's great oceans. "The belief that all things are created solely for the utility of man," he proclaims, "when admitted without restriction, has stained with many errors that most noble part of Physics which deals with the ends of things."

Aepinus considers at length in his book the orientation of a compass needle with respect to the Earth or any other large magnet. Traditionally, this had been explained in terms of the streams of magnetic fluid that circulated around the magnet forcing the needle to align the channels passing through it with the direction of flow. A similar explanation had been offered for the well-known patterns formed by iron filings sprinkled on a sheet of paper held over a magnet; indeed, these patterns had often been presented as an obvious visual proof of the existence of the supposed streams of magnetic matter in the vicinity of a magnet. Aepinus, however, succeeds in reducing both effects to the joint action at a distance of the two poles of the magnet in question. He in fact defines a series of curves, "lines of force" as we would call them, in the vicinity of a magnet, such that "no matter where the center of the needle is located on this curve, the needle disposes itself of its own accord so as to lie on the tangent to this curve" (§ 297). Since iron filings, when scattered near a magnet, themselves become magnetic by induction, they too will align themselves with these curves, and so reveal the shape of them by the patterns they form. To illustrate the argument, Aepinus assumes for a moment that the magnetic force law is one of simple inverse proportionality, and that the force of each half of the magnet is concentrated at a single point somewhere within it. He shows that in these circumstances the curves will be circles passing through the points at which the magnetic power is concentrated. The patterns actually formed by iron filings around a magnet are not exact circles, he says, because the force law he has assumed is in fact not the one that operates in nature; but "for the rest, it is evident that everything else will apply, whatever the function for the magnetic actions."

Aepinus is, of course, well aware that both the magnetic declination and the dip vary with time at any point on the Earth's surface. He mentions Halley's attempt to account for this, but holds that Halley's theory implies that isogonic lines will always retain the same form and will merely move slowly around the surface of the Earth. Aepinus maintains that there is as yet insufficient evidence to decide the matter, but he is clearly very skeptical. Later, he speculates that the variation might have a quite different cause: "perhaps," he suggests, "the principal cause of the variations observed in the position of the magnetic needle should be sought in the generation and destruction of magnetic ores, and in the other mutations which can occur to these" (§ 364). With this in mind, it is hardly surprising that he never attempted to construct a detailed mathematical theory of the variation of the kind developed at about the same time by both Euler and Tobias Mayer.[89] Aepinus does not mention Euler's theory, and since Mayer's was never published, he remained unaware of most of its ramifications. But Aepinus, like most other eighteenth-century investigators concerned with the practicalities of magnetic behavior, clearly and not unreasonably considered the variation of the Earth's magnetic field as an errant and non-law-like phenomenon, and therefore not open to mathematical treatment.

Aepinus does consider at some length the way iron can acquire magnetism simply by virtue of its position in the Earth's field. That it should do so is plain from his theory. However, Aepinus is again able to go further than this, and reduce the effect to calculation. He is also led to ascribe the power of the lodestone to its having been held fixed with respect to the Earth's field for a sufficiently long time. This seems to him much more plausible than to suppose that mines from which such stones are dug actually tap the Earth's magnetic core. He likewise invokes the magnetism of the Earth to explain why iron becomes magnetized, and bodies that are already magnets have the direction

---

[89] Euler, *Mém. Acad. Berlin,* XIII, 1757, 175–251; XXII, 1766, 213–64. Eric G. Forbes, ed., *The Unpublished Writings of Tobias Mayer,* Göttingen, 1972, III.

of their magnetism altered, when they are struck either by an electrical discharge (whether a bolt of lightning, or a discharge from a Leyden jar) or repeatedly with a hammer. Either cause, Aepinus argues, will agitate the particles of the iron. This, he says, will open up the pores between them and enable the magnetic fluid therein to move more freely. Thus freed, the fluid will move in whichever direction is dictated by the overriding field in that location, namely that of the Earth, and hence poles will be produced that are aligned with that field. The effect is purely mechanical, and certainly does not reveal, for example, any "hidden nexus" between electricity and magnetism, such as had incautiously been supposed by some in the light of Franklin's experiments on the subject.

Aepinus's account of the origin of the lodestone's power raises a different kind of problem. If, as Aepinus asserts, the stone acquires its magnetism from the Earth's field, why do not other samples of iron ore extracted from the Earth display similar properties? Since his answer is that only certain kinds of iron ore are suitably susceptible to magnetization, the question leads Aepinus into a somewhat extended discussion of the mineralogy of iron and its ores (§§ 348–56). His account is fairly orthodox for the time, assuming a chemistry of metals that takes them to be compounds of various "metallic earths" with phlogiston. Aepinus suggests that the more perfect the "metallic character" of an ore, that is, the more phlogiston it contains, the more susceptible it is to magnetization. Interestingly, this view of the matter brings him to a new method of increasing the strength of natural magnets. He argues that if a lodestone is heated in a strong fire or, better still, cemented "in a strong fire after the addition of a large supply of phlogiston," it ought to become not only more metallic but also more apt to receive an increased magnetic force. Sure enough, when he puts his idea to the test, the experiment succeeds.

We have already noted that Aepinus saw his *Essay* primarily as a contribution to the science of magnetism rather than to that of electricity. We are now in a position to see why. Great though his contributions were to electrical theory and experiment, in the magnetic realm they were even greater. In Aepinus's hands, the study of magnetism took on a completely different form. From a subject that had until then been almost totally immersed in qualitative discussions of the interactions of Cartesian-style circulations of subtle matter, he fashioned a rigorous and at least semi-quantitative account based on forces acting at a distance between particles of matter. Admittedly, his theory had weaknesses, some of which were later overcome by Coulomb. But weaknesses and all, it was a vast improvement on anything that had gone before. Also, it burst upon the scientific stage almost completely unheralded. Aepinus's approach was so novel that few of his ideas were foreshadowed in the writings of others. Michell alone seems to have been heading in the same direction; but his published work failed to do more than hint at the radical nature of his thinking.

Up to a point, at least, we know precisely the genesis of Aepinus's reform of magnetic science—his revolutionary new theory of magnetism sprang directly from his perception of the analogy between a magnet and an electrified tourmaline crystal. Aepinus first developed a semi-mathematical action-at-a-distance theory of electricity, and then transferred it, lock, stock, and barrel, to the magnetic case, and showed that it served there equally well. However, his readiness to transfer the theory in this way surely implies that Aepinus was in a frame of mind receptive to new ideas about magnetism even before he began work on the tourmaline. He must at least have been thoroughly skeptical about the supposed circulating streams of subtle matter, and ready to think in terms of some specialized kind of action at a distance.

Beyond this, the evidence lets us down. Though we can make some fairly obvious inferences concerning the development of Aepinus's ideas, there is nothing that takes our understanding any further. In particular, the notebook that Aepinus kept during the crucial years just prior to the publication of his *Essay* is even less helpful here than it was in the electrical case.[90] Unless more of his manuscripts come to

---

[90] AAN, raz. v, op. 3–7, no. 1. The only references to magnetism in the entire notebook are as follows:

(i) on l. 30: "lex quam sequitur ortus punctorum consequentium in ferro magnete adfricto.

"lex secundum quam vires diversis ~~corp~~ frustis ferri similibus aut dissimilibus communicantur.

"Lex qua attrahuntur diversa ferri frusta."

(ii) further down the same page, accompanied by the diagram below: "suspenda-

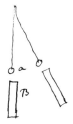

Figure 5. Aepinus's notebook illustration of a possible method of measuring magnetic forces.

light, therefore, we must once again resign ourselves to recognizing not only the remarkable nature of his achievement but also our very limited understanding of the thought processes from which it emerged.

> tur parvum ferri frustum a ex filo tenui, ac huic pendulo magnes approximetur. ex angulo ad quem elevari potê pendulum, magnetis vis determinari poterit."
>
> (iii) on l. 37: "Eine aufgeh. eiserne Nadel, cui approximatur magnes, ob sie nicht erstl. gegen ein groß Stück Eisen, prope quod pendet angez. werde, ob sie in athmosph. unius poli, nicht post contactum alterius poli inimici repellirt wird. "Theoria magnetis
>
>> fluidi magnetici partes se repellunt, a ferro attrahuntur.
>>
>> fluid. ~~elect~~ magn. ϱ ferrum se momento $=$liter distribuit.
>>
>> ferrum in eum statum redigi potest, ut fl. magn. non sit permeabile."

CHAPTER FIVE

A Place in History

When, in June 1759, Aepinus sought approval from his colleagues in the St. Petersburg Academy of Sciences for the printing of his great *Essay on the Theory of Electricity and Magnetism,* he also sought and obtained their permission to have the work published in the form of a supplement to the Academy's *Commentarii* for that year.[1] As a consequence, the book is so described on its title page. Aepinus's intention was not simply to insure that the printing costs were met out of the Academy's purse rather than his own. He was also concerned about the remoteness of St. Petersburg from the rest of the learned world, and he wanted to take advantage of the established channels through which the annual volumes of the *Commentarii* were distributed. This is made clear by the fact that fifteen months later he brought the matter up again, at another meeting of the Academic Conference. On this occasion, "the question arose, whether books, printed at the Academy's expense in the form of a supplement to the *Commentarii,* should be sent together with the *Commentarii* to foreign members as a gift."[2] Apparently, something had gone wrong with Aepinus's original plan, and he was now seeking to put things right. The other academicians accepted his point and resolved that representations should be made to the chancellery of the Academy. Accordingly, G. F. Müller, the secretary of the Academic Conference, wrote to the chancellery a few days later, setting out the academicians' views, and offering to send out copies of Aepinus's book with the next volume of the *Commentarii* if a sufficient number were made available to him. In the end, however, the chancellery released only twelve copies for distribution, not nearly enough to go around, given that the Academy boasted over forty foreign members at the time.[3]

The problem of distribution was a very real one. St. Petersburg was a long way from the scientific centers of western Europe, and com-

[1] *Protokoly zasedaniĭ konferentsii Imperatorskoĭ Akademii Nauk s 1725 po 1803 goda,* II, 428.

[2] Ibid., II, 455.

[3] Müller to the Chancellery of the Academy, 19 September 1760; AAN, fond 3, op. 1, no. 257, 1. 9. The Chancellery's decision is noted on the back of the letter. The Academy's foreign membership has been determined from B. L. Modzalevskiĭ, *Spisok chlenov Imperatorskoĭ Akademii Nauk 1725–1907,* St. Petersburg, 1908.

munication could sometimes be unreliable. As we shall see, however, Aepinus did manage by one means or another to insure that copies of his book reached the west. Nevertheless, his work failed for a time to have much impact. If the remoteness of St. Petersburg contributed to this, it did so less directly. The leading scientists of Paris and London undoubtedly tended at times to look down on the work of their colleagues from other parts of the world and therefore sometimes gave it less attention than it deserved. There is some evidence that Aepinus in particular suffered in this way. Yet there were more important reasons than this for the initial failure of his book to create a stir, which we will explore. First, however, it must be established that Aepinus's contemporaries had access to his work, and thus could have profited from it had they seen fit to do so.

It has been largely impossible to determine the identity of the fortunate twelve who received copies of Aepinus's book from Müller. Leonhard Euler may have been on the list, since some three months after the St. Petersburg academicians discussed the matter, Euler reported to Müller that he had presented to the Berlin Academy in Aepinus's name the copy of the book that had been sent to him. Yet even here, the suspicion lingers that the copy might have been one sent to Euler by Aepinus himself for presentation to his old Academy, rather than one sent to Euler personally on behalf of the St. Petersburg Academy.[4]

The ordinary mechanics of the eighteenth-century book trade would have insured at least a limited distribution of Aepinus's work. Toward the end of the 1760s at the latest, books published by the St. Petersburg Academy were available on a regular basis from booksellers in Frankfurt-am-Main, Leipzig, Berlin, and Breslau, as well as in St. Petersburg itself.[5] It appears, too, that the Academy sent shipments to the principal German book fairs, though the Seven Years' War no doubt interrupted this trade for a time.[6] Aepinus's book would, of course, have been listed regularly in any catalogue of the Academy's publications; indeed, a hundred years later it was still listed and advertised for sale at the Academy's agents in St. Petersburg, Riga, and Leipzig![7] A

[4] Euler to Müller, 10 January 1761; Yushkevich et al., *Die Berliner und die Petersburger Akademie,* I, 166.

[5] J. A. Euler to S. Canterzani, – October/November 1769; I. I. Lyubimenko, ed., *Uchenaya korrespondentsiya Akademii Nauk XVIII veka,* Moscow, 1937, p. 159.

[6] G. Heinsius to J. von Stählin, 28 November 1767, ibid., p. 92.

[7] *Catalogue des livres publiés en langues étrangères par l'Académie Impériale des Sciences de St. Pétersbourg, en vente chez ses commissionaires* . . . , St. Petersburg, 1867.

lengthy review in the Leipzig journal *Commentarii de rebus in scientia naturali et medicina gestis* early in the 1760s would have encouraged German scholars to seek out the work,[8] and evidence from later in the century suggests that some of them did so.[9]

From the book fairs and the Academy's various accredited outlets, a few copies of Aepinus's *Essay* would also have penetrated to more distant parts of Europe. Henry Cavendish found and presumably bought a copy of the work in Nourse's, the well-known eighteenth-century London bookstore, probably in June 1766.[10] The price, however, may have discouraged potential buyers less wealthy than Cavendish; though Aepinus's book is not specifically mentioned in this context, at this period the St. Petersburg Academy's publications were generally regarded in western Europe as outrageously expensive.[11]

As authors usually do, Aepinus himself took a hand in distributing copies of his work. In addition to the copy he may have sent to Euler for presentation to the Berlin Academy of Sciences, it would be extraordinary if he did not send a copy to Wilcke, by now installed as lecturer in experimental physics at the Royal Swedish Academy of Sciences in Stockholm. Aepinus's election in July 1760 to foreign membership of the Swedish Academy may even have been prompted by the arrival of a copy of his book; in any case, the work was listed in a catalogue of the Academy's library published only a few years later, in 1768.[12] In London, a copy of the book, along with a copy of Aepinus's *Sermo academicus de similitudine vis electricae atque magneticae* and two copies of his *Cogitationes de distributione caloris per tellurem,* was presented to the Royal Society at its meeting on 4 November 1762. The presentation was made by Benjamin Wilson, F.R.S., with whom Aepinus was corresponding at this time and to whom he also sent a personal copy of the *Essay*.[13]

[8] *Commentarii,* Supplement, Decade 1, Leipzig, 1763, pp. 681–93.

[9] Cf. below, pp. 209–12.

[10] Cavendish to John Canton, n.d.; Royal Society of London Library, Canton Papers, II, 31. For a tentative dating of this letter, see R. W. Home, "Aepinus and the British electricians: the dissemination of a scientific theory," *Isis,* LXIII, 1972, 190–204.

[11] Lyubimenko, *Uchenaya korrespondentsiya,* p. 47.

[12] *Förtekning på K. Vetensk. Academiens Bok-Samling, År 1768,* Stockholm, 1768, p. 21.

[13] Royal Society of London, Journal Book, entry for 4 November 1762. Cf. Wilson, *Phil. Trans.,* LIII, 1763, 436–66. Also Wilson to Torbern Bergman, 29 June 1763, in G. Carlid and J. Nordström, eds., *Torbern Bergman's Foreign Correspondence,* Stockholm, 1965–, I, 423. Wilson's presentation copy of the *Essay* is preserved in the Silvanus P. Thompson Collection at the Institution of Electrical Engineers, London.

At least one copy of Aepinus's book was already circulating in London before this, however, since Wilson was able to report to Aepinus the views of several of his learned friends concerning the work at a time when he himself had still not yet seen it and still did not know its exact title.[14] This copy may have belonged to John Canton, who would very likely have been on Aepinus's original distribution list. At any rate, Canton somehow or other acquired a copy quite early, and it is duly listed in a catalogue of his library drawn up by the executors of his estate in April 1772, shortly after his death.[15] This was probably the copy that Priestley used in preparing his famous *History and Present State of Electricity* in 1766. We know, at least, that when Priestley began work on his book, Aepinus's *Essay* was one of the first works he sought out, and it was from Canton that he sought it.[16] Later, Tiberius Cavallo used this same copy while preparing his popular expositions of electrical and magnetic science.[17]

In addition, Aepinus sent a copy of his *Essay* to Franklin. The volume, inscribed on the fly-leaf "B. Franklin from the Author," is preserved in the library of the American Philosophical Society. It reached Franklin following his return to Philadelphia in November 1762 from his first, five-year sojourn in London as agent for the colony of Pennsylvania.[18] (Aepinus would not have known that Franklin had previously been living in London, and so would have sent the book directly to Philadelphia.) Franklin promptly told his philosophically minded fellow colonials Cadwallader Colden and Ezra Stiles about it,[19] and within a few months he arranged for both Stiles and John Winthrop, professor of mathematics and natural philosophy at Harvard, to read it. Franklin had intended that his friend James Bowdoin should also see it, but apparently he did not do so.[20]

Given the extent to which Musschenbroek dominated the science of magnetism at this period, Aepinus would certainly have wanted him to have a copy of the *Essay*. However, there is no firm evidence

---

[14] Wilson to Aepinus, n.d. (draft); British Library, Add. MS 30094, f. 91. Aepinus is not identified explicitly as Wilson's correspondent, but from the contents of the letter it could be no one else.

[15] Royal Society of London, Canton Papers, II, 97.

[16] Schofield, *A Scientific Autobiography of Joseph Priestley*, p. 16.

[17] Royal Society of London, Canton Papers, II, 113.

[18] See Franklin's somewhat belated letter of thanks to Aepinus, dated 6 June 1766; N. N. Bolkhovitinov, "V arkhivakh i bibliotekakh S.Sh.A.: Nakhodki, vstrechi, vpechatleniya," *Amerikanskiĭ ezhegodnik*, 1971, p. 331.

[19] Franklin to Colden, 26 February 1763, in Labaree, *The Papers of Benjamin Franklin*, x, 204. Franklin to Stiles, 29 May 1763, in ibid., x, 266.

[20] Ibid., x, 351, xi, 22, 230, 246, 254.

that a copy was actually sent to Leyden, and it may well be that news of the Dutch professor's death on 19 September 1761 reached St. Petersburg in time to prevent its dispatch. The work is not listed in the catalogue of the auction at which the bulk of Musschenbroek's library was later sold.[21] The first Dutchman known to have read the book is Jan Hendrik Van Swinden, then a young professor at the University of Franeker, who frequently referred to it in his *Tentamina theoriae mathematicae de phaenomenis magneticis*, published in Leyden in 1772.[22]

The *Essay* evidently reached Italy at about the same time. The Milanese mathematician and one-time writer on electricity, Paolo Frisi, received a presentation copy directly from Aepinus, but the inscription makes it plain that the copy was not sent before the late 1760s.[23] Beccaria's reference to the work in his *Elettricismo artificiale*, published in Turin in 1772, was probably the earliest mention by an Italian author, but this ardent Franklinist had met Aepinus's book for the first time only a short time before.[24] Soon afterward, a copy came into the hands of Carlo Barletti, professor of experimental physics at Pavia. Barletti, who had previously had to rely on Priestley's *History* for his knowledge of Aepinus's work, now came to appreciate better the power of the German's approach, and so impressed was he by what he read that he proceeded to publish a series of extracts from the *Essay* as an appendix to his own book on electricity. To this he added an open letter to Volta in which he again drew attention to Aepinus's novel approach to electrical theory.[25] Barletti's message was not lost on Volta, who was by this means himself made familiar with Aepinus's theory for the first time.[26]

In one way or another, then, copies of Aepinus's book became available in most major scientific centers within a few years of its publication. (Paris, however, was probably an exception, to which we shall return later.) In particular, Aepinus's book became known to nearly

---

[21] Personal communication from the Town Archivist, Gemeentelijke Archiefdienst Leiden.

[22] Van Swinden testified still more directly to the impact Aepinus had on his thinking in a letter he sent to St. Petersburg at about this time: Van Swinden to Aepinus, 5 September 1773 (draft); Bibliotheek der Rijksuniversiteit te Leiden, MS BPL 755.

[23] See above, p. 116.

[24] Beccaria, *Elettricismo artificiale*, p. 175. In his *Experimenta atque observationes quibus electricitas vindex late constituitur, atque explicatur*, Turin, 1769, Beccaria showed no awareness at all of what Aepinus had done in his book.

[25] Barletti, *Dubbj e pensieri sopra la teoria degli elettrici fenomeni*, Milan, 1776.

[26] Cf. Heilbron, "Volta," *Dictionary of Scientific Biography*, XIV, 72.

all the world's leading Franklinist electricians. Yet by no stretch of the imagination can it be said to have had a dramatic impact on their thinking. Some of the ideas Aepinus promulgated soon won acceptance, it is true, but they were accepted separately and for themselves, and not on account of their place in the broader structure of Aepinus's theory. The merits of the theory when seen as a whole, and especially its mathematical rigor and consistency, were simply not appreciated at first.

One of Aepinus's ideas that quickly won acceptance was his denial of Franklin's doctrine that glass was absolutely impermeable to the electric fluid. Franklin's own belief in the uniqueness of glass had already been weakened some years earlier, and in a letter to his friend Ebenezer Kinnersley he had suggested that a wooden board, if properly dried and heated, might take the place of the glass in the Leyden experiment.[27] Yet it was Aepinus's discussion that persuaded him to abandon his previous position entirely in favor of Aepinus's view that impermeability was a property common to a greater or lesser extent to all electrics *per se*. "I think with him," Franklin wrote to John Winthrop in 1764, "that Impermeability to the El[ectric] Fluid, is the Property of all El[ectric]s per se; or that, if they permit it to pass at all, it is with Difficulty, greater or less in different El[ectric]s per se." [28] By 1767, when Priestley wrote his *History*, Aepinus's opinion was the new orthodoxy among British electricians, and Priestley reported it in a very matter-of-fact way as correcting what "Dr. Franklin once supposed." [29]

In discussing the Leyden experiment, Aepinus introduced a further modification of Franklin's account when he showed that, as a result of the finite thickness of the glass or other dielectric, the charges on the two plates could not be exactly equal. Priestley's reaction to this was probably typical of the Franklinist electricians of the period; he unhesitatingly accepted the results of the experiment Aepinus devised

---

[27] Franklin to Kinnersley, 28 July 1759, in Labaree, *The Papers of Benjamin Franklin*, VIII, 417.

[28] Franklin to Winthrop, 10 July 1764, in ibid., XI, 254. Cf. Cohen, *Franklin and Newton*, p. 540.

[29] Priestley, *History*, p. 244. Since whether or not glass was impermeable to the electric fluid had been one of the central points at issue in the debate between Nollet and the Franklinists, the Frenchman's supporters naturally seized on this apparent weakening of their opponents' position. Mathurin-Jacques Brisson's editorial remarks in his French translation of Priestley's *History* are a notable instance of this (Priestley, *Histoire de l'électricité*, Paris, 1771, II, 498). Yet had Brisson and his fellows taken the trouble to read Aepinus's book, they would, of course, have found no comfort in it.

to verify his conclusion, but that was as far as he went. Apparently blind to the deeper ramifications of Aepinus's discussion, he made no mention either of Aepinus's reasons for doing the experiment or of the mathematical analysis that underpinned his argument.[30]

A much more far-reaching amendment to Franklin's theory was Aepinus's categorical denial that there existed any such things as the electrical atmospheres Franklin invoked to account for electrical attractions and repulsions, and later used to explain inductive effects as well. We have seen that most of Franklin's followers failed to understand exactly what he had in mind with regard to these atmospheres, and that, as a result, a number of variant "atmosphere" theories were developed during the 1750s to account for the effects in question. None of Franklin's followers before Aepinus, however, was prepared simply to abandon the idea of atmospheres, and to rely on unexplained actions at a distance instead.

No trace has survived of Franklin's own attitude toward this revolutionary feature of Aepinus's account; even Priestley's *History* is uninformative here, since Priestley restricts his discussion of Franklin's views concerning the crucial induction experiments to the period before the appearance of Aepinus's *Essay*.

Franklin's supporters reacted to Aepinus's notion in various ways. Benjamin Wilson, a Franklinist only insofar as he accepted that "plus" and "minus" were two different modes of electrification, certainly read the *Essay*, but he simply ignored Aepinus's arguments and continued to espouse unaltered his own peculiar theory of atmospheres.[31] John Canton did not discuss the subject in print, but we learn from Priestley's *History*, a work written with Canton's encouragement and assistance and hence unlikely to misrepresent his views, that his reaction was more positive.

Priestley, after describing Canton's induction experiments of 1753, stressed how "at the time these experiments were made, Mr. Canton retained the common idea of electric atmospheres." Priestley then pointed out that, as a result of the work of Wilcke and Aepinus, it could now be seen that in reality the experiments "tend to refute the common opinion, and are much easier explained upon the supposition, that the portion of fluid belonging to any electrified body is constantly held in contact, or very nearly in contact, with the body; but acts upon the electricity of other bodies at a certain distance."[32] Priestley then gave a brief account of Franklin's contributions to the

[30] Priestley, *History*, p. 273.
[31] Wilson, *Phil. Trans.*, LIII, 1763, 436–66.
[32] Priestley, *History*, p. 236.

subject, followed by a long account of the work of Wilcke and Aepinus, drawn largely from the latter's *Essay*, together with a cursory (and mistaken) reference to Beccaria's views.[33] Immediately afterward, he returned to Canton's opinions and reported a change of heart. "It is now also Mr. Canton's opinion," he said, "that electric atmospheres are not made of effluvia from excited or electrified bodies, but that they are only an alteration of the state of the electric fluid contained in, or belonging to the air surrounding them, to a certain distance." [34] At the same time, Priestley was careful to attribute the initial denial of the existence of atmospheres to Aepinus alone, even though he had until this point made no attempt to discriminate between Aepinus's contributions and those of Wilcke. Priestley's implication seems clear; Canton's reading of Aepinus's *Essay* had led him to abandon his earlier views about atmospheres.

Priestley also made it clear that he himself favored Aepinus's doctrine. He doubted that the supposed atmospheres had any real existence, and accepted the idea of action at a distance.[35] Yet he also continued to speak of electrical atmospheres; and although in most cases he was clearly adopting Aepinus's usage, and using the phrase simply to refer to "spheres of influence" around electrified bodies, an air of ambivalence still pervades his account. In particular, to speak of "the mutual repellency of similar electric atmospheres," as Priestley did at one point,[36] seems to restore to the atmospheres a physical reality earlier denied them.

Canton was the most esteemed of the English electricians throughout the 1760s. It is therefore noteworthy that although he was, according to Priestley's report, persuaded by Aepinus's arguments to abandon his previous conception of real atmospheres of electric matter surrounding charged bodies, he remained reluctant to reduce everything to simple actions at a distance as Aepinus had done. Canton now envisaged a different kind of "atmosphere" surrounding an electrified body. The presence of a charge of either kind on a body would, he suggested, modify the electrical state of the surrounding air. Priestley described the idea as follows: "Excited glass . . . repels the electric fluid from it, and consequently, beyond [a certain] distance makes it more dense; whereas excited wax attracts the electric fluid existing in the air nearer to it, making it rarer than it was before." Induction was then explained in terms of the immersion of a second

[33] See above, p. 105.
[34] Priestley, *History*, p. 246.
[35] Ibid., pp. 236, 237.
[36] Ibid., p. 262.

body in such an "atmosphere": "When any part of a conductor comes within the atmosphere of [excited glass], the electric fluid it naturally contains will be repelled by the dense atmosphere, and will recede from it. But if any part of a conductor be brought within the atmosphere of [excited wax], the electric fluid it naturally contains will be attracted by the rare atmosphere, and move towards it. And thus may the electric fluid contained in any body be condensed or rarified." [37]

These ideas, or something like them, subsequently enjoyed some favor among British writers on electricity. Sometimes, as in the general failure adequately to appreciate Henry Cavendish's important contributions to the subject, this had unfortunate consequences. Cavendish, in 1771, proposed a theory of electricity which though very similar to Aepinus's was developed independently. The notion of an extended atmosphere of electric matter around a charged body had no more place in Cavendish's theory than it did in Aepinus's. For Cavendish, as for Aepinus, electrical charges simply exerted their powers in some unexplained manner at a distance.[38] Yet Cavendish's countrymen gave his theory a similar reception to the one they earlier gave Aepinus's; and this despite the fact that Aepinus's work had in the meantime considerably weakened their previous commitment to the doctrine that real atmospheres of electric matter surrounded charged bodies. Even after Cavendish's work, most British electricians continued to discuss electrostatic induction, in particular, in terms of electrical atmospheres. Nevertheless, the phrase seldom retained its original meaning. In some cases, what was meant was simply "spheres of influence," as in Aepinus's (and Priestley's) usage. More often, however, something like Canton's idea seems to have been intended.

Canton's position received powerful backing some years afterward from the work of Beccaria. In his important *Elettricismo artificiale* of 1772, Beccaria quietly discarded many of the central tenets of the elaborate theory he had earlier expounded concerning the mechanics supposedly underlying electrical attractions and repulsions, and the formation of electrical atmospheres. In place of his former views, Beccaria, like Canton, now maintained that any charged body was surrounded by a zone of air in an electrically modified state. It was to this zone that Beccaria now applied the term "electrical at-

---

[37] Ibid., pp. 246–47. Cf. Canton to Priestley, 5 April 1766, Royal Society of London, Canton Papers, II, 65.

[38] Cavendish, *Phil. Trans.*, LXI, 1771, 584–677. Cavendish expressly declared that the charge on a body did not extend sensibly beyond its surface in his early draft, "Thoughts concerning Electricity" (*Electrical Researches of the Honourable Henry Cavendish*, pp. 94–95).

mosphere"; and he now supposed that induction occurred on account of the increased "expansive force" or "tension" of the electric matter that other bodies encountered when immersed in such a zone. Nor can one help suspecting that Beccaria's change of heart was prompted, like Canton's, by Aepinus's work, because the latter's *Essay* is cited at the very point in Beccaria's book where he denies the existence of extended atmospheres of electric fluid around charged bodies.[39]

It is also difficult not to see Aepinus's influence in Beccaria's sudden abandonment of the "hydrodynamical" account of electrical attractions and repulsions that had occupied so prominent a place in his earlier writings. In contrast to what he had once said, Beccaria now drew the same careful distinction that Aepinus drew in his *Essay* between those phenomena that arise from an actual transfer of electric fluid and those, namely attractions and repulsions, that depend, rather, on the forces exerted by static accumulations of it.[40] However, Beccaria still supposed these forces to be mediated by electrical atmospheres (of his newly modified variety) surrounding the bodies in question. He was no more prepared than Canton to follow Aepinus's example and have recourse to simple actions at a distance in this matter.[41]

Another influential Franklinist electrician to invest the phrase "electrical atmosphere" with a new meaning was Charles Mahon (later third Earl Stanhope). Mahon, like Canton and Beccaria, used the term to refer to a zone of electrically modified air that, in his view, necessarily surrounded any charged body. Unlike the others, however, Mahon held that the electrical condition in question amounted merely to the acquisition of a charge by the particles of air, through contact with the central charged body. "When Bodies are charged with Electricity," he said, "it is the *Particles of (circumambient) Air being electrified,* that constitutes the *electrical Atmosphere* which exists around those Bodies." [42]

One of the more remarkable features of Aepinus's electrical theory is his doctrine that the particles of ordinary matter exert mutually repulsive forces on each other. When Coulomb took up Aepinus's ideas a generation after they were first published, he found this particular proposition impossible to accept, and he therefore settled for

[39] Beccaria, *Elettricismo artificiale,* Turin, 1772, p. 175. I have also used the English translation, *A Treatise upon Artificial Electricity,* London, 1776.

[40] Ibid., pp. 1–2.

[41] Ibid., p. 393. Cf. Beccaria to Franklin, n.d. (1776?), Antonio Pace, *Benjamin Franklin and Italy,* Philadelphia, 1958, pp. 376–77.

[42] Mahon, *Principles of Electricity,* London, 1779, p. 11.

the so-called "two-fluid" theory. According to this, ordinary matter is normally saturated with not just one but a neutral combination of two subtle electric fluids, and electrification consists in separating the two, so that one body acquires a superabundance of one, while a second acquires a superabundance of the other. Instead of bodies becoming charged "plus" or "minus," they become charged "vitreously" or "resinously." Beyond this, the theory is formally equivalent to Aepinus's; in fact, the only difference is that the mutual repulsion that Aepinus had ascribed to the particles of ordinary matter is now transferred to the second electric fluid.[43]

Because Coulomb invoked the two-fluid theory explicitly to circumvent Aepinus's discomfiting proposition about ordinary matter, it has sometimes been assumed that the theory was actually invented for this purpose. Such, however, was not the case. The theory was developed contemporaneously with Aepinus's by the Scot Robert Symmer, in complete ignorance of Aepinus's work and in the context of a quite different set of problems from those with which Aepinus was concerned.[44] Symmer was persuaded by some curious experiments on the mutual electrification of pairs of black and white silk stockings "that what is called *negative* electricity is, in reality, a positive active power; and that electricity, in general, consists not of one alone, but of two distinct, positive powers, acting in contrary directions, and towards each other." "It is my opinion," he proclaimed, "that there are two electrical fluids (or emanations of two distinct electrical powers) essentially different from each other."[45] Further experiments in which he discharged Leyden jars through his body, or through

---

[43] Coulomb, *HAS*, 1788, Mém. pp. 671-73.

[44] Symmer, *Phil. Trans.*, LI, 1759-60, 340-89; also published separately under the title *New Experiments and Observations concerning Electricity*, London, 1760. The invention of the two-fluid theory has often been ascribed to Du Fay rather than to Symmer. When Du Fay announced his famous distinction between vitreous and resinous electricity, however, he was not advocating, as Symmer did, a theory of electricity that assumed the existence of two distinct subtle electric fluids or powers. On the contrary, he was referring merely to his experimental discovery that there are two different modes of electrification possible in nature, one seemingly associated particularly with glass, the other with sulphur and various resinous materials (Du Fay, *HAS*, 1733, Mém. pp. 457-76). A shift in the meaning of the word "electricity" appears to have led many authors astray here. In Du Fay's day, electricity was an attribute, not a substance, an effect, not a cause. The word referred to a peculiar set of properties that certain classes of matter acquired when rubbed. Only later did it come to refer to the substance (or substances) whose redistribution was supposed to endow matter with these properties. For an account of Symmer's work, see J. L. Heilbron, "Robert Symmer and the two electricities," *Isis*, LXVII, 1976, 7-20.

[45] Symmer, *New Experiments*, pp. 38, 47.

quires of paper, had convinced him of the correctness of this doctrine. Nowhere, however, did he reveal the slightest awareness of the problem in Franklin's theory that Aepinus uncovered, and that later caused Coulomb to adopt the two-fluid alternative.

The same may be said of Beccaria's nephew, Gian Francesco Cigna, who took Symmer's work as his starting point in an important paper published soon afterward, in the transactions of the infant Academy of Sciences of Turin. In addition to considerably extending Symmer's experimental investigations, Cigna also entertained seriously the Scot's novel approach to the theory of the subject, and eventually concluded that his ideas fitted the phenomena as well as Franklin's did. Cigna was careful with his footnotes, and at the appropriate places in his account he cited both Aepinus's paper on the tourmaline and his analysis of the experiment devised by the Jesuit Fathers in Peking. Nowhere, however, did Cigna mention Aepinus's *Essay* or the arguments it advanced concerning the foundations of Franklin's theory. On the contrary, his discussion of the two-fluid theory was couched in terms very similar to those used by Symmer.[46]

Though Aepinus's modified version of Franklin's theory was unknown to Symmer and Cigna, it was well known to another early proponent of the two-fluid alternative, Aepinus's friend and former student, Johan Carl Wilcke. As was the case with Symmer and Cigna, Wilcke's principal reason for favoring the two-fluid theory was his becoming convinced that so-called "negative" electricity involved some positive agency, and not a mere deficiency of fluid.[47] He also mentioned "negative repulsions" as a phenomenon that presented difficulties for Franklin's theory. Yet no more than the others did Wilcke specify, as a reason for discounting the one-fluid theory, Aepinus's troubling proposition about the mutual repellency of particles of ordinary matter.

Priestley, "convinced . . . of the usefulness of various theories, as suggesting a variety of experiments, which lead to the discovery of new facts," expounded the two-fluid theory in his *History and Present State of Electricity,* notwithstanding his own preference for Franklin's views.[48] Priestley claimed, indeed, to do more justice to the theory than its advocates, even Symmer himself, had done. Like Wilcke, and in contrast to Symmer and Cigna, Priestley was aware of Aepinus's postulate concerning the mutually repulsive character of ordinary

---

[46] Cigna, *Miscellanea Taurinensis,* II, 1762-65, 31-72. A decade later, Cigna had apparently still not seen the *Essay,* and relied instead on Priestley's *History* for his knowledge of what Aepinus had done (ibid., v, 1770-73, 97-108).

[47] Wilcke, *Abhandl. Schwed. Akad. Wiss.,* 1763, pp. 207-26.

[48] Priestley, *History,* pp. 441-50.

matter. Unlike many subsequent historians, he also correctly reported Aepinus's reason for adopting it: "He thinks," Priestley said, "that all the particles of matter must repel one another: for that, otherwise . . . it could not happen, that bodies in their natural state, with respect to electricity, should neither attract nor repel one another." [49] Yet Priestley's very next sentence was his notorious blanket condemnation of Aepinus's reasonings, which suggests that he was not much persuaded by the argument he had just set out. Had Priestley taken Aepinus's notion seriously, he would surely have at least mentioned the conceptual difficulties involved in reconciling it with the Newtonian doctrine of universal gravitation. But he did not, and in his subsequent presentation of the two-fluid theory there is not the slightest suggestion that this was being put forward as a possible alternative to what Aepinus had proposed. It was presented, rather, in a spirit of pure eclecticism, because he "thought it had been, hitherto, too much overlooked." [50]

So far as Aepinus's new postulate was concerned, Priestley went somewhat further than most other Franklinist theorists of the age simply by reporting it. Most authors ignored it altogether. Yet Aepinus had shown unequivocally that, without it, Franklinian electrical theory was inconsistent with the simplest of observations. The general failure to take Aepinus's argument seriously is perhaps the most striking evidence we have of the complete lack of understanding, on the part of almost all electricians in the 1760s and 1770s, of what Aepinus was trying to do in his electrical theory.

To Nollet and his supporters, of course, Aepinus was speaking what amounted to a foreign language, so that his work would have had little or no impact on their thinking even in the best of circumstances. But Franklinists everywhere were equally uncomprehending. Though happy to let Aepinus persuade them—up to a point, at least—in cases where he was able to marshal direct experimental backing for his position, they invariably considered each proposition or group of propositions in isolation. They failed to see that, in Aepinus's hands, these formed part of a coherent and mathematically consistent whole. Cavendish was the sole exception in this regard. He did appreciate the rigor and systematic character of Aepinus's analysis, since these were features he constantly strove for in his own work.[51] But his ideas fell on equally stony ground.

Eighteenth-century experimental physicists were unused to requir-

[49] Ibid., p. 436.
[50] Ibid., p. 450.
[51] Russell McCormmach, "Henry Cavendish: a study of rational empiricism in eighteenth-century natural philosophy," *Isis*, LX, 1969, 293–306.

ing of their theories the degree of rigor demanded of Franklin's theory by Aepinus and later by Cavendish. "Physics" still retained much of its ancient aura of being, by its very nature, at best approximative and concerned, according to the traditional Aristotelian dichotomy, with quality and not quantity. Though Priestley could blithely remark that "Æpinus has lately given us an excellent specimen of what use mathematics, and especially algebraical calculations, may be of to an electrician," [52] others among Aepinus's readers were opposed to his efforts to bring mathematics to bear on their subject. Benjamin Wilson reported the views of others besides himself when he wrote in patronizing terms to Aepinus that "the introducing of algebra in experimental philosophy, is very much laid aside with us, as few people understand it; and those who do, rather cho[o]se to avoid that close kind of attention: tho' I make no doubt but I dar[e] say you had a very good reason for making use of that method." [53]

Others again, less self-opinionated than Wilson, simply failed to see the point of what Aepinus was doing and hence tended to see his work as adding little to what Franklin had done already. John Winthrop clearly fell into this category, despite the fact that he was professor of both mathematics and natural philosophy at Harvard and thus might have been expected to see the value of Aepinus's calculations. He spoke of Aepinus's work in a letter to Ezra Stiles: "Aepinus is a man of clear ideas & a comprehensive as well as inquisitive mind. His book contains many curious things. He has set the theory of Magnetism in a new light; & his hypothesis agrees so well with the various, puzzling phenomena of that mysterious Power, as renders it in a considerable degree probable; tho many of 'em might be accounted for in a different way. But he has the right German prolixity, even to a degree of tediousness; & I freely confess, if I had had the leisure, I should not have had the patience, to follow him minutely thro all his formulas, which fill a 4to of almost 400 pages; — a Vol. big enough to contain a System of Philosophy. As to Electricity, I don't find that he has advanced much, if at all, beyond our excellent Countryman, to whom he renders great & just honors; & I scarcely recollect any thing he has said on this subject, but what I had learn'd before, & with greater pleasure, from Dr. Franklin." [54]

Perhaps the most hostile reaction to Aepinus's introduction of mathematics into what had been, until then, the exclusive preserve of

---

[52] Priestley, *History*, p. 475.

[53] Wilson to Aepinus, n.d., British Library, Add. MS 30094, f. 91.

[54] Winthrop to Stiles, 21 February 1764, Stiles Papers, Beinecke Rare Book and Manuscript Library, Yale University.

the experimentalists came from the instrument maker George Adams. "Æpinus," Adams wrote, "by a mathematical theory of electricity, has closed the door on all our rese[a]rches into the nature and operations of this fluid." Aepinus's work manifested, so far as Adams was concerned, all the evils attendant upon the failure of mathematicians to attend closely enough to the phenomena: "when neglecting these, calculations are made to serve an hypothesis; the more elegant and beautiful they are, the more detrimental they become to science."[55] Yet Adams's discussion in the same work of the phenomena of magnetism reveals that he was ignorant of Aepinus's important discoveries in this field. His harsh judgment of the latter's book appears to have been based on hearsay alone!

Even those readers who approved Aepinus's new theory of magnetism while rejecting his contributions to electrical theory often did so with reservations such as those expressed by Winthrop. Franklin accepted Aepinus's basic idea more freely than most. As early as 20 February 1762 he wrote of a magnet as "a body containing a subtle moveable fluid, in many respects analogous to the electric fluid,"[56] and since he always thereafter credited this doctrine to Aepinus, it would appear that he had already seen the *Essay* by then. Yet even Franklin felt constrained to admit "some Difficulties in his Doctrine that as yet I do not see how to solve."[57] These centered on the appearance of a new pair of poles whenever a magnet was broken in two; Franklin appears not to have been persuaded by Aepinus's discussion of this point.[58]

Despite his misgivings on this matter, Franklin was instrumental in bringing Aepinus's ideas about magnetism to the notice of a wider audience. He drew attention to this aspect of Aepinus's work in letters to his fellow Americans Colden, Stiles,[59] and Winthrop. He included in the definitive edition of his papers, published in 1769, his earlier letter to Kinnersley alluding to the theory Aepinus had proposed. Most important of all, he set out the elements of Aepinus's magnetic theory in a letter to Barbeu Dubourg that the latter included in his French edition of Franklin's works, published in 1773.[60] It was

[55] Adams, *Lectures on Natural and Experimental Philosophy*, 2nd ed., London, 1799, IV, 304.

[56] Franklin to Kinnersley, 20 February 1762, *Benjamin Franklin's Experiments*, p. 366.

[57] Franklin to Cadwallader Colden, 26 February 1763, in Labaree, *The Papers of Benjamin Franklin*, x, 204.

[58] Franklin to Winthrop, 10 July 1764, in ibid., XI, 254.

[59] Franklin to Stiles, 29 May 1763, in ibid., x, 266.

[60] *Oeuvres de M. Franklin*, ed. Barbeu Dubourg, Paris, 1773, I, 277–79.

probably this brief but admirably clear summary that first directed Coulomb to Aepinus's work, with significant consequences for the future development of the subject.

Other readers in the early 1760s reacted with less approval to Aepinus's theory of magnetism than did Franklin. In the letter from Benjamin Wilson to Aepinus that has been quoted with reference to the place of mathematics in experimental physics, Wilson reported, in the same patronizing and at this point downright insulting manner, that Aepinus's "hypothesis" had not found favor with his colleagues: "Those of my learned friends who have had the satisfaction of perusing your work, think it a very ingenious performance, but do not allow of your hypothesis. And add, that it appears you are a stranger to many very curious works upon those subjects, which if you had read before, would have afforded new lights, and probably saved you a great deal of trouble." [61]

In addition to being unfair, this is not even an altogether accurate report, since Franklin was undoubtedly one of the "learned friends" to whom Wilson referred. Yet it helps to explain why Aepinus's ideas about magnetism had so little impact in England at this time. It suggests a general feeling that Aepinus's theory, though ingenious, was rendered untenable by certain phenomena he was thought to have overlooked, presumably those same experiments in which magnets were broken in two, the consideration of which had tempered even Franklin's enthusiasm for his account.

This in turn explains the peculiarly equivocal tone of Priestley's brief discussion of Aepinus's hypothesis, which he did not even bother to prepare himself, but took instead from a summary drawn up for his use by Richard Price.[62] In this account, though the chief elements of Aepinus's analogy between electricity and magnetism are carefully set out, it is never expressly stated how Aepinus intended to build a theory of magnetism upon them. In particular, the analogy between Aepinus's theory of magnetism as a whole and Franklin's theory of electricity is not mentioned. As a result, while someone who had read Aepinus's book would certainly recognize his doctrines in Price's summary, the converse would not have held; that is, someone unfamiliar with Aepinus's ideas would not have gained a clear picture of them from Priestley's book.

In the 1770s, the traditional concept of magnetic circulations received new support with the publication of Euler's immensely popular *Letters to a German Princess*. British writers of the period, such as

[61] Wilson to Aepinus, n.d.; British Library, Add. MS 30094, f. 91.
[62] Priestley, *History*, pp. 406–408.

Oliver Goldsmith, tended still to support the same idea and hence, by implication, to reject Aepinus's theory.[63] In a new edition of Chambers's *Cyclopaedia*, published in the 1780s, the traditional views still predominated.[64] Likewise, though George Adams described both Euler's theory of magnetism and Aepinus's in the *Essay on Magnetism* that he appended to his longer *Essay on Electricity* of 1784, he made it clear that he, too, preferred Euler's account.[65] The earliest major work in English that took Aepinus's ideas on magnetism as its starting point was Tiberius Cavallo's *Treatise on Magnetism, in Theory and Practice, with Original Experiments* (London, 1787).

During the 1760s, two of Germany's leading scientists announced the results of major investigations concerning the magnet. The first of these, Tobias Mayer, did his work too early to be influenced by Aepinus's ideas, but he nevertheless needs to be considered here because his report provoked a direct response from Aepinus.

Mayer set out the results of his magnetic research in two papers he read to the Royal Society of Sciences in Göttingen, the first on 7 June 1760, the second on 16 January 1762, only a month before he died. Unfortunately, neither paper was published in full at the time, although brief summaries did appear in the *Göttingische Anzeigen von gelehrten Sachen*.[66] Only in recent times have the complete texts become available.[67] Aepinus's response to Mayer's work was based on a but partial knowledge of what the Göttingen professor had done, and was in fact directed at only the first of the two printed summaries.[68] However, it is unlikely that a closer acquaintance with Mayer's work would have altered Aepinus's opinion of it.

Both Mayer's initial paper and the published summary of it began with a vigorous attack on the theory of magnetic circulations. Thereafter, without concerning himself further with possible causes of the phenomena he was studying, Mayer turned to analyzing the action of

---

[63] Goldsmith, *A Survey of Experimental Philosophy*, London, 1776, I, 26-32.

[64] Chambers, *Cyclopaedia*, London, 1779-89, art. "Magnetism" (in vol. III, publ. 1781).

[65] Adams, *An Essay on Electricity . . . With an Essay on Magnetism*, London, 1784; 3rd ed., London, 1787, pp. 379-81, 433-41. Adams did not mention Aepinus's name in his discussion; instead, he gave as his source for the doctrine in question "Oeuvres de Franklin, tom. 1, p. 277," that is, the letter from Franklin to Barbeu Dubourg discussed already. As noted above, Adams appears to have had no direct acquaintance with Aepinus's book.

[66] The summaries appeared in the issues for 16 June 1760 (pp. 633-36) and 6 February 1762 (pp. 377-79), respectively.

[67] Forbes, *The Unpublished Writings of Tobias Mayer*, III.

[68] Aepinus, *Novi commentarii Academiae Scientiarum Imperialis Petropolitanae*, III, 1766-67, 325-50.

one magnet on another, with the ultimate aim of establishing a mathematical theory of the terrestrial declination and dip. His approach contrasted markedly with Aepinus's, and was in many ways typical of the rational mechanicians of the age. With his ultimate aim firmly in view, Mayer paid little heed to the detailed behavior of actual magnets in the laboratory. What he sought was a simple law of force that matched the phenomena reasonably well. This he proceeded to use, despite the manifest physical implausibilities associated with it, as the foundation of his larger calculations.

Mayer restricted his analysis to uniform bar magnets of negligible thickness. He assumed that these would have their magnetic "centers" or neutral points located at and fixed in their geometrical equators, and that elsewhere within them the magnetic intensity would be directly proportional to the distance from this equator. In the published summary, it was simply stated, without any indication of how Mayer arrived at the notion, that, in addition, "the power, with which each particle of the one magnet acts on a particle of the other magnet, is in conformity with the distance between the two particles in such fashion that it is inversely proportional to the square of the distance." On the basis of these two "laws," it was said, "he finds the most perfect accord between his calculation and experiments." [69]

From the full text of Mayer's paper, we gain a better appreciation of his argument at this point. Mayer's initial justification of his assertion that the law of force was inverse square in character came, we discover, from some very neat experiments in which he timed the oscillations of a compass needle placed on the extension of the axis of a bar magnet.[70] Mayer recognized, however, that his experimental arrangement was such that his results gave no more than an indication of the true form of the law, and certainly did not constitute a proof. For this, he turned to a direct comparison between the measured forces between two magnets at different distances and the forces calculated according to the principles he had adopted. Assuming an inverse square law, Mayer obtained excellent agreement between theory and experiment. He did so, however, only by making a further and highly implausible assumption, namely: "Each part [of a magnet] is thought of as attracted not . . . by all the unlike parts of the other magnet, but by one only, and similarly repelled by one only of the like parts; and the parts which act thus on one another are those which in similar magnets are the same distance from their centres or whose distance from their centres is proportional to the

[69] *Göttingische Anzeigen von gelehrten Sachen*, 1760, p. 635.
[70] Mayer, *Unpublished Writings*, III, 38–39 (English trans. pp. 70–71).

distance between their poles." [71] If, on the contrary, Mayer made the physically much more reasonable assumption that any part of one magnet was attracted by all the unlike parts of the other, and repelled by all the like parts, theory and experiment were wildly divergent.[72]

Even when spelled out in detail, then, Mayer's "proof" of the inverse square law was far from conclusive. Aepinus, with only the bald statement of Mayer's conclusions to go on, found his analysis most unsatisfactory. Some of the experiments Aepinus had described in his *Essay* directly contradicted Mayer's principles. He pointed out that, according to his experiments, the magnetic center of a magnet was only very rarely located in its geometrical equator. He argued that Mayer's postulated linear variation of magnetic intensity along a magnet was certainly false whenever consequent poles were present (as, he maintained, they usually were), and was highly unlikely even if the magnet had only a simple pair of poles. More particularly, he pointed to a number of phenomena that he had treated as effects of induction that were, in his view, incompatible with Mayer's principles. These included the alteration of the repulsion between two like poles into an attraction if the two were brought sufficiently close together, and the migration of the magnetic center along a magnet even in less extreme circumstances than this. Above all, he insisted that, because of inductive effects, the magnetic intensity at different points in a magnet itself varied whenever a second magnet was brought near, the extent of this variation depending, like the force itself, on the distance between the two.

Mayer made no reference to the problem posed for his theory by the phenomenon of consequent points, nor to that due to the uncertainties about magnetic centers. He was aware that the strength of a magnet might be altered by the presence of another magnet nearby, but suggested that this could readily be fitted into his analysis merely by using different values for the intensity on different occasions.[73] He failed to consider the possibility that a continuous dependence of the intensity distribution on distance between the magnets might need to be included in his formulae, and that if this were done, an experimental test of those formulae would offer even less prospect than it did already of yielding an unequivocal answer concerning the form of the force law.

Ultimately at issue here, it is clear, was the question of how far one could legitimately abstract from the complications and confusions of

[71] Ibid., pp. 45, 51–55 (English trans. pp. 76, 82–86).
[72] Ibid.
[73] Ibid., p. 56 (English trans. pp. 87–88).

the real world in order to obtain a valid mathematical description of it. Aepinus obviously felt that Mayer had considerably overstepped the bounds of the permissible. Insofar as Mayer's theory purported to apply at all to interactions between real magnets on a laboratory scale, Aepinus's objections were not at all unreasonable; Mayer's analysis *was* unsatisfactory from a physical point of view. And yet the fact remains that, within a limited set of circumstances, Mayer was able to attain excellent agreement between his theoretical calculations and experiment. Were this the sole criterion to be taken into account, his work would have to be judged successful.[74]

The second major work on magnetism to emerge from Germany in this period came from the pen of Johann Heinrich Lambert, and appeared in the form of two papers in the 1766 volume of the *Mémoires* of the Berlin Academy of Sciences.[75] Lambert's primary aim was, like Mayer's, to develop a mathematical description of the interaction between two magnets. No more than Mayer did he concern himself with discovering the cause of the interaction, though at times his choice of phrase suggested an underlying acceptance of the circulation theory. He took even less account than did Mayer of physical complexities of the kind Aepinus pointed to in his critique of the latter's work. On the other hand, Lambert's experimental procedures were better suited than Mayer's to minimize the impact of these, and for this reason his conclusions are less open to objection. These conclusions somewhat resembled Mayer's. Lambert, too, found the force acting between individual particles of the two magnets to depend on the "absolute force" of these particles and inversely on the square of the distance between them. The force also depended, he said, on the sine of the angle enclosed by the axes of the two magnets. He maintained that the "absolute force" of a particle increased with distance from the center of the magnet, and he inclined to Mayer's supposition that the relationship involved here was a linear one.

Lambert's experiments, and more particularly his analyses of the numerical data they yielded, were most impressive. Despite Lambert's having done the work in Berlin, however, as a fellow member with Euler of the Academy of Sciences there, he shows not the slightest awareness of what Aepinus had written on the same subject a few years earlier. The Berlin Academy had a copy of Aepinus's *Essay*,

[74] For assessments, more favorable to Mayer, of Aepinus's objections, see Christopher Hansteen, *Untersuchungen über den Magnetismus der Erde*, Christiania, 1819, pp. 288–94, and Forbes's introduction to Mayer's magnetic papers (Mayer, *Unpublished Writings*, III, 9–11).

[75] Lambert, *Mém. Acad. Berlin*, XXII, 1766, 22–48, 49–77.

and we might have expected Euler to have drawn it to Lambert's attention if he did not know of it already. Yet Lambert wrote as if Aepinus's book had simply never been written! His ignorance of the work was the more unfortunate, because even though his own study was a truly virtuoso performance, it nevertheless lacked a sufficiently strong physical framework to support it and enable it to be taken further, and this Aepinus's analysis might well have provided.

Lambert was not the only author on magnetism in the mid-1760s to write in apparent ignorance of Aepinus's work. The important and influential *Tentamina philosophica de materia magnetica* of the Franeker professor Anton Brugmans, published in Franeker in 1765, revealed no greater acquaintance with Aepinus's *Essay* than did Lambert's two memoirs. Yet Brugmans was, with Aepinus, one of the first finally to abandon the traditional circulation theory of magnetism. In its place, Brugmans developed a theory somewhat analogous to, but more fully articulated than Symmer's two-fluid theory of electricity. Iron, Brugmans supposed, is normally saturated with a neutral combination of two distinct, highly elastic fluids. Magnetization consists in the separation of these, so that one end of a piece of iron acquires a surplus of one fluid, while the other end acquires a surplus of the other. Because each fluid is highly elastic, such accumulations tend to spread out into the surrounding air. Each pole of a magnet is therefore surrounded by an atmosphere of its particular magnetic fluid. Whenever such an atmosphere interacts with another or with the neutral combination of fluids in an unmagnetized piece of iron, there is, Brugmans said, a strong tendency toward "equilibrium," by which he seems to have meant a state in which the two fluids neutralize each other to the maximum possible extent and in which the "elastic force" or pressure in any residual atmospheres is minimized. Brugmans emphasized repeatedly that magnetic effects did not involve circulations of the fluids he had invoked, but were brought about by static distributions of these. He apparently saw them as pressure effects of some kind; for example, in a very interesting passage in which he attempted to explain the patterns formed by iron filings around a magnet, Brugmans supposed that the loops of gradually increasing "radius" formed by the filings represented strata of gradually increasing "elastic force" in the atmosphere surrounding the magnet.[76]

During the 1770s, Aepinus's ideas became somewhat better known in both Germany and Holland, as is evidenced by the entries submitted to a prize competition conducted by the Bavarian Academy of

---

[76] Brugmans, *Tentamina*, pp. 140–43.

Sciences. The subject of the competition, first announced in 1774, was the question, "Is there a true physical analogy between the electrical and magnetic forces?" The Academy, prompted perhaps by news from Vienna of Franz Anton Mesmer's curious ideas, added to this the subsidiary question, "And if there is one, in what manner do these forces act on the animal body?" Dissatisfied at first with the standard of the entries received, the Academy renewed the competition in 1776. Even at the second judging no entry was deemed worthy of the full prize, but on this occasion the Academy did bestow a gold medal valued at twenty ducats upon the work submitted by one of Brugmans's fellow professors at Franeker, Van Swinden. At the same time, the entry submitted by the Benedictine Georg Christoph Steiglehner, a teacher at the St. Emmeran seminary in Regensburg who was soon to become professor of mathematics, experimental physics, and astronomy at the University of Ingolstadt, was awarded a lesser medal worth ten ducats. The Academy subsequently published both award-winning essays in its transactions, along with an unsuccessful entry submitted by the otherwise undistinguished Jesuit professor, Lorenz Hübner.[77]

It is striking that both the award-winning entries in this competition took Aepinus's *Essay* as their starting point. Indeed, Steiglehner's answer to the first of the questions posed by the Bavarian Academy amounted to little more than a detailed presentation of Aepinus's basic analyses, slightly extended at one point to deal with that newly discovered device, the electrophorus.[78] Van Swinden, though full of praise for Aepinus's work, was at the same time more critical of it. In fact, he denied that any valid parallels could be drawn between magnetism and electricity, and he devoted much effort to refuting both Aepinus's arguments and those put forward by Cigna in favor of an analogy rather different from that favored by Aepinus.[79]

[77] *Neue philosophische Abhandlungen der baierischen Akademie der Wissenschaften*, II, 1780, 1–384. I have also used the French translations of these essays published by Van Swinden in his *Recueil de mémoires sur l'analogie de l'électricité et du magnétisme, couronnés & publiés par l'Académie de Bavière . . . augmentés de notes & de quelques dissertations nouvelles*, La Haye, 1784.

[78] Van Swinden, *Recueil*, II, 3–114.

[79] Cigna, *Miscellanea Taurinensis*, I, 1759, 43–67. At this period, Cigna seems to have accepted a theory of electricity rather like Canton's "dynamical" version of Franklin's doctrine, and a traditional circulating-fluid theory of magnetism. Within this context, he saw a magnet as analogous to an electrified globe, and iron as the magnetic analogue of an electrical conductor, with the proviso (p. 44) that "a glass sphere does not emit electric fluid nor a resinous one receive it, unless the sphere is rubbed, [while] a magnet at any time whatsoever without any preparation receives and emits."

This was not the first occasion on which Van Swinden had taken issue with Aepinus's doctrines. He tells us that, as early as 1770, he wrote down for his own benefit "un examen très-détaillé" of Aepinus's work, in order to discover why this left him unconvinced despite its many excellent features.[80] Then, in his *Tentamina theoriae mathematicae de phaenomenis magneticis,* published in 1772, he publicly challenged some of Aepinus's ideas about magnetism. Both in this work and in his later award-winning essay, however, Van Swinden coupled his criticisms with high praise for what Aepinus had contributed to the subject.

In his prize essay, Van Swinden confessed that he found Aepinus's arguments much more difficult to rebut than Cigna's. The latter's claims could be isolated and countered one by one, whereas in Aepinus's case, his principles were only suppositions: "they cannot be refuted directly . . .: it is only by examining the consequences, more or less remote, that this physicist has drawn from them, that one can evaluate them."[81] Van Swinden's criticisms of Aepinus's work are unconvincing. They depend for the most part on a very narrow conception of what constitutes an acceptable analogy in science. The only kinds of analogy Van Swinden would admit were those between particular experimental facts. In his view, broad structural analogies between theories, such as the one Aepinus wished to draw, smacked too much of the old *esprit de système* that experimental science had long since outgrown: "I acknowledge," he said, "that I am not attached to systems in my work on this subject, but to the facts alone."[82] Yet we have seen that Aepinus provided, in the course of his analysis, a great deal of experimental backing for his position. Hence Van Swinden's criticism amounts to a rejection of the *way* Aepinus used experimental evidence to support his theory. Van Swinden, like others in the eighteenth-century tradition of experimental physics, wanted to use only well-attested experimental facts as his starting point in constructing his theories. Aepinus and other mathematical physicists in increasing numbers in the years that followed allowed themselves much greater licence in developing their theories, requiring instead, as their empirical control, that these led to experimentally confirmed consequences.

Van Swinden's arguments were criticized, and the alternative approach of Steiglehner and his master, Aepinus, upheld, in an extended review of the relevant section of the Bavarian Academy's transactions

[80] Van Swinden, *Recueil,* I, 157.
[81] Ibid.
[82] Ibid., I, 246.

that appeared in 1781 in the *Rheinische Beiträge zur Gelehrsamkeit*.[83] The writer was J. J. Hemmer, secretary of the local learned society at Mannheim, and author of several works on electricity at this period. Hemmer, like Steiglehner, was familiar with Aepinus's work at first hand. So too was J.S.T. Gehler, who gave actual page references to the *Essay* in the articles on electricity and magnetism in his excellent *Physikalisches Wörterbuch*, first published during the years 1787–1791.[84] Few other German writers on electricity were so well read, however, and most knew Aepinus only by vague repute, usually as the person who first denied the existence of real atmospheres of electric matter around electrified bodies, and thereby found the key to a proper understanding of induction effects such as the Leyden jar and the electrophorus.[85] A second extended review of the *Essay* itself in the Leipzig-based journal *Commentarii de rebus in scientia naturali et medicina gestis*,[86] a full quarter-century after the first, must have helped to make the work better known. Yet even after this date, would-be readers could sometimes find it hard to lay their hands on a copy.[87]

Meanwhile, in Holland, Van Swinden was not the only one to be influenced by Aepinus's ideas. When Martinus Van Marum and Paets van Troostwyk presented a memoir on the electrophorus to the Batavian Society for Experimental Philosophy in Rotterdam in 1783, they accepted without further explanation the additional force of repulsion between the particles of matter that Aepinus had introduced into Franklin's theory, and cited Aepinus's *Essay* as they did so. Unfortunately, Van Marum was swayed soon afterward by an entirely fallacious argument of Van Swinden's purporting to show that Aepinus had committed an elementary blunder in the analysis that led him to introduce his new force. He therefore abandoned Aepinus's views, and turned instead to Beccaria's version of the theory of atmospheres.[88]

---

[83] *Rheinische Beiträge*, 1781, Bd. I, pp. 428–59.

[84] Gehler, *Physikalisches Wörterbuch, oder Versuch einer Erklärung der vornehmsten Begriffe und Kunstwörter der Naturlehre*, 2nd ed., Leipzig, 1798–99, III, 99; IV, 801.

[85] E.g. A. Socin, *Anfangsgründe der Elektricität*, 2nd ed., Hanau, 1778, p. 54; J.C.P. Erxleben, *Anfangsgründe der Naturlehre*, 5th ed., Göttingen, 1791, pp. 497, 513, 518, 543 (notes by G. C. Lichtenberg); K. G. Kühn, *Die neuesten Entdeckungen in der physikalischen und medizinischen Elektrizität*, Leipzig, 1796–97, I, 2.

[86] *Commentarii*, xxx, 1788, 313–23.

[87] G. C. Bohnenberger, *Beyträge zur theoretischen und praktischen Elektrizitätslehre*, Stuttgart, 1793, I, 11.

[88] W. D. Hackman, "Electrical Researches," in R. J. Forbes et al., eds., *Martinus Van Marum: Life and Work*, Haarlem, 1969–74, III, 337. Cf. Van Marum and Paets

So far we have said little about the reception accorded Aepinus's *Essay* in Paris, despite the fact that this was the scientific capital of the world during the eighteenth century. It is high time we took up that part of our story, since it was the eventual adoption of Aepinus's ideas by the French school of mathematical physicists that began to emerge in the 1780s that finally secured him his place in the mainstream of electrical and magnetic history. Until then, however, progress was very slow: Nollet's effluvialist ideas about electricity dominated French thinking on the subject well into the 1770s.[89]

Around 1765, one of Franklin's few French supporters, Jean Baptiste Le Roy, wrote to Benjamin Wilson in London complaining that Aepinus's book, which Wilson had told him about, was unobtainable in Paris. Le Roy asked Wilson to bring a copy with him during his forthcoming visit to the French capital.[90] When the visit actually took place, however, it was not the leisurely affair that Wilson had intended but a desperate chase after a confidence trickster who had duped him of his life's savings. Le Roy's modest request would have been one of the last things Wilson thought about in his anxiety to track down his money. It is therefore unlikely that the Frenchman gained access to a copy of Aepinus's work at this period, at least from Wilson. In any event, he did not refer to it in papers on electricity that he read to the Paris Academy of Sciences in the early 1770s.[91] Nor did Sigaud de la Fond, who promulgated Franklinist ideas about electricity in the courses of public lectures he offered regularly in Paris in the late 1760s.[92] On the other hand, in 1773, Barbeu Dubourg, in his French edition of Franklin's works, included Aepinus in a select band of the century's foremost electrical investigators.[93] Obviously Dubourg, at least, was acquainted with and had been impressed by some of Aepinus's work. Yet even here, we cannot confidently conclude that it was Aepinus's *Essay* that had so impressed him, since it could equally well have been the work on the tourmaline. Indeed, the fact that

---

van Troostwyk, *Verhandelingen Bataafsch Genootschap Proefondervindelijke Wijsbegeerte*, VII, 1783, 214.

[89] R. W. Home, "Electricity in France in the post-Franklin era," *Actes du XIVe Congrès International d'Histoire des Sciences, Tokyo-Kyoto, 1974*, Tokyo, 1974–75, II, 269–72.

[90] Le Roy to Wilson, n.d.; British Library, Add. MS 30094, f. 122. The letter is postmarked 17 September, but the year is not specified. Internal evidence suggests that it was written in 1765.

[91] Le Roy, *HAS*, 1770, Mém. pp. 53–67; ibid., 1772, Pt. I, Mém. pp. 499–512; ibid., 1773, Mém. pp. 671–86.

[92] Sigaud de la Fond, *Traité de l'électricité*, Paris, 1771.

[93] *Oeuvres de M. Franklin*, I, 335.

Aepinus's name stood next to Benjamin Wilson's in Barbeu Dubourg's list suggests that the latter possibility is the correct one.

The edition of Franklin's works published by Barbeu Dubourg also included a letter in which the American outlined Aepinus's adaptation of the theory of "plus" and "minus" to the magnetic realm.[94] Here, and in the French edition of Priestley's *History and Present State of Electricity*, published a little earlier,[95] some of the ideas developed in Aepinus's *Essay* were brought to the attention of French readers for the first time. But no Frenchman seems to have taken up Aepinus's work in any serious way prior to Coulomb, in the prize-winning memoir on magnetism that he submitted to the Paris Academy in 1777.[96]

Coulomb may have learned of Aepinus's work from Le Roy, who was an old family friend.[97] Alternatively, his interest may have been aroused by Franklin's brief discussion in his *Oeuvres*. (Priestley is a less likely influence, since Priestley's account of Aepinus's magnetic theory was so inadequate that from it the basic structure of the theory could scarcely be made out at all.)[98] Wherever Coulomb first learned of Aepinus's work, it is clear that he subsequently read the *Essay* itself; otherwise he could scarcely have alluded with such evident familiarity to Aepinus's modified version of the "double touch" method of magnetizing iron.[99] By contrast, although a series of questions on electrical theory inserted by the Abbé Rozier in his journal *Observations sur la physique* early in 1777 clearly derived ultimately from Aepinus's denial of the Franklinian doctrine of atmospheres, their immediate inspiration seems to have been not a reading of the *Essay* but Cigna's successful repetition of the air condenser experiment.[100] And Cigna in turn relied on Priestley's *History* for his knowledge of what Aepinus had done in his great work! As late as 1781, Sigaud de la Fond's acquaintance with Aepinus's book was obviously still only second-

---

[94] Ibid., I, 277–79.

[95] Priestley, *Histoire de l'électricité*, Paris, 1771.

[96] Coulomb, "Recherches sur la meilleure manière de fabriquer les aiguilles aimantées," *Mémoires de mathématique et de physique présentés à l'Académie Royale des Sciences, par divers savans*, IX, Paris, 1780, 185, 258–59, 262.

[97] C. S. Gillmor, *Coulomb and the Evolution of Physics and Engineering in Eighteenth-Century France*, Princeton, 1971, pp. 7–8.

[98] Priestley, *Histoire*, II, 438–40.

[99] Coulomb, "Recherches," p. 185.

[100] *Observations sur la physique, sur l'histoire naturelle et sur les arts*, IX, 1777, 6, 81, 161. Cf. Cigna, *Miscellanea Taurinensis*, V, 1770–73, 105–108.

hand,[101] and in this he was probably more typical than Coulomb of French scientists of the day.

Coulomb was one of the greatest engineers of the age. His training at the French Army's engineering school at Mézières included a sizable mathematical component and also thoroughly familiarized him with the eighteenth-century tradition of *physique expérimentale*. Coulomb became both a highly skilled experimenter and a very competent mathematician. He was, in fact, one of the principal founders of the new tradition of mathematical physics that developed in France in the years after 1780.[102] Aepinus's *Essay* was an important forerunner of this new tradition and was, indeed, one of the earliest instances of such an approach to appear in print. Coulomb was greatly impressed by the work, which became the starting point of his own thinking on the subjects of electricity and magnetism.

Coulomb's 1777 prize essay on magnetism marked his first foray into this area. In it, his general approach to the subject was already clearly defined, and it was along the lines marked out by Aepinus. So far as the theory of magnetic action was concerned, Coulomb sought above all to combat the old theory of circulating streams of fluid and to replace it by a mathematical analysis based on the concept of attraction. Coulomb laid much stress on this point in his paper, arguing, in fact, that he had provided a formal proof that the directive property of the compass needle could not be explained in terms of the pressure of a torrent of fluid.[103] His concern bears witness to the continuing strength of the circulation theory in his day.

Duhamel and Antheaulme, the two leading French authors on magnetism in the 1750s, had accepted the circulation theory unhesitatingly. Nollet, in the sixth and final volume of his influential *Leçons de physique expérimentale* (1764), summed up the situation at that time by saying: "Although savants have embraced various opinions on the causes of magnetism, and have followed different paths in order to explain the phenomena thereof, they have always agreed on one point for, as it were, the basis of their systems; there is scarcely one amongst them who does not admit, around each natural or artificial magnet, a subtle and invisible fluid which circulates from one pole to the other, and which has been given the name, magnetic matter. This supposition is entirely probable, and can scarcely be denied in view

---

[101] Sigaud de la Fond, *Précis historique et expérimental des phénomènes électriques*, Paris, 1781, p. 555.
[102] Gillmor, *Coulomb*, passim.
[103] Coulomb, "Recherches," pp. 172–73, 256–57.

of the following experiment." [104] The experiment he gives is, as usual, the scattering of iron filings in the vicinity of a magnet. Yet Nollet himself had hesitations about the theory, even though he found the alternative of action at a distance unthinkable. "It is true," he said, "that one is as if forced to admit this cause in general, because one cannot perceive another; but when one compares it with its effects, the mind rebels, and conceives only with much difficulty that so many marvels could come about from such an apparently infertile source." [105] The Abbé Paulian went further and rejected the circulation theory altogether.[106] However, the alternative he proposed was even less satisfactory, since it merely replaced the mysterious power of a magnet by an equally mysterious action of a surrounding atmosphere of particles that were themselves little magnets. He therefore deservedly remained entirely without influence. In the 1770s, with the publication of Euler's *Lettres à une princesse d'Allemagne,* the traditional theory gained further support, and it continued to be generally accepted until Coulomb threw out his challenge to it.

Coulomb attributed his alternative theory based on the notion of attraction to "plusieurs Physiciens," of whom he identified three, Aepinus, Brugmans, and Wilcke. Aepinus he correctly regarded as the champion of the one-fluid theory developed by analogy with Franklin's theory of electricity. Wilcke he coupled with Brugmans as an advocate of the two-fluid theory on the basis of some brief remarks set out in 1766, within a few months of the appearance of the Dutch professor's book.[107] Coulomb recognized that the two theories were formally identical: "These two hypotheses," he said, "explain all the magnetic phenomena equally well, and in the same manner." In contrast to his later writings, where he clearly preferred the two-fluid account, Coulomb here seems to incline, if anything, toward the one-fluid version, despite his explicit acknowledgment of the point that later swayed him the other way, the need to postulate a force of repulsion between solid particles of iron.[108] Coulomb was less concerned with making a choice between the two theories, however, than with pointing out that in his view neither could adequately explain what happened when a magnet was cut in half. To account for this, Coulomb suggested that the quantity of fluid transferred from one end

---

[104] Nollet, *Leçons,* VI, 212.

[105] Ibid., pp. 223-24. Nollet had admitted to very similar reservations almost twenty years earlier (*HAS,* 1745, Mém. p. 112).

[106] Paulian, *Dictionnaire de physique portatif,* Avignon, 1760, art. "Aiman."

[107] Wilcke, *Abhandl. Schwed. Akad. Wiss.,* 1766, p. 316.

[108] Coulomb, "Recherches," pp. 262, 258.

of a steel bar to the other during magnetization might be very small compared to the total quantity it contained. In one of his later papers, he introduced, as a rather more satisfying alternative, the notion that magnetic fluid is confined within each molecule of iron, rather than being free to redistribute itself throughout a given sample.[109] This idea quickly came to be generally accepted by magnetic theorists. Nevertheless, leading nineteenth-century investigators such as Poisson and William Thomson still found it convenient to base a number of their calculations on supposed macroscopic distributions of fluid, as Aepinus did, rather than on Coulomb's more "correct" molecular dipole distributions.[110]

In his great series of researches on electricity and magnetism in the 1780s, Coulomb used his newly invented torsion balance to establish at last the laws according to which electrical and magnetic forces vary with distance. In the magnetic case, a number of earlier workers had already tried to establish the form of the force law, but their efforts had for the most part fallen down because they tried to measure the force between a magnet as a whole and a nearby piece of iron, rather than the force between two magnetic poles. Coulomb recognized very clearly that it was the force between two poles that he needed to measure. In his experiments, he made a point of using long (25 *pouces*, or approximately 2 ft.) thin magnetized needles, and the first thing he sought to establish was that, with them, "one can without sensible error suppose that the magnetic fluid is concentrated within 10 *lines* [i.e. 10/12 inch] of its end."[111] (In an aside, Coulomb pointed out that this result controverted the assumption that "certain authors" had made, that the density of magnetic fluid increased linearly from the center to the extremity of a magnet.[112] We have seen that an assumption equivalent to this was central to the arguments of Mayer and Lambert; Coulomb probably also had Van Swinden in mind, as a third proponent of the idea.) Thereafter, he was able to confine his attention to a single force, the one acting between the two poles relatively close to each other in his experiments. This he could measure with the torsion balance, confident that the forces due to the more distant poles of the two needles could be safely neglected.

In the electrical case, prior to Coulomb's work fewer attempts had

[109] Coulomb, *HAS*, 1789, Mém. p. 488.
[110] Poisson, *Mém. Acad. Roy. Sci. (Paris)*, v, 1821-22, 492ff.; Thomson, *Phil. Trans.*, CXLI, 1851, 250ff.
[111] Coulomb, "Second mémoire sur l'électricité et le magnétisme," *HAS*, 1785, Mém. p. 593.
[112] Ibid., p. 600.

been made to determine the form of the force law. Around 1760, Daniel Bernoulli had used a suitably modified hydrometer to measure electrostatic forces directly and had concluded that they followed an inverse-square law with respect to distance. His experimental arrangement had left much room for error, however, and, perhaps because he recognized this himself, he had never published a formal report of his investigations. Only a brief account had ever appeared in print, in an article published in the *Acta Helvetica* of the Physico-Medical Society of Basel by one of Bernoulli's colleagues at the University of Basel.[113] Some years later, Priestley had inferred, from Franklin's observation that cork balls suspended within a metal cup were unaffected by any electricity given to the cup, that the law must be inverse-square in character. However, he himself seems not to have been particularly impressed by his argument, because immediately afterward he added, in increasing numbers in later editions of his book, alternative explanations more in keeping with contemporary ideas about electrical atmospheres.[114]

Not long after this, John Robison, while still a young lecturer at Glasgow University, discovered the inverse-square form of the law with some experiments "conducted in the most obvious and simple manner, suggested by the reasonings of Mr. Æpinus." So, at least, he stated in a work written many years later, adding that he "had read an account of his investigation in a public society in 1769." [115] Of this account, however, no trace has survived, and the experiments to which Robison lays claim seem to have been entirely without influence. The same may be said for Henry Cavendish's famous and very precise determination of the law in 1772, since this, too, remained entirely unknown until long afterward.[116] As in the magnetic case, it was upon Coulomb's work rather than anybody else's that the scientific community of the day based its acceptance of the inverse-square law.

The chief factor limiting Aepinus's computations in his *Essay* had been his ignorance of the precise forms of these force laws. Coulomb's work made possible the transformation of both electrostatics and magnetostatics from the semi-mathematized condition in which

[113] A. Socin, *Acta Helvetica*, IV, 1760, 214–30.

[114] Priestley, *History*, London, 1767, p. 732. Cf. 2nd ed., London, 1769, pp. 711–12, and 4th ed., London, 1775, pp. 690–91.

[115] Robison, "Electricity," *Encyclopaedia Britannica*, 3rd ed., *Supplement*, vol. I, Edinburgh, 1801; 2nd ed., Edinburgh, 1803, p. 578.

[116] Cavendish's account of his experiments was first published in his *Electrical Researches*, ed. J. Clerk Maxwell, Cambridge, 1879, pp. 104–12.

Aepinus had left them into rigorously deductive mathematical sciences. Coulomb began the process himself, once he had the forms of the laws firmly settled. It was completed a generation later, in the work of Poisson and Green.[117]

Coulomb's admiration for Aepinus is evident throughout his papers on electricity and magnetism, as is his respect for Musschenbroek as an authority on experimental magnetism. But, as Gillmor points out,[118] Coulomb cites few other authors. Most notably, he mentions Franklin only once, and then only to report that his theory of electricity had been adapted by Aepinus to the magnetic case.[119] It is evident that, so far as Coulomb was concerned, Franklin's qualitative discussion belonged to an older tradition of physical inquiry that was rapidly being outdistanced by the newer mathematical approach based unequivocally on the notion of action at a distance that Aepinus had pioneered.

In his writings after 1780 Coulomb opted for two-fluid theories of both electricity and magnetism, rather than Aepinus's one-fluid varieties. Nevertheless, he still saw the two alternatives as operationally indistinguishable, and as a matter of convenience he was perfectly happy, in the analyses that constitute the bulk of his later papers, to write consistently of "the electric fluid" and "the magnetic fluid," as if there were only one of each rather than two. Coulomb's preference for the two-fluid accounts derived simply from his distaste for Aepinus's postulate that the particles of ordinary matter mutually repelled each other, but he was never dogmatic on this point.

Almost certainly as a result of Coulomb's repeated insistence upon the merits of Aepinus's work, and perhaps also inspired by Volta's energetic promulgation of similar views about electrical atmospheres during a visit to Paris in 1782,[120] other French writers during the 1780s began at last to pay heed to what Aepinus had done. It was at this time that Gaspard Monge, for example, prepared a brief summary of the *Essay* for his own use.[121] Later in the decade, Buffon constantly cited Aepinus's book and Musschenbroek's *Dissertatio de*

---

[117] S. D. Poisson, *Mémoires de la classe des sciences mathématiques et physiques de l'Institut Impérial de France*, 1811, Pt. 1, pp. 1–92; ibid., Pt. 2, pp. 163–274; *Mém. Acad. Roy. Sci. (Paris)*, v, 1821–22 (publ. 1826), 247–338, 488–533. George Green, *An Essay on the Application of Mathematical Analysis to the Theories of Electricity and Magnetism*, Nottingham, 1828.

[118] Gillmor, *Coulomb*, p. 175.

[119] Coulomb, "Recherches," p. 258.

[120] Volta, *Epistolario*, Bologna, 1949–55, II, 84–85, 96. A later letter from De Luc to Volta (ibid., p. 163) suggests that Volta's visit did not have much lasting impact on the French.

[121] René Taton, *L'oeuvre scientifique de Monge*, Paris, 1951, pp. 326, 386.

magnete as his chief authorities on magnetic practice in the volume on the magnet in his *Natural History of Minerals*.[122] Buffon, however, was an old man by this time, and he failed to profit from Aepinus's theoretical analysis, or for that matter from Coulomb's. He advocated instead the highly idiosyncratic notion that the Earth's magnetism was caused by *electric* matter, exhaled from the hot equatorial regions of the globe, streaming both northward and southward to re-enter the Earth at the poles. Similar streams, he imagined, also surrounded smaller magnets.

Much more significant was the publication by the Abbé René-Just Haüy in 1787 of an excellent epitome of Aepinus's *Essay*.[123] Even though Haüy made it clear in his introduction that, like Coulomb, he preferred the two-fluid alternatives to Aepinus's doctrines, he presented a very faithful account of what the latter had done. His book therefore served to make Aepinus's ideas much more accessible to French readers than they had ever been before, while the publication of an extract from Haüy's book in the journal *Observations sur la physique* [124] brought them to the attention of a still wider public.

Significantly, when the Paris Academy of Sciences set up a committee to report on Haüy's book, none of the old-style electricians was included on it. Instead, the committee—consisting of Coulomb, Laplace, and the mathematicians Legendre and Cousin—had a strong mathematical bias.[125] Coulomb departed on a visit to England soon afterward, so that he was unable to append his signature to the committee's report. There is no doubt, however, that he agreed entirely with the sentiments expressed by his colleagues, who wrote in terms of the highest praise of Aepinus's achievement, stating, indeed, that his book constituted a veritable "epoch in the history of the sciences." [126]

Thereafter, Aepinus's standing in the eyes of the French physics community was secure. When, in August 1787, a vacancy appeared in the select ranks of the Foreign Associates of the Paris Academy of Sciences, Aepinus's name was one of those brought forward by the commissioners appointed to prepare nominations. When the matter was put to the vote, however, Sir Joseph Banks was elected, with Joseph Black as second choice. Yet the mere fact that Aepinus's name

---

[122] Buffon, *Histoire naturelle des minéraux. Tome V. Traité de l'aimant et de ses usages*, Paris, 1788.

[123] Haüy, *Exposition raisonnée de la théorie de l'électricité et du magnétisme, d'après les principes de M. Æpinus*, Paris, 1787.

[124] *Observations*, XXXI, 1787, 401–17.

[125] Gillmor, *Coulomb*, p. 176.

[126] Extract from the *Registres* of the Académie Royale des Sciences, Paris, published by way of a preface to Haüy, *Exposition raisonnée*, p. [xxix].

was put forward in such august company is evidence of the esteem in which he was held by this time. Coulomb was one of the nominating commissioners on this occasion, and one cannot help suspecting that it was he who suggested Aepinus as a suitable candidate. A further vacancy occurred in 1789, Coulomb was not a commissioner, and Aepinus's name was not brought forward; yet following Franklin's death in 1790, when Coulomb was again appointed a commissioner, Aepinus's name reappeared on the list. His candidature was no more successful this time, however, than on the previous occasion; the naturalist P. S. Pallas was appointed instead.[127]

A further indication of Aepinus's new standing in France at this period comes from a treatise on magnetism published in 1788 by the Genevan professor Pierre Prévost. Prévost's chief concern in his book was to expound a novel theory about the ultimate causes of magnetic forces. Concerning the manner in which these forces manifested themselves in the world, he simply referred to Aepinus's *Essay* and Haüy's epitome. As to Aepinus's importance in the history of the subject, he had no doubts at all. "Mr. Æpinus," he wrote, "has produced in the theory of magnetism a revolution that can be compared to the one that Newton brought about in general physics."[128]

Similar statements recurred frequently in the French scientific literature in the years that followed. Delambre, in his *éloge* of Coulomb, remarked that "Aepinus is the first who submitted them [i.e. electricity and magnetism] to analysis."[129] Arnault's *Biographie nouvelle des contemporains* restated the theme at greater length in a passage that brought out very clearly the basis of Aepinus's new-found reputation. Aepinus, we read, is "one of the men who apply mathematics to physics with the most utility; rigorous in his deductions, endowed with a rare sagacity in his experiments, he caused several important steps to be made in the natural sciences. To submit to exact calculation that electrical power so rich in phenomena, so rapid, so mysterious and so little known, as well as the magnetic force, whose relations with electricity are still a secret; that was one of the most audacious efforts of human patience and science: Æpinus dared to attempt it."[130]

---

[127] Gillmor, *Coulomb*, pp. 246, 250; Institut de France, *Index biographique des membres et correspondants de l'Académie des Sciences* . . . , Paris, 1968; personal communication from the Permanent Secretaries, Académie des Sciences, Paris.

[128] Prévost, *De l'origine des forces magnétiques*, Geneva/Paris, 1788, preface.

[129] Delambre, "Eloge historique de M. Coulomb," *Mémoires de l'Institut National des Sciences et Arts (sciences mathématiques et physiques)*, VII, 1806, Hist. p. 212.

[130] A. V. Arnault et al., *Biographie nouvelle des contemporains, ou dictionnaire historique et raisonné* . . . , Paris, 1820–25, art. "Æpinus, François-Marie-Ulrich-Théodore."

Later in the nineteenth century, readers of *La grande encyclopédie* were told that Aepinus's "profound mathematical knowledge imprinted on his demonstrations a remarkable character of rigor and clarity." His works were, they were assured, "generally very remarkable."[131]

In France, therefore, the belated recognition of Aepinus's merits is to be dated to the 1780s, and is intimately bound up with the successes that Coulomb achieved at that period. Thereafter, Aepinus was widely hailed as the founder of that mathematical approach to the sciences of electricity and magnetism that Coulomb had subsequently done so much to foster. Although Aepinus's work was known outside France much earlier than this, few if any of those who knew it would have placed it on such a high pedestal as the French now did. Nevertheless, as the remarkable power of the new mathematical approach to physics that Aepinus had pioneered came to be more widely appreciated, physicists elsewhere came to esteem his *Essay* equally highly.

A German edition of Haüy's epitome, followed perhaps by a second edition a few years later, doubtless contributed to Aepinus's still expanding reputation in that country.[132] In Britain, the article on electricity in the third edition of the *Encyclopaedia Britannica*, published in the 1790s, still showed no awareness of Aepinus's work. (The article on magnetism, being based largely on Cavallo's *Treatise*, was somewhat better informed.) The real breakthrough came shortly afterward, with the publication of a two-volume supplement to this edition of the *Britannica*, because included in these additional volumes were long articles by John Robison on electricity and magnetism, incorporating extensive summaries of Aepinus's arguments.[133]

Robison indicates that he himself knew Aepinus's work as early as 1769. Later, he spent some years in St. Petersburg as secretary to Admiral Sir Charles Knowles, who was on loan from Britain to the Russian navy. During that time Robison came to know Aepinus personally, and was presented by him with a copy of the *Essay*.[134] We

---

[131] A. Berthelot, ed., *La grande encyclopédie: inventaire raisonné des sciences, des lettres et des arts*, Paris, 1882–1902, art. "Æpinus, Franz-Ulrich-Theodor."

[132] Haüy, *Darstellung der Theorie der Elektricität und des Magnetismus nach den Grundsätzen des Herrn Aepinus*, Altenburg, 1801. Cf. Appendix below, item **66C**.

[133] *Encyclopaedia Britannica*, 3rd ed., *Supplement*, Edinburgh, 1801, arts. "Electricity" (vol. I) and "Magnetism" (vol. II). These articles were reprinted in toto in Robison's posthumously published *System of Mechanical Philosophy*, Edinburgh, 1822, IV.

[134] Robison to James Watt, – December 1796, Eric Robinson and Douglas McKie, eds., *Partners in Science: Letters of James Watt and Joseph Black*, London, 1970, p. 248. Robison's annotated copy of the *Essay*, marked on the title page "Joh

may presume that, following his taking up the Chair of Natural Philosophy at the University of Edinburgh, Robison included some discussion of Aepinus's work in his lectures. Certainly we know that his lectures were unpopular because they were regarded as too severely mathematical in their approach,[135] a criticism that would be entirely consistent with his having used Aepinus's work rather than, say, Franklin's, as the basis for what he had to say about electricity. In any event, Robison's articles in the *Encyclopaedia Britannica* would for the first time have brought the full power and extent of Aepinus's analysis, "unquestionably one of the most ingenious and brilliant performances of this century," before a wider British audience to whom it had been until then, so Robison lamented, "little known."

By this time, Aepinus's name had already begun to appear more frequently in the writings of British electrical and magnetic investigators. As in Germany, the acquaintance with his work was often second-hand, and came by way of Priestley's *History*. Sometimes, however, as in Cavallo's *Treatise on Magnetism* (1787), the knowledge of what he had achieved came from the *Essay* itself. An article in *Nicholson's Journal* in 1800 took as its starting point the extract in the *Observations sur la physique* prompted by the appearance of Haüy's epitome.[136] And one suspects Robison's guiding hand behind a paper read by Gaspard de la Rive to the Royal Medical Society in Edinburgh in 1797 on the theme, "Does the theory of Æpinus account in a satisfactory manner, for the principal phenomena of Magnetism?" Echoing what Prévost had written some years before, or perhaps even quoting him directly, de la Rive began his paper with the unequivocal assertion, "Æpinus has produced in the theory of Magnetism a Revolution, which may be compared to that which was operated by Newton in Physics." [137] The final seal of approval, so far as Aepinus was concerned, came with the publication of Thomas Young's *Course of Lectures on Natural Philosophy and the Mechanical Arts* (London, 1807). As in Robison's *Encyclopaedia Britannica* articles, the theories of electricity and magnetism set out in this work were those of Aepinus,

---

Robison 1772" and inscribed "Ex dono illustr.[m1] Auctoris," is preserved in the Ronalds Library, Institution of Electrical Engineers, London.

[135] J. B. Morrell, "The University of Edinburgh in the late eighteenth century: its scientific eminence and academic structure," *Isis*, LXII, 1971, 158–71, p. 161.

[136] George Miller, "Observations on the theory of electric attractions and repulsions," *Journal of Natural Philosophy, Chemistry and the Arts*, IV, 1800–1801, 461–66.

[137] De la Rive, *Royal Medical Society, Edinburgh: Dissertations*, XXXVII, 1797–98, 249–66. Cf. Paul A. Tunbridge, "Faraday's Genevese friends," *Notes and Records of the Royal Society of London*, XXVII, 1973, 263–98.

with the inverse-square force law added.[138] Thereafter, Aepinus was cited as a matter of course by British as well as French authors, as one of the principal founders of the modern theories of electricity and magnetism.

It is noteworthy that Robison and Young, authors of the first extended expositions of Aepinus's work to be published in English, were, like their French counterparts Coulomb and Haüy but unlike most earlier British students of electricity and magnetism, competent mathematicians. The merits of Aepinus's approach were never obvious to those working in the older experimentalist tradition of physical inquiry. Nor was his mathematics sufficiently challenging in itself to interest men such as Euler and d'Alembert, who tended to see physical problems merely as excuses for developing new and ever more sophisticated methods of mathematical analysis. Only when mathematics and experiment combined in the last years of the eighteenth and the first years of the nineteenth century to form a rich new tradition of genuinely mathematical physics did Aepinus at last find an audience that could fully appreciate what he had tried to do in his book. But once this new tradition developed, those working within it found in Aepinus's work a remarkably early and also a remarkably well-developed example of the kind of thing they themselves were trying to achieve. It was not without reason, therefore, that they hailed his *Essay* as marking "an epoch in the history of the sciences." In a very real sense, it did.

[138] Young, *Course*, I, 658–67, 685–95. Some years earlier, Young had for a time maintained an alternative theory of aethereal electrical and magnetic atmospheres (Geoffrey Cantor, "The changing role of Young's ether," *British Journal for the History of Science*, v, 1970, 44–62).

AN ESSAY ON THE THEORY OF
ELECTRICITY
AND
MAGNETISM

Together with two dissertations, of which the first explains a certain electrical, and the second, a certain magnetic phenomenon.

BY

F. U. T. AEPINUS

Member of the Imperial Academy of Sciences of St. Petersburg, the Royal Berlin Academy of Sciences and the Electoral Academy of Mainz at Erfurt.

In the form of a Supplement to the Memoirs of the Imperial Academy of St. Petersburg.

ST. PETERSBURG

AT THE PRINTERY OF THE ACADEMY OF SCIENCES.

TO THE MOST ILLUSTRIOUS
AND
EXCELLENT

CYRIL

COUNT RASUMOWSKI,

HER SACRED MAJESTY'S

HETMAN

OF LITTLE RUSSIA TO BOTH BANKS OF THE
BORYSTHENES, AND OF THE LEGIONS
BEYOND THE CATARACTS,

HER MAJESTY'S

CHAMBERLAIN ORDINARY,
PRESIDENT

OF THE IMPERIAL ACADEMY OF SCIENCES,

PROTRIBUNE

OF THE ISMAILOVSKII PRAETORIAN GUARD,

KNIGHT

OF THE ORDERS OF ST. ANDREW, THE WHITE EAGLE,
ST. ALEXANDER AND ST. ANNE,

MEMBER

OF THE ENGLISH ROYAL SOCIETY, AND OF
THE BERLIN ACADEMY OF SCIENCES
AND BELLES-LETTRES

TO MY MOST GRACIOUS LORD

Most Illustrious Count
Most Gracious Lord [1]

Among those who have scrutinized nature, some have outstripped the rest in their rapid progress. Such a man did Newton show himself to an astonished world: a mighty river, overflowing its banks, bursting through levees, flooding the plains with its ferocious waters, submerging everything far and wide as far as the eye could see.

What raised such immortal men to the pinnacle we admire is quite clearly an unusual vigor of mind received from nature, which was generous to them. But does not the man who dares to ascribe everything to their genius alone, attribute to this more than is proper?

Suppose that a man endowed with such great gifts of mind lacks the stimuli to attempt outstanding deeds in order to win the first place among mortals. Will he seek the Sublime? Or are greater things to be expected from a man of moderate genius who is inflamed by a strong incentive? A little globe driven by compressed vapors can smash an obstacle which a rock hurled by a weak hand hardly makes tremble.

Members of the Academy which venerates PETER THE GREAT as its Founder and ELISABETH, heir of her father's renown, as its Patron, lack nothing which can elevate the mind. To You, to whom her HIGHNESS has committed the care of our Society, the Academy owes it that we lack no opportunity and no means of satisfying the ardor which the great Name of our Founder and that of our Patron excites; through that grace of Yours, with which you are accustomed to watch our labors, you fire the heart of each of us more and more every day to strive for excellence.

Driven by such incentives, I dared to test whether I could approach certain mysteries of nature and discover what causes aid nature to produce its marvelous phenomena. That power of bodies which is called Electricity is newly discovered and insufficiently explored; but that it is common to several bodies, and can produce most violent effects, must be taken as a sign that it should be numbered among

---

[1] Count Kyril Razumovskiĭ (1728–1803), younger brother of the Empress Elisabeth's favorite, Alexeĭ Razumovskiĭ, was president of the St. Petersburg Academy of Sciences, 1746–1765.

the prime driving forces of nature, and we must not consider as fallacious the opinion of those who suspect that this force plays a role in important phenomena of nature.

Therefore why should not this force be considered worthy of provoking the constant attention of investigators of nature, since when it has been sufficiently explored, it is to be hoped that it will help us to unlock the very secrets of nature? I was much attracted by the theory of this force proposed by Franklin, *for I frequently found that his theory satisfied the phenomena better than any other. Yet I seem to have discovered some defects in that excellent theory; these I have striven to emend, and with the help of these improvements to bring the theory into complete agreement with the phenomena. These attempts, I dare to hope, have not been completely unsuccessful and they have lit my way to beginning various and, I think, new discussions; this Book which I dedicate to You, contains the most important of these.*

Magnetic force, the twin of electric force, was known earlier to investigators of nature and was explored with perservering zeal. Although it is not found in all bodies, but is a property of iron and of minerals of iron alone, still it is worthy of being strenuously toiled over in diligent investigation because it is as marvelous as, and has to date offered far greater uses to the human race than, the electric force, yet the true causes of its phenomena seem more hidden than the causes of the effects arising from the electric virtue of bodies. A skillful examination of this force is the chief task I have proposed to follow in my work.

I offer jointly a theory of electric and magnetic virtue and will perhaps amaze some who are unaware of the reasons which prompted me to act in this way. I demonstrated in a lecture I gave last year on the day dedicated to the Name *of the* AUGUST EMPRESS, *celebrated by a public gathering of the Academy, that there is such a great, I do not say similarity, but congruence in the phenomena of electrical and magnetic force that it can hardly be doubted that the causes, by means of which nature achieves the effects of the two forces, are very similar.*[2] *The observations I am speaking of led me to detecting the theory expounded here and I consider that I could only give a clear explanation of it if I maintained a constant comparison between the electric and magnetic forces.*

The theory led to some new discoveries and I owe to it, as well per-

---

[2] Aepinus, *Sermo academicus de similitudine vis electricae atque magneticae*, St. Petersburg, 1758.

DEDICATION

haps to other things of lesser moment, my happening on a method of increasing the strength of natural magnets to an unusual degree and my capability of making the methods of constructing artificial magnets used by Physicists to date easier, more convenient, and more efficient. With its help I was able to remove the doubts weighing heavily on the opinion of those doctors of natural Philosophy who assert that the globe of the earth contains, as a nucleus, a magnet of great mass. Finally, I strove to perfect the magnetic needle, an instrument of basic necessity for the human race to live a more convenient life. I give here only the fundamentals of my improvements and keep a more detailed exposition to a book on this subject alone.[3]

To You, Most Illustrious Count, *I offer a book which attempts to demonstrate what I have recounted and I beg with all fitting veneration that you deign graciously to accept it. I do so with the intention, first, of giving an account of the labors in which I have been involved since* You *called me to St. Petersburg and of showing that, though my strength failed, I never lacked the desire, by satisfying* Your *zeal for the increase of the Sciences, of fulfilling the goal of the GREAT FOUNDER of the Academy; and finally I do so in order to provide myself an opportunity of giving* You *with devoted heart most humble thanks for the outstanding benevolence which you have deigned to bestow on all those who cultivate the Sciences and very often on me in particular.*

F. U. T. AEPINUS

[3] This projected work was never published.

Preface

I write the preface of a work which is published in an imperfect state. For, at the conclusion of my labor, I see that some things could have been expressed more harmoniously, more distinctly and more methodically than they are. Though I was much inclined to refine the whole work, there was neither the time available, nor did I think it advisable.

An illness which kept me bed-ridden for quite a long time at the start of the year prevented me from the very beginning from making the book as perfect as I wanted. When I recovered my health there was hardly enough time to revise it. The Academy wanted me to hand it to the printer so that the learned world might know the work being done in our Academy on the theory of the magnet, before it decided the prize proposed on the subject of artificial magnets.[4]

I did not want to work longer on my book. For perhaps the theory I propose is false, in which case I thought I could do better than waste time uselessly polishing an erroneous hypothesis. If my hypothesis is contrary to nature I want it to be consigned to eternal oblivion along with other figments of rank geniuses. But if those who are learned about nature judge more mildly and think my theory is consonant with the truth, I hope they will readily condone certain defects in a book and an author which open a new path to the secrets of nature. So in this case too I did not judge it necessary to remove some of the remaining faults in the book.

I have included certain analytic, though short and simple, calculations, forced in part by the nature of the subject and in part to escape, as I have mentioned regularly in the work itself, the prolixity of the usual language. I trust no one will take this badly. For I hardly think that anyone who longs to be called an investigator of nature will be so unacquainted with the mathematical sciences that he will be troubled by the easy calculations employed here.

To readers who are skilled in natural Philosophy, I leave free judgment concerning this work of mine. If the theory I have proposed is found, after skilled examination, to stray from the truth, I confess I have missed my primary aim, but I think that, nevertheless, my readers

[4] See above, p. 36.

will find in my book some not unuseful things. For if I judge rightly there are certain new things in it which are not so dependent on my theory that they ought to collapse if the theory falls. Thus I was the first to observe that unelectrified bodies, whilst they remain such, are neither attracted nor repelled by other electrified bodies, just as ferreous bodies are acted upon by a magnet only after they have acquired magnetic force. I have set out various noteworthy phenomena which are usually observed when electricity is produced by rubbing two glass plates together. I have given a clear description of the method I invented some years ago [5] of producing the electric commotion, as it is usually called, without the help of glass. I have shown why the globe of the earth, even though it is a magnet, exerts no magnetic attraction or repulsion on iron or on a magnet, and thus I have dismissed a serious doubt which impeded the opinion, familiar to investigators of nature, that either the globe of the earth itself is a magnet of great mass, or it at least encloses one. I have proposed a method of increasing the strength of natural magnets, or rather of increasing their aptitude for receiving a greater degree of magnetic force, by means of cementation. I have improved *Michell's* method of the double touch; I have supplied a more efficient way than was known before of constructing artificial magnets with the help of a natural magnet; and I have demonstrated several others things which do not need to be set out in detail.

I have something further to say about the final two chapters. A treatise written in French, bearing the epigram *Ab una scintilla augetur ignis,* was delivered, among others, to the Academy, in the hope of winning the prize, and I see that its author had in his possession before I did an improved method of the double touch similar to mine, and that he is accustomed to use almost the same method as my own in, so to speak, exciting the first seeds of magnetic force without the help of a natural magnet.[6] So lest I appear to claim what is not mine, I declare that the treatise of that author must be considered as the first account of these methods, and I gladly yield this praise to him.

I have added two Dissertations at the end. The first supplies an explanation, drawn from *Franklin's* Theory of electricity and improved by me in certain heads, of an electrical phenomenon detected by that outstanding and diligent investigator of nature, the famous

---

[5] Cf. Aepinus, "Mémoire concernant quelques nouvelles expériences électriques remarquables," *Mém. Acad. Berlin*, XII, 1756, 119–121.

[6] Antheaulme, *Mémoire sur les aimants artificiels,* Paris, 1760.

Richmann.[7] The second tries to give the causes of a paradoxical magnetic phenomenon whose discoverer is uncertain. I trust that both will be not unpleasing to my readers.

St. Petersburg, October, in the year of the Holy Redeemer 1759.

[7] G. W. Richmann, *Novi commentarii Academiae Scientiarum Imperialis Petropolitanae,* IV, 1752–53, 324.

Introduction

1) After the discovery of the marvelous phenomena and laws of electricity, a force almost unknown to the ancients, some physicists soon began to contemplate a comparison between this newly discovered force and the magnetic force, which is equally marvelous but already known in ancient times. For at first sight they were presented, though obscurely, with an analogy between the phenomena depending on the two forces, and this stimulated those who were well aware of the extent nature loves to act according to analogical laws to work on this point. In this task, as in other matters, the most worthy *Musschenbroek*[8] showed himself highly diligent; he published a full catalogue of the similarities and dissimilarities of the two forces.[9] A few years later, if I recall rightly, the *Bordeaux Academy* of sciences publicly proposed a prize for research into the similarity between electricity and magnetism; but I do not remember, nor have my very careful inquiries informed me, which dissertation won the prize, nor whether it was published and made publicly available.[10] There were several others also who indicated their suspicion of a similarity or nexus obtaining between electricity and magnetism; but it is pointless to list them. We must indeed confess that before *Franklin's* theory of electricity it was scarcely possible to develop the analogy that was being sought. For apart from the extremely meager resemblance, that both electrified bodies and those imbued with magnetic force exercise various attractions and repulsions (and even then in doing this they seem to follow quite diverse laws), hardly any other similarity is apparent; so it had to be considered that, in producing the phenomena of these forces, nature was following paths that were altogether different rather than similar or analogous.

2) Although I had already at times in the past pondered this same problem, I saw hardly any glimmer of hope of uncovering anything remarkable and I refused to apply myself more diligently to a task I considered useless. But toward the end of 1756 and the beginning of

---

[8] Aepinus regularly spells Musschenbroek's name with only one 's.' Also, 'Franklin' frequently becomes 'Francklin.' Both errors have been corrected wherever they occur.

[9] P. van Musschenbroek, *The Elements of Natural Philosophy*, trans. J. Colson, London, 1744, I, 210–11.

[10] The winning author was the Jesuit Laurent Béraud, and his essay *was* published; Beraut (sic!), *Dissertation sur le rapport qui se trouve entre la cause des effets de l'aiman, et celle des phenomenes de l'électricité*, Bordeaux, 1748.

'57, when I was still living in Berlin, I was unexpectedly recalled to considering it. I was occupied at that time in studying the marvelous properties of the *Tourmaline* stone and in producing a description of my discoveries which I offered to the illustrious Berlin Academy of Sciences; this is printed in the Commentaries of that illustrious Academy for the year 1756.[11] I was struck at the time, and not only me but also those with whom I shared my discoveries about the Tourmaline, by the utmost similarity between this stone and a magnet. This is so obvious that I have no doubt that anyone who has read what I then wrote about the Tourmaline has thought of it without prompting. So, spurred on by this opportunity and by a brighter gleam of hope, I began afresh, and more diligently, to explore the similarity of the magnetic and electrical forces. I was indeed making outstanding progress, but since I was preparing to depart, my time was too short to lend sufficient maturity to my thoughts, and for this reason I made no report to the Academy, though I informed some friends of my discoveries; I told them I believed that the causes of magnetic and electrical phenomena were extremely similar and that the effects of a magnet were analogous to the effects of a Leyden jar.

3) After my arrival in St. Petersburg, I felt my labor should not be wasted, since I had begun to hope that through its help I could gain entrance to the secrets of nature. I do not think that the diligence I employed in probing this matter turned out unhappily. For after I had examined the analogy of electricity and magnetism for some time, as much by theory as through the help of experiments (for the setting up of which the collection of very powerful artificial magnets in the possession of our Imperial Academy afforded outstanding opportunity) I clearly perceived at last that there is a fundamental analogy which is not open to doubt. Indeed after I had begun to transfer the *Franklinian* theory of electricity, emended on certain heads, to magnetism, I soon became convinced that a theory of magnetic phenomena could be constructed, quite similar to Franklin's theory of electricity, by which the things observed round a magnet can be equally well explained, in the same way that the various electrical phenomena are derived *ex posteriori*. This fortunate success spurred me on and by diligently applying myself to these researches I arrived, within a year or so, at the theory which I am publishing here.

4) Meanwhile the Imperial Academy conferred on me the honor of giving a public address to the solemn assembly celebrated on 7th September 1758. The Academy was not displeased by the material

---

[11] Aepinus, *Mém. Acad. Berlin*, XII, 1756, 105–21.

INTRODUCTION 239

I proposed on the similarity of electricity and magnetism, which I thought new and pleasing for the public, so at its request I spoke on this topic.[12] But since a diligent and careful exposition of the theory and its application to the phenomena seemed on the one hand to lead to somewhat deeper and tedious researches, requiring as a result a great deal of persistence; and on the other to demand being written out, since it would trespass far beyond the limits of a public lecture, I abandoned the plan of explaining the theory itself at that time and limited myself to the history of this matter, to an enumeration, that is, of the similarities of the two forces as they become known through experiments. I added briefly at the end an account of the general principles on which I consider a universal magnetic theory depends. But since I did not believe that I had completed my work, and judged that I ought to give to the world of learning a finalized and fully worked out exposition of the whole theory, I thought I was obliged to compose this treatise which I now issue publicly.

5) I am fully aware that it cannot be certainly concluded from the agreement of an hypothesis with the phenomena that we have reached the true cause. Although the theory I propound here satisfies the majority of magnetic phenomena, I prefer to proceed more modestly than confidently, and to put forward my proposition as probable rather than as certain. I hope meanwhile that my hypothesis will be proved outstandingly to my readers, by its full agreement as much with the magnetic phenomena as with the laws of nature recognized in other spheres. I add to this finally that, because it is quite likely that nature produces analogous phenomena by analogous means, I think that a considerable weight attaches to my opinion. It is established by experiment that magnetic phenomena are like those of electricity, and, as those who have read my treatise will surely admit, this is not to be called into doubt. So if everything is thus considered, I trust that my hopes for my theory are not empty, and that it will find approval among investigators of nature so they will rank it as being likely to the same degree that the experts admit they are convinced of the truth of *Newton's* explanation of the phenomena of the system of the world. For in support of its truth, Newton's theory can claim only all kinds of agreement with the phenomena and with laws of nature recognized in other fields.

6) To make them explicable, I here consider all magnetic phenomena in terms of a fluid whose parts mutually repel each other but are attracted by particles of iron, and my procedure in this whole

[12] Aepinus, *Sermo academicus de similitudine vis electricae atque magneticae*, St. Petersburg, 1758.

treatise is none other than showing how, from these few assumed principles, quite conformable to the analogy of nature, innumerable marvelous magnetic phenomena are as if spontaneously derived. I do not however inquire into the source of these primitive forces of attraction and repulsion with which I suppose the particles of iron and magnetic matter to be endowed. I leave that to the further discussion of those who are happy to spend time in investigations of that kind. I perceive that in a way I follow the custom of the geometers; if they have to solve a problem depending on the quadrature of the circle, they are satisfied to reduce it to this quadrature and to show how the problem depends on it. They do not require that the quadrature of the circle be taught at the same time. I thought that they are all the more deserving of imitation because I would conjecture that the primitive and fundamental forces, whose assistance nature uses, are much the same thing in physics as transcendental quantities in geometry. The eminent *Newton* proceeded by this method. He demonstrated how the motion of the heavenly bodies depends on universal gravity; but he did not spend energy on rooting out the source of this universal gravity. My plan of action will doubtless be castigated by those who cannot bear talk of attraction and repulsion in Natural Philosophy. To satisfy these, I declare that I am fully convinced of the existence of forces of attraction and repulsion, but I do not go so far as to hold, as do some incautious disciples of the great *Newton,* that these are forces innate in bodies, and I do not approve of the doctrine which affirms the possibility of action *in distans.* For I hold as an indubitable axiom the proposition that a body cannot act where it is not. If ever it is proved that an attraction or repulsion does not depend ultimately on some external pressure or impulse, then I judge that we are reduced to the point where we are forced to ascribe the execution and production of movements of that kind, not to corporeal forces but to spirits or beings with understanding of the things which act, and I cannot be induced to believe that this happens in the world. So the whole of my opinion amounts to the fact that I consider the attractions and repulsions I have spoken about as phenomena whose causes so far lie hidden, but from which other phenomena depend and are derived. I think that considerable progress can be made in the analysis of the operations of nature by the scholar who reduces rather complicated phenomena to their proximate causes and primitive forces, even though the causes of those causes have not yet been detected. I trust that this declaration of my opinion will satisfy the most rigid censor.

ST. PETERSBURG in the year of the Holy Redeemer 1758.

CHAPTER I

General Principles of the Theory of Electricity and Magnetism

1) The scholar who knows *Franklin's* very elegant theory of electricity, which agrees astonishingly with the phenomena, will recognize that it can be reduced to these few universal propositions.

α) There is a certain most subtle, truly elastic fluid producing all electrical phenomena, called electric on account of them, whose parts sensibly repel each other mutually even over rather large distances.

β) The particles of this fluid are attracted by the matter from which all bodies known to this time are made.

γ) There is a remarkable diversity in the way in which other bodies act on the particles of electric fluid; there are some bodies so constituted that the electric matter moves in their pores with the greatest of ease, and meets no resistance in passing freely through their pores in any direction; other bodies, on the contrary, are of such a nature that they admit its movement only with difficulty, and prevent this from happening freely; bodies of the former kind are usually called *non-electric per se;* of the latter kind, *electric per se.*

δ) Electrical phenomena are of two kinds. Some of course arise from the actual transit of electric matter from one body to another as it crosses from a body which contains a greater supply of this fluid to another which holds less, to which class especially pertain electric sparks and other phenomena concerned with the electric light; others occur without the actual movement of fluid, or its transit from one body to another, to which class must be particularly referred the attractions and repulsions which are usually exercised by electrified bodies.

2) I consider that these propositions contain all the heads of the theory we need for the explanation of electrical phenomena. Franklin indeed joins to these what he calls the impermeability of glass, which he thinks he needs to explain the phenomena of the Leyden jar and the experiment of commotion.[13] But I am astonished that the sagacious man did not observe that this impermeability or power of strongly resisting the transmission of electric fluid constitutes the essence of bodies electric *per se,* and is not proper to glass, but is common to all

[13] *Benjamin Franklin's Experiments,* ed. I. B. Cohen, Cambridge, Mass., 1941, pp. 180-81, 227.

bodies electric *per se*. For since, as the famous man excellently observed, bodies electric *per se* are such because the electric matter cannot pass through their pores without difficulty,[14] it is immediately obvious that it is very much the same whether a body is called electric *per se* or impermeable to the electric fluid. And so, while I am establishing the general foundations of the theory of electricity, I consider it altogether superfluous to mention the impermeability of glass since it cannot be considered as a peculiar property and is not proper to glass.

3) Since it is the aim of this work to set up a constant comparison between electricity and magnetism and to reveal in this way the similarity that obtains everywhere in the phenomena of both forces, and in the causes of the phenomena as well, I shall not now linger over deriving the theory of electricity from these assumed principles, but pass immediately to establishing the foundations of magnetic theory. It is my supposition that all the phenomena of the magnet can be derived from the following few principles, quite similar to those on which *Franklin's* theory is built.

α) There is a fluid producing all the phenomena of magnets, therefore called magnetic, which is most subtle, able to permeate the pores of all bodies, and whose parts, just like those of the electric fluid, mutually repel each other.

β) This fluid suffers no action from most other bodies found in the world, and is neither attracted nor repelled by them.

γ) There is though a certain kind of body whose parts attract magnetic matter and are in turn attracted by it; and the body endowed with this property is iron primarily, and also all bodies called ferreous, to the extent they contain iron, such as various minerals of iron—the very well-known stone called Magnet, a kind of black sand found in Virginia and frequent on the shores of the Baltic Sea,[15] and others of that kind.

δ) There is the greatest similarity between iron and ferreous bodies and bodies that are electric *per se*. For just as electric fluid moves with difficulty in the latter, so does magnetic fluid in the former. What is more, it seems that magnetic fluid experiences a still greater difficulty in crossing through the pores of iron than electric fluid usually finds in the pores of the bodies so far known to us that are electric *per se*.

ε) There is so far no known body which acts on magnetic matter,

---

[14] Ibid., p. 247.

[15] This black sand, evidently a form of haematite, was frequently referred to in eighteenth-century writings on magnetism. See, in particular, Musschenbroek, *Phil. Trans.*, XXXVIII, 1733–34, 297–302.

GENERAL PRINCIPLES 243

analogously to bodies that are non-electric *per se*. There is of course no body whose parts attract magnetic matter and through whose pores there lies open a free and unimpeded transit to magnetic fluid. Meanwhile in iron itself a certain gradation seems to obtain in this respect. For the softer the iron is, whether it is pure or mixed with heterogeneous parts, as in minerals, the freer is the movement of magnetic fluid in its pores, while the more hardened the iron is, the more difficult is the movement, so soft iron approaches in a way the analogy with bodies non-electric *per se,* more so at least than hardened iron.

4) As a result it is clear that I in no way consider the magnetic and electric fluids as one and the same thing, as do those who toil to derive the phenomena of both electricity and magnetism, and many other things, from one single extremely subtle fluid, namely the aether. It is my supposition that these fluids are endowed with very different properties that are not compatible with a single underlying material; of these the most significant is that electric matter is attracted by all bodies known to date, but that, on the contrary, magnetic fluid is completely immune from all action of all bodies, save iron alone. I am not really afraid of the objection, familiar at other times, that I am inventing a new subtle fluid without necessity and am thus departing from the harmony which nature is constantly accustomed to observe in its own operations, so that it gains many ends with few means. For since the effects of the magnet can only be happily explained if properties are attributed to the magnetic fluid which differ completely from those with which electric matter is endowed, it is not without reason, but guided by the contemplation of nature itself, that I propose here the total diversity of the two fluids.

5) From what is supposed in § 1 and 3, by comparing the principles of magnetic and electrical theory, it can easily be perceived what is analogous in the electrical and magnetic phenomena and what is not. In general, it is easily obvious that electricity must be significantly richer in the variety of its phenomena than magnetism. For there are innumerable bodies which act on electric fluid, and it in turn on them, and in which electric matter moves freely and without any difficulty; yet there are no bodies in which the same obtains in respect of magnetic matter. And so in the whole magnetic doctrine, there are no phenomena of bodies, so to speak, non-magnetic *per se,* though on the contrary those that occur in bodies non-electric *per se* are fairly numerous. And so magnetism supplies no phenomena analogous to those phenomena which are proper to a non-electric *per se,* but the analogy only exists in particular with phenomena which occur in bodies electric *per se*. For the principles from which magnetic phenomena

flow scarcely differ from those from which an explanation is to be sought of the things which occur in the electric *per se*. Yet it must not be thought that a full congruence is to be found even here, for iron, at least hardened iron, differs notably from the bodies electric *per se* known to us, in that the latter admit the movement of electric fluid in their pores more easily than the former admits the movement of magnetic fluid. Since therefore among the phenomena of electricity, apart from the phenomena which concern bodies electric *per se*, things do not occur that might be analogous to magnetic bodies, scarcely anything dealing with the theory of electricity can be expected here, except what concerns the phenomena of electrics *per se*. It would be completely foreign to the scope of this essay to bring in here the other phenomena of electricity, since my intention is to consolidate the theory of magnetism rather than that of electricity.

6) Let there be a body A, Fig. I, containing a certain quantity of fluid, whose parts are attracted by the body A, but which mutually repel each other, and it is plainly obvious that a double action is then exercised on a particle of this fluid such as B clinging around the surface of the body A. For this particle B will be attracted by the body A but will be repelled by the fluid enclosed in the pores of the body A. These contrary forces are either equal or unequal, and in the latter case, either repulsion or attraction prevails. If we assume that these forces are equal, it is clear that there is no action on the particle B, for the forces acting on it, and striving to produce contrary effects, destroy each other. But if a supply of fluid is enclosed in the body A such that the repulsion overcomes the attraction, the particle B will then have to yield to the former force; it must move quickly away and be separated from the body A. When the same obtains for the other particles of the fluid too, it is clear that a continuous flow of fluid must take place from the body A, unless there is some foreign or external cause preventing it. Because the quantity of fluid enclosed in the body A is continuously diminished, and as a result the repulsive force continuously decreased, the flow of fluid lasts until the repulsive force is reduced to equality with the attractive force (which meanwhile remains unchanged) and at this point all efflux must cease. If finally it is assumed that the attraction exercised on the particle B is greater than the repulsion, it is clearly established that the contrary ought to take place. For in this case particles of fluid will enter the body from every part until, because of the continuously increasing supply of fluid, the repulsion finally increases to the point where it becomes equal to the attraction, and then all further influx must cease.

7) It is clear from this that a fixed quantity of fluid can always be

GENERAL PRINCIPLES

assigned; if it is contained in the body A, there is neither efflux nor influx of fluid. But if a change takes place, fluid either escapes of its own accord, if its quantity has been increased, or if this has been diminished it increases until the equilibrium is restored. It seems that this quantity can be fittingly called *natural,* since a body left to itself always returns of its own accord to the state where it contains neither a lesser nor a greater quantity of fluid than is exactly sufficient to produce an equilibrium between the attracting and repelling forces. So if we apply this to the electric and magnetic fluids, the conclusion must be that any body left to itself always comes of its own accord to the state where it contains exactly the quantity of electric fluid sufficient to produce equilibrium between the attracting and repelling forces, and similarly that any ferreous body of its own accord always reaches the state where it contains a like quantity of fluid. Therefore readers will easily comprehend what I later on intend to call the *natural quantity* of electric or magnetic fluid.

8) Moreover it is established by experience that if two bodies A and B, Fig. II, containing a natural quantity of either electric or magnetic fluid, approach each other, these bodies exercise no mutual action on each other which could be attributed to electricity or magnetism, and they neither attract nor repel each other. To understand the truth of the matter more distinctly, let us examine more carefully the way in which two bodies filled with a fluid such as the magnetic or electric ones, act on each other. It is readily evident that attraction as well as repulsion occurs between these bodies. For the fluid contained in A, repelling that in B, repels at the same time this whole body to which the fluid clings. On the contrary, the matter from which the body A is made, by attracting the fluid enclosed in B and clinging to it, at the same time entices the whole body to itself, and vice versa the body B acts on the body A in the same way. So these contrary forces acting between the two bodies are either equal and destroy each other, or one of them prevails over the other either by attraction or repulsion. But this equality of the contrary forces obviously depends on the supply of fluid contained in any of these bodies. The question then arises whether, if each of the bodies contains the natural quantity of fluid, repulsion and attraction destroy each other or not. The answer to this is easy with the help of experiment. For it is established from § 7 that any body left to itself always comes of its own accord to the state where it contains exactly the natural quantity of fluid; but we know equally from experiment that all bodies left to themselves exert on one another no action that could be attributed to electricity or magnetism, and if through some ex-

ternal action they are disturbed from this state, if nothing foreign impedes, they return of their own accord to their former state where they do not mutually act on one another. It is evident then that the state of bodies where they contain the natural quantity of electric or magnetic fluid plainly coincides with that where the attraction and repulsion destroy each other, and where they do not act on one another.

9) Since therefore bodies do not exercise on one another any action that depends on electricity and magnetism so long as they contain the natural quantity of fluid, we must look especially at those cases where the natural quantity of fluid has been changed. It is evident, generally speaking, that it can be changed in two ways; electricity or magnetism is produced by increasing the supply of electric or magnetic fluid beyond the natural, or by diminishing it below the natural. In each case it is self-evident that the equality of the attraction and repulsion cannot subsist any longer. *Franklin* called the electricity which is produced by augmenting the electric matter *positive,* and that produced by diminishing it, *negative.* I transfer these words in the same sense to magnetism. So I understand that there are two different kinds of magnetism, as well as of electricity, one of which will be called *positive* magnetism, the other *negative.* It follows from this that there are three ways in which a body can be endowed with magnetism or electricity. For a body may be either totally positively or negatively magnetic or electric, or finally it may be reduced to a state where, while one part of it is positively electric or magnetic, the other, at the same time and the same moment, is negatively electric or magnetic. Let us therefore consider a little more exactly these three states of bodies endowed with electricity or magnetism.

10) Let there be a body A, Fig. I, constituted first in the natural state, that is it contains the natural quantity of fluid, which we shall call Q; and if we suppose that B is a particle of either electric or magnetic fluid clinging round the surface of the body, the particle B will be attracted by the whole body A, but will be repelled by the fluid with which its pores have been filled. Let us call this attractive force $=a$, and the repulsive force which is exercised on $B=r$, and the force by which the particle B is attracted toward the body A will be $=a-r$. But since the natural state is supposed to exist, the body A contains precisely as much fluid as its attractive force can retain, and so the whole action on the particle, or $a-r, =o$. Let us suppose that, however this happens, there is added to the fluid contained in the body a definite quantity of it, uniformly distributed through the whole body, which is to the natural quantity of the fluid as α is to

Q. Then the repulsion [16] of the particle B $=\dfrac{(Q+\alpha)r}{Q}$, but the attraction is the same as before, so the particle B will be attracted toward the body A by a force $= a - r - \dfrac{\alpha r}{Q}$. But since $a - r = 0$, the force which draws the particle B towards A $= -\dfrac{\alpha r}{Q}$, or the particle B will be repelled from the body A by a force $\dfrac{\alpha r}{Q}$.

11) So unless the particle B is impeded by some external cause from obeying the force acting on it, it will have to separate from the body A and escape. But since the same is true about any other particle of fluid clinging on the surface, there must occur a spontaneous efflux of abundant matter lasting until the body A has returned to its natural state again. For although the supply of fluid contained in the body A, and as a result α, continuously diminishes during the efflux, and the force producing the efflux, which $= \dfrac{\alpha r}{Q}$, continuously decreases at the same time, yet this efflux cannot stop before the force $\dfrac{\alpha r}{Q}$ has vanished, that is, before α has become $= 0$, or the body has returned to the natural state. But if there is an impeding cause able to resist this efflux, then the efflux will be slowed, but will not be completely stopped until after the body A has reached the natural state again.

12) These arguments concerning the case where fluid abounds in a body are easily transferred to the other, where the quantity of fluid is diminished below the natural. For it, to fit the formula discovered above to this case, instead of assuming as in § 10 that fluid abounds in the body A beyond Q by a quantity $+\alpha$, it is now supposed that it abounds by a quantity $-\alpha$, or what is the same thing, is lessened by a quantity α, then the particle B clinging on the surface of the body A will be attracted toward it by a force which $= \dfrac{\alpha r}{Q}$ and, if nothing stands in the way, will enter the body. So a continuous influx of fluid into the body will occur, and will last until α has become $= 0$, that is, until the body A has recovered its natural state.[17] But if there is some impediment resisting this influx, a longer time will be needed before the body can return to its natural state, but the influx will not be able to

[16] Aepinus here, as elsewhere, makes the unstated assumption that the force exerted is proportional to the quantity of fluid involved. The assumption is made explicit in § 30 below.

[17] This argument only holds, of course, provided the presence of some external source of fluid is assumed.

cease altogether until after the body has returned completely to the natural state.

13) If we apply what has been proposed so far to electricity, it is soon obvious that the state of positive or negative electricity is not durable, but must always be quickly destroyed, unless some outside cause preserves it for some time. We can imagine two such causes which can make for the conservation of both positive and negative electricity; one of these is external to the electrified body, the other internal. If, first, the body A is surrounded in all directions by bodies electric *per se,* the particle B is repelled by the state of positive electricity, but because it can only with difficulty enter the pores of the contiguous bodies (which, as electrics *per se,* do not permit free entry to themselves), the efflux of the electric matter abounding in A is significantly retarded; so the state of positive electricity ought to last much longer in this case than if this impediment is not there. But if the body A is negatively electric, since then particles of electric fluid cannot enter the body A unless they have left the pores of the bodies contiguous to its surface and crossed over, the afflux too ought to be slower, and as a result this state should also be preserved longer.

14) It is further possible to posit an internal impediment to the destruction of electricity if the body A itself is electric *per se.* For in this case, the effluent matter cannot easily extricate itself from the pores of the body, nor can affluent matter easily enter them, so both positive and negative electricity is destroyed with more difficulty in a body electric *per se* than in one non-electric *per se.* Experiment agrees excellently with this reasoning. For we know that if electricity is produced in a body non-electric *per se* and this contacts another body non-electric *per se* which allows very free transit to the electric fluid, then the electrified body returns to the natural state in a moment. But if the same body is supported by a body that is electric *per se,* e.g. glass, and surrounded by dry air, which is of course electric *per se,* it has been found that an appreciable time is then required before the electricity is completely destroyed and the body returns again to the natural state. It is similarly established that the electricity of electrics *per se* is not as quickly destroyed as with those non-electric *per se,* if the same conditions hold for both. It seems that an exception must be made of the case where such a body electric *per se* is touched along the whole of its surface, after it has been electrified, by a body non-electric *per se,* e.g. if it is submerged in water. Here experiment testifies that the whole of the electricity is suddenly destroyed even in bodies of this kind. But it is not difficult to assign a cause for this. For while a body electric *per se,* e.g. a glass globe A, Fig. III, is electrified,

the abundance or deficiency never reaches the inside of the body, but subsists in the parts closest to the surface. If the electric matter on the surface of the globe A is increased, because of the difficulty with which it passes through the pores of this body it is forced to subsist in the parts closest to the surface and cannot reach to the inside. If the surface is emptied of electric fluid, then the parts occupying the interior of the globe tend toward the surface, but cannot free themselves and reach it because of the difficult movement in the pores. So negative electricity in a body of this kind never extends far beyond the surface. In each case, it is in our power to electrify only the outermost surface *abcd* of the globe and the interior nucleus of the body always continues in its natural state. But only the electricity that clings on the surface can be completely destroyed at high speed by the surrounding water. Yet I have no doubt that if a glass globe of appreciable size could be electrified to such a degree that the electricity penetrated the whole, and it were then submerged in water and quickly withdrawn, it would not lose all its electricity. The surrounding water would destroy the electricity clinging on the surface but this destruction would not be able to reach right to the nucleus. In a similar fashion, when a heated iron globe is immersed in cold water and soon withdrawn, its surface is cold to the touch at first, but after a short interval it grows appreciably warm again. But it is scarcely possible that the matter can be put to the test in this fashion. On another occasion I shall relate certain experiments which will clearly show that a body electric *per se* tenaciously preserves its own electricity, even though a body non-electric *per se* is in close contact with it over its whole surface.[18]

15) Consider also the third case, which is particularly necessary for the foundation of our magnetic theory. Suppose that a body A, Fig. IV, is divided into two equal parts AB, AC and imagine first that this body is constituted in the natural state, and that the natural quantity of fluid pertaining to the part AC as well as to $AB=Q$. Suppose further that D and E are two particles of either electric or magnetic fluid clinging round the surface of the body. Suppose that each of these is attracted by the body A by a force $=a$, and let E be repelled by the natural quantity of fluid contained in AB by a force $=r'$ and by that pertaining to the other part AC by a force $=r$. It is obvious that, when all circumstances in respect of the particles D and E are the same, the particle D is also attracted by a force $=a$, but is repelled by the fluid contained in AC by a force $=r'$, and by

[18] See below, pp. 275–76.

that included in AB by a force $=r$. Therefore the attraction by which the particles D and E are pushed toward the body A will be $=a-r-r'$, and because the body A is suposed to be constituted in the natural state, this will be $=$ o. If now the supply of fluid in AC is considered increased by a quantity which is to the natural as α is to Q, but in BC diminished by a quantity which is to the natural as β is to Q, the supply of fluid contained in AC will be $Q+\alpha$, and in $AB=Q-\beta$. But if we further suppose that these quantities are uniformly distributed through the parts AC and AB respectively, the particle E, which is attracted by the body A but repelled by the fluid to be found in both AC and AB, will be attracted toward the body A by a force which $=a-\dfrac{(Q+\alpha)r}{Q}-\dfrac{(Q-\beta)r'}{Q}$ (§ 10. 12); because $a-r-r'=$ o, this becomes $\dfrac{\beta r'-\alpha r}{Q}$, or what amounts to the same thing, the particle E will be repelled from the body by a force $\dfrac{\alpha r-\beta r'}{Q}$. But if a calculation is made in a similar fashion for the particle D, the force drawing it toward the body A will be $=a-\dfrac{(Q+\alpha)r'}{Q}-\dfrac{(Q-\beta)r}{Q}=\dfrac{\beta r-\alpha r'}{Q}$.

16) If we further imagine a particle of fluid F clinging to the middle of the body near A, and if again we first consider the body A to be wholly constituted in the natural state, then it is obvious that this particle is attracted by the parts of which the body is composed, but that this attraction is equal in all directions, and so can be considered zero. But this particle is also repelled by the fluid contained in AC as well as AB, and indeed by each, as is easily shown, by a force $=r$; but these repelling forces, being equal and opposite, mutually destroy each other. So as long as the body A is constituted in the natural state, the particle F tends no more to one part than to the other, and does not try to move toward either B or C. If we assume that the supply of electric fluid is increased in the part AC, but diminished in AB, and if we take over the same letters we used in the previous §, then it is clear that the particle F is pushed toward the emptied part AB by a force $=\dfrac{\alpha r+\beta r}{Q}$. For since the body A does not act on the particle F, only the fluid masses contained in AC and AB must be taken into account here for the repulsions which they exert. The fluid contained in AC pushes the particle F toward B with a force $=\dfrac{(Q+\alpha)r}{Q}$ (§ 10) and that contained in AB in the opposite direction toward C with a force equal to $\dfrac{(Q-\beta)r}{Q}$ (§ 10). Therefore the whole force pro-

GENERAL PRINCIPLES   251

pelling the particle F toward B will be $=\dfrac{(Q+\alpha)r-(Q-\beta)r}{Q}=\dfrac{\alpha r+\beta r}{Q}$.

17) It is obvious that here there is a force by which the particle F is urged to leave the part AC, and it follows from this that there must be a continuous flux from the part AC to AB, ceasing only when the fluid is uniformly distributed through the whole body A and is of equal density at all points of the body. So if there is no impeding cause—as is the case with bodies which admit an easy movement of fluid in their pores, such as bodies non-electric *per se* and to some extent soft iron—the outcome must be this: the state cannot last long, unless it is preserved through some external cause. But if movement of fluid in the pores of the body takes place with difficulty, then the state can subsist for some time; for rather a long time in fact if this difficulty is great, but shorter if it is small. For the actual transit of the particle F into the part AB cannot take place unless the force pushing it can overcome the resistance. So this state happens only in bodies electric *per se*, or, if I may use the term, magnetic *per se;* for in other bodies it can be preserved only momentarily without the help of some external force, the fluid distributing itself in a moment through bodies which are left to themselves.

18) Further, it is clear from § 15 that a particle of fluid E clinging round C is urged to leave the body and escape by a force less than it would be if the part AB were filled with fluid to the same degree as AC. For when that is the case, the force enticing the particle E to leave is $=\dfrac{\alpha(r+r')}{Q}$, as is obvious from § 10, and it is very plain that this quantity is greater than the force by which the particle is driven if the part AB of the body is emptied. For it is self-evident that the quantity $\dfrac{\alpha r-\beta r'}{Q}$ is less than the quantity $\dfrac{\alpha(r+r')}{Q}$. Similarly, too, the force which entices the particle D clinging round B to enter the emptied part of the body A, which we found in § 15 $=\dfrac{\beta r-\alpha r'}{Q}$, is less than it would be if the whole body were filled with fluid. For then the force urging the particle D to flow in would be $=\dfrac{\beta(r+r')}{Q}$ (§ 12), and this quantity is altogether greater than $\dfrac{\beta r-\alpha r'}{Q}$.

19) It is easily concluded from the preceding that once this state has been induced in the body A it can be destroyed by either of two mechanisms. For either the fluid crosses from the part AC where it abounds to the other AB where it is deficient until it is uniformly distributed through the whole body A, or it flows out from the part AC

and in toward the part AB until both parts have returned to the natural state. In bodies electric or magnetic *per se* to a sufficiently high degree, that is, in bodies in which the fluid experiences a sufficiently great difficulty in passing through the pores, the former mode of destruction either has no place or at least it occurs only at a quite slow rate. So therefore if we suppose that the state we are considering here is induced in a body electric *per se* or magnetic *per se* to a high degree, we need not look for the former mode of destruction of this state. If we look only for the latter, though, it is clear from the preceding paragraphs that, other things being equal, this state is preserved much longer and is more durable than if fluid is either abundant or deficient in the whole body A and in every part of it. For although here too there are forces tending to the destruction of this state, yet we showed just previously that these forces are significantly less if the fluid abounds in one part of the body A and is at the same time deficient in the other than it would be if the fluid either abounded equally in the whole body or was deficient in the whole.

20) All the reasonings I have developed so far are also valid for two bodies A and B, Fig. V, close to each other, which, although they are such as to admit the free movement of fluid in their pores, have had thrust between them a third body C which can place a barrier against the movement of the fluid from one body to the other. In this case, if the supply of fluid is increased in one of them, e.g. A, and diminished in the other, B, the particle of fluid F tries to cross from A to B but is prevented by the intermediary body C from actually crossing. The particle E is also enticed to leave the body A, though by a force less than if the body B were not present, and in a similar way, the particle D is urged to enter the body B, but by a force less than there would be if the body A were not present. If a calculation is undertaken for this case in a manner similar to that in § 15 above, the same formulae as discovered there for a body electric *per se,* one part of which is filled with fluid and the other emptied of it, will do service. So also, the things subsequently deduced from these formulae will be equally valid for both cases. This is so obvious that it does not seem worthwhile to repeat the reasonings employed previously and adapt them to this case.

21) Since α and β are indeterminate and can receive any value at all, values can be assigned to these letters such that the forces urging the particles D and E, Fig. IV, (whose analytic expression has been found in § 15 above) become either equal to zero, or greater or less than it. So if first you want the force enticing the particle E to leave to be zero, let $\beta r' - \alpha r = 0$, or $\beta = \frac{\alpha r}{r'}$. When this value is substituted in

GENERAL PRINCIPLES

the formula for the force acting on D, that force [19] will be $=\frac{\alpha(r^2-(r')^2)}{Qr'}$. But if the force urging the particle D to enter is to disappear, $\beta r - \alpha r'$ will $=0$, hence $\beta=\frac{\alpha r'}{r}$ and the force [20] repelling the particle E from the body $A=\frac{\alpha(r^2-(r')^2)}{Qr}$. It follows from this

1) that the force on the two particles D and E can never disappear at the same time, but only one or other of them. For them to disappear at the same time, it would be necessary that $r^2-(r')^2=0$, or that r be equal to r'. But this is impossible since r' is constantly less than r. For since the supply of fluid natural to AB is equal to the quantity natural to AC, but the former quantity acts on the particle E from a greater distance than the latter, its action on that particle must necessarily be less than the action of the latter, and the same obtains for the particle D. Although we do not so far know the function according to which particles of either the electric or the magnetic fluid mutually repel each other, yet by analogy with what nature usually follows at other times and in conformity with innumerable experiments instituted on the magnet and on electrified bodies, it has been concluded that the intensity of the repulsive force in those bodies decreases when the distance is increased.

2) That if there is no force on D, then the force acting on E and attracting this particle toward the body A will be negative and hence repulsive.

3) That if, on the contrary, the force on E disappears, then the force enticing the particle D toward A will be positive and hence attractive.

These latter two heads follow immediately since, as has just been shown, r' is always less than r.

22) If further we want the force by which the particle E is attracted to be greater than zero, it must be assumed that $\beta r' - \alpha r$ is a positive quantity, or that $\beta r'$ is greater than αr whence $\beta > \frac{\alpha r}{r'}$. If it is assumed that μ indicates a positive quantity, β can be put $=\frac{\alpha r}{r'}+\mu$. If we substitute this value in the formula for the force attracting the particle D, that force will be $=\frac{\alpha(r^2-(r')^2)}{Qr'}+\frac{\mu r}{Q}$, whence if the force acting on the

---

[19] The text has $\frac{a(r^2-(r')^2)}{Qr'}$.

[20] The text has $\frac{\alpha((r')^2-r^2)}{Qr}$.

particle E is positive, the other force, acting on D, is also positive and greater than $\frac{\alpha(r^2-(r')^2)}{Qr'}$. But if the force urging the particle E is to be a repulsive one, $\beta r' - \alpha r$ must be negative, whence we obtain $\alpha r > \beta r'$, and $\beta = \frac{\alpha r}{r'} - \mu$. With this value substituted in the formula for the force acting on D, that force becomes $\frac{\alpha(r^2-(r')^2)}{Qr'} - \frac{\mu r}{Q}$, and this quantity is always less than $\frac{\alpha(r^2-(r')^2)}{Qr'}$, but according to the various magnitudes of the quantity μ it is now positive, now negative, and now equal to zero.

23) If we consider the force on D in a similar way and we want it first to be attractive, then $\beta r - \alpha r'$ must be positive, whence is deduced $\beta = \frac{\alpha r'}{r} + \mu$. When this value is substituted in the formula for the force acting on E, this becomes $= \frac{\alpha((r')^2 - r^2)}{Qr} + \frac{\mu r'}{Q}$. According to the various magnitudes of the quantity μ, this force can be equal to zero or positive or negative, as the positive quantity $\frac{\mu r'}{Q}$ is either equal to the negative one $\frac{\alpha((r')^2 - r^2)}{Qr}$ or greater or less than it. If finally the force acting on D is to be negative and repulsive, $\beta r - \alpha r'$ must be negative, whence is obtained $\beta = \frac{\alpha r'}{r} - \mu$. With this value substituted in the formula for the force acting on E, it becomes $= \frac{\alpha((r')^2 - r^2)}{Qr} - \frac{\mu r'}{Q}$ and since $(r')^2 - r^2$ is negative this quantity is always negative.

24) By collecting all the cases we have considered in the preceding paragraphs, we see that there can be various states in which a body such as we are considering here can be constituted with one part emptied of fluid, while fluid abounds in the other part; yet the cases can be reduced conveniently to these four classes.

1) One part neither attracts nor repels fluid, while the other clearly either attracts or repels it.
2) The body repels fluid from both parts.
3) The body attracts fluid to both parts.
4) Finally, the body repels fluid from one part, and attracts it to the other.

Readers will find examples that are worth noting of all these cases in what follows. But the things that I have demonstrated here about

GENERAL PRINCIPLES 255

a single body in one of whose parts fluid abounds while it is deficient in the other are valid also for two bodies combined together in the way I have described in § 20. This is sufficiently obvious and needs no special discussion.

25) Before we move on, it will be convenient to note that the hypothesis we assumed in § 15 above for convenience of calculation, namely that while fluid abounds in one part of the body A and is deficient in the other the fluid is uniformly distributed in each one of the parts, hardly ever happens in nature, or at least rarely. It is obvious that there are infinitely many ways in which fluid may abound or be deficient in each part AC or AB as a whole, without being of the same density at any two points of these parts. It is impossible to enter on a calculation for these cases; for the law of the variation of the density of the fluid in various places can be changed in infinitely many ways, and until the discovery of the function according to which repulsions or attractions of electric or magnetic fluid take place, even if there is a law of density, it is not right to draw any conclusion. So I have considered only that case which is suited to calculation. Readers will meanwhile easily perceive how if fluid abounds in the part AC and is deficient in AB, even if it is not uniformly distributed, the conclusions to be drawn will still be similar to those I have discovered from the fictitious hypothesis of uniform distribution. So, as I have demonstrated above, they can be transferred without fear of error to those cases where fluid is not uniformly distributed. Thus it is also obvious that even if the parts AB and AC are not equal, as I supposed above, but unequal, things similar to what is valid if these parts are equal among themselves should still obtain.

26) If further a body is constituted in a state where it can possess each electricity or each magnetism at the same time, it is not necessary, as I have so far supposed, that fluid should abound in one half of it, while it is deficient in the other. Rather there are innumerable other possible cases. Let there be a body AB, Fig. VI, divided into any number of parts e.g. five, and let fluid abound in AC, be deficient in CD, abound again in DE, be deficient once more in EF, and so on. Once this state has been induced in the body AB, if this body is electric *per se* or magnetic *per se,* it can be preserved, or at least can last for an appreciably long space of time. It is evident that it cannot be destroyed without an actual movement of fluid through the pores of the body, and since this happens with difficulty under the assumed hypothesis, it is altogether possible for this state to be preserved for an appreciable time.

27) Since some cases pertaining to this are dealt with in what fol-

lows, let us strive to discover the general rule for the action which a body constituted in this state exercises on neighboring particles of fluid. So let G be a particle of fluid found round the end A and granted that the whole body AB is first constituted in the natural state, let the particle G then be repelled by the fluid contained in AC by a force $=r$, by the fluid in CD by a force $=r'$, by the fluid in DE by a force $=r''$ etc., and let the natural quantity of fluid in each part AC, CD, DE etc. $=Q$. Let it further be posited that fluid abounds over and above Q in AC by a quantity a, in CD by a quantity b, in DE by a quantity c etc. and if we determine the forces by which each part attracts the particle G one by one, using the method of § 10, and collect all these forces into a total, the whole force by which the body AB attracts the particle G will $= -\dfrac{(ar+br'+cr'' \ldots \text{etc.})}{Q}$. In the application of this formula to special cases it must be noted that, if the fluid does not abound in a certain part but is deficient, a negative value must be attributed to the letter that designates the quantity abounding in that part. It comes to the same thing of course if I say that a quantity $+a$ is deficient in a certain part, or that a quantity $-a$ abounds in a certain part. So e.g. this general formula is suited for the cases handled in § 15, if for the action on E we put $a = +\alpha$, $b = -\beta$, $c = o$ etc., and for the action on D, $a = -\beta$, $b = +\alpha$, $c = o$ etc.,[21] for then the same formulae as those evolved above are produced.

28) Now that we have abundantly considered the three states which bodies can attain in respect of electricity or magnetism, we will have to toil at discovering the laws by which bodies constituted in this or that state act mutually on each other. Therefore take two bodies A and B, Fig. II, close to each other, and suppose them constituted first in the natural state. To determine the mutual action of those bodies, let us separate the whole action of the body A on the body B into two parts. For it is evident that the body A, saturated with fluid, or filled to the natural state, acts first on the fluid enclosed in the body B, but then also on the body B itself or that proper matter from which the body B is made. The first of these actions according to the hypotheses we fixed in § 1. 3 above must be separated afresh into a double action. The fluid contained in A repels that contained in B, but the body A itself attracts the fluid; whence if the former action is indicated by r, and the latter by a, this total action will $=a-r$, or the fluid contained in B is attracted toward the body A by a force which $=a-r$. But if, further, we look at the action on the body B itself, or on the matter from which the body B is made, it is already determined by the hypotheses

---

[21] The text has $\gamma = o$ rather than $c = o$ on both occasions.

GENERAL PRINCIPLES

posited above that the fluid contained in A attracts it, and we must suppose this to happen by a force$=$A. But it can be further asked whether the proper matter of the body A must also be said to act on the proper matter of the body B. Whatever may be the case, let us assume that the proper matter of the body A attracts that from which the body B is made by a force$=x$, and then the action which the proper matter of the body A exercises on the body B and on the fluid contained in it$=$A$+x$; whence the total force by which the body A attracts the body B and vice versa will be$=a-r+$A$+x$.

29) Since we introduced x at will into the calculation, the value to be attributed to it will have to be decided, whether zero or positive or negative. If we find the first of these to be the case, then it will be obvious that there is no mutual action of the proper matter of the bodies on each other; but if x is found positive, it will have to be concluded that the bodies A and B, or rather their proper matter, attract each other, and if negative that they repel each other. Now we proved above (§ 6. 7.) that while a body is constituted in the natural state, or contains the natural quantity of fluid, it plainly does not act on other particles of fluid existing outside the body; whence, since we have supposed that the body A is constituted in the natural state, we obtain the equation $a-r=o$. We then indicated (§ 8.) that experiment shows that two bodies, each of which contains the natural supply of magnetic or electric fluid, plainly do not act on one another, and least of all do they exercise a mutual action on each other which can be attributed to electricity or magnetism. So in our case $a-r+$A$+x=o$. But if we combine this equation with $a-r=o$ found above, we will obtain A$+x=o$, or $x=-$A.

30) From this we can immediately deduce various noteworthy conclusions. It is clear first that A$=a$. To make this evident, let the mass of the body A$=$M, the natural quantity of fluid contained in A$=$Q, and similarly the mass of the body B$=m$, and the natural quantity of fluid pertaining to that body$=q$. Since corporeal actions always happen in the ratio of the masses, a will be to A, in the composite ratio of M : m and q : Q, whence is obtained the analogy a : A$=$Mq : mQ, and A$=\dfrac{amQ}{Mq}$. But since both A and B are constituted in the natural state, it is obvious that M : $m=$Q : q, or M$q=m$Q, and that when this equation is combined with the equation just found for A, A$=a$. Then it follows further from what was demonstrated in the preceding paragraphs, that the four quantities a, r, A, and $-x$ are all equal to each other. For since I have given the demonstration that $a=r$, and A$=-x$, and since it is already proved that A$=a$, it is clear that $a=r=$

$A = -x$. It may be further concluded from what has already been demonstrated,

1) That the proper matter of all bodies known to us to date is mutually repulsive. For the value of x (the letter by which we indicated this action above) has been found in the preceding paragraphs to be negative.

2) Since it is established through experience that two bodies constituted in the natural state exercise no mutual action that could be attributed to electricity or magnetism, whether these bodies are close together or removed from each other by some distance, it follows that the forces a, r, A and $-x$ always remain equal, whatever the distance between the bodies. Since, therefore, these forces are changed when the distance is changed, it follows that each of them suffers the same change from a change in distance, since otherwise their equality could not persist at all distances. So the forces often mentioned not only follow the same function of the distances but are also always of the same intensity at equal distances. I assert only generally here the similarity of the functions according to which the actions I have discussed are exercised, but so far I dare not specify what the law of those functions is; though I confess that, if given the option, there is some probability in the belief that they follow the inverse square law of distances, for the analogy of nature seems to militate on behalf of this law.

31) Of the propositions I have evolved here, the one enunciating that the proper matter of any bodies at all is mutually repulsive will perhaps seem to my readers too hard to be readily tolerated. I do not deny that when it first came to me right at the beginning of my thinking about Franklin's theory of electricity, I was somewhat horrified at it. But after I began to consider that it contained nothing contrary to the analogy of the operations of nature, I easily got used to it. For if we are willing to admit the truth, we observe innumerable cases in nature where we find attractive or repulsive forces inherent in bodies, so there is no need to be afraid of adding this new force discovered here to the number of primitive and fundamental forces of nature. At the beginning I was particularly worried because this proposition seemed diametrically opposed to another universal law of nature detected by *Newton,* by virtue of which all bodies are said to attract each other in the composite ratio of their masses directly and the square of the distances inversely. But I satisfied myself without great difficulty. For since the repulsion I have spoken about is reduced to precisely nothing by the magnetic or electric fluid in bodies containing the natural quantity of those fluids, the remaining corporeal actions, whatever they are, are clearly not disturbed by it, but remain altogether

the same as they would be if there were no repulsion of that kind. It follows that this force can always be considered as not existing at all, except only when the discussion is about phenomena arising from electricity or magnetism. So it is clear that *Newtonian* attraction can be saved even though repulsion of this sort is admitted. But if any one thinks there is a contradiction latent in our assumption that corporeal matter is endowed at the same time with two opposing forces, that is repulsive and attractive, I should like him to note that I do not consider either the repulsive force discovered here, or the attractive force called universal gravity, to be forces inherent in matter or essential to its make-up. So it is clear that I cannot be judged guilty of a contradiction, for I do not assume that these opposed qualities inhere in the same subject; but although I do not know what means nature uses to produce these forces, I am certain that each force arises from some extrinsic cause. It is no contradiction to say that the same body at the same time and at the same moment is enticed by two extrinsic and opposed forces.

32) Let us proceed further in evolving the laws of action of bodies endowed with electricity or magnetism. Assume that of the two bodies A and B, Fig. II, B is still constituted in the natural state, but A is replete with fluid beyond the natural quantity, and indicate the natural quantity of fluid for the former body by q, for the latter by Q. Since in the body A the quantity of fluid abounds, let that be put $=Q+\alpha$. But if we now examine the mutual action of the bodies on each other, or, since we know from the law of nature that the action is mutual, if we examine the action of the body B on the body A and the fluid contained in A, it is obvious that it is made up of the action of the fluid to be found in B on the fluid contained in A as well as on the proper matter of the body A, and of the action of the proper matter of the body B on the proper matter of the body A as well as on the fluid enclosed in the body A. If therefore we retain the denominations assumed in § 28, but use the letter R in place of x, and then begin calculating, it is obvious that the forces which act in the case we are dealing with are the following:

1) The fluid contained in B, whose quantity $=q$, attracts the proper matter of the body A by a force $=a$.

2) This same fluid repels the fluid contained in A, or $Q+\alpha$, by a force $=\dfrac{Qr+\alpha r}{Q}$. Since the attracting and repelling forces of bodies are always in ratio to their masses, Q will be to $Q+\alpha$, as r is to the force by which $Q+\alpha$ is repelled.

3) The proper matter of the body B attracts the fluid contained in A by a force $= \dfrac{QA+\alpha A}{Q}$.

4) The proper matter of the body B repels the proper matter of the body A by a force $= R$.

By collecting the four expressions for these forces, the total force by which the body B attracts the body A and vice versa $= a + \dfrac{QA+\alpha A}{Q} - \dfrac{Qr+\alpha r}{Q} - R$. But since from § 29, $a+A-r-R=0$, this attracting force will be $= \dfrac{\alpha A}{Q} - \dfrac{\alpha r}{Q}$. Further $A=r$ (§ 30), so the force $= \dfrac{\alpha r}{Q} - \dfrac{\alpha r}{Q} = 0$. So the force by which the bodies A and B attract each other is zero, and these bodies plainly do not act on one another. If on the contrary we assume that the body B is still constituted in the natural state, but rather than fluid abounding in A, the supply is reduced below the natural quantity, the formula found for the preceding case may be adapted to this one provided α is made negative. The force by which the bodies attract each other is then found $= \dfrac{\alpha r}{Q} - \dfrac{\alpha A}{Q} = \dfrac{\alpha r}{Q} - \dfrac{\alpha r}{Q} = 0$. So in this case too there will be no mutual action of the bodies A and B on each other. And the same is found if the attractive force of the bodies A and B is sought by the direct method we used for the previous case.

33) All this leads to the following propositions:

1) As long as a body contains the natural quantity of electric fluid, or as long as it is constituted in the natural state, it is neither attracted nor repelled by other bodies whether positively or negatively electric or constituted in the natural state, and no action on a body of that kind is to be attributed to electricity.

2) Iron or any other ferreous body, so long as it possesses no magnetic force itself, is neither attracted nor repelled by a magnet; or a magnet only acts on other magnets or on iron which has already been magnetized.

These propositions will doubtless seem absurd to the majority of Physicists, especially to certain disciples of *Franklin* who do not explore the objections to *Franklin's* system sufficiently deeply and who, with the celebrated *Franklin* himself, assume it as a rule established by experience that bodies either positively or negatively electric always attract any other bodies not imbued with electricity.[22] But although

---

[22] Cf. *Benjamin Franklin's Experiments*, p. 185.

these assertions of mine can seem to be discordant with and opposed to everyday experience, they must be considered completely true. For not only do they flow from the fundamental principles of *Franklin's* theory so necessarily that if they are not true the whole of *Franklin's* theory must collapse, but also they agree so well with experience that physicists should have arrived at these laws through this alone if they had attended sufficiently to all the circumstances. I shall later apply the general principles of the theory evolved here to the phenomena of electricity and magnetism, and that will be the time to expound these things more carefully; but so readers are not too offended because right at the beginning of this commentary they meet propositions which cannot but seem contradictory to experience, I give notice that I will later prove by evidence that no body constituted in the natural state can approach another body either positively or negatively electric without being disturbed from its natural state and becoming electric.[23] And so it is not to be wondered at that those who instituted electrical experiments should have been deceived initially. For when they moved a body constituted in its natural state to an electrified body, they observed that it was attracted. And since they did not know that this body cannot but be deeply electrified itself in approaching another electrified body—they did not even harbor a suspicion of this—they could hardly avoid fixing the incorrect rule that bodies constituted in the natural state are attracted by electrified bodies. How nicely experience agrees with my theory, and how true my paradoxical assertions are, my readers will be able to guess when I show below that the action of electrified bodies is less on bodies that, because of either their own nature or an external impediment, become electric with difficulty as they approach the electrified body, and that the action is greater if the bodies approaching the electrified body can easily be electrified; in fact that, as required by the theory, the attraction is completely checked if the production of electricity in the body approaching the electrified body is thoroughly impeded.[24] Readers will also see below that there is a certain phenomenon concerning electrified bodies that would be judged absolutely contradictory and contrary to the first principles of reasoning if my assertions are not allowed, and that there is no difficulty at all once they are assumed. The case of magnetism is completely similar. For iron constituted in the natural state close to a magnet immediately becomes a magnet and is only attracted after it has gained magnetic force. Furthermore, the same is entirely valid here as for electricity. For bodies that become

[23] See below, p. 312.
[24] See below, pp. 274, 315.

magnetic with difficulty are attracted rather weakly, those that acquire magnetism easily are attracted more strongly; and if the production of magnetism in iron near a magnet is completely inhibited, then there is also no attraction. Finally that same phenomenon I mentioned above as observed in electrified bodies occurs also in the magnet and is absolutely inexplicable without the use of my propositions.

34) This will remove readers' scruples to some extent, and I promise those who perhaps still nourish some doubt that later they will be completely satisfied. Let us now pass to the further development of the laws of action of electrified and magnetic bodies. Let us retain the letters used in § 29. 32, and suppose that a supply of fluid beyond the natural abounds in the body A as well as in the body B, Fig. II. In this case, the quantity of fluid contained in A, since it exceeds the natural, must be written $Q+\alpha$, and similarly the quantity of fluid in B is $q+\delta$, and if we determine the four forces from which the whole mutual action of bodies A and B on each other is composed (§ 32), we will obtain the following expressions for them:

1) the fluid in B attracts the mass of the body A by a force
$$=\frac{qa+\delta a}{q},$$

2) this same fluid repels the fluid contained in A by a force
$$=\frac{(Q+\alpha)(q+\delta)r}{Qq},$$

3) the mass of the body B attracts the fluid contained in A by a force $=\dfrac{QA+\alpha A}{Q}$,

4) the mass of the body B repels the mass of the body A by a force $=R$. So the whole force by which the bodies attract each other $= a+A-r-R+\dfrac{\delta a}{q}+\dfrac{\alpha A}{Q}-\dfrac{qar+Q\delta r+\alpha\delta r}{Qq}$; since $a+A-r-R=o$ (§ 29) and the quantities a, A, r, R, are equal (§ 30), this reduces to $-\dfrac{\alpha\delta r}{Qq}$. Since this formula is negative, it shows that the attraction becomes a repulsion, and the bodies A and B mutually repel each other if fluid abounds in both.

35) If the fluid in both the bodies A and B is supposed to be diminished below the natural quantity, the formula just constructed fits this case if both α and δ are made negative. Thus, the force by which the bodies A and B mutually attract each other is found as clearly as before $=-\dfrac{\alpha\delta r}{Qq}$, whence also two bodies, in both of which fluid is lacking, mutually repel each other. But if finally fluid abounds

GENERAL PRINCIPLES

in the body A and is deficient in B, with δ taken as negative in the formula found in the preceding paragraph, and α positive, we shall obtain a formula corresponding to this case. The force by which the bodies attract each other in this case $= +\dfrac{\alpha \delta r}{Qq}$, and since this is positive, it must be concluded that the bodies attract each other. The same conclusions are reached, if formulae for the cases handled here are sought by the direct method, as was done in the preceding §.

36) So the following are the general rules by which the magnetic or electrical actions of bodies on one another are determined:

1) If both bodies, or only one of them, is constituted in the natural state, there is no mutual action of the bodies on one another.

2) If fluid either abounds or is deficient in both bodies, then those bodies mutually repel one another.

3) If fluid abounds in one of the bodies, but is deficient in the other, those bodies mutually attract one another.

These rules suffice for determining the actions of bodies, if in the whole of them there is either an abundance or a deficiency of fluid. So it remains for us to examine also the action of bodies when either one or both bodies is constituted in a state where fluid abounds in one part, but is deficient in another. Because these cases prepare the way for the explanation of the majority of magnetic phenomena, they particularly deserve our consideration.

37) In this discussion let us retain the notation assumed in § 15. If we first consider that f, a particle of the matter of which the body A itself is made, is placed in the middle of the body A, Fig. IV, it is indeed repelled by both parts AC and AB (§ 30), but by both equally, whence these forces, exerted from this direction and that, destroy each other and reduce to nothing. But a particle f of this kind is attracted by the fluid contained in AC by a force $= \dfrac{(Q+\alpha)r}{Q}$ (§ 30), and by the fluid contained in AB by a force $= \dfrac{(Q-\beta)r}{Q}$, whence it is pressed toward B by a force $= \dfrac{(Q-\beta)r - (Q+\alpha)r}{Q} = -\dfrac{\alpha r + \beta r}{Q}$. Since therefore this force is negative, it is obvious that a particle of the body A is drawn toward C by a force which is equal to $\dfrac{(\alpha+\beta)r}{Q}$. So it is evident that a particle f of the body is urged by the same force by which the particle F of the fluid (§ 16) is drawn, and that there is no other difference than that the force on the particle f is the negative of that exerted on the particle F.

38) Further, if a body G, whose natural quantity of fluid$=q$, is moved up to the part AC, the magnitude of the action of the body A on this body is quite easily calculated. If the quantity by which fluid abounds in the body G beyond the natural quantity q is called δ, the body G will be attracted toward the part AC alone by a force $=-\dfrac{\alpha\delta r}{Qq}$ and toward the part AB alone by [25] a force $=+\dfrac{\beta\delta r'}{Qq}$ as immediately follows from § 34. 35. So the whole attraction of the body G toward A will [26] be$=\dfrac{(\beta r'-\alpha r)\delta}{Qq}$. By a completely similar argument, the force by which the part AB attracts the body H close to itself is found $=+\dfrac{\beta\delta r}{Qq}$, and the force by which the part AC attracts it $=-\dfrac{\alpha\delta r'}{Qq}$, whence the whole force by which H is attracted by the body A$=\dfrac{(\beta r-\alpha r')\delta}{Qq}$.

39) If the body G, or H, be supposed to contain only the natural quantity of fluid, δ will$=0$; so under this hypothesis the attractions which the body A exercises on the body G and on the body H, found to be $=\dfrac{(\beta r'-\alpha r)\delta}{Qq}$ and $\dfrac{(\beta r-\alpha r')\delta}{Qq}$ respectively, disappear. And so neither by its part AC in which fluid abounds nor by its part AB in which fluid is deficient does the body A exert any action on other bodies constituted in the natural state. Finally let the quantity of fluid in the body G be deficient, and by making δ negative in the formula found in the preceding §, the force by which the body G is attracted toward A in this case will become $=\dfrac{(\alpha r-\beta r')\delta}{Qq}$. Similarly by putting δ negative in the formula for the action on H, the formula is converted to the case where the quantity of fluid in the body H is deficient, and so this force is equal to $\dfrac{(\alpha r'-\beta r)\delta}{Qq}$.

40) Since the ratio between α and β is indeterminate, it will be proper to determine it at will and to deduce various consequences from the preceding. We have shown above (§§ 21. 22. 23. 24.) for various determinations of the ratio between α and β how both the part AC and the part AB of the body A act on particles of fluid, either

---

[25] The text has $+\dfrac{\alpha\delta r'}{Qq}$.

[26] The text has $\dfrac{(\beta r-\alpha r)\delta}{Qq}$.

GENERAL PRINCIPLES

not at all, or to attract them or repel them. So the question arises how the body A in any of these states acts on bodies close to itself. In reply, it is evident, first, from the preceding §, that whatever ratio is assumed between α and β, and in whatever state the body A is, its action on the bodies G and H, if they have been constituted in the natural state, is nothing. Since this case is easily explained, we shall have to look particularly at the other cases we handled in § 38. 39.

41) The general formulae found there are easily fitted to each case we must examine here, provided the ratio which is assumed to obtain between α and β is introduced into these formulae. If this is done, we obtain the following formulae:

I

For the part AC in which fluid abounds

The state of the body A	The state of the body G	The force by which G is attracted	
If the part AC does not act on the fluid, in which case $\beta = \frac{\alpha r}{r'}$ (§ 21)	If fluid abounds in G	$= 0$ $- - - -$	0
Ditto	If fluid is deficient in G	$= 0$ $- - - -$	0
If the part AC attracts fluid, in which case $\beta = \mu + \frac{\alpha r}{r'}$ (§ 22)	If fluid abounds in G	$= \frac{+\mu \delta r'}{Qq} - -$	positive
Ditto	If fluid is deficient in G	$= \frac{-\mu \delta r'}{Qq} - -$	negative
If the part AC repels fluid, in which case $\beta = \frac{\alpha r}{r'} - \mu$ (§ 22)[27]	If fluid abounds in G	$= \frac{-\mu \delta r'}{Qq} - -$	negative
Ditto	If fluid is deficient in G	$= \frac{+\mu \delta r'}{Qq}$	positive

[27] The text has $\beta = \frac{\alpha r}{r'} + \mu$.

II

For the part AB in which fluid is deficient

The state of the body A	The state of the body G	The force by which G is attracted	
If the part AB does not act on the fluid, in which case $\beta = \dfrac{\alpha r'}{r}$ (§ 21)	If fluid abounds in H	$= o$ – – – –	o
Ditto	If fluid is deficient in H	$= o$ – – – –	o
If the part AB attracts fluid, in which case $\beta = \dfrac{\alpha r'}{r} + \mu$ (§ 23)	If fluid abounds in H	$= \dfrac{+\mu \delta r}{Qq}$ – – –	positive
Ditto	If fluid is deficient in H	$= \dfrac{-\mu \delta r}{Qq}$ – – –	negative
If the part AB repels fluid, in which case $\beta = \dfrac{\alpha r'}{r} - \mu$ (§ 23)	If fluid abounds in H	$= \dfrac{-\mu \delta r}{Qq}$ – – –	negative
Ditto	If fluid is deficient in H	$= \dfrac{+\mu \delta r}{Qq}$ – – –	positive

42) By comparing the consequences deduced so far and the formulae which I have expounded in the two tables in the preceding §, the truth of the following rules is already obvious.

1) The body A plainly does not act on other bodies constituted in the natural state.

2) If one part of the body A is constituted in that state where it plainly does not act on particles of fluid, then this part of it acts as if it had been constituted in the natural state (§ 36).

3) If a part of the body A, whichever it is, repels particles of fluid, it acts on other bodies as a body in which fluid abounds usually acts.

GENERAL PRINCIPLES

It then attracts bodies in which fluid is deficient, but repels those in which fluid abounds (§ 36).

4) If the other part of the body attracts particles of fluid, this part acts on other bodies as a body in which fluid is lacking usually acts (§ 36), that is, it repels bodies in which fluid is lacking and attracts those in which fluid abounds.

43) Finally, we must begin a discussion of the mutual action of two bodies on each other when each is in that state where fluid abounds in a certain part of it while it is deficient in the other. To avoid the need to descend to particular cases let us immediately develop a general formula that we can subsequently fit to any case which might offer itself. Imagine two bodies AD and EH, Fig. VII, close to each other and each divided into three equal parts. It is sufficient to evolve a formula for this case alone, since cases other than those which can be easily explained from this do not occur in what follows. Let the natural quantity of fluid in the parts AB, BC, CD $=Q$, but let the natural quantity of fluid of the parts of the other body EF, FG, GH be called $=q$. Let it be further supposed that while each body is constituted in the natural state,

The fluid in AB acts on that contained in EF by a force $=r$
– – – BC – – – – – – $=r'$
– – – CD – – – – – – $=r''$
– – – AB – – – – – FG by a force $=\rho$
– – – BC – – – – – – $=\rho'$
– – – CD – – – – – – $=\rho''$
– – – AB – – – – – GH by a force $=\mathfrak{r}$
– – – BC – – – – – – $=\mathfrak{r}'$
– – – CD – – – – – – $=\mathfrak{r}''$

Finally let the quantity of fluid beyond the natural abound

in AB, by a quantity $=a$
– BC – – – $=b$
– CD – – – $=c$
– EF – – – $=e$
– FG – – – $=f$
– GH – – – $=g$.

If by the help of § 34 we determine the action by which any part of the body AD acts on any part of the body EH, we find

The action of the part BA on the part EF $= -\dfrac{aer}{Qq}$

- - - - BC - - - - $= -\dfrac{ber'}{Qq}$

- - - - CD - - - - $= -\dfrac{cer''}{Qq}$

- - - - AB - - - FG $= -\dfrac{af\rho}{Qq}$

- - - - BC - - - - $= -\dfrac{bf\rho'}{Qq}$

- - - - CD - - - - $= -\dfrac{cf\rho''}{Qq}$

- - - - AB - - - GH $= -\dfrac{ag\mathbf{r}}{Qq}$

- - - - BC - - - - $= -\dfrac{bg\mathbf{r}'}{Qq}$

- - - - CD - - - - $= -\dfrac{cg\mathbf{r}''}{Qq}$

By collecting these expressions together, the total force by which the bodies AD and EH attract each other [28]

$$= \frac{-(ar+br'+cr'')e - (a\rho+b\rho'+c\rho'')f - (a\mathbf{r}+b\mathbf{r}'+c\mathbf{r}'')g}{Qq}$$

We shall be able to fit this formula to any case without difficulty, if we substitute for the letters a, b, c, e, f, g the values due to them; these are negative if the fluid in a certain part does not abound, but is deficient.

44) The general principles of our universal magnetic and electrical theory, now that they are thus settled, will have to be applied to the phenomena of each kind. Happily the explanation for very many electrical phenomena, and especially those of the *Leyden* or *Musschenbroekian* jar, can be first sought among them. My procedure throughout will not be to linger over the explanation of all these phenomena; but remembering the scope of my writing, I will only inquire here into the more outstanding phenomena and those that have not yet perhaps been as distinctly explained by others as can be done with the aid of the principles developed here, and especially into those which will make for a more distinct exposition of magnetic theory. This is the reason I begin with the experiment of commotion and a consideration of the Leyden jar.

---

[28] The text has $- \dfrac{(ar+br'+cr'')\,e - (a\rho+b\rho'+c\rho'')f - (a\mathbf{r}+b\mathbf{r}'+c\mathbf{r}'')g}{Qq}$

GENERAL PRINCIPLES

45) Since the shape of the glass used for the experiment of commotion does not affect the success of the experiment, let us suppose in the following that we use a flat glass plate; for the experiment succeeds equally with this as with a vessel in the shape of a jar. Therefore let ABEF, Fig. VIII, be a glass plate covered on both sides as is usual by metal plates IK and CD. Let the plate IK be touched by a body non-electric *per se*, such as the chain LM in the figure, not supported by bodies electric *per se* but connected with the globe of the earth itself through other bodies non-electric *per se*. Let electricity be communicated in the usual way to the leading plate CD with the help of a chain or wire TS which brings electricity to it. If positive electricity is communicated to the chain TS in some way by means of either a globe or an electrificatory tube, the plate CD will also be filled with electric matter. Let the quantity of electric fluid accumulated in this way in the plate CD beyond the natural quantity $=\alpha$, and, if we retain the notation assumed in § 15, the particles of electric fluid which cling to the plate IK will be repelled by a force $=\dfrac{\alpha r'}{Q}$ (§ 10). Since there is no impediment, these particles ought to pass into the chain LM and disperse through it, and this efflux will not cease until after the plate IK is so emptied that the force acting on the particles clinging round it disappears. Since, if the quantity of fluid which has flowed from the strip IK is put $=\beta$, this force is $=\dfrac{\beta r - \alpha r'}{Q}$ (§ 15), the efflux will not stop until after β has become $=\dfrac{\alpha r'}{r}$, at which stage the force by which the particles clinging round the plate CD are repelled will be $=\dfrac{\alpha(r^2 - (r')^2)}{Qr}$ (§ 21).[29] So if electrification is continued, by this reasoning more and more electric matter will be continuously accumulated in the plate CD, but so much will be emptied from the opposite plate IK that the force on particles of electric fluid next to the strip IK is always reduced to 0. Yet this accumulation cannot proceed to infinity. For since the [30] force $\dfrac{\alpha(r^2 - (r')^2)}{Qr}$ increases continuously as α increases, this force enticing particles out of the plate CD must finally become so great that the resistance which the encompassing air offers against the dissipation of electric fluid—air as an

---

[29] The text has $\dfrac{\alpha((r')^2 - r^2)}{Qr}$.

[30] The text again has $\dfrac{\alpha((r')^2 - r^2)}{Qr}$.

electric *per se* admits fluid into its pores with difficulty—no longer suffices for retaining these particles, so then the fluid must escape again of its own accord, because it is accumulated beyond this limit.

46) Let it be assumed on the contrary that negative electricity is communicated to the plate CD and on account of the consequent evacuation of the plate CD, it is evident that the particles of electric fluid clinging round the plate IK are attracted toward it, whence they must enter the chain LM and be accumulated in it. Let the quantity of electric fluid evacuated from the plate CD$=\beta$, and the force attracting the particles of fluid toward the plate IK will be$=\dfrac{\beta r'}{Q}$ (§ 12). The afflux of electric fluid into the plate IK cannot cease until so much has entered this plate that the force producing the afflux has disappeared; that is, if the quantity of fluid which entered the plate IK is called $=\alpha$, since that force is equal to $\dfrac{\beta r' - \alpha r}{Q}$ (§ 15), until β has become$=\dfrac{\alpha r}{r'}$. Then the force by which the particles clinging round the other plate CD are attracted$=\dfrac{\alpha(r^2-(r')^2)}{Qr'}$ (§ 21). For the rest, similar things obtain as in the preceding case. The evacuation of the plate CD can indeed continue on and on, but not to infinity. For the attractive force of the plate CD, $\dfrac{\alpha(r^2-(r')^2)}{Qr'}$, must become so great that the air impeding the entrance of electric matter into the plate finally becomes unequal to performing this task. When this happens, as much electric fluid will always spontaneously enter the plate CD as is taken away by electrification.

47) I have considered here only the metal sheets covering the two sides of the glass plate, and I have not looked at the glass plate itself. It is evident that in respect of its surfaces the same obtains as I have demonstrated is the case with the plates CD and IK. For while electric fluid accumulates in one of the plates and is evacuated from the other, because of the close contact of the metal with the surfaces of the glass plate electric fluid must also enter one surface of the latter and leave the other. Although glass is electric *per se,* it is not perfectly so, and electric fluid is not altogether immobile in its pores, though it moves only with appreciable difficulty. So it is impossible for electric fluid to accumulate in a metal plate, e.g. CD, without some of it entering the surface of the glass plate, and the plate IK cannot be evacuated without some electric fluid moving from the surface of the glass plate into IK. It is therefore easily seen that, in the case we dealt with in § 45, fluid accumulates in the surface AE of the glass

GENERAL PRINCIPLES

plate but is evacuated from the other BF; and in the case we considered in § 46, on the contrary, the surface AE of the plate is evacuated of fluid while the opposite one BF is filled with it. What has been demonstrated in § 45 and 46 about the metal plates DC and IK, and what we shall demonstrate later about them from principles drawn from there, are equally valid for the surfaces of the glass plate corresponding to each metal plate, and can be transferred to them without fear of error.

48) From this it follows first that electric fluid can accumulate on the surface AE of a Leyden jar and on the metal plate CD covering it, or be evacuated from it, to a far greater degree than can occur if *Musschenbroek's* apparatus is not used, and electrification is instituted without it. For if the plate CD alone is electrified, and the quantity of electric fluid accumulated in it, if positively electrified, $=\alpha$, the force driving the particles of fluid from the plate will be $=\frac{\alpha r}{Q}$ (§ 10). But if *Musschenbroek's* apparatus is used, this repelling force will be $=\frac{\alpha(r^2-(r')^2)}{Qr}$ (§ 45). Suppose in the former case that this repelling force is the greatest possible, at which point α is made $=\gamma$, and in the latter when the *Leyden* jar is called into use, that this terminal force is obtained when α has increased to the point where it has attained a value $=\delta$. Since each force can overcome the same impediment (§ 45), they will be equal, so $\frac{\gamma r}{Q}=\frac{\delta(r^2-(r')^2)}{Qr}$, whence is obtained $\delta=\frac{\gamma r^2}{r^2-(r')^2}$. Therefore, since $\frac{\gamma r^2}{r^2-(r')^2}$ is greater than γ, it is clear that a far greater accumulation of electric fluid is possible in the plate CD if the *Leyden* jar is used, than if electrification is instituted without it. This same reasoning is so easily transferred to the case where negative electricity is communicated to the plate CD that we do not need a specific application to this case. So if negative electrification is instituted it can similarly produce evacuation of the plate CD to a far greater degree by using *Musschenbroek's* apparatus than can happen if the plate CD alone is electrified.

49) Experience agrees extremely well with these reasonings deduced from the theory. For it is very well known to anyone who has ever taken part in electrical experiments, that if chains or beams are suspended, and the whole electrical apparatus is electrified without using *Musschenbroek's* jar, three or four turns of the electrificatory globe are quite enough to communicate the greatest degree of electricity of which it is capable to the apparatus, and further turns serve

not to increase the degree of electricity but only to preserve it. But if on the contrary a *Leyden* jar is employed, three or four turns of the globe are a long way from producing the maximum degree of electricity. In fact it can still be increased even after a thousand turns, especially if the jar used has a rather ample surface and thin walls.

50) I say if the jar is rather thin walled, for it can easily be shown that, other things being equal, if the plate ABEF is rather thin a greater accumulation of electric fluid is possible than if it has a greater thickness. For although it is clear from § 48 that δ is always greater than γ, it is easily evident that δ gets proportionally greater than γ as $r^2 - (r')^2$ decreases, or as r' approaches equality with r. But it is obvious from § 21 above that r' differs less from r to the extent that the distance between the plates CD and IK is less, that is, the less is the thickness of the intervening plate. This is also why a much more violent commotion can be produced with a thinner jar than with one made from thicker glass, as experience long ago taught investigators of nature. I shall show clearly later on that the violence of the electrical commotion depends particularly on the degree of condensation of the electric fluid.[31]

51) Certain phenomena are usually observed in the *Leyden* jar which seem at first sight paradoxical and hardly reconcilable with *Franklin's* theory of this experiment. They are the following:

1) After the *Leyden* jar has been filled with electricity the exterior plate IK gives no signs of electricity. It does not provoke sparks from other non-electrified bodies and it does not, as is usual at other times with electrified bodies, attract light bodies nearby. Yet each of these things should happen if the experiment has been carried out in the fashion indicated in § 45, since according to *Franklin's* theory the plate IK is then evacuated of electric fluid and is strongly negatively electric.

2) Although according to *Franklin's* theory the strip CD is positively electric to the highest degree, even more strongly indeed than is ever possible without the use of a jar, nevertheless in respect of the degree of electricity it possesses it gives only weak signs of electricity and, if the electrification has not been continued for quite a long time, hardly as much as it would give off without the use of *Musschenbroek's* apparatus. For it does not throw out sparks over longer distances and it attracts light bodies only quite weakly.

52) Yet the reason for these phenomena can easily be assigned from our theory. For since the force by which the plate IK attracts

---

[31] See below, p. 280; and also above, p. 116.

GENERAL PRINCIPLES

electric fluid outside itself is reduced to nothing (§ 45), its condition is completely as if it were constituted in the natural state, whence no force is exerted to entice the electric fluid enclosed in the pores of other bodies close to itself. So fluid leaving a body of this kind cannot leap into the plate IK and give out sparks. For the same reason it cannot exert either electrical attractions or repulsions, as is sufficiently clear from § 42 no. 2.

53) As for the plate CD, although it is positively electric itself in a most violent degree, yet it can give out only weak signs of electricity. For since the force by which it repels its abundant electric fluid, a force which $=\dfrac{\alpha(r^2-(r')^2)}{Qr}$, is always less than it would be if the plate CD had been filled to the same degree with electric fluid without *Musschenbroek's* apparatus $\Big($ in which case that repelling force would be $=\dfrac{\alpha r}{Q}\Big)$, it is evident that electric fluid leaps out and gives off a spark with more difficulty in the former case than if the same should happen in the latter. It is easily shown that weaker electrical attractions and repulsions ought to be exerted by the plate in the former than in the latter case. For, if *Musschenbroek's* apparatus is used, the repulsion by which the plate AC acts on another body in which the quantity of fluid exceeds the natural by a quantity $=\delta$ is from § 38, $=\dfrac{(\alpha r-\beta r')\delta}{Qq}$, whence by substituting the value which β has in this case (§ 45), or $\dfrac{\alpha r'}{r}$, the repulsion in our case $=\dfrac{\alpha(r^2-(r')^2)\delta}{Qqr}$. But if we do a calculation for the case where the *Leyden* jar is not used, the repulsive force of the place CD will be $=\dfrac{\alpha\delta r}{Qq}$ (§ 34). It is self-evident that, since $\dfrac{r^2-(r')^2}{r}$ is less than r, the latter force is significantly greater than the former.

54) Although the plate CD gives out only feeble electrical effects, the point cannot be reached where it plainly gives out none. As is clear from § 21. 38. 42, this cannot happen unless r' becomes $=r$, and it is necessarily always less than this (§ 21). Meanwhile the force by which the plate CD repels electric fluid and acts on other bodies can become so small that it is indetectable and escapes the notice of the senses. This happens if there is very little difference between r' and r, which is the case if the plates CD and IK and the intervening glass plate ABEF are quite thin (§ 21). In the *Leyden* jar it can hardly ever happen that the electrical action from both parts becomes completely indetectable, for between the two metal plates or the bodies

used in their place there must of necessity be interposed a glass plate, and this cannot be thin enough for such a point to be reached. I have come across certain other relevant cases where, though a strong enough electricity is produced, it does not reveal its presence. These experiments, in which I gained different results from other scholars, seem to warrant a more careful description.

55) Take two plates of clear glass of the same shape, AB, EG, Fig. IX, each with a surface of a few square inches. Glue a handle KIH, made up of glass cylinder HI of moderate thickness and three inches long and a wooden haft IK, to either one of them by means of sealing wax. Wipe both the plates and the cylinders properly; if the air is humid and full of vapors at the beginning of the experiment dry and warm them at a fire. Prepare a pendulum NO made of a single well dried silk thread, such as is normally woven from bombyx, and a small bob of cork about the size of a lentil (this little bob must be slightly damp), and suspend the pendulum freely. Then rub the glass plates mentioned above against each other a few times. After this rubbing, if the plates are not separated from one another but remain in contact and are brought close to the pendulum NO, they give no sign of electricity and plainly do not act on the pendulum. But if one of the plates is separated from the other, and then either one of them alone is brought close to the pendulum, it will be found to be electric, at first strongly attracting the pendulum and afterward repelling it. When the plates are placed together or joined again, the electricity is immediately extinguished, but later, when they are separated once more, it is revived in both; this can be repeated on almost innumerable occasions. Although both plates becomes electric through this mutual friction, they never acquire the same kind: one of them gains positive electricity, the other negative.

56) After these plates have become electrified by mutual attrition, separate them and bring one of them close to the pendulum NO. It will attract the pendulum, and after it has communicated some of the electricity it possesses to the little bob O, it will repel it. Then keep the plate ABCD in the position shown in Fig. X, so that the pendulum, repelled at its lower end, comes to the position N*o*. Further, hold the other plate EFGH in a position parallel to the former at a distance of three or four inches. If the plate EFGH is moved one or two inches closer to ABCD, the pendulum NO will begin to descend. When the plate EFGH is drawn back, it will ascend again to its former height. In this way, the pendulum may be forced either to ascend or descend at will, as the plate EFGH is removed further from ABCD, or is brought closer to it.

GENERAL PRINCIPLES 275

57) If the often mentioned plates are first rubbed together in the manner described, and then all electricity is removed from one of them, e.g. the one that gained a positive electricity, by contact with a body non-electric *per se,* and then they are rubbed against each other once more, when placed together and made to touch there will not be an absence of all activity as before, but they will act as if they possess negative electricity. The contrary of this happens if the electricity was removed from the plate which possessed a negative electricity, for then the joined plates show a positive electricity. Nevertheless if in each case the plates are separated from one another, one of them, as in the previous experiment, will be positively electric, the other negatively.

58) In carrying out these experiments there is no need for two glass plates specifically. Any two thin plates at all, of whatever matter they are made, produce exactly the same results. The experiment succeeds equally well if both plates are of glass, or one is made of sulphur or metal, or both or one of them is made from sulphur or sealing wax or some other material. The experiment can be varied in infinitely many ways and there is no fear of failure provided care is taken that both plates, or at least one of them, is made from some matter electric *per se.* For if neither plate is electric *per se,* it is to be expected that electricity will never be obtained by mutual attrition; it is in fact evident that in this case, even if electricity is supposed to be produced, it will be immediately destroyed in a moment. If one plate, e.g. ABCD, Fig. IX, is made to gain a positive and the other HGEF a negative electricity, while the two plates touch, the fluid abounding in ABCD will be urged to cross to the evacuated second plate EFGH, as is evident from § 16. But if both plates were non-electric *per se,* the electric fluid, finding no difficulty in moving freely in the pores of such bodies, ought to obey this force and hence ought to cross from ABCD into EFGH, whence the instantaneous destruction of the electricity in both plates follows of necessity.

59) I have obtained similar phenomena by still other means. I had a dish AB, Fig. XI, shaped like a segment of a sphere, made from thin tin or brass; to it was joined a hollow cylinder FE by means of cement. I placed this dish on the stand EFGH, consisting of a square wooden foot GH and a glass cylinder IF.[32] I poured liquefied sulphur into the dish and when it congealed I glued to it a handle of the type I described in § 55. As long as the sulphureous segment was joined to the dish, no trace of electricity was observed. As soon as the sulphur was separated

---

[32] The text has EF.

from the dish, both dish and sulphur were found to be vigorously electric, and this electricity was at once completely extinguished when the sulphureous segment was put back in the dish, and then revived if these two bodies were once more separated. In this experiment the sulphur and the dish showed contrary electricities, but sometimes the sulphur possessed a positive, at other times a negative electricity. If it were the former, and I touched the dish with my finger after removing the sulphur, I destroyed in this case the negative electricity it had gained; when the sulphur was replaced, the whole apparatus acted as if it were positively electric, although when the sulphur and dish were separated the former was positively electric, the latter negatively. If the sulphur had initially gained a negative electricity, and I proceeded in the experiment in a similar fashion, the effect was the same, only now when the sulphur and the dish were joined, the whole apparatus acted as if it were negatively electric.[33]

60) All the phenomena reviewed here agree remarkably well with the theory expounded up to this point. For if the plates described in § 55 are rubbed, a certain part of the electric fluid crosses from one into the other; call this γ. Since the quantity that abounds in one of the plates e.g. ABCD is equal to the quantity that is deficient in the other EFGH, if we use the notation of § 15, the force by which the plate ABCD repels fluid while the plates are held parallel will be $=\frac{\gamma(r-r')}{Q}$, and the force by which EFGH attracts it, equal to the former, will likewise $=\frac{\gamma(r-r')}{Q}$ (§ 15). If the plates are moved closer to one another, r' gets closer and closer all the time to equality with r, whence the force by which they both act on fluid must decrease. But if they are completely touching, then on account of the close contact of the two plates, r' does not sensibly differ from r, so the force by which the two plates act on fluid disappears.

61) If further the natural quantity of electric fluid in the bob O of the pendulum is called q, and fluid is considered abundant in it by a quantity δ, the force by which the plates held in a parallel position act on the pendulum, if the plate that is positively electric is closer to the pendulum, will be $=\frac{\gamma\delta(r'-r)}{Qq}$, and if the negatively electric plate is closer to the pendulum, will be $=\frac{\gamma\delta(r-r')}{Qq}$ (§ 38. 39). Whether δ is positive or negative, it is evident that these actions are

[33] This experiment has sometimes been seen as anticipating Alessandra Volta's discovery in 1775 of the electrophorus. See above, pp. 129–30.

GENERAL PRINCIPLES

less if there is very little difference between r' and r, and that they disappear altogether, if r' becomes completely equal to r. The reason for the phenomenon described in § 56 is immediately clear from this, and as easily discerned now is the reason why, when two plates touch each other, they plainly exercise no electrical attractions or repulsions (§ 55). I consider this last phenomenon quite noteworthy. For we learn from it that electricity is in fact produced rather often in nature, without being manifested by any sign. So it is right to believe that electricity acts in very many phenomena where we have not the least suspicion of its presence.

62) If the experiment described in § 57 is performed, and electricity is removed from one of the plates after rubbing (let this plate be first of all that which possessed positive electricity), then with the first rubbing the effect was that in this plate the quantity of electric fluid was changed to $Q+\gamma$, and in the other plate to $Q-\gamma$. When the electricity in the first plate is destroyed, the plate is reduced again to the natural state, and the quantity of electric fluid contained in it becomes again $=Q$. With the second rubbing, a certain amount of electric fluid passes afresh from the negatively electric plate to the other; let this quantity be called ϵ; the quantity of electric fluid in the former plate will now be $=Q+\epsilon$, in the latter $=Q-\gamma-\epsilon$. Therefore when the plates are held in a parallel position, the force by which the positively electric one attracts fluid [34] will $=\dfrac{\gamma r'+\epsilon r'-\epsilon r}{Q}$ but the negatively electric one attracts it with a force $=\dfrac{\gamma r+\epsilon r-\epsilon r'}{Q}$ (§ 15). Further, the force by which the former plate attracts the pendulum is equal to $\dfrac{(\gamma r'+\epsilon r'-\epsilon r)\delta}{Qq}$, and that by which the latter does is $\dfrac{(\gamma r+\epsilon r-\epsilon r')\delta}{Qq}$ (§ 38). But it is evident that if r' becomes $=r$, these forces do not disappear, but the former two become $=+\dfrac{\gamma r'}{Q}$ and $+\dfrac{\gamma r}{Q}$, while the latter two $=+\dfrac{\gamma\delta r'}{Qq}$ and $+\dfrac{\gamma\delta r}{Qq}$, and will be positive if δ is positive, negative if δ is negative. The reason is already clear why the joined plates act completely as if their sum total were negatively electric (§ 12. 35).

63) The calculation for the other case is similar. If the electricity of the negatively electric plate is destroyed, let the quantity of electric fluid in it, after it has been reduced again to its natural state, $=Q$.

---

[34] The text has $\dfrac{\gamma r'-\epsilon r'-\epsilon r}{Q}$.

When a certain amount of electric fluid crosses from it to the other plate through a new rubbing (call this quantity ϵ), the quantity of fluid in the negative electric plate becomes $Q-\epsilon$, in the positively electric plate $=Q+\gamma+\epsilon$, so if the plates are held in a parallel position the force by which the positively electric one attracts fluid $=\dfrac{\epsilon r' - \gamma r - \epsilon r}{Q}$ and that by which the negatively electric plate attracts it $=\dfrac{\epsilon r - \gamma r' - \epsilon r'}{Q}$ (§ 15). Further, the forces by which they attract the bob of the pendulum NO are for the positively electric plate $=\dfrac{(\epsilon r' - \gamma r - \epsilon r)\delta}{Q}$ and for the negatively electric one $=\dfrac{(\epsilon r - \gamma r' - \epsilon r)\delta}{Qq}$. When r' is made $=r$, the two former forces become $=-\dfrac{\gamma r'}{Q}$ and $-\dfrac{\gamma r}{Q}$, while the two latter become $=-\dfrac{\gamma \delta r'}{Qq}$ and $-\dfrac{\gamma \delta r}{Qq}$, and these are negative if δ is positive, and positive if δ is negative. A consequence of this is that the joined plates then act as if their aggregate were positively electric (§ 10. 34. 35).

64) These statements are easily applied to the experiments described in § 59, so there is no need for a specific discussion of them. Readers may harbor doubts though because the sulphureous segment is not thin but quite thick, whence it seems that in this case r' can never become sensibly fully $=r$. But it must be noted that this sulphureous segment gains electricity only in that surface which touches the dish, and its electricity does not penetrate much beyond this surface. For the electric fluid abounding in it cannot penetrate deeply into the very substance of the sulphur in proportion as it is outstandingly electric *per se,* and if fluid is deficient in the surface it cannot cross to there from the interior parts of the sulphur. So the electricity of a piece of sulphur subsists only on its outermost surface, and this is sufficiently thin that in contact with the dish, r' can be assumed to be $=r$.

65) The phenomenon occurring in the *Leyden* jar is also noteworthy, since its electricity lasts tenaciously and much longer than if the jar is not used. Electricity lingers on for several days, for a whole week in fact, when in bodies electrified without the use of *Musschenbroek's* jar it hardly lasts fifteen or thirty minutes. The reason for this phenomenon springs so clearly to the eyes that it hardly seems worthwhile to expound it. For in part, as was demonstrated in § 48, a great quantity of electric fluid is accumulated in the plate CD, such as cannot be forced into a body other than by using a *Leyden* jar;

and in part the force which tends to expel the abundant quantity of fluid while the *Leyden* jar is being used is less than that which strives to dissipate the fluid if electrification has been instituted without the use of the jar, as I proved in § 48. Both of these must give rise to a longer conservation of electricity, for it is self-evident that it must last longer where there is a greater quantity of fluid abounding in a body and where at the same time the force tending to expel it is less. It is obvious from this at the same time that, all things being equal, the continued existence of electricity in the *Leyden* jar, and in the other cases relevant here, ought to be protracted longer, the less r' differs from r. For the less this difference is, the greater are both the causes of the continued existence. So electricity will last the longest in the two cases I described in § 55. 59, because there the difference between r' and r completely escapes the senses. Experience agrees exceptionally well with this reasoning, and with the help of it not only I myself but long ago several others understood that, under the adduced circumstances, electricity is hardly ever destroyed. Thus e.g. if liquefied sulphur is poured into a dry glass and is taken out only after very many months, whole years even, electricity is found still existing in it.[35]

66) Although the continued existence of electricity in a *Leyden* jar left to itself is quite long, yet it can be destroyed by a single act in a single moment. This happens when that equally marvellous and noteworthy experiment of the electrical commotion is performed—the experiment which gave a special opportunity to investigators of nature to inquire more carefully into the phenomena of the *Leyden* jar. To understand properly how the outcome of this experiment depends on the theory we have so far explained, suppose that, after the jar has been filled with electricity and established in the state I described in § 45 above, a curved metal wire PQR, Fig. VIII, or any other body non-electric *per se,* touches the plate IK with one of its ends P. It is clear first of all that no change can take place from this contact, for the plate IK is established in the state where it plainly does not act on electric matter. But if R, the other end of the wire, is moved closer to the other plate CD, and right up to contact with it, then a spark is drawn since the force of the plate CD for propelling electric fluid, though weakened, is not altogether reduced to nothing (§ 48). At the same moment, as some of the electric fluid begins to cross from the plate CD to the end of the wire R, the force of the plate IK for attracting fluid no longer remains zero, and it immediately begins to attract electric fluid again. For if the quantity of fluid which crosses from the plate CD into the wire is called ϵ, the force by which the plate IK attracts

---

[35] This was first reported by Stephen Gray, *Phil. Trans.,* xxxvii, 1732, 285–91.

fluid will $=\frac{er'}{Q}$, as is readily obvious from § 15. So the fluid which entered the end R of the wire must immediately cross into the plate IK. So there must arise a continuous influx of the surplus fluid from CD into IK through the wire PQR, and this cannot stop until both plates are equally filled with fluid, that is, until the state of the jar described in § 45 above is completely destroyed.

67) This transfer should occur with considerable speed and violence. For

1) At the same moment as the electric fluid begins to cross from the plate CD to the end R of the wire its other end is evacuated of fluid. Both make for a significant acceleration of the movement of electric fluid along the wire from R toward P. For the fluid that entered the end R ought to tend toward P, as is shown clearly in the previous discussion, in as much as that part of the body would contain a smaller quantity of electric fluid, even if the end P had not been evacuated. But since at the same moment it becomes emptied of fluid, this tendency must be greater, so the force urging the fluid to flow through the wire PQR from the plate CD to the plate IK is as if double what it would be, if the end of the wire did not touch the plate IK.

2) Since these causes entice the fluid to cross as long as the plate CD contains a greater quantity of fluid than the other plate IK, from the continuous action of these forces, the flow will of necessity accelerate appreciably. And especially

3) The electric fluid is appreciably condensed, even to a thousand times greater degree than is possible if *Musschenbroek's* apparatus is not used. So since this great quantity of fluid crosses in a very short time, in a single moment as it were, from the former plate to the latter, it is clear that extremely violent effects must necessarily occur as a result.

4) Finally electric fluid incited to the swiftest movement will be appreciably compressed, and, as it is very elastic, will be excited into very fast vibrations.[36] In this last seems to consist the particular cause of the pain, and the extremely violent concussion which a man feels when his body touches the wire PQR, and allows the electric matter to cross through it.

68) Other extremely violent effects accompany the crossing, but the scope of this work prevents my discussing them, and it seems hardly worthwhile to begin since he who wants to apply what has so far been

---

[36] Cf. Helmholtz, "Über die Erhaltung der Kraft" (1847), in his *Wissenschaftliche Abhundlungen,* Leipzig, 1882–95, I, 46.

expounded to particular phenomena, can easily understand the explanation of innumerable phenomena flowing from it almost spontaneously. Meanwhile it seems that some mention ought to be inserted of the explosion and crackling noise which accompanies the transfer just described of electric matter through the wire. I think that nothing else than a strong compression of fluid in the plate CD, and a sudden cessation of the compressing force, can be ascribed as the reason. For the elements of Physics show abundantly that, in an elastic fluid such as the electric one, these things are sufficient to produce a violent explosion and an appreciable crashing.

69) I have so far considered only the case where the plate CD has been positively electrified when the jar was imbued with electricity, so the consequences deduced concern only the case I considered in § 45. But all these conclusions can easily be transferred to the second case, dealt with in § 46, where the plate CD was negatively electrified. For one needs only to stay strictly on the path I have followed in deducing conclusions up to now. So readers will not require from me a special consideration of that case, since that would be completely superfluous. Finally I issue a warning once more that although I have considered here only the action of the metal plates CD, IK and have looked at the intervening glass plate only insofar as it impedes the actual transfer of electric fluid from one of the plates to another, I do not consider the glass plate itself as altogether inert. In § 47 above I have already expounded my reasons for thinking that it plays some part in producing the electrical commotion.

70) A certain gentleman particularly expert in *Franklin's* theory [37] has informed me that he had hoped he could sometime, guided by his contemplation of this theory, produce the electrical commotion without the help of the glass which most people think necessary to the success of this experiment. He tackled the matter in the following way. He suspended next to each other from silk cords two iron bars AB and CD, Fig. XII, and electrified the former positively, the latter negatively. Then at the same moment he touched the bar AB with one hand, and the bar CD with the other; but success did not answer his hope, for by this method he was unable to obtain any sensation like the electrical commotion. But if we compare this experiment with the theory so far expounded, it is clear that

1) Of the causes producing the violent transit of electric matter in the experiment of commotion (reviewed above in § 67), the first and second are in a way applicable here, provided both bars are touched at the same moment, and a spark is enticed from each at

[37] Presumably J. C. Wilcke.

the same time, but this I think is difficult in practice and can hardly ever be obtained. But if we look at the other causes, it is evident that

2) the third cause of the violent explosion, and the fourth, which depends on it, are not applicable here. For electric fluid is not strongly compressed in the bar AB, and the bar CD is not evacuated of that fluid to a very great degree. For from the way the experiment was set up, the electric fluid could not be more condensed in the bar AB and could not be evacuated from CD to a degree greater than if the electrification was carried out in the ordinary manner without the help of the jar.

So it is not surprising that the electrical commotion could not be obtained in this fashion since, with the primary cause absent, the violence with which electric matter passes through the experimenter's body in this test was considerably, even a thousand times, less than usually happens when *Musschenbroek's* jar is employed.

71) But if two bodies are combined so that, as in the *Leyden* jar, a similar condensation of fluid is obtained in one part, and a similar rarefaction in the other, there is no doubt that the test would then succeed as desired. I actually did this in the experiment in which I first produced the electrical commotion without the help of the glass which most people thought clearly necessary for the success of the experiment of commotion. Prevented by lack of time, I mentioned this briefly in the paper containing the description of the phenomena of the *Tourmaline* stone which I delivered to the illustrious Academy of Sciences at *Berlin,* and which was printed in the Commentaries of the Academy for the year 1756, where I described the things required for the success of the experiment.[38] This extremely noteworthy experiment gave me the opportunity of making a significant correction to the theory of electricity handed down by *Franklin,* which was spoiled by some quite improbable hypotheses, and it seems to warrant an account of certain things relating to it.

72) When *Musschenbroek's* experiment first became known, most Physicists thought the special cause of the violent phenomena of the *Leyden* jar ought to be attributed to the peculiar nature of the water generally used for this experiment. But once it was established that any other body non-electric *per se* could be substituted for water with equal or even greater success, anyone could easily have come upon a test such as mine. For the suspicion might easily have arisen that any other body electric *per se* could play the role of glass, and since it was more than sufficiently understood that air is electric *per se,* it was an

---

[38] The experiment was described very briefly in the last two and a half pages of the paper, i.e. *Mém. Acad. Berlin,* XII, 1756, 119–21.

GENERAL PRINCIPLES

obvious move to test whether air could play the role of glass in this experiment. As far as I know, it never occurred to anyone,[39] and I myself did not come to it this way; I was led to it by *Franklin's* theory itself. A friend occupied with electrical experiments was showing me a certain test which in its essentials coincided completely with the one *Franklin* described in the English Transactions Vol. LIX, Part I "For the Year 1755", pp. 302. 303. 304; [40] he asked me to work out how it could be reconciled with *Franklin's* theory, something which seemed difficult to him. Turning over in my mind the principles of this theory, I immediately fell upon an explanation which clearly agreed with that which *Franklin* had assigned to this phenomenon. I perceived from the theory I had formed that, if a certain positively electric body approaches another body constituted in the natural state, from which fluid can escape freely through the mediation of other bodies, this latter body must become emptied of fluid because of the repulsion which the abundant electric matter in the former exerts; but, after this has been done, the air flowing between these two bodies must prevent electric matter from crossing from the former into the latter, and rectifying the shortage caused by the repulsion. Since I saw that all this corresponded exactly with what, to *Franklin's* mind, happens in the *Leyden* jar, I came away persuaded that the electrical commotion could be produced even without the aid of glass, by using air in its place.

73) My friend was doubtful because he imagined another explanation not altogether consonant with *Franklin's* theory for the phenomenon whose cause I had discovered. So thinking that an experiment ought to be set up quickly to convince my doubting friend and to confirm the truth of my explanation, and since other apparatus suitable for this experiment was not available, my friend helped me to suspend a square iron bar and below it, about one and a half inches away in a parallel position, a tube made from soldered brass plates. I touched the tube with my right hand, and electrified the iron bar in the usual manner. When I considered it was endowed with a sufficiently strong electricity, I tried to entice a spark from it with my left hand, but was unable to get any sensation similar to that which usually

---

[39] Nollet, who first established that other conductors could be substituted for the water in the Leyden experiment, also tried substituting other substances—though not air—in the place of glass. He reported, however, that when he did this, the experiment no longer succeeded, except, with some differences in the effects obtained, with porcelain (Nollet, *HAS*, 1746, Mém. pp. 1–23).

[40] Aepinus gives the wrong volume number. The correct reference is Franklin, *Phil. Trans.*, XLIX, 1755, 302–304; reprinted in *Benjamin Franklin's Experiments*, pp. 304–305.

accompanies the electrical commotion. But since I was thoroughly convinced of the truth of my reasoning, I sought a reason for my lack of success and I came upon one which seemed very probable. It was easy to perceive from *Franklin's* theory (and readers will easily understand this from what has been discussed previously) that the commotion will always be less violent, firstly to the extent that the surfaces, in one of which the fluid is condensed and in the other evacuated, are smaller, and secondly to the extent that these surfaces are a greater distance apart, since the cause of the greater rarefaction and condensation decreases when the distance is increased (§ 50). I easily recognized that both causes of debility occurred here. For in the usual method of instituting *Musschenbroek's* experiment, when a jar or a glass plate is used, the surfaces are hardly half a line apart, here they were more than thirty-six times that. Moreover the bodies I was using, the quadrilateral prism and the cylindrical tube, turn toward each other quite small surfaces, and a repulsion between them could only be acting as it were on a line almost devoid of breadth. It was not in my power to remove the first of these causes, for air does not impede the transit of electric matter as powerfully as does glass, so the electric matter leaps from one body to another in the form of a spark over greater distances if air rather than glass is the intermediary; and I was sure from elsewhere that another body could not be moved closer than one or one and a half inches from the tube without enticing a spark. I was readily able to remove the second debilitating cause by using bodies with large surfaces. I hoped too that the surfaces might be so great that the resulting increase could overcome the debility created by the greater distance, and compensate as it were.

74) I had two plates about one foot broad and one and a half feet long made from iron strips covered with tin, and I used them in the test in the manner described previously. But not even this could gain the desired effect. For the sensation which the leaping sparks produced had very little similarity to that which usually accompanies the electrical commotion. Meanwhile the following phenomena made me keep hoping.

1) If I was not touching the lower plate and then drew a spark from the upper positively electrified one with my left hand there was hardly a more violent sensation than from any ordinary spark of reasonable strength.

2) If I kept my right hand applied to the lower plate, and then drew a spark from the upper one with my left, a violent and painful sensation spread through almost the whole left hand, and at the same time a similar pain was felt in the fingers touching the lower plate.

These phenomena clearly argued for a transit of electric fluid from the upper plate to the lower, so the reason I had not obtained a true and sufficiently violent commotion should be laid at the door only of insufficient size in the plates I was using.

75) To obtain greater surfaces, I had wooden plates made of rather ample size, about eight square feet in surface, covered with metal leaf, and I suspended them close to each other in a parallel position at a distance of one and a half inches; immediately I carried out the experiment in the usual manner I obtained a strong commotion altogether like that usually produced by the *Leyden* jar. Moreover all the other phenomena obtaining in the jar could be demonstrated from my apparatus, but there is no need to review them.

76) The particular benefit I immediately obtained from this experiment was that I received a distinct idea of the impermeability of glass which *Franklin* had proposed and *Nollet* had contested. To solve the phenomenon of the impermeability of glass, *Franklin* had put forward quite improbable hypotheses, and felt that he had to take refuge in the peculiar structure of glass.[41] Since one cannot suppose a similar or somewhat equivalent structure in air, which is a most fluid body, and since, all the same, air can play the role of glass in the *Leyden* experiment, I readily perceived that the celebrated *Franklin* had showed human weakness. By examining the matter more deeply, it was soon obvious to me that impermeability was involved in the essential property making bodies electric *per se;* it is by virtue of this that they admit the movement of electric fluid only with difficulty through their pores. I could not help marvelling that the celebrated *Franklin,* who had so outstandingly expounded the nature of bodies electric *per se,* had not perceived that the immediate consequence of the essential feature of bodies electric *per se* is that property which he gave the name impermeability. Moreover this experiment also helped me to a more correct understanding of the universal economy of the experiment of commotion, because it is sufficiently evident that it is very well illustrated here. This test was also most useful in gaining a proper understanding and explanation of the effects of lightning. For if one of my plates described in the preceding paragraph is taken to be a cloud heavy with electricity or thunder, the other to be the surface of the terrestrial globe threatened by the cloud, it is clear that this experiment exhibits all the things that happen during lightning. Since this is foreign to the scope of this work, to mention it is sufficient.

77) Let us rather return once more to a discussion of certain of the phenomena of the *Leyden* jar. For there are still things which de-

[41] *Benjamin Franklin's Experiments,* pp. 229–30.

serve rather close attention since they are either quite noteworthy in themselves, or they will help in making a comparison between the phenomena of magnetism and electricity. In founding the theory of the *Leyden* jar, *Franklin* made a mistake which we have an opportunity of correcting here. The celebrated man considers that when the jar is laden with electricity, precisely as much electric fluid is expelled from the exterior surface of the jar, as is accumulated on the interior, or, if we use the method employed above for expressing these quantities, he supposes α always equals β.[42] But we have proved in § 45 above, that $\beta = \dfrac{\alpha r'}{r}$, and, because $r > r'$, this quantity is necessarily less than α, and is never equal to it unless r' becomes completely or at least sensibly $= r$; and this case cannot occur in a *Leyden* jar, whose walls are never thin enough for the difference between r and r' to disappear. Although it can occur in the experiments reviewed in § 55. 59, it is established that *Franklin* had no idea of it. This means that *Franklin's* hope of filling innumerable jars with electricity to the same degree as a single one with the same labor, in the way which the Celebrated Man described,[43] is completely undermined. For if the experiment is set up in the way indicated by *Franklin*, since the quantity expelled from one jar and entering the next is less than that which is accumulated in the former jar in the ratio r' to r, it is clear that while all the jars are filled with electricity, the electricity in each succeeding jar is less than in the nearest preceding one in the ratio r' to r.

78) To prove the equality between α and β, the Celebrated Man points to the experiment in which a man, supported on pitch or glass and holding a jar in his hand, applies his other hand to a wire lowered into the jar and discharges the jar, and yet after he has sustained the commotion, no sign of electricity either positive or negative can be detected.[44] It must be confessed that at first sight *Franklin's* quite ingenious deductions seem to follow completely from this. For if it is supposed that $\alpha = \beta + m$, and the natural quantity of electric fluid in the man supported on pitch performing the experiment $= q$, the sum of the natural quantity of fluid contained in the outer and inner surfaces of the jar and in the experimenter before electrification is sustained $= 2Q + q$. But if the jar is filled with electricity, the quantity of electric fluid in the internal surface of the jar, under the hypothesis assumed above, will $= Q + \beta + m$; in the external surface $= Q - \beta$; and

[42] Ibid., pp. 180–85.
[43] Ibid., p. 190.
[44] Ibid., p. 185.

in the man $=q$. So the total in all three $=2Q+q+m$. If this total is equally distributed through the whole apparatus as happens when the jar explodes (§ 66), the quantity of electric fluid found in the total apparatus will $=2Q+q+m$, which differs from the natural quantity $2Q+q$ by the quantity m. So it is evident that, if m were positive, the man together with the jar would come out positively electric after the explosion, and if m were negative, negatively electric. After the explosion the natural state will not be found either in the experimenter or in the jar unless m were $=o$, or unless α were $=\beta$.

79) Although the Celebrated Man conducts his reasoning properly, it can be shown that he applied this reasoning incautiously to the experiment proposed. For if we compute the forces by which the whole apparatus must act on electric fluid and other electrified bodies after the equable distribution of fluid, the first is found to be $=\dfrac{mr}{2Q+q}$, and the latter $=\dfrac{mr\delta}{(2Q+q)q'}$ (§ 10. 34).[45] But it is evident that if m is very small with respect to $2Q+q$, this force can become so small that it is plainly insensible. If we look at *Franklin's* experiment, it is clear that the quantity m was sufficiently small; for

1) In this experiment he used a jar of only rather small mass; this was all he could use otherwise the commotion would have been too violent for a man to endure easily.

2) For the same reason the jar could not be loaded with electricity to the highest degree but only moderately.

3) The jar *Franklin* used was thin-walled, so since r' is little different from r, because $\beta=\dfrac{\alpha r'}{r}$, it departs too little from equality with α. On the contrary,

4) The mass of the man, and the natural quantity of electric fluid found in the man (which is proportional to this mass), are quite great.

One must conclude from all this that m was significantly less than q, or $2Q+q$, so it is not surprising that after the explosion of the jar the electricity in the man was so small as not to betray itself by any indication.

80) If the experiment is set up so that m is sufficiently large with respect to the quantity $2Q+q$, the outcome will evidently show that the quantity α is not equal to β. To test this, I tackled the matter in the following way. I suspended two plates of rather ample surface

---

[45] Where q' is the charge on the second body. The text has $\dfrac{mr\delta}{2Q+q}$.

such as I described in § 75 about two inches apart and, supported on pitch, I sustained the commotion in the usual manner; if I had electrified the leading plate positively I always gained positive electricity and if I had electrified it negatively, I always gained negative electricity. In this way the experiment achieved what it could not achieve if it were performed according to *Franklin's* method, since

1) the surfaces, in one of which the fluid was condensed, and in the other rarefied, were quite large, and

2) these surfaces were rather distant from each other, so, because r' differed greatly from r, β or $\dfrac{\alpha r'}{r}$ was significantly less than α if the leading plate had been positively electrified, and β or $\dfrac{\alpha r}{r'}$ was significantly greater than α if the leading plate had been negatively electrified (§ 45. 46).

81) I have already noted in § 47 above that when producing the electrical commotion one must not look solely to the metal plates covering the glass plate, but that the glass plate also plays some part in producing it. I showed, and experiments confirm this, that the glass plate itself is reduced to the state required for producing the commotion, so that electric fluid is accumulated in one part and rarefied in the other. From this readers could infer the conclusion that in producing the electrical commotion, there is clearly no need of metal plates or even of other bodies performing the same function. For since the glass plate itself can be reduced to the state necessary for the experiment of commotion, it seems evident that the experiment ought to succeed even if the glass plate is used alone. Those who conclude this, however, will wonder that experience is never found to agree with this reasoning, for the *Leyden* experiment never succeeds unless the jar is covered on both surfaces by some matter non-electric *per se*. To stem the resulting doubt it will be necessary to explain distinctly the uses of the metal sheets in this experiment; this task is all the more urgent since from its results certain phenomena of magnetism can also be understood. Metal sheets, apart from what they contribute of themselves to the commotion, also play the following two roles.

1) They are necessary for imbuing the glass plate with electricity, and without their assistance or that of other bodies performing the same function, the plate itself can never be imbued with electricity. For if we suppose that the two sheets are taken away and the experiment is set up on the bare glass, and the electrifying chain ST, Fig. VIII, is joined immediately to the leading surface of the plate, and the chain LM through which electric matter ought to escape is joined

GENERAL PRINCIPLES 289

immediately to the back surface of the plate, at the points S and L respectively, electric matter is brought to the point S, and from this point it distributes itself to the neighboring points of the surface AE. Since the distribution of electric matter through glass, being an electric *per se,* is performed with difficulty, in sum only in the region around S is some sort of accumulation of electric matter obtained; it can hardly, if at all, reach the parts spread out on the surface; thus a notable condensation of fluid can be obtained only at the point S and the neighboring region, but not on the total surface AE. But if the surface AE is covered by the metal sheet CD, the electric fluid distributing itself immediately through the whole sheet is at once carried to every point of the surface, so a significant condensation of fluid also spreads itself through the whole surface. *Mutatis mutandis* the same is true for the surface BF which, for the same reasons, without the help of the metal sheet can be evacuated of fluid only around the point L.

2) When the explosion of the jar occurs, the metal sheets alone are strong enough to produce the violent explosion and commotion (as is sufficiently obvious from what was expounded above), even if we completely take away from the effect that which the plate contributes to render it more violent, and we consider the plate as altogether inert. But at the same time the metal sheets cause the interposed glass plate itself to join in increasing the force of the explosion, since they help the transit of electric matter from the leading surface of the plate into the back one. For since they touch the whole of these surfaces tightly at every single point, electric fluid can immediately cross from any point of the surface AE into the leading sheet CD, can in turn be brought along the wire RQP to the back sheet IK and be distributed immediately through it, and can then immediately cross to any point of the surface BF.

82) Again, if we are thinking of taking away these metal sheets CD and IK and leaving the glass plate bare after it has been imbued with electricity with their help, then even if the curved metal wire RPQ is applied in the fashion described above in § 66, no explosion is to be expected. For the electric matter accumulated round the point R will enter the wire and pass through it to the point P of the back surface and fill the neighboring region previously evacuated, but the rest of the electric matter accumulated in points located away from the point R can only cross into the wire quite slowly and it will not be energetic enough to distribute itself quickly through the parts of the surface BF further removed from the point P. So there cannot be a very rapid transit of electric fluid from the leading surface into the back one, but it must occur at such a slow pace that a sensible effect

will arise from it only with difficulty. It is readily obvious that this transit will be slower to the extent that the plate used in the experiment is more greatly electric *per se,* and that it will be completely zero if a body could be used which was perfectly electric *per se,* or if electric matter was completely immobile in its pores; so it is clearly obvious that neither an explosion nor a destruction of electricity can be obtained with the help of a curved wire of this kind or of any other body playing its role.

83) I have treated the phenomena of the *Leyden* jar here at greater length than I had planned but I am confident my readers will not take it amiss. There occur among what I have discussed very many things which are not outside the scope of this work, and they will be much used subsequently. Moreover I think I have expounded the theory of this experiment more distinctly than anyone else has done, and have revealed much which clears a path for the explanation of innumerable electric phenomena; my readers, I hope, will not be irked at having paid some attention to them. Let us now at last proceed to developing a magnetic theory.

84) From what I expounded at the beginning of this commentary (§ 9), three kinds of magnets can be imagined. For it follows from this that there may be magnets in which magnetic fluid abounds, others in the whole of which the quantity of this fluid is diminished below the natural, and finally others which in one part possess a positive magnetism, if it is proper to speak thus, and in the other a negative magnetism. Conformably to our theory, in the case of bodies imbued with electricity it is more than sufficiently known that there are these three kinds. It will not be unknown to my readers that it cannot be shown by experience that the same situation occurs in the magnet, and this may provide a handle for doubting my theory as a whole. For it is established that all magnets produced either by nature or by art generally have two poles (sometimes more), which do not act according to the same laws, but in some way by opposite laws, and that no magnet has been found with one single pole only, or which exercised its actions through its whole substance and from all parts of it according to the same laws. It is easy to conclude from this that all magnets produced to date by either nature or art have always been imbued not with a single kind of magnetism but with both. So I judge that I am obliged to remove this scruple from my readers and to show why, although they are possible according to theory and in themselves, there are no magnets which possess only one single kind of magnetism. I may be allowed to digress from my discussion for a while however and deal first with magnets endowed with both kinds of magnetism, for thus the way will be prepared for resolving the difficulties.

GENERAL PRINCIPLES

85) Let us imagine a body A, Fig. IV, capable of magnetism, made either purely of iron or with some heterogeneous parts mixed in, as in minerals, and let us suppose that however this is achieved, magnetic matter is distributed unequally through the body so that it abounds in the part AC and is deficient in the part AB. It is then evident that everything I demonstrated in general terms at the beginning of this chapter (§ 15–27, and § 37–44) can be brought in here, and that if we keep the values assigned to the letters used above, all the formulae developed and the conclusions deduced from them can be transferred here. So from § 15 we have formulae expressing the action of a magnet on particles of fluid close to itself and from § 38. 39 its action on other ferreous bodies whether imbued with magnetism or not. Let us not waste time in applying these to the magnet since these discussions will find their right place below. We have shown above, § 16, that a particle of fluid F, located on the border between the part AC filled with fluid and the evacuated part AB, is urged to leave the part AC and cross to AB, by a force $=\frac{\alpha r + \beta r}{Q}$. So the magnetic fluid in any magnet will be urged by this force $\frac{\alpha r + \beta r}{Q}$ to leave the part AC, cross to the part AB, and distribute itself uniformly through the whole of the body A.

86) Thus in any magnet left to itself, there is a force tending continuously to the destruction of the magnetism, so the force of any magnet must be continuously weakened unless there is some external conserving cause; and indeed as long as the force $\frac{\alpha r + \beta r}{Q}$ is sufficiently great to overcome the difficulty with which magnetic fluid moves in the pores of the body A, the magnet will be weakened speedily enough. If however the force $\frac{\alpha r + \beta r}{Q}$ is less, the flow will stop completely, or at least take place at such a slow pace that a sensible weakening could only occur after a rather long time. From this it is first of all obvious why the magnetic force in any magnet, whether artificial or natural (and this is verified by constant experience) decreases and finally is almost completely destroyed, unless it is conserved by certain artifices which I shall have the opportunity of discussing later. From this it follows also that any given ferreous body is capable of a given degree of magnetism only; if it becomes magnetic to a greater degree this excess is quickly destroyed again. As is evident from the previous discussion, the quantity of this degree of saturation, so to speak, depends on the difficulty with which magnetic fluid moves in the pores of the body to be magnetized. If there is no resistance, the degree of

saturation will be nil, so such a body, even if strongly magnetized, loses all its magnetism once it is left freely to itself. If this resistance is infinitely great the degree of saturation will be infinite, and given such a body, whatever degree of magnetism is once received, it will be lost only after an infinite time, that is it will always be preserved unaltered without diminution. So if a resistance of a degree intermediate between these two extremes is supposed, those bodies in which magnetic matter moves with some difficulty will always be able to assume a greater degree of magnetism, or rather conserve it longer, and those where it moves more easily, a less degree.

87) Since magnetic matter moves with more difficulty in ferreous bodies the more hardened they are, the degree of saturation in hardened iron will always be greater than in that which is softer. Experience is in outstanding agreement here. For it is very well known to those who labor at making artificial magnets that a considerable degree of magnetism can easily be communicated to soft iron, but that this labor is to some extent frustrated since the magnetism impressed in a body of this kind usually disappears at once, completely if the iron is very soft, or nearly so if it is hardened only a little. As a result the hardest steel is best for making artificial magnets. Anyone who wants can also be convinced by the following experiment. Let two parallelepipeds be made exactly alike, DC and dc, Fig. XIII, one of which, DC, is of quite soft iron, and the other, dc, of moderately hardened steel; be careful too that they acquire no magnetism before the experiment is begun. Then take a rather strong magnet AB and place the parallelepiped DC made of rather soft iron at one of its poles A; it will be observed that it receives a significant magnetism. For by its end D it will be able to lift an appreciable weight of magnitude E. Next remove this parallelepiped from the magnet; not only will it no longer be able to lift the weight E, but the magnetism will be diminished at once to such a degree that it will hardly suffice to pick up one piece of iron filing. If the other parallelepiped dc made of hardened steel is substituted for DC, while it is applied to the magnet it cannot lift as great a weight as DC, but after it has been removed it will not lose its magnetism completely as the previous parallelepiped did, but will still lift an appreciable ferreous mass by its end d.

88) Apart from the cause of the destruction of magnetism which we have considered here, it is obvious from § 15 that there is still another. A particle of magnetic fluid E, clinging round C, is enticed by a force $= \dfrac{\alpha r - \beta r'}{Q}$ to leave, but a particle D, found round B, is enticed

GENERAL PRINCIPLES 293

to enter by a force $=\dfrac{\beta r - \alpha r'}{Q}$. In the *Leyden* jar, and in other similar cases pertaining to electricity, there is an external cause preventing these forces from being able to produce their effect, namely the surrounding air, but we cannot imagine such a cause with a magnet. For since all the bodies surrounding the magnet neither attract nor repel magnetic fluid, they do not help the forces tending to the destruction of magnetism, and equally they cannot impede them from gaining their effect. So magnetic fluid would be able to flow freely out of the part AC and flow into the part AB, until the magnetism was totally destroyed, without the bodies surrounding the magnet preventing it, were there not some internal cause standing in the way. This is the difficulty, which I have previously introduced, with which magnetic fluid moves through the pores of ferreous bodies. It is evident, of course, that this opposes itself not only to the efflux but also to the influx of magnetic matter, which naturally cannot take place without an actual movement through the pores. And so the same cause which puts an obstacle to the force tending to the destruction of magnetism previously considered (§ 85) performs the same function in respect of the forces we introduced at the beginning of this paragraph. From this it is clear that what I said in § 86. 87 about the destruction of magnetism and about the various degrees of saturation in various bodies applies also in respect of the forces we are considering here.

89) So it is possible for the state described in § 85 to persist in a body of a ferreous kind, and be destroyed only after quite a long time. These things are valid if the magnet is left to itself, but the question arises whether, just as electricity is destroyed in the *Leyden* jar (to which the magnet has a significant resemblance) (§ 66), magnetism can be destroyed in the magnet in the same manner. If the discussion of the relevant § is applied here, it can be seen that the same ought to happen in a magnet as in *Musschenbroek's* jar. For if a curved iron wire is made to touch the part AC of the magnet filled with fluid, Fig. IV, and the evacuated part AB at the same time, for the same reasons as with the *Leyden* jar, a transit of magnetic matter ought to take place from the filled part to the emptied one, and not stop until the magnetism is completely destroyed. But anyone who was willing to attend to my commentary on the *Leyden* experiment § 81. 82, will easily understand that, for the same reason that a sudden destruction of the electricity does not take place when the metal sheets covering the walls of a *Leyden* jar are taken away and the bare glass is revealed, in a magnet too a destruction of magnetism cannot be brought about in this fashion. For since magnetic fluid moves with difficulty in the

pores of ferreous bodies, especially those that are hardened, with much more difficulty in fact than does electric fluid in the pores of glass, neither will the particles which cling round the surface C of the magnet, and much less those which occupy the interior of the part AC, be able to extricate themselves from the pores and cross into the iron wire, nor, similarly, will magnetic matter be able to flow across from the other end of the wire touching the evacuated part AB into the surface B. Therefore it is clear that a destruction of magnetism, such as usually occurs through an explosion in the *Leyden* jar, cannot happen with a magnet, just as it cannot happen in the said jar except with the help of the metal sheets with which it is covered. Because we are unable to produce in a magnet an explosion and a destruction of magnetism similar to that in the *Leyden* jar, no reason for this needs to be imagined other than that there are absolutely no bodies non-magnetic *per se*, or analogous to those non-electric *per se*; that is, through whose pores magnetic matter could move freely.

90) I affirmed above that magnets endowed with only a single kind of magnetism could not be found, and that all magnets possess both magnetisms at the same time. Yet I was not of the opinion that the case I have so far considered, where one part of the body imbued with magnetism possesses a positive magnetism, the other a negative, is the only one which can occur. Rather, theory and experience show that there are several other cases. Thus we quite often observe that in iron rods imbued with magnetic force, there are not two but three, four, five, indeed many poles. If e.g. AG, Fig. XIV, is an iron rod, or a somewhat longer steel rod, it is possible for the part AB to be a north pole, BC a south, CD a north again, DE a south, and so forth. It is customary to call opposite poles succeeding each other alternately in this way consequent points; unless I am mistaken the celebrated *Musschenbroek* first used this term.[46]

91) We easily gather from our theory that a state of an iron rod of the kind I have just described is altogether possible. For if we assume the rod to be divided into equal parts and we make the quantity of fluid abounding beyond the natural Q in AB=α, in CD=α', in EF=α'' etc.; the quantity of fluid less than the natural in BC=β, in DE=β', in FG=β'' etc.; and the forces by which the natural quantity of magnetic fluid in the part AB repels particles of fluid are, on $b=r$, on $c=r'$, on $d=r''$, on $e=r'''$, on $f=r^{IV}$, on $g=r^{V}$, etc.; and if we calculate according to § 27,[47] the force enticing the particle a to leave

---

[46] Aepinus *is* mistaken here. The term was introduced not by Musschenbroek but by Brook Taylor (*Phil. Trans.*, XXXI, 1720-21, 207).

[47] The text has § 34.

GENERAL PRINCIPLES

$$= \frac{\alpha r - \beta r' + \alpha' r'' - \beta' r''' + \alpha'' r^{IV} - \beta'' r^{V} - \text{etc.}}{Q},$$

the force pushing the particle b toward the part BC

$$= \frac{\alpha r + \beta r - \alpha' r' + \beta' r'' - \alpha'' r''' + \beta'' r^{IV} - \text{etc.}}{Q},$$

the force pushing the particle c toward CB

$$= \frac{-\alpha r' + \beta r + \alpha' r - \beta' r' + \alpha'' r'' - \beta'' r''' - \text{etc.}}{Q},$$

the force pushing the particle d toward DE

$$= \frac{\alpha r'' - \beta r' + \alpha' r + \beta' r - \alpha'' r' + \beta'' r'' - \text{etc.}}{Q},$$

the force pushing the particle e toward ED

$$= \frac{-\alpha r''' + \beta r'' - \alpha' r' + \beta' r + \alpha'' r - \beta'' r' - \text{etc.}}{Q},$$

the force pushing the particle f toward FQ

$$= \frac{\alpha r^{IV} - \beta r''' + \alpha' r'' - \beta' r' + \alpha'' r + \beta'' r - \text{etc.}}{Q},$$

and so forth.

These formulae are easily obtained, for one need only compute the actions of each part AB, BC, CD etc. taken in order on any of the particles a, b, c, d, e, etc. and take the sum of all those forces, where it must be noted that those forces which act in the opposite direction must be subtracted from the total, or must be taken to be negative.

92) As is easily perceived, all these forces tend to the destruction of the state of the rod and of the magnetism, so everything I said above § 86. 87 applies here too. Provided these forces are so tempered that they cannot easily overcome the resistance which magnetic fluid meets to its moving through the pores of ferreous bodies, once that state has been induced in a rod in some way it will be able to last for a considerable length of time; so there is clearly the possibility of consequent points in magnets. How a rod may come to be endowed with the state where consequent points are to be found in it will be discussed later when we are dealing with the communication of magnetism.[48]

93) When we dealt previously with the phenomena of the *Leyden* jar, we assumed a completely indeterminate ratio between α and β

---

[48] See below, pp. 353–55.

and we assumed that this could be fixed in various ways according to the circumstances. But the matter is different in regard to the magnet. For on account of the very great difficulty with which the fluid moves in the pores of ferreous bodies, it is scarcely possible for any magnetic matter to enter a body of this kind from outside by any means, or for it to be evacuated from it. So all the ways by which either nature or art can impress magnetism on ferreous bodies, are almost reduced to the point where the magnetic fluid contained in the body to be magnetized, is, as it were, transferred and forced to cross from one part of the body to the other; this will appear more clearly later when we discuss the communication of magnetism. Thus e.g. if an iron rod A, Fig. IV, is magnetized, nothing else happens except that a certain part of the fluid is propelled from the part AB into the part AC. So the amount of magnetic fluid to be found in a magnet is always constant and remains the same as it was before magnetization was instituted. So since, according to the notation we have assumed, the amount of magnetic fluid contained in the magnet before magnetization $= 2Q$, and after the magnetic force has been impressed in it, it $= 2Q + \alpha - \beta$, $2Q + \alpha - \beta$ will $= 2Q$ and α will $= \beta$. I will later indicate some cases which must perhaps be excepted from this rule.

94) But if we introduce this ratio found between α and β into the formulae discovered in § 15, the force by which the part AC repels magnetic fluid $= \dfrac{\alpha(r-r')}{Q}$, and the force by which the other part AB attracts the fluid will likewise $= \dfrac{\alpha(r-r')}{Q}$. Since this force is equal to the former, it is clear that the three possibilities I expounded in § 21. 22. 23 cannot happen in a magnet although they do occur in the *Leyden* jar and in other similar electrical experiments. So a magnet can never be reduced to the state where one of its two parts plainly does not act on magnetic fluid, or where both parts either attract or repel fluid; rather, any magnet must be found constantly in a state where it repels magnetic fluid in one part and attracts it in the other.

95) I previously promised my readers a more exact exposition of the reasons why there are no magnets endowed with a single kind of magnetism; I have decided to fulfil this promise. It is first of all clear from § 18 that, if a body possesses both magnetisms at the same time, the force tending to the destruction of magnetism is significantly less than if this body is supposed to be imbued with a single kind of magnetism. So if we imagine either that magnetic fluid abounds in the whole of a ferreous body, or that it is deficient in the whole of it, it is clear that magnetism of this kind cannot last for as long a time.

GENERAL PRINCIPLES

I dare not assert however that it ought to disappear in an instant though this would otherwise solve the whole difficulty. Although since bodies surrounding a magnet do not act on magnetic fluid, it is not correct to imagine in a magnet an external cause that can inhibit the free efflux of magnetic fluid from a magnetized body, or the free influx into it, it has already been established from what was introduced earlier that there is an internal cause which can impede this free efflux or influx. It is this impediment, more often called a resistance, which stands in the way of the free movement of magnetic matter in the pores of iron; unless the force urging the fluid to either efflux or influx can overcome this, destruction of magnetism cannot occur. Our theory therefore requires us to lay down that, provided this state can be once induced in a ferreous body, so that it becomes either wholly positively or wholly negatively magnetic, it can last for some time. And it is clear that it ought to be the more durable, the greater is the difficulty with which fluid moves in the pores of the body used in the experiment; since this difficulty increases with hardness, this state, all things being equal, will be conserved longer in quite hard steel than in significantly soft iron.

96) So the theory does not state that magnets with only one kind of magnetism are impossible, provided we can contrive a method of inducing this state in a ferreous body. But for reasons in part already expounded in § 95, in part still requiring further development later when I shall be dealing with the communication of magnetism, it seems quite difficult to reach the point where magnetic fluid either abounds or is deficient in a body as a whole. For as will become obvious later, neither art nor nature knows ways of impressing magnetism on a ferreous body other than those that are plainly useless for our proposition, because these ways consist in the mere propulsion of magnetic fluid from one part of the body to the other. Except for the following, I have been unable to find a way to obtain a body which one could with any likelihood hope might be imbued with a single kind of magnetism. Let a body A, Fig. XV, be imbued with magnetism, and let fluid abound in the part AC but be deficient in AB. Suppose a part nC be cut from AC and a part mB be cut from the other end AB; it is clear that when the former part nC is taken away from the magnet it must be completely imbued with a positive magnetism, and the other part mB must be imbued with a negative magnetism.

97) My readers will be aware of the outcome of this experiment, for it is discussed at length by several who have described experiments on

the magnet.[49] But before I expound what I experienced in this matter myself, let us inquire more accurately, according to our theory, into the necessary or possible consequences of this test. So that we can apply the theory to this experiment, let us note first that in the magnet A, magnetic matter is never uniformly distributed in such a way that e.g. at any point of the part AC where fluid abounds or at any point of the other part AB where fluid is deficient there is an equal density. I have so far supposed that this is the case, but I had no intention of asserting that it actually occurs in nature. I only assumed it for ease of calculation, and I could assume it as long as I only deduced conclusions that are little altered whether or not the fluid is uniformly distributed through the parts AC and AB; these are the kinds of conclusions I have drawn from that supposition to date. As I shall show later, the assumption should rather be that the densities of magnetic fluid are shown by a curve DAE whose ordinates np, mq are proportional to the difference between the density which the magnetic matter actually has at any point of a magnet and the natural density, which is the density at any point when the body is constituted in the natural state. I take the ordinates drawn above the axis as positive and those below it as negative.

98) Consider that a part Cn, Fig. XV and XVI, is cut from the magnet, first of all from that part of it AC which is positively magnetized, and fluid will abound in this whole part, but will not be distributed equally through it; but the part Dp of the curve of densities, corresponding to the part nC, will show the density at any point. It is further evident that as long as this part nC is joined to the magnet from which it was cut, the force tending to dissipate the fluid contained in it is less than if it were completely removed from the magnet. Consequently, the moment the part nC is removed from the magnet some part of the fluid accumulated in it can be dissipated and can flow out from any point of the part nC. If this happens, all the ordinates of the curve of densities, or rather of the part pD pertaining to Cn, will be reduced by a definite quantity. The size of this reduction cannot be determined from our theory alone, and can only be known through experience. From our theory alone, though, it is not difficult to perceive that this decrease can be so large that certain of the ordinates expressing the density of magnetic fluid finish up negative. For suppose that the ordinate np corresponding to the point n decreased till it became zero, Fig. XVII, and it is evident that,

---

[49] Cf. William Gilbert, *On the Magnet* (1600), trans. Silvanus P. Thompson, New York, 1958, p. 101. Few of Aepinus's readers were persuaded by his discussion of this point.

because of the magnetic matter still abounding round C and exercising a repulsive force, there is a force tending to expel the fluid from the point n, even though only the natural quantity of fluid is found at that point. But if this force is still capable of overcoming the difficulty with which fluid moves in the pores of iron, some fluid will still trickle from the point n and neighboring points; thus the ordinate of the curve of densities corresponding to the point n becomes negative, Fig. XVIII, and the body nC becomes negatively magnetic in the region of n.

99) All this is also valid if a part is cut from the negatively magnetic part AB, Fig. XV and XVI. Let the part cut off be Bm, and let the part of the curve of densities Eq indicate the density of fluid at any point; so the whole part Bm is negatively magnetic, though to a greater degree round the point B than in the other part round m. It is further evident that, as long as the part Bm is joined to the magnet, the force which urges the magnetic fluid to enter it is less than if this part is completely removed from the magnet. So at the very moment it is separated from the magnet a certain part of the magnetic fluid will be able to enter any point; if this happens, all the ordinates of the curve of densities will be diminished.[50] Although the amount of this reduction cannot be defined *a priori*, it is readily obvious that, by an argument similar to the preceding case, this decrease of the negative ordinates can become so great that some of them move in the opposite direction and become positive. For if the ordinates are first supposed to have decreased till the one corresponding to the end point m, which is the smallest of all, becomes zero, Fig. XVII, nevertheless, because the other end B is still evacuated, there is a force enticing further magnetic fluid to flow in. So if this force is still capable of overcoming the difficulty with which fluid moves in the pores of the body Bm, some magnetic fluid will still enter round the end m; therefore the ordinates round m become positive, Fig. XVIII, and the body mB becomes positively magnetic in the part m.

100) Before I proceed further I must explain the significance of a certain term I use in what follows. If we imagine a magnet, one part of which, AC, Fig. IV, is positively magnetic and the other AB negatively magnetic, then a point can always be assigned that separates the positive magnetic part from the negative one, and at which for that reason the density of fluid is equal to the natural, as A here. In what follows, I shall call this point the *magnetic center*. It is true that this limit separating the positive magnetic part from the negative one is not a

---

[50] Here Aepinus again assumes, though much less plausibly on this occasion, the presence of some external source of fluid.

point but rather a surface, so the name *center* does not seem really suitable for it. Meanwhile readers will easily tolerate this word, since I have been unable to find one more fitting straight away.

101) If we now suppose that the case I mentioned in § 98 occurs with the part nC cut from the positive part AC of the magnet, Fig. XVI,[51] so that fluid is dissipated to the extent that the ordinates round n become negative, it is evident that the part nC gains its own magnetic center. For after so much fluid is dissipated that the ordinate at the point n is zero, Fig. XVII, the magnetic center falls right at the point n. But if still more fluid trickles out, the magnetic center is transferred from the point n to another point f, Fig. XVIII, closer to the center of the body nC, and if the efflux of fluid continues still further, the magnetic center is also transferred still further, and recedes from the end n toward the center of the body. All this is easily transferred to the case treated in § 99. If we suppose a piece mB cut from the negative part AB of the magnet, Fig. XV, all this is equally valid in its case, and, before the afflux completely stops, it gains a magnetic center that recedes the more from the end m toward the middle the longer the afflux of magnetic matter continues. Since therefore the efflux and afflux of fluid stops faster the harder the magnetic matter finds it to move through the pores of the bodies nC and mB, it is clear that in each the magnetic center will be closer to the end m or n, and further from the middle of the body, if steel or reasonably hard iron is used in the experiment, than it will if the iron is soft.

102) Now that I have explained what we consider, according to the theory, necessary or possible in the test, I shall describe here certain appropriate experiments that will make it clear that, far from there being any contradiction between theory and experiment, there is excellent agreement between the theory and the phenomena. I had a rod made from pure steel, square in cross-section, 2 lines broad and 10 inches 3 lines long, and I made sure that it was hardened to the maximum degree. First of all, I broke this rod into two parts; the longer, which I shall call A, was 5 in. 9 lin., the shorter, called B, 4 in. 4 lin. long.[52] I joined these two pieces on a wooden plate at the point of fracture, and I pressed them strongly together so that they touched one another exactly and could be regarded as a single body.

[51] The text has Fig. XIV.

[52] Aepinus normally uses the standard eighteenth-century units of length, the inch and the line, where 1 line $=\frac{1}{12}$ inch. In this discussion, however, for no apparent reason, he uses the "decimal line" (*linea decimalis*), equal to $\frac{1}{10}$ inch.

GENERAL PRINCIPLES

I then magnetized this whole rod composed of two parts by *Canton's* method, which I shall describe later.[53] After this was done, I found that so long as the rod was not removed from its position, it had only two poles, one of which *md*, Fig. XIX, was north, the other *ma* south and, as was proper, a single magnetic center situated in the part A at a point *m* such that *am* = 5 in. 2 lin.; so the magnetic center was placed practically in the middle of the whole rod.[54]

103) I then separated the pieces A and B from one another and I found that the piece B, which previously possessed only a single kind of magnetism, had now gained two poles. It still showed north magnetism in the part *d*, but in the part *c*, south. Thus the piece B had its own magnetic center at the point *n*, Fig. XIX; its position was not in the middle, but was found closer to the end *c*, so that *cn* = 1 in. 8 decimal lin. So it was still 4 lin. distant from the middle. On examining the piece A, to which, as readers will easily perceive, all the reasoning of § 98. 99. 101 can be applied, I observed that its magnetic center, which was previously 7 lin. distant from the end *b*, had changed its position and had moved forward toward the middle of the rod, with now *bm* = 1 in. 4 lin., so it was still 1 in. 5½ lin. distant from the middle.

104) These phenomena are not at all contrary to the theory, but rather they quite conform to the consequences previously deduced from it, as is readily obvious to anyone. There was however in both pieces of the rod, as I have just indicated, a magnetic center not placed exactly in the middle, but still appreciably distant from it; this can be attributed to the difficulty with which magnetic fluid either enters or departs from the pores of hard steel, as I showed in § 101. I began to hope therefore that after some time both centers would approach still closer to the middle. For the afflux and efflux of magnetic fluid does not stop completely, as I showed previously, but is only significantly slowed down. Because of this I put the piece A on one side to test how the phenomena agreed with this theory, and after about half an hour I subjected it to examination once more. I found that the result was as I had hoped. For the distance of the magnetic center from the point *b*, previously 1 in. 4 lin., was now 1 in. 5½ lin.

105) Since I had discovered elsewhere what I shall give my attention to proving below, that when iron is struck its pores open up, as it were, so that during the concussion magnetic fluid moves more freely

---

[53] See below, pp. 368, 373. Cf. Canton, *Phil. Trans.*, XLVII, 1751–52, 31–38.

[54] We learn from elsewhere that Aepinus used iron filings patterns to locate the magnetic centers of his magnets (Aepinus, *Acta Acad. elect. Mogunt.*, II, 1761, 258–59).

through them, I conceived the hope that the magnetic center would move closer to the middle if the piece of rod were struck violently. So I hurled the piece B about 50 times with considerable violence against the stone pavement, and after it had sustained these 50 strong concussions, I again found that the result agreed completely with the theory. For the distance of the magnetic center from the end c, which had previously been 1 in. 8 lin., was now 2 in. 0 lin. The same thing happened with the piece A, where, after 24 throws against the pavement, the magnetic center was 1 lin. closer to the middle, so that bm now $= 1$ in. $6\frac{1}{2}$ lin.

106) At another time it occurred to me, and the experiments discussed below will prove it with evidence, that, when steel is fired or at least heated considerably, the movement of magnetic matter becomes appreciably freer. A consequence of this seemed to be that if I had exposed to the fire the pieces whose phenomena I have explained so far, the magnetic centers ought to have moved still closer to their centers. But since I had not thought of this until I had completed the experiment described, the whole test had to be undertaken again. So I took a well hardened steel rod, length 5 in. 9 lin., of the same thickness as I described before. I broke it into two pieces A and B, Fig. XIX; A was 4 in. 3 lin. long, B 1 in. 6 lin. When these were put together as described above and magnetized, a single magnetic center was found, positioned at m; its distance from the end a was 2 in. 8 lin., so ad was not much distant from the middle of the total aggregate. When the pieces were separated, the magnetic center in A moved closer to its center, so that now $mb = 1$ in. 7 lin.; it had been only 1 in. 5 lin. before. I then exposed this piece to the fire and heated it until it turned a blue color. After it was completely cooled the magnetic center was 2 in. $\frac{1}{2}$ lin. distant from the end b, so it was located almost in the middle. Similarly, in the other rod, length 4 in. 2 lin., which likewise had its magnetic center 1 in. 7 lin. distant from its center, I discovered at another time that, after firing it, the center had moved forward almost exactly to the middle, for its distance from the end b was $= 2$ in. 1 lin.

107) The tests I undertook in this matter which I have recounted here were not single tests; I repeated them very many times. Since they had approximately similar results, it is not worth describing them at greater length. We can therefore be rightly convinced that in a piece cut from a magnet nothing happens which is contrary to the theory and which does not agree well enough with it. The agreement of the phenomena with our reasoning reveals itself particularly in that the magnetic center is found to move ever closer to the middle of

GENERAL PRINCIPLES

the rod, on the one hand of itself, if a longer delay is permitted, on the other if the magnetic fluid wins a freer movement in the pores. Setting up the same experiments on rods made of softer steel, I found that immediately after the separation of the pieces the magnetic center moved almost exactly to the middle of each piece, as is required by the reasoning expounded in § 101.

CHAPTER II

Concerning Electrical and Magnetic Attraction and Repulsion

107A) [55] The attraction of two bodies close to one another, both abounding beyond the natural quantity (by quantities α and δ) with a fluid endowed with the properties we have ascribed to the magnetic and electric fluids, has been found above, § 34, to be generally $=\dfrac{-\alpha\delta r}{Qq}$. Since this formula embraces all the cases which can occur (§ 35), and since it disappears whether δ, or α, or both of these quantities, is made $=o$, what we have demonstrated above (§ 32) is obvious here: these two bodies not only plainly do not act on one another if both are constituted in the natural state, but this also remains true if just one of them is found in the natural state, even though fluid either abounds or is deficient in the other. So it is a most certain consequence of our theory that any body constituted in the natural state is neither attracted nor repelled by any bodies whether positively or negatively electrified. I am not now going to consider magnetism for it will be convenient to deal especially with this later on.

108) Thus we should not expect any electrical action on other bodies constituted in the natural state, provided these bodies preserve their natural state while they are approaching an electrified body. It follows from the principles evolved above that the contrary of this always happens and that any body close to an electrified body ought itself to become electric, even though it was constituted in the natural state before it came close. Remember that this is clear from the things we demonstrated in the previous chapter about the action of electrified bodies on particles of electric fluid found outside themselves. For we proved in § 10 that electric fluid found outside a body A, Fig. XX, is attracted by a force $=\dfrac{-\alpha r}{Q}$, or, what comes to the same thing, is repelled by a force $=\dfrac{+\alpha r}{Q}$, if fluid abounds in this body by a quantity α. But if a body B constituted in the natural state approaches the body A, the electric matter found in this body will be repelled by a force $=\dfrac{\alpha r}{Q}$. It will thus be forced to leave the part CDGH in the vicinity of A and move toward the other end GHEF.

[55] Two successive paragraphs are numbered 107 in the text.

ATTRACTION AND REPULSION

109) The fluid will obey this force as far as is permitted by those impediments which may be present. Such an impediment can arise from the difficulty with which electric fluid moves through the pores of the body B. If of course we imagine that the body B is perfectly electric *per se,* since electric matter then sticks so fast in its pores that it cannot be moved by any force, the force propelling it will have no effect, and the body B will preserve its natural state. But if the body B is only imperfectly electric *per se,* that is, if it provides not an infinite but only some resistance to the movement of fluid, the force $\frac{\alpha r}{Q}$ will not be completely without effect, even though it cannot have its full effect; and fluid will obey the propulsion to a lesser degree the more the body B is perfectly electric *per se,* and to a greater degree the more the body B is imperfectly electric *per se.* Finally, if the body B is completely non-electric *per se,* then because the resistance is nil, the force $\frac{\alpha r}{Q}$ must gain its full effect.

110) The first of these cases does not, so far as we know, occur in nature, for no body has so far been found which can be said to be perfectly electric *per se.* But if there was a body of this kind, or if at some time one was found, there is no doubt that its natural state would always be preserved in approaching electrified bodies, and as a result it would never be attracted by bodies, whichever way they were electrified. So the second and third cases concern us particularly. Let us first consider the last of them a little more carefully. Here the electric matter, obeying the repulsion exercised on it, will cross from the end CD toward the more distant parts of the body B, but at this point two possible cases can arise. For either the body B is surrounded on all sides by bodies electric *per se,* as would be the case, for example, if it were suspended from a silk thread, and surrounded by dry air; or electric fluid can flow freely from it, as happens if it is touched by a body non-electric *per se.* In the first case, when the electric matter is propelled, it will not be able to flow out of the body B, so it will of necessity accumulate around the end EF. And the body B will then be reduced to the state considered in § 15; part of it, CDGH, facing the body A, will become empty of electric fluid, while the part EFGH, furthest from A, will be filled with it.

111) So that we can judge what mutual action there is between the bodies A and B, suppose that the natural quantity of electric fluid in each of the parts CDGH and EFGH $=Q$, and in the body $A=q$; further, let fluid abound in the body A by a quantity δ, and let the quantity of fluid propelled from CDGH into EFGH be put $=\beta$. If, in

the general formula already evolved in § 43, a is made $=-\beta$, $b=+\beta$, $c=0$, $e=\delta$, $f=0$, $g=0$, it will be adapted to the case we are considering here, and the mutual attraction of the bodies A and B will $=\frac{-(-\beta r+\beta r')\delta}{Qq}=\frac{\delta\beta(r-r')}{Qq}$. But because $r>r'$, this force is positive, and hence, under the circumstances assumed so far, the body A first electrifies and then attracts a body B close to it.

112) It is self-evident that the closer the body B moves to the body A, the greater becomes the repelling force of the latter. Since a greater force ought to produce a greater effect and a smaller force a smaller effect, it is clear that the evacuation of the part CDGH and the filling of the part EFGH ought to increase to a greater extent, the closer the bodies move to one another. So it is evident that β becomes greater when the distance decreases and smaller when it increases; thus also the attracting force $\frac{\delta\beta(r-r')}{Qq}$ is less at greater distances and greater at smaller distances. We can get this far with the help of our theory alone, but what precisely the law is according to which the attracting force changes when the distance is changed is not evident and so far I dare not determine it. I hope nevertheless that, with the help of the principles of electrical theory dealt with here, combined with suitable experiments, an entry into this mystery of nature can be gained, and later I shall devote all my energy to discovering it.[56]

113) In order to cast more light on what was said in the previous §, suppose that the body B has become electric by coming near the body A and that, as we have supposed, a quantity of fluid $=\beta$ has been propelled from CDGH into EFGH. Consider further that a certain particle of fluid, I, has been positioned near GH and let us determine the force by which that particle is agitated. It is obvious at once that the force enticing the particle I toward the interior of the body B is a threefold one. If we retain the denominations employed previously, let the force by which the particle I is agitated by the natural quantity of electric fluid contained in the body A be put $=\rho'$, and the force by which the natural quantity of fluid in each of the parts CDGH and EFGH acts on it $=\rho$; the particle I will be repelled from the body A toward EF by a force $=+\frac{\delta\rho'}{q}$ (§ 10), and from the part EFGH toward CD by a force $=\frac{\beta\rho}{Q}$, but it will be attracted by the part CDGH toward

---

[56] No evidence survives of any subsequent efforts by Aepinus to determine the form of the law.

CD by a force $=\frac{\beta\rho}{Q}$. Thus the total force by which it is propelled toward EF $=\frac{\delta\rho'}{q}-\frac{2\beta\rho}{Q}$. So electric fluid ought to flow across from the part CFGH into the part EFGH. But since β is always increased by this flow, the force producing it becomes continuously smaller. The transit of fluid will not however be able to stop completely until β has increased so much that $\frac{\delta\rho'}{q}-\frac{2\beta\rho}{Q}$ has become $=o$. So the greater is $\frac{\delta\rho'}{q}$, the greater will be β, the quantity of fluid propelled from CDGH into EFGH; thus since ρ' increases as the distance between the bodies A and B diminishes, the closer the bodies move to one another, the greater is β, as I supposed in the previous §.

114) To turn to the other case, if the body B is touched by some other body C, Fig. XXI, not electric *per se,* and not supported on other bodies electric *per se,* so that fluid can freely flow out, the accumulation of electric fluid in the part EFGH which happened in the previous case does not take place. For if we compute the force by which a particle K located near the surface FE is urged toward the body C, if the fluid expelled from the part CDGH is added to the part EFGH, and the particle K is taken to be repelled by the natural quantities of fluid contained in the body A and in the parts CDGH and EFGH by Forces ρ'', ρ' and ρ respectively, it will be found that the particle is repelled by the body A toward C by a force $=\frac{\delta\rho''}{q}$, and from the part EFGH by a force $=\frac{\beta\rho}{Q}$, while by the part CDGH it is drawn in the opposite direction by a force $=\frac{\beta\rho'}{Q}$. Therefore the total force propelling the particle K toward C will be $=\frac{\delta\rho''}{q}+\frac{\beta(\rho-\rho')}{Q}$. Since ρ is greater than ρ', it is clear that this force is always positive and cannot become $=o$. So there cannot be an accumulation of the kind that occurred in the previous case, and fluid propelled from CDGH into EFGH ought to cross, at least in part, into the body C.

115) To understand correctly what happens in this case, and to know the state to which the body B will be reduced, let us assume that the quantity of electric fluid abounding in EFGH, which we have already proved cannot be equal to β, is $=\gamma$, and if in a similar manner to the previous § we compute the force by which the particle K is urged

toward the body C, it will be [57] found $=\frac{\delta\rho''}{q}+\frac{\gamma\rho-\beta\rho'}{Q}$. It is evident here that the body B will reach the state where this force is zero, for until this has happened, electric matter, being still repelled, ought to cross continuously into the body C. For the formula $\frac{\delta\rho''}{q}+\frac{\gamma\rho-\beta\rho'}{Q}$ to become zero, it is necessary for the quantity $\frac{\gamma\rho-\beta\rho'}{Q}$ to be negative, for unless it is, it will not be able to cancel out the positive term $\frac{\delta\rho''}{q}$. It is clear that it is possible for the entire formula $\frac{\delta\rho''}{q}+\frac{\gamma\rho-\beta\rho'}{Q}$ to disappear, provided $\frac{\gamma\rho-\beta\rho'}{Q}$ is negative.

116) There are various cases in which the quantity $\frac{\gamma\rho-\beta\rho'}{Q}$ can become negative. This happens if
  1) γ itself is negative, in which case the part EFGH too will be negatively electric.
  2) If $\gamma=o$, in which case the part EFGH is constituted in the natural state.
  3) If γ is positive, and hence also the part EFGH is positively electric, but $\gamma\rho<\beta\rho'$. To obtain this it is necessary that $\beta>\frac{\gamma\rho}{\rho'}$.

But since ρ is always greater than ρ', $\frac{\rho}{\rho'}$ is greater than 1, and hence $\frac{\gamma\rho}{\rho'}$ is greater than γ, so that γ must be significantly less than β.

117) If we evolve formulae indicating the force with which the bodies A and B attract each other in each of these cases, by substituting the appropriate values in the general formula discovered in § 43, we obtain
In the first case, where γ is negative,
$$\frac{+(\beta r+\gamma r')\delta}{Qq}.$$
In the second case, where $\gamma=o$,
$$\frac{+\delta\beta r}{Qq}.$$

[57] The text has $\frac{\delta\rho''}{Q}+\frac{\gamma\rho-\beta\rho'}{Q}$.

In the third case, where γ is positive,

$$\frac{+(\beta r - \gamma r')\delta}{Qq}.$$

But in this last case $\beta > \gamma$; if therefore we make $\gamma = \beta - \mu$, the formula will become $\dfrac{+\delta\beta(r-r') + \delta\mu r'}{Qq}$.

All these formulae are positive. Thus for this case the same things are valid as for the preceding one (§ 111). For as the body B approaches the body A, it first becomes electric and then is attracted by it. It must further be noted, however, that even though the same things happen in this case, the attracting force is greater than in the preceding. For all the formulae found here are greater than $\dfrac{\delta\beta(r-r')}{Qq}$,[58] and it is this formula which, according to § 111, expresses the attractive force acting in the former case.

118) If we now consider the case where the body B is electric *per se* but only imperfectly so, it is clear that all the reasoning we have used previously can be employed here, and there must be a similar result on all points. Everything said previously (§ 110. 117) applies here, and the only difference is that the electric fluid moves with difficulty in the pores of bodies of this kind; hence the force propelling the fluid achieves a less complete effect. But the only definite outcome of this is that the quantity of electric matter driven from CDGH into EFGH is less than it would be if the body B were perfectly non-electric *per se*. Generally, bodies like this become electric themselves while approaching a positively electrified body, but since they are not disturbed from their natural state to such a great degree as bodies non-electric *per se*, and since β is thus less in them than in bodies non-electric *per se*, so too the force by which bodies electric *per se* are attracted will be more sluggish than the force by which bodies non-electric *per se* are agitated under the same circumstances. For it is easily seen that the formulae expressing the attractive forces become less if β decreases.

119) We have so far examined the case where electric fluid abounds in A. But if we turn to the other, where the quantity of fluid in that body is diminished below the natural, it is evident from § 12 that the electric fluid is attracted toward the body A by a force $= \dfrac{+\alpha r}{Q}$. Since this force is positive, the fluid contained in the body B is not repelled as before but is attracted toward the surface CD, so it must leave the

---

[58] The text has $\dfrac{\delta\beta(r-r')}{Q}$.

part EFGH and move toward CDGH. Assuming that we must inquire what ought to be the consequences of this, it is readily clear that we can pursue our reasoning in a completely similar manner as in the preceding case.

120) If, first of all, the body B, Fig. XX, is non-electric *per se* but is surrounded by other bodies electric *per se* which can inhibit the free efflux or afflux of fluid, then, retaining the notation employed above by calling the quantity of fluid deficient in $A = \delta$, and the quantity of fluid which crosses from EFGH into $CDGH = \beta$, and calculating just as in § 111, the mutual attraction of the bodies will be found to be, exactly as above, $= \dfrac{\delta\beta(r-r')}{Qq}$.[59] So everything proved in § 112. 113 is valid for this case too. For, by calculating just as in § 113, the force by which a particle I located near GH is urged to flow across from the part EFGH into CDGH is found $= \dfrac{\delta\rho'}{q} - \dfrac{2\beta\rho}{Q}$. So the greater $\dfrac{\delta\rho'}{q}$ is, that is, the closer the bodies A and B approach each other, the greater β must be to reduce the force urging the particle I to zero. And the closer the bodies A and B become, the more fluid is evacuated from EFGH and accumulated in CDGH; at the same time there is an increase in the force by which the bodies attract each other according to a certain function of the distance so far unknown.

121) If electric matter can flow freely into the body B, as happens if it is touched by a body C, Fig. XXI, non-electric *per se*, then, calling the quantity of fluid deficient in $EFGH = \gamma$, a particle of fluid k located close to the surface EF is urged to enter the part EFGH by a force $= \dfrac{\delta\rho'}{q} + \dfrac{\gamma\rho - \beta\rho'}{Q}$, because, in an obviously similar way to § 115, the influx cannot stop until this force has been reduced to zero. Since this can only happen in the three ways evolved in § 116, it is evident that γ ought to become either negative, or $= 0$, or positive but by a quantity less than $\dfrac{\beta\rho'}{\rho}$. In the first case, the part EFGH will be positively electric, in the second it will be constituted in the natural state, and in the third it will be negatively electric, but to a lesser degree than CDGH. If we next evolve formulae expressing the attractive force of the bodies in a similar fashion as in § 117,

In the first case, where γ is negative, it will be

$$\dfrac{+(\beta r + \gamma r')\delta}{Qq}.$$

---

[59] Again the text has $\dfrac{\delta\beta(r-r')}{Q}$.

In the second case, where $\gamma = 0$

$$\frac{+\delta\beta r}{Qq}.$$

In the third case, where γ is positive

$$\frac{+(\beta r - \gamma r')\delta}{Qq}.$$

But since in this case $\beta > \gamma$, let $\beta - \mu$ be substituted for γ and we shall get

$$\frac{+\delta\beta(r-r') + \delta\mu r'}{Qq}.$$

So for this case it is likewise true that a body B brought close to a negatively electric body first becomes electric and is then attracted. The things I mentioned at the end of § 117 concerning the magnitudes of the forces are clearly valid for this case too. Finally, if the body B, close to the negatively electric A, is electric *per se* not in the highest degree but only imperfectly, by applying the reasoning employed in § 118, we come to the same conclusions.

122) We have so far considered only those cases where the whole of the body A is reduced to a state where it is either positively or negatively electric. But a third case can obtain, where the body A possesses positive electricity in one part and negative in the other, and the case of the *Leyden* jar is pertinent here. After what has been dealt with so far, this case is easily handled, for it has been shown above, § 21. 24, that there can be three possible states of the body A when it has been electrified in the way described. For either the part NOIK, Fig. XXII turned toward the body B plainly does not act on electric fluid, or the body A attracts fluid with that part, or, finally, it repels it. In the first case, the body A does not act on the electric fluid contained in B. So the body B moved to the body A preserves its natural state and there is no mutual action of the bodies on one another. But if the body A repels fluid from the part IK, then it is easily understood that all the reasoning I used in § 113. 118 can be employed here. For since in this case the body A acts on electric fluid found outside it just like a positively electric body does, it is obvious that what has been demonstrated earlier about a positively electric body is likewise valid here too. But if the body A attracts fluid with the part turned toward the body B, it can be considered negatively electric, since it acts in exactly the same way. As a result, what was said in § 119. 121 can be applied to this case also. All this is sufficiently evident for us to be able to dispense with the rather lengthy calculation required for evolving

formulae valid for these cases. If anyone meantime wants these formulae, he will be easily able to work them out from § 43.

123) We must now consider how the theoretical reasonings proposed so far agree with experience; we shall in fact find complete agreement. First of all, it is commonly known that a body constituted in the natural state, when moved to an electrified body, is attracted by it, as the theory states. So in this direction, the propositions evolved here do not need further confirmation through experiments. But I asserted with the guidance of theory that this attraction does not happen except insofar as and until the body attracted has itself become electric. So the crux of the matter is that we ask whether it is proved by experience that any body constituted in the natural state, on being moved to an electrified body, is itself disturbed from the natural state and becomes electric. Although this is so clear from *Franklin's* discoveries that we hardly need new experiments, I judge it worthwhile to fit those experiments to the theory so that there can be no residue of doubt about the truth of my propositions.

124) Place a metal rod AB, about a foot long, on glass supports CD, EF, Fig. XXIII, and on the end A place a metal piece GL about an inch and a half long fitted in the middle with a little hook M to which has been attached a well-dried silk thread HM. Then take the electrificatory glass cylinder IK, and, after it has been electrified by rubbing, move it to the end A of the rod, to a distance of about an inch, and hold it there motionless. Lift the metal piece GL by means of the silk thread HM and place it on the glass support NO.[60] If the body GL is examined it will be found to be electric and indeed negatively so. In the second experiment, let everything happen in the fashion described, and move the glass tube IK once more to the end A, with the body GL placed on the end B of the rod. If everything is done as in the preceding experiment, the body *gl* or GL placed on the support NO will again be electric, but contrary to before, it will have a positive electricity. If a sulphur cylinder is used in place of the glass tube, there will be a similar outcome throughout, only in the contrary order: GL placed on the end A will become positively electric, and applied to the end B will become negatively electric. Quite evidently these four experiments prove the truth of what I have expounded above, and this is so obvious that I can dispense with further explanations.

125) Afterward when the body GL is again placed on the end A of the rod, move the electrified glass tube or sulphur cylinder once more

[60] The text has N*o*.

ATTRACTION AND REPULSION

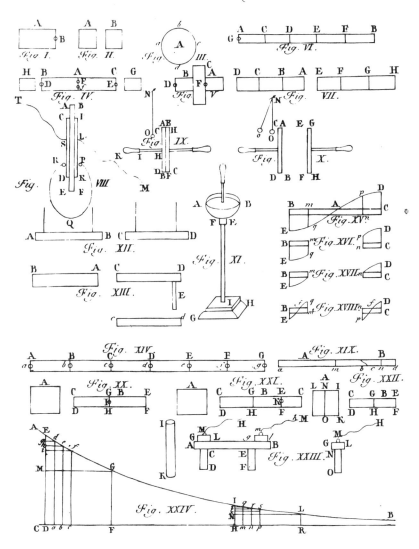

Figure 6. Facsimile Plate I from the *Essay*.

toward A, but now let the rod be touched at B by a body non-electric *per se,* through which electric matter can flow freely in or out. Then the body GL raised and placed on the support NO is found to be electric once more, positively so if the sulphur cylinder is used and negatively if the glass tube is used. Finally, if the body GL is placed on the rod close to B, and for the rest everything happens as before, when using a glass tube, the body GL will be found at times to have gained a negative electricity as before, at times none, and at times a positive one, though to a weak degree. By substituting a sulphur cylinder for the glass tube, GL is likewise at times found to be positively electric, at times constituted in the natural state, and at times negatively electric although only weakly. The reason for these phenomena is quite obvious from § 115. 121, with which these phenomena are in such agreement that they plainly demonstrate the truth of the reasonings produced there.

126) Although these experiments can sustain a comparison with the theory, and can in my opinion remove all doubts that perhaps arose concerning the paradoxical proposition that I brought forward previously (§ 33), that a body constituted in the natural state, as long as it is such, is neither attracted nor repelled by any electrified body, yet I judge it worthwhile to add certain things still which can shed further light on this matter. It is easily perceived that, if what I have so far taught about the attraction of bodies constituted in the natural state is correct, a consequence of it is that bodies ought always to be more strongly attracted, the more strongly they are rendered electric by the approach of an electrified body, and that this action will be weaker, the more weakly and the greater the difficulty with which they become electric under the circumstances described. So since it is evident that when bodies electric *per se* are brought to an electrified body, because of the difficulty with which electric matter moves in their pores, they always become electric to a lesser degree than happens under the same circumstances with bodies non-electric *per se* (§ 118. 121)—this agreeing very well with experience, which testifies to the same thing—it follows from our theory and from the method proposed here for explaining electrical attraction that, if all the circumstances are the same, bodies electric *per se* ought to be more weakly attracted than those non-electric *per se,* and that among themselves the more electric *per se* ought to be more weakly, the less electric *per se* more strongly attracted. It is very well known to all those who have given their efforts to performing electrical experiments that the theory is here consonant with experience. Since experience alone shows that all bodies moved toward an electrified body become electric and are attracted, though those observed to be-

come more weakly electric are more weakly attracted, even without the help of the theory there is a reason for stating that if a body does not itself become electric when moved toward an electrified body, it will certainly also not be attracted.

127) There is a further phenomenon mentioned above (§ 33) which would have to be recognized as an outstanding paradox if it could not be reconciled with our theory; but the reason for it is obvious from the theory even at first sight. Rub two glass mirrors ABCD, EFGH, Fig. IX, on one another as indicated in § 55. If they are separated from one another after the rubbing, either of them on its own attracts the globe NO of the pendulum. So if they are joined, it might be expected that, since the anterior and the posterior alone attract the pendulum, the combination of the plates ought to exert a still stronger attraction than when only one is employed. For since the pendulum is already urged by two forces tending to the same effect, one of which is exercised by the plate ABCD, the other by the plate EFGH, according to the first principles of reasoning it seems a necessary conclusion that there ought to be a double effect from a doubled force. But, as I mentioned above in § 55, experience points to the exact opposite of this. For as soon as the plates are joined, the attractive force is quite inactive and completely extinguished. This most difficult phenomenon will always be a salve to those unwilling to admit our theory. But, that conceded, the reason for the phenomenon is easily given. For if the plate ABCD is e.g. positively electric, exerting a repulsion on the electric matter contained in the globe O, it exerts an effort which tends to bring this globe to the state where, in the part facing the plate, it becomes negatively electric, and in the part facing away from the plate, positively electric. On the contrary, when the plate EFGH, which is negatively electric, is joined to it, this tends to produce a directly opposite effect. For since it attracts the electric fluid to be found in the globe, it tends to make the part of the globe facing the plates positively electric, but the part facing away negatively electric. So since these contrary forces are equal (§ 60), neither of them achieves any effect, but the globe O will preserve its natural state unchanged, and, conformably with our theory, it will not be attracted either by the positively electric plate ABCD or the negatively electric one EFGH when they are in combination. Evidently then, this experiment shows that provided the production of electricity in a body is prevented there is no action of electrified bodies on it.

128) We must now turn to a more careful examination of the phenomena which occur if both the bodies A and B close to one another, Fig. XXII, possess either positive or negative electricity. It could

seem that this case is covered by the things dealt with above, § 36; but it must be noted that we examined there the mutual action of bodies on one another in as much and for as long as they possessed the state which was assigned to them. We did not look at the change of state which happens in those bodies as they approach one another, and so a special examination must be considered particularly warranted here. Let us suppose first that two bodies A and B, both positively electric, approach one another, and let us ask what will happen to these bodies. Imagine that in the body A electric fluid is accumulated beyond the natural quantity $(=2q)$ by an amount α, and in B beyond the natural quantity $(=2Q)$ by an amount β.

129) It is readily apparent from what has been dealt with above that here the body A exerts a repulsion on the electric fluid contained in B and that in turn the body B exerts a repulsion on that contained in A. So since electric fluid is propelled from the ends CD, IK toward EF and LM, electric fluid cannot remain uniformly distributed as it was before the bodies approached one another, but some part of it must leave the parts NOIK, CDGH and move to LMNO, GHEF. Let the quantity of fluid which crosses from IKNO into LMNO $=\eta$, and that which flows out from CDGH into GHEF $=\theta$, and if in the formula already evolved in § 43, we make $e=\frac{\alpha}{2}-\eta$, $f=\frac{\alpha}{2}+\eta$, $g=0$, $a=\frac{\beta}{2}-\theta$, $b=\frac{\beta}{2}+\theta$, $c=0$, it will express the force by which the body A and the body B mutually attract each other.

130) Since, as is self-evident, the magnitudes of the quantities η and θ depend on the magnitudes of the repelling forces which the bodies B and A exercise, and these forces in turn depend as much on the quantity of fluid abounding in each body as on the mutual distance of the bodies from one another (as is evident from the formula derived in § 10), one can easily see the possibility of adjusting the quantity of fluid abounding in A or B or in both bodies, as well as the distance between these bodies, in such a way that either η is greater than $\frac{\alpha}{2}$, or θ is greater than $\frac{\beta}{2}$, or both occur. So it is possible for either a alone or e alone or both quantities to become negative. As long as all four quantities a, b, e, f, are positive, it is self-evident that, with the substitution of positive values, the formula $\dfrac{-(ar+br')e-(a\rho+b\rho)f}{Qq}$ always remains negative.[61] But if it is assumed that either a, or e, or both

---

[61] The text has $-\dfrac{(ar+br')e-(a\rho-b\rho')f}{Qq}$.

ATTRACTION AND REPULSION   317

quantities, can acquire negative values, then the formula yields respectively [62]

$$1) \quad \frac{+(ar-br')e+(a\rho-b\rho')f}{Qq}$$

$$2) \quad \frac{+(ar+br')e-(a\rho+b\rho')f}{Qq}$$

$$3) \quad \frac{+(br'-ar)e-(b\rho'-a\rho)f}{Qq}.$$

It is obvious that any of these formulae could be either positive or zero, or could even become negative. So various things can happen if two positively electric bodies are moved close to one another; yet these can be conveniently reduced to three. For either η and θ have such values that the attractive force of the bodies A and B will be negative, and then they will repel each other, or the force may be positive and the bodies will attract each other, or finally there is no attraction and the two bodies will clearly not act on one another.

131) From this there arise certain noteworthy phenomena which may seem to contain something paradoxical and alien to the analogy of the other operations of nature. Thus, for example, it is possible that two bodies fairly distant from one another may repel each other, yet if they are brought closer together by some external force, they may begin to attract each other. For as long as the bodies are rather distant from one another, the repelling force by which each of them urges the fluid contained in the other is weak, so the quantities η and θ are quite small. It is possible that then $\eta<\frac{\alpha}{2}$ and $\theta<\frac{\beta}{2}$, and so a and e have positive values, and the bodies A and B repel each other (§ 130). But since the repelling forces, and thus η and θ, increase when the distance is decreased, as the bodies are brought closer together it is possible that then either the two quantities $e=\frac{\alpha}{2}-\eta$ and $a=\frac{\beta}{2}-\theta$, or at least one of them, becomes negative, and the total force by which the bodies repel each other may similarly gain a negative value (§ 130).

132) It is further possible that, so long as the body A is only weakly electric, it repels the body B; but if it is electrified to a stronger degree, it begins to attract the body B. For if the body A is weakly electric, the

[62] Consistent with the error noted earlier in the paragraph, and with that which appeared in the initial statement of the general force formula (§ 43), in the text the leading "+" sign in each of these formulae is placed in front of the entire expression. Having had to make the adjustment in earlier cases because the leading sign was a "—," we have elected to continue doing so here and in several other places in the next few paragraphs, without drawing attention to each case individually.

repulsive force which it exerts on the fluid contained in B is small, and therefore θ will also be small. It is then possible for the quantity $a = \frac{\beta}{2} - \theta$, and all four quantities a, b, e, f, to be positive. But if the supply of electric fluid abounding in A is increased, θ increases, so then $\frac{\beta}{2} - \theta = a$ can become negative, and the force by which the bodies A and B mutually act on one another can switch and become positive, or attractive.

133) These phenomena can seem contrary to the analogy of nature because, whereas in nature at other times it is usual for a force to increase as the distance decreases, in the first of these (§ 131), on the contrary, it decreases beyond zero and becomes a force in the opposite direction, and in the second (§ 132), when the mass is increased, the force does not likewise increase but diminishes and passes through zero into one in the opposite direction; this is quite contrary to the custom of nature. These phenomena would then contain enormous paradoxes were not their source and their friendly agreement with the fundamental laws of nature clear from our own theory. At the same time there is not a little probability in our theory, for when it is assumed, these paradoxical phenomena agree so aptly with the fundamental laws of nature; and I doubt very much whether this could as happily be the case with any of the other hypotheses developed to date.

134) The same things clearly occur when both the bodies A and B are negatively electric. For in this case the body A attracts the electric matter to be found in B and the body B in turn attracts that contained in A; a consequence of this is that the electric fluid does not remain uniformly distributed as it was before the bodies were moved near each other. Rather, some part of the fluid should move from EFGH into CDGH and from LMNO into IKNO. So if we suppose that the quantity of fluid which is deficient in A=α and in B=β, and we say further that the quantity which crosses from LMNO into IKNO=η, and that moving from EFGH into CDGH=θ, and then we put into the general formula (§ 43), $e = -\frac{\alpha}{2} + \eta$, $f = -\frac{\alpha}{2} - \eta$, $g = 0$, $a = -\frac{\beta}{2} + \theta$, $b = -\frac{\beta}{2} - \theta$, $c = 0$,[63] or, which comes to the same thing, if we put a, b, e, f into the formula (§ 43) as negative quantities, and g and $c = 0$ (in which case we get $\frac{-(ar + br')e - (a\rho + b\rho')f}{Qq}$, which is the

---

[63] The text has $a = -\frac{\beta}{2} + \eta$, $b = -\frac{\beta}{2} - \eta$, $c = 0$.

ATTRACTION AND REPULSION

same formula as is obtained if a, b, e, f, are positive), and then let $e=\frac{\alpha}{2}-\eta$, $f=\frac{\alpha}{2}+\eta$, $a=\frac{\beta}{2}-\theta$, $b=\frac{\beta}{2}+\theta$,[64] we shall get a formula which expresses the mutual action of the bodies A and B on one another in this case.

135) It is clear that here, as in the previous case (§ 130), because η and θ can be allotted various values, it is possible for the attractive force of the bodies A and B to be either negative or positive or even at times zero; thus in this case the same holds as in the preceding one. Negatively electric bodies A and B close to one another can either repel or attract each other, or they may not act on each other at all. The phenomena reviewed in § 131 and 132 arise here also. For it is possible for two negatively electric bodies which repel one another at a greater distance to begin to attract each other when moved closer together, and also for a body A to repel a body B when weakly negatively electric, but to attract B once it is more strongly electrified. Since all these things are demonstrated in precisely the same fashion as in § 131. 132, there is clearly no need for further explanation here.

136) Finally, the last case, where one of the two bodies that have been moved toward one another, A for example, is positively electric, while the other, B, is negatively electric, is also explained by a similar argument. In this case, the body A repels the electric matter contained in B, but the body B attracts that found in A, whence some of the electric matter crosses from the part LMNO into NOIK and some of it is propelled from the part CDGH into EFGH. If we call the quantity of fluid abounding in A, α, and that deficient in B, β, and the quantity which crosses from LMNO into IKNO, η, and that propelled from CDGH into EFGH, θ, we easily obtain the formula for the force by which the bodies A and B attract one another. For if in the general formula of § 43 we make $e=\frac{\alpha}{2}+\eta$, $f=\frac{\alpha}{2}-\eta$, $g=0$, $a=-\frac{\beta}{2}-\theta$, $b=-\frac{\beta}{2}+\theta$, $c=0$ (or if in the general formula we make $g=0$, e and f positive, a and b negative, and $c=0$, when it becomes $\frac{+(ar+br')e+(a\rho+b\rho')f}{Qq}$;[65] and then e becomes $=\frac{\alpha}{2}+\eta$, $f=\frac{\alpha}{2}-\eta$, $a=\frac{\beta}{2}+\theta$, $b=\frac{\beta}{2}-\theta$), we will obtain the formula we were looking for.

---

[64] The text has $a=\frac{\beta}{2}-\eta$, $b=\frac{\beta}{2}+\eta$.

[65] The text has $+\frac{(ar+ba')e+(a\rho+b\rho')f}{Qq}$.

137) Since η and θ increase with an increase of the repulsive force of the body A and of the attractive force of the body B, it is evident that, with the different degrees to which the bodies A and B may be electrified and with the different distances they may be from one another, it is possible for the quantities $\frac{\alpha}{2}-\eta$, or f, and $\frac{\beta}{2}-\theta$, or b, to be sometimes positive, sometimes zero, and sometimes negative. As long as f and b remain positive, it is self-evident that the formula [66] $\frac{+(ar+br')e+(a\rho+b\rho')f}{Qq}$ also remains positive, and as a result the bodies A and B attract each other. And it is equally easily proved that if either b alone or f alone becomes negative the formula still remains positive. To make this clear, let us first of all note that even though b becomes negative, it still cannot ever be greater than a. If $b=\frac{\beta}{2}-\theta=-\gamma$, then $a=\frac{\beta}{2}+\theta$ will be $>\gamma$. For let $b=\frac{\beta}{2}-\theta=-\gamma$, and θ will $=\frac{\beta}{2}+\gamma$, hence $a=\beta+\gamma$, and this value is clearly greater than γ. In the same way, it is quite clear that, even if f becomes negative, $=-\delta$, e, which then $=\alpha+\delta$, would still be greater than δ. If we suppose

1) b alone becomes negative, in which case our formula transforms into $\frac{+(ar-br')e+(a\rho-b\rho')f}{Qq}$,[67] it is clear that, because $r>r'$, $\rho>\rho'$ and $a>b$, both $ar-br'$ and $a\rho-b\rho'$ are postive, so that the whole formula is positive. But if

2) f alone becomes negative, our formula becomes [68] $\frac{+(ar+br')e-(a\rho+b\rho')f}{Qq}$ or [69] $\frac{+(er-f\rho)a+(er'-f\rho')b}{Qq}$, where again, because $r>\rho$, $r'>\rho'$ and $e>f$,[70] the value of the whole formula is positive.

138) But the third and remaining case, where b and f become negative at the same time, is more difficult and can hardly be expounded through the theory alone as long as we do not know the function according to which the particles of electric fluid repel each

---

[66] Again the text has $+\frac{(ar+ba')e\cdot+(a\rho+b\rho')f}{Qq}$.

[67] The text has $+\frac{(ar-br')e+(a\rho-a\rho')f}{Qq}$.

[68] The text has $+\frac{(ar+br')e-(a\rho-b\rho')f}{Qq}$.

[69] The text has $+\frac{(er-f\rho)a+(a\rho-b\rho')f}{Qq}$.

[70] The text has $f>e$.

other. In this case the attractive force of the bodies A and B $=$ $\dfrac{+(ar-br')e-(a\rho-b\rho')f}{Qq}$ where it is clear from the preceding § that $a>b$, and $e>f$. Therefore let it be assumed that $a=b+\mu$, $e=f+\nu$, $(r-r')=m$, $\rho-\rho'=n$, and [71] then $ar-br'=bm+\mu r$ and $a\rho-b\rho'=bn+\mu\rho$, and the force by which the bodies A and B attract each other $=$ $\dfrac{+fb(m-n)+f\mu(r-\rho)+\nu bm+\nu\mu r}{Qq}$, where the last three terms are positive. So unless $fb(m-n)$ or $m-n$ can become negative, it will also be impossible for the whole formula ever to become negative; on the other hand, if it is possible for $m-n$ to be negative, there will possibly be cases where the attractive force of the body becomes negative and transforms into a repulsive force. So the question is reduced to the point where it must be determined whether n can become greater than m, that is, whether there can be any cases wheret $r-r'<\rho-\rho'$.[72] And no satisfactory reply can be given to this question until we understand the function according to which electrical repulsion is exerted.

139) Although at this point theory has to be abandoned and recourse had to experience, let us add certain considerations to cast some light on the matter. We may rightly assume that the function according to which the particles of electric fluid repel each other is such that, when the distance is increased continuously, the force decreases more and more, for this is completely consonant not only with the analogy of nature, but also with experience. For experience shows that everything deduced from our supposition takes place in nature. An immediate conclusion from this is that, if the curve AB, Fig. XXIV, is the graph of electric repulsions, or is arranged such that, with abscissae CD taken as distances, the corresponding ordinates DE are proportional to the repulsive forces, this curve would be asymptotic.[73] If this were not so, a continuous decrease in repulsion with increasing distance would not be possible, but there would be a terminus at some point beyond which the repulsion would not decrease even if there was an increase in distance. This asymptotic curve can be of two kinds, for either it is convex and approaches the asymptote CR continuously, or at some point or other it has a point of inflection, and is at one time concave to the asymptote and at another time convex to it.

140) If we assume that the first of these occurs, and that the ab-

[71] The text has $ar - br' = +\mu r$.
[72] The text has $r - r' > \rho - \rho'$.
[73] M. C. Shields, "The early history of graphs in physical literature," *American Physics Teacher*, v. 1937, 68–71, draws attention to the infrequency with which graphical representations were used in eighteenth-century physics. He does not mention Aepinus's use of them.

scissa$=x$, the ordinate$=y$ and dx is constant, dy corresponding to this dx will decrease continuously the more x is increased, so that any dy corresponding to a smaller x is greater than a dy which corresponds to a larger x, yet dy does not become zero until x has become infinite. The truth of this is verified from Geometry, where it is clear that if the curve AB does not have any inflection, ddy never becomes$=$o. So then ddy will be either always positive or always negative and will never cross from positive to negative or vice versa. Therefore in a curve of that kind, dy either increases perpetually or decreases perpetually. The latter is the case here. For since the curve is asymptotic, at $x=\infty$, dy will eventually become equal to o, so dy ought necessarily to decrease perpetually. Thus if in such a curve two pairs of ordinates DE, FG and HI, KL are taken so that the distances DF and HK are equal, DE$-$FG$=$EM will be greater than HI$-$KL$=$IN. For since the lines DF and HK are equal, let us consider them divided into an equal number of infinitely small parts Da, ab, bc ... etc. and Hm, mn, np ... etc., and by drawing the ordinates ad, be, cf, ... etc., and mq, nr, ps ... etc., EM will be$=$E$g+dh+ei$ ... etc., IN$=$ I$t+qv+rx$ ... etc. Therefore since E$g>$It, $dh>qv$, $ei>rx$ ... etc., E$g+dh+ei$ ... etc.$=$EM will also be greater than I$t+qv+rx$ ... etc.$=$IN.

141) If the graph of repulsions, AB, Fig. XXV, is conceived to have a point of inflection, ddy corresponding to this point L will be$=$o, and dy will change from increasing to decreasing or vice versa. The first of these possibilities is the case here, since at $x=\infty$, dy ought to become$=$o. Therefore since dy changes from increasing to decreasing, dy will be a maximum at the very point of inflection. So the dy which most closely precedes this maximum dy is less than it, and if two pairs of equidistant ordinates are taken, the first of which HI, KL, is close to the point of inflection L,[74] and the second DE, FG lies some short distance nearer the origin of the curve, it is easily understood by the same reasoning as in the previous § that it is now possible for ED$-$FG$=$EM to be less than HI$-$KL$=$IN.

142) Since the letters r, r', ρ, ρ' denote four repulsive forces so arranged that the difference of the distances relevant to r' and r is equal to the difference of the distances relevant to ρ' and ρ, (§ 43), these quantities can be represented by the four ordinates of the graph of repulsions DE, FG, HI, KL, so arranged that DF$=$HK.[75] If therefore the curve AB is such that it has no point of inflection, $r-r'$ will be

---

[74] The text here refers to the point of inflection as C, but no such point is shown in the figure.
[75] The text has DE $=$ FG.

perpetually $> \rho - \rho'$, so the attractive force of the bodies A and B can never become negative or repulsive (§ 140 and 138). But if the curve AB has a point of inflection, it is possible for $r - r' < \rho - \rho'$,[76] and for the bodies A and B to mutually repel one another (§ 141 and 138). So the whole question we are discussing here reduces to deciding whether the graph of electrical repulsions is endowed with a point of inflection or not. But an answer to this question must be left to experience. And if we consult experience, there can hardly be any doubt remaining that the often-mentioned graph of repulsions ought never to have a point of inflection. For from innumerable experiments it has never happened with me, or with others as far as I know, that bodies possessing different kinds of electricity have been observed to repel one another. So we can state with the highest degree of probability that the function according to which electrical repulsions occur, though we do not yet know what it is exactly, is nevertheless of such a nature that the graph of repulsions has no point of inflection.

143) Experience agrees excellently with everything I have mentioned to date; and it will be convenient to show this briefly. It is very well known that two bodies, one of which is positively electric, the other negatively electric, always attract each other. This can easily be tested by suspending a little globe or some other light body from a silk thread and communicating electricity to it from a sulphur cylinder, and then bringing an electrified glass tube or some other body which has received electricity from a glass tube close to it; or vice versa, by bringing a sulphur cylinder close to a little globe which has been electrified by means of a glass tube. It will be observed that the pendulum is always attracted and is never repelled, even if the experiment is repeated a thousand times under circumstances varied in any way whatever. But what happens if bodies possess the same kind of electricity is not sufficiently understood; therefore relevant experiments deserve a more careful exposition.

144) Suspend a small globe of cork, B, about the size of a pea, from a silk thread AB, Fig XXVI. Below the pendulum AB, place a metal cylinder E about an inch in diameter on glass supports FG, KL, and arrange the supports and the pendulum AB so that when the latter is vertical the globe B does not quite touch the cylinder E. To the globe B tie the silk thread *bcd* and pass it over the little hook *c*; fix to the cylinder E an iron wire HI five or six feet long, properly supported on bodies electric *per se,* by means of which electricity can be conducted to it. Electrify the globe B in the normal manner with the

---

[76] Again the text has $r - r' > \rho - \rho'$.

help of the glass tube and afterward, by moving the rubbed glass tube to the end I of the iron wire, make the cylinder E electric too. It will then be observed that the globe B is repelled and the pendulum is raised to the position Ab. Then pull the thread bcd and force the globe B closer to the cylinder E, and after it has approached to a distance of 2, 3 or 4 lines, it will suddenly be observed that the repulsion has changed to an attraction, and the pendulum comes to the vertical position AB; if it is disturbed a little from this position by pulling the thread bcd forward a half or whole line, it spontaneously returns again to its previous position. This experiment succeeds as well if a sulphur cylinder is substituted for glass.

145) Further, restrict the pendulum AB by winding the thread bcd a few times round the hook c so that it cannot move further than about 2 lines from the cylinder E. Electrify the globe B and the cylinder E again with the help of a glass tube in the manner described in the preceding §, but both of them initially only weakly; the pendulum will be raised to a position Ab, but will only recede from the cylinder E to the extent permitted by the thread bcd. Then electrify the cylinder E strongly, and the repulsion will be suddenly changed to an attraction, and the pendulum will move to the vertical position AB, and if it is disturbed a little from this position it will return spontaneously to it (§ 144). In this experiment too, the sulphur cylinder can be used in place of glass with the same success.

146) Although these experiments are outstanding proof of what was stated in § 130. 135, yet doubt might still be encountered in those who ponder all the circumstances correctly. It is very well known that air is imperfectly electric, and so it continuously reduces the electricity of the globe B which it surrounds; thus it might seem that the sudden change of repulsion to attraction in the experiments reviewed in § 144. 145 should be ascribed to a different cause from the one I alleged. For air can lower the electricity of the globe B to such a degree that either all or the greatest part of it is destroyed, and if this happens the globe B is as if not electrified and so ought to be attracted by the cylinder E. This is altogether possible, but that it did not happen in fact in our experiments can easily be proved. In the experiment of § 144, after the pendulum has reached the vertical position, pull the thread bcd until the pendulum AB has reached the position Aβ, and it will be observed that it is repelled once more by the cylinder E. In the second experiment (§ 145), after the pendulum has similarly arrived at the vertical position, take all the electricity from the cylinder E, then electrify it once more but only weakly, and it will again repel the pendulum. Neither of these things could happen if all the

electricity had been taken away from the globe B by the surrounding air, for then in each case the repulsion would not take place, but rather the pendulum should always be attracted. My final advice is that these experiments may easily fail to succeed if they are not set up with the greatest care on all points, with particular attention to the dryness of the air.

147) There occur among writers dealing with electricity the rules that differently electrified bodies mutually attract one another, and those possessing the same kind of electricity repel one another.[77] As for the first rule, it is clear from § 143 that it can be accepted; but the latter, as it is usually promulgated, must be considered completely false. We have proved (§ 130. 135. 144. 145) that if both bodies possess the same electricity, they can at times repel one another, as the rule states, at other times show no action on one another, and at other times attract one another. Meanwhile we can generally admit the converse of the rule as true, if it is accepted in the sense that, if two bodies repel one another, then the quantity of electric fluid in both bodies at the same time is either greater than or less than the natural. For it is easily established that only in this case can bodies mutually repel one another, since in other cases, as I have sufficiently shown already, they always attract each other and never repel.

148) The action of a magnet, both on iron not endowed with magnetism and on another magnet, follows exactly the same laws as are found in electrical attraction and repulsion. Indeed it is readily obvious that the almost universal reasoning which I used for electricity is equally valid for magnetism if only a few things are changed; but since the similarity of the causes of magnetic and electrical phenomena may be greatly elucidated as a result, it seems worthwhile to expound and confirm by experiments the theory of magnetic attraction and repulsion. Therefore consider that a magnet A, Fig. XXVII, is filled with magnetic fluid in its part AC, but evacuated in the part AD, and let the quantity of fluid which has passed from AD into AC$=\alpha$. This magnet then repels magnetic fluid found outside itself from the part A, but attracts it from the part B, by a force which in each case $=\dfrac{\alpha(r-r')}{Q}$ (§ 94). First move to the part AC an iron body B, constituted in the natural state and so far not imbued with magnetic force. If this body continues in the natural state, it is evident from § 41. 42 that the action of the magnet on it will be zero, and so it can neither be attracted nor repelled as long as that condition lasts. But since the magnetic

---

[77] Du Fay, *HAS*, 1733, Mém. pp. 457–76.

fluid found in B is propelled by the magnet by a force $=\frac{\alpha(r-r')}{Q}$, some part of it must cross from BE into BF, so the iron B becomes negatively magnetic in its part BE and positively magnetic in the part BF. In order that we may determine the action of the magnet on the iron B, let the natural quantity of fluid appropriate to each of the parts BE and BF$=q$, and the quantity of fluid which crosses from BE into BF$=\beta$, and if in the general formula of § 43, $a=\alpha$, $b=-\alpha$, $c=0$, $e=-\beta$, $f=\beta$, $g=0$, the action sought will be $=\frac{\alpha\beta(r-r')-\alpha\beta(\rho-\rho')}{Qq}$.

149) This quantity will always be positive, and will never be able to become negative, unless $\rho-\rho'$ becomes $>r-r'$. So what ought to happen here cannot be decided from the theory alone, but as in the case of electricity (§ 139ff), recourse must be had to experience. Of course, everything said in § 139–142 is valid and can be immediately transferred to here. Since experience teaches that the magnet constantly attracts non-magnetized iron and that ever since Physicists began to inquire into the nature of magnetism experimentally, there is no recorded example of a magnet which has repelled iron, we can state with the greatest probability that the graph of magnetic repulsion, just like that of electrical repulsion, must be a curve without a point of inflection.

150) So the force by which a magnet attracts iron is always positive, but it increases when β increases and likewise it decreases when β decreases. Therefore, since the size of β itself depends on the size of the repelling force $\frac{\alpha(r-r')}{Q}$ (§ 148), the attraction of the iron must be greater if this force is greater. But it increases

1) If the distance decreases. For when the distance is reduced, it is evident that, because the graph of magnetic repulsion does not have a point of inflection, the quantity $r-r'$ must be increased (§ 149. 140).

2) If r greatly exceeds quantity r'. Other things being equal, the longer the magnet A is, the greater is this difference. For then the part AD is much further from the body B than the part AC, so of necessity r' must be much less than r.

3) If α increases, whence, other things being equal, the magnet must attract nearby iron with a greater force, the greater is the degree to which it is magnetic. A fourth case must be added to these.

4) The magnitude of β depends on the greater or lesser facility with which magnetic fluid moves through the pores of the body B. For if this difficulty is infinitely great, β will be zero, and the body B will preserve the natural state and will not be attracted. In the remaining

cases, β will always be the greater, the more easily magnetic fluid moves in the pores of the body B, and always the less, the greater the difficulty with which magnetic fluid moves in the pores of the body B. So the attraction between iron and a magnet will be greater in the former case, and less in the latter.

151) The way is now prepared for an explanation of some quite noteworthy magnetic experiments. First, it is evident that the attraction of iron is always greater, the closer it is moved to the magnet, and so, with decreasing distance, the attraction ought to increase according to a certain fixed function. I do not determine what that function is, for the principles of our theory do not suffice to unravel it. *Musschenbroek* and several others have worked hard on this matter, which can only be discovered *a posteriori* and through experience, but they have not had the success they sought.[78] Although I have found this matter difficult enough, I am still not without hope that, if an experiment is properly set up with the help of artificial magnets, and if the heads of the theory explained here are called in to help, there is a possibility that the law itself will finally be unearthed; but other tasks have so far prevented my inquiring into this matter.[79]

152) It is clear from no. 2 § 150 that, other things being equal, a magnet will always be weaker the shorter it is, and stronger the longer it is. This agrees excellently with the observation of Physicists who declare unanimously that if part of a magnet is broken off in such a way that its axis becomes shorter, it suffers a far greater loss than if some is taken off around the sides. I also consider that this provides the reason, which need not be looked into here, for artificial magnets generally possessing so significantly stronger a force than unarmed natural magnets. For rather long steel rods are usually chosen for artificial magnets while, on the contrary, natural magnets are usually quite short. My opinion is strengthened by certain experiments I have performed recently, though meanwhile I do not deny that other causes are also present; I shall have more opportunity for discussing these later.

153) Finally, from no. 4 § 150 it follows that soft iron is more strongly drawn by a magnet than is hardened steel. This is so obvious

[78] See above, pp. 169–73.

[79] The poles of artificial magnets were usually further apart than those of natural ones. Hence if artificial magnets were used, the force being measured could be expected to approximate much more closely to the ideal case of the force between two isolated poles. Aepinus's explicit mention here of the need to use artificial magnets in the experiments he is proposing indicates that he, unlike most earlier investigators, appreciated the need to reduce things to this ideal case. Unfortunately, there is no record of his subsequently taking these inquiries any further.

that it has been known for a long time through experience, and as a result the iron bars attached to magnets (called *Supports* in French) from which hang the weights which the magnet is supposed to carry, are usually made from the softest iron.

154) All this can be applied to the case where the negatively magnetic part of a magnet is turned toward the iron. For since the magnet A, Fig. XXVII, then attracts the fluid contained in the iron b by a force $= \dfrac{\alpha(r-r')}{Q}$ (§ 94), the fluid is urged to cross from bf into be, so the iron b becomes magnetic, and be becomes positively magnetic with bf negatively magnetic. So if the quantity of fluid which crosses from bf into be is called β, it will be found from § 43, by making $a=-\alpha$, $b=\alpha$, $c=o$, $e=+\beta$, $f=-\beta$, $g=o$, that the force by which the magnet A acts on the iron $b = \dfrac{\alpha\beta(r-r')-\alpha\beta(\rho-\rho')}{Qq}$. But since this formula agrees entirely with that discovered above (§ 148), what was said in § 149 is equally valid for this case too. And in the same way, the other things dealt with in § 150–153 are all equally valid for this case and may be transferred to it without difficulty.

155) After the explanation of the things determined by the theory concerning the action of a magnet on iron, it remains for experiments to prove the truth of my reasoning. That a magnet draws by either of its poles any iron that is not imbued with magnetic force, and that it draws soft iron more strongly than hardened iron, is generally known and can easily be tested. So what I have to do in particular is demonstrate by experiments that iron moved to a magnet is disturbed from its natural state and becomes a magnet itself. This can be done with very little trouble. For if a piece of iron B or b is moved to either pole of the magnet, Fig. XXVII, whether it touches this pole or is held a short distance from it, it will be found to be magnetic and will pick up another iron weight of appreciable size. More than that, if it is removed from the magnet some magnetism still remains in it, provided the body is not made of the softest iron, but has some hardness. However, we have explained sufficiently in § 87 above why, in the former case, the body B or b becomes magnetic in the vicinity of the magnet but immediately loses all its magnetism on being taken away.

156) Although this single experiment is abundantly sufficient to prove the agreement of our theory with experience, I shall add a description of certain other phenomena depending on what has been expounded here. For in part they seem sufficiently noteworthy in themselves, and in part they are excellent proof of the truth of the principles expounded here. Set up a bifurcated body ABC as is shown

ATTRACTION AND REPULSION

in Fig. XXXVIII. Move A to either pole of a magnet AE, and the body ABC will become magnetic and its end B will raise an iron weight F. Then take another magnet CD and put it at the other end C in such a way that the pole C is the contrary of the pole A. When this is done, the magnetism in the end B will be destroyed immediately and the body ABC will no longer be able to pick up the weight F, or, in fact, any other much smaller weight, with this end.

157) If in this experiment it is assumed that the pole A is positive, by exerting the usual repulsion when it is joined to this leg A, it will urge magnetic fluid to leave the end A and move towards B. But if the negative pole of the magnet CD alone were moved to the leg C, it would attract magnetic matter and urge it to leave the part B empty and cross over into C. Since therefore the magnets AE, CD tend to produce directly contrary effects, if they are of about the same strength and are moved to the legs A and C simultaneously, they will produce no effect and the end B of the iron ABC will remain constituted in its natural state. But if only one of the magnets is joined to either end, it produces its particular effect and renders the end B either positively or negatively magnetic.

158) Further, suspend EF, a piece of rather thin and softened iron wire some inches long, from a linen thread GE, Fig. XXIX, and move either one of the poles of a magnet A, say AC, to a distance of half an inch or an inch below EF. Then bring HI, another iron bar, unmagnetized, to the wire EF from the side, and it will be observed

1) If the body HI moves toward the lower end F of the wire EF, this is repelled by the body HI. But if

2) the iron HI is later raised so that it is opposite the upper end E of the wire, this will be attracted by the body HI.

In this experiment the lower end F of the wire, which is closer to the magnet, gains, according to the principles of the theory, a magnetism contrary to that possessed by the pole of the magnet AC, but the other end E acquires the same magnetism as the pole AC; this is manifest from § 148. 154. The same is true for the body HI; its end I which is closer to the magnet acquires for the same reasons a magnetism contrary to that of the pole AC. A reason can now be given without difficulty for the phenomena of the experiment reviewed. For I shall demonstrate later what is sufficiently known from elsewhere, that bodies which turn ends imbued with different kinds of magnetism toward one another attract each other, and that ends possessing the same magnetism, when turned to face each other, ordinarily repel one another. When these laws are supposed, there is nothing in the experiment for which the reason is not self-evident.

159) Although it is already sufficiently clear from what has been

330 THEORY OF ELECTRICITY AND MAGNETISM

discussed that our theory teaches things entirely consonant with experience, yet it seems worthwhile to remove altogether a doubt which could grow in readers because of my assertion that a magnet does not act on iron in as much as and for as long as it is constituted in the natural state. I can meet this doubt most aptly by demonstrating experimentally that when the production of magnetism in an iron body is inhibited for any reason whatever, a magnet does not act on it. It is first of all quite evidently clear, and experience testifies, that the weaker is the magnetism produced in a body moved to a magnet, the weaker too is always the force by which that body is attracted. For it is very well known that even if all the circumstances are supposed to be the same, yet hardened steel receives a lesser force than does softened iron. So it is proper to conclude that if iron could be hardened to such a degree that no magnetism at all could be produced in it, there would then also be no attraction. But the truth of my assertion is more clearly obvious still from what follows.

160) Prepare two quite similar and equal magnets A and a, Fig. XXX, of as equal force as possible, and both quite thin. In this experiment I myself used artificial magnets, namely two thin steel plates impregnated with magnetic force. Place the magnet A on a wooden table so that AC, one end of it, protrudes a little beyond the table, and from this end hang a sufficiently light weight F, much less than the weight the magnet can pick up, such as the thin iron wire gFh,[80] 5 or 6 inches in length. Afterward place the magnet a on A in the reverse position, so that the poles with opposite forces face each other and touch. This done, the force of the pole AC will all seem to vanish for it cannot by any means lift the weight F or any other weight, no matter how small, unless the magnet a is taken off again.

161) To the person who refuses to recognize the truth of my assertions, the outcome of this test ought to seem highly contradictory and paradoxical. For since both the magnet A and the magnet a attract the iron F and are able to carry it, the first principles of reasoning lead us to conclude that when united, as in our experiment, their forces ought to produce a still greater effect. But all the difficulties which can arise over this test are easily removed through my theory. If, for example, the pole AC of the magnet A is positive, it tries to propel the magnetic matter from the part gF into the part Fh, and, on the contrary, the negative pole ac of the magnet a tends to accumulate magnetic fluid in the part gF but evacuate it from Fh. Since these contrary forces are equal (for we have supposed that each magnet is of

---

[80] The text has gEh.

equal force and sufficiently slender for the actions of each on the iron to be almost the same) they can produce no effect. So the body F continues in the natural state and, in conformity with the theory, is not attracted. It is readily apparent that the same ought to happen in the case where AC is a negative pole and *ac* positive.

162) Let us not leave this experiment, which so outstandingly merits our attention, but let us inquire further into its phenomena and evolve their causes clearly. Suppose that the experiment is performed in the manner described (§ 160), but that the magnet *a* is not placed on the magnet A but is held some distance from it in a parallel position. Suppose further that the quantity of magnetic fluid abounding in AC is α, and the quantity evacuated from *ac* is β. If the calculation is worked out in a manner similar to § 15, it appears that the magnet A alone, or rather the part AC alone (there is no need to look now at the action of the other part AD), will repel the magnetic fluid contained in F with a force $=\frac{\alpha r}{Q}$, and the part *ac* of the other magnet will attract it by a force $=\frac{\beta r'}{Q}$. So when the magnets are held in parallel positions, the total force by which the fluid is propelled from gF into Fh will be $\frac{\alpha r - \beta r'}{Q}$. But it is immediately clear from this formula, first of all, that the force $\frac{\alpha r - \beta r'}{Q}$ exerted by the magnets joined in the manner described is less than $\frac{\alpha r}{Q}$, the force with which the magnet A alone acts; next, that as the magnets are moved closer to one another, because r' continuously increases, the repelling force $\frac{\alpha r - \beta r'}{Q}$ is diminished; and further, if the magnets are of equal strength (in which case $\alpha = \beta$) it almost completely disappears when the magnets are placed immediately on one another, for then r' becomes very nearly $=r$. It follows from this not only that a magnet will carry a smaller weight when another magnet is moved to it in the manner described than it is able to carry at other times, but also that the closer the other magnet is moved to it, the smaller is the weight it can lift, and in particular, under the conditions adduced previously, when the second magnet is completely joined to it, it gives almost no further sign of attraction, as happened in the experiments described in § 160.

163) Any person can easily convince himself that these things completely conform to experience if he is prepared to carry out the test. Hence there is clear confirmation of what I asserted above (§ 152),

that, other things being equal, any magnet will be weaker the shorter is its axis, or, the less is the distance between its poles. Imagine further that α is less than β; in this case the magnet A will be weaker than the magnet a and if we put $\alpha = \beta - \mu$ the force by which the magnets A and a, held in parallel positions, repel the magnetic fluid in F will be $= \frac{\beta(r-r') - \mu r}{Q}$. If then $r - r' = 0$, which is the case or is at least very nearly so when the magnets are actually in contact, the force by which the fluid found in F is repelled will be $= \frac{-\mu r}{Q}$, whence it is then negative and attractive; but if the magnet a is quite a long distance away from the other magnet A, because then r' will be almost evanescent, the force by which the fluid in F is repelled $= \frac{(\beta - \mu)r}{Q}$, and, because $\beta - \mu = \alpha$, hence positive, this quantity is always positive. Since while the magnet a is being moved from a greater distance up to contact with the magnet A, the force repelling the fluid in F is initially positive and finally becomes negative, there will be a certain distance between the magnets at which this force completely vanishes and becomes zero.

164) The explanation of the following phenomena must be sought from these reasonings. Take two similar and equal magnets such as described in § 160. Let the magnet a be much stronger than the magnet A. Hang from the latter an iron wire F, Fig. XXX, and then slowly move the second magnet a, from a distance, closer and closer to A. When the magnets are close to each other, at a distance of some inches, the attraction suddenly ceases and the wire F, no longer held by the magnet A, will fall spontaneously. If the magnet a is held motionless in the position in which this happens, for example $\gamma\alpha\delta$, hardly any trace of magnetic force will be observed left in A. Move the magnet a still closer, and the magnet A will recover its attractive force by degrees, and, even if the magnets are in fact in complete contact, it will carry the wire F once more.

165) From § 163, it is readily evident why these phenomena happen in this way. If the magnet a is held in position $\gamma\alpha\delta$, the action on the fluid contained in the wire F is zero, so then the wire remains in its natural state and is not attracted; but if the magnet a is at a greater distance, the fluid in F is repelled, the part gF becomes negatively magnetic as the part Fh becomes positively magnetic, and the wire F it attracted. Finally, if on the contrary the magnet moves closer, the magnetic fluid in F will be attracted, so conversely the part gF becomes positively magnetic and the part Fh negatively magnetic, and the wire F is similarly attracted (§ 154). It is further obvious from our

theory, and it will be convenient to add this here, that, as long as the magnet a is distant from A beyond the null point γαδ, the lower part Fh of the wire F ought to gain the same magnetism as the pole AC; but once the magnet a has been completely joined to the magnet A, the opposite magnetism is produced in the end Fh, the same, that is, as pole ac possesses. If an examination is set up in the customary manner with the help of a magnetic needle, and a test is made of the kind of magnetism the part Fh possesses, experience testifies that the fact is just the same.

166) What has been expounded so far, from § 160, is equally valid if the poles of the magnet are changed around; if, that is, rather than AC being taken as positive and ac negative as we supposed, the pole AC is now thought to be endowed with a negative magnetism and ac with a positive magnetism. But this is so obvious that no difficulty can arise for readers if we leave them to adapt what has been said so far to that case. Let us rather move on to other things. Place the magnet CAB, Fig. XXXI, as before on a board so that part of it projects beyond the board. Find the greatest possible size of an iron weight F which the magnet can carry with its end C. Then hold a second magnet a below the weight F hanging from the magnet A, at a distance of half an inch or an inch, so that ac is the opposite of the pole AC. It will be seen that the magnet A can carry a far larger weight than when the magnet a is removed. But if the pole ac is similar to the pole AC, the magnet will be seen to grow weaker, and will not be able to lift as great a weight as before. We can now give the cause of these phenomena without difficulty. Let for example AC be the pole of the magnet A in which fluid abounds, and ac that in which fluid is deficient. Since AC repels the fluid contained in F, it tends to produce that state in which the part Fg of the iron is emptied of fluid but the part Fh is filled. Since further the negative pole ac of the second magnet a attracts the fluid in F, this too tends to produce the same effect. For by enticing the magnetic material to itself it causes the part gF to become negatively magnetic and Fh positively magnetic. So since the forces exercised by both magnets conspire to the same effect, they necessarily achieve a greater effect than if the magnet A were acting alone. And so the part gF is evacuated to a greater degree, and Fh is filled to a greater degree than happens at other times, so a greater attraction ought to arise in this case. But if the pole ac is similar to the pole AC, in our case, the magnet A, for example, where fluid abounds, still tends to produce the same effect as before, but the other magnet now exerts an opposite effort. For since it repels fluid it tends to make the part gF positively magnetic and Fh nega-

tively magnetic. So in this case, the action of the magnet A is at least in part destroyed; therefore the iron F becomes magnetic to a lesser degree and is attracted to a lesser degree also. All this reasoning is very readily adapted to the other case where AC is not the positive but the negative pole of the magnet A.

167) In this experiment, substitute in place of the magnet a a piece of soft iron of a rather large size, possessing no magnetic force. Held in the position where the magnet a was previously, because of the proximity of the magnet A, it likewise becomes magnetic, in such a way that its pole ac is opposite to AC, but ad is similar to it. (§ 148. 154). So the situation is the same as if a magnet were held in this position, whence the same phenomena should also happen. If an iron body of the same kind is moved to the magnet A in the fashion described, the magnet A will carry a greater weight than if the iron is removed (§ 166). The celebrated *de Reaumur* was the first to observe this, and he found that a magnet can easily lift a mass of iron which has been placed on iron, e.g. on an anvil, which it could not lift if it were put on a hand or on a wooden board.[81]

168) From the pole AC of the magnet A, Fig. XXXII, hang an iron body F, and hang weights to find the force by which it is attracted. Hold the magnet a close to the iron F in the position shown in the figure, so that the pole ac is the contrary of AC. When the experiment is performed, it will be observed that now the magnet A attracts the iron F with a much smaller force. If, e.g., the pole AC is positive, by repelling magnetic material, it induces in the body F a state such that the part gF is evacuated of fluid, and Fh is filled with it. Since when the magnet a is moved to the side, it attracts fluid with its part ac, it is easily understood that the effect produced by the magnet A will be completely destroyed and some part of the fluid caused to cross back from Fh into gF; anyone willing to give the matter a little attention will readily understand this. Since in this case, F is prevented by the magnet a from becoming as magnetic as at other times, it ought to be attracted altogether more weakly.

169) The same phenomenon occurs if a non-magnetic body, similar and equal to F, is used in place of the magnet a. For this immediately changes into a magnet whose pole ac is opposite to the pole AC. Join the iron bodies a and F together and it is evident that the part gF will still prevent the adjacent part ac from becoming as greatly magnetic as at other times, and the part ac will do the same in respect of gF. Therefore it is readily clear that a body of double thickness ought to be attracted not by an exactly double force but by a force

[81] Réaumur, *HAS*, 1723, Mém. p. 126.

somewhat smaller than this. Further, the same reasoning is easily fitted to the case where 3, 4, 5, or more such bodies coalesce into one, and it is generally clear from this that if the thickness of the body is increased m times, the attraction on that account does not increase m times, but remains somewhat less than this.

170) Thus the attraction changes according to some fixed law as the thickness is increased; but this law cannot be determined as long as we do not know the function according to which the repulsion of magnetic fluid is exerted. Meanwhile it may be seen without difficulty that there is a certain thickness for which the attraction is a maximum. Let the body F, Fig. XXXIII, be attracted by a force $= A$, and its thickness be increased by an infinitely thin plate cad which, if alone, would be attracted by a force $= a$. This plate joined to the body F will lessen the magnetism of the body F, and at the same time lessen by an infinitely small amount, $= \gamma$, the force by which it is attracted. Similarly the body F will prevent the plate a from becoming magnetic to the same degree as at other times. And so after the plate a is joined to the body F, it will no longer be attracted by a force $= a$, but by a smaller force. Let us suppose that the force by which the plate a is attracted is diminished by the action of the body F by a quantity $= \epsilon$, and the total force by which the body F is attracted, after its thickness has been increased by the plate a, will be $= A + a - \gamma - \epsilon$. It is evident that the quantity a is constant, if the plates increasing the thickness of the body F are assumed to be of constant thickness, but that the quantity ϵ depends on the magnetic force of the part gF and increases as that increases, while the quantity γ which depends on a constant force a can likewise be held as constant. If then it is supposed that the magnetic force of the part gF increases with the increased thickness, at the same time ϵ will continuously increase and come finally to the point where $\epsilon + \gamma$ becomes equal to a. But if this happens, $a - \gamma - \epsilon$, or the increment in the force by which the body F is attracted, will be $= 0$, and the thickness of the body F will then be that which produces the maximum attraction.

171) We are forced to stay with these rather general considerations, for what precisely is the thickness which corresponds to the maximum attraction cannot be determined until we know the function according to which the magnetic fluid acts, and this cannot be determined by theory alone. It is very well known however that what the theory has taught us actually takes place. For it is common knowledge that some of those who founded the study of magnetic phenomena have observed this phenomenon and have been struck by it.[82]

[82] Cf. Musschenbroek, *Dissertatio physica experimentalis de magnete,* Leyden, 1729, p. 45.

172) Now that we have sufficiently examined the action of a magnet on iron not previously imbued with magnetic force, the laws of action of two magnets on one another remain to be discovered. Let us suppose first of all that the magnets are opposed to one another, so that the positive pole of the former, AC, Fig. XXXIV,[83] faces the negative pole BE of the latter. Then the magnet A repels the magnetic material contained in B, whence the evacuated part EB ought to be evacuated to a still greater degree, and similarly the magnet B causes magnetic material to be accumulated in AC to a greater degree than before, by attracting the fluid to be found in A. Let us indicate the natural quantity of magnetic fluid appropriate to the parts AC and AD of the magnet A by Q, and that appropriate to the parts BE, BF of the other magnet B by q. Further, let the quantity of magnetic fluid which abounded in AC and was deficient in AD before the magnets were moved to one another be called α, and that deficient in BE and abounding in BF be called β. Finally, let the quantity of fluid forced to cross from AD into AC through the action of the magnet B, when the magnets have been moved toward one another, be $=\eta$, and that which the magnet A forces to cross from BE into BF be $=\theta$. If now in the general formula of § 43 we put $a=\alpha+\eta$, $b=-\alpha-\eta$, $c=0$, $e=-\beta-\theta$, $f=\beta+\theta$, $g=0$ or, which comes to the same thing, we put $a=b=\alpha+\eta$ and $e=f=\beta+\theta$ in the formula $\dfrac{+(ar-br')e-(a\rho-b\rho')f}{Qq}$, we get the formula which expresses the force by which the magnets A and B attract one another, and so it $=\dfrac{(\alpha+\eta)(\beta+\theta)((r-r')-(\rho-\rho'))}{Qq}$.[84]

173) Provided $(r-r')$ is always $>(\rho-\rho')$, it is self-evident that this force will always be positive. Therefore, since I have proved (§ 149) that this undoubtedly is always the case, it follows that magnets whose opposite poles are turned toward one another always attract each other. But although this is generally true, as a result certain phenomena can occur which have an unexpected outcome. Thus e.g., at first sight, there would hardly be anyone who would hesitate to assert, if all the other circumstances were similar, not only that the stronger the magnet B was, the greater ought to be the force by which it was attracted, but also that if the body B were destitute of all magnetism, it ought to be attracted by a smaller force than a magnet like and equal to itself would be. Although these things can seem self-evident,

---

[83] The text has Fig. XXXV.

[84] The text has $\dfrac{(\alpha+\eta')(\beta+\theta)((r-r')-(\rho-\rho'))}{Qq}$.

nevertheless not only have I myself rather often observed the contrary (and I find that it has also been noted by others) [85] but also, if we look more closely and more carefully at the matter, we are easily convinced that these things are false. For the strength of the force by which the magnets attract each other depends to a large extent on the greater or lesser facility with which magnetic fluid moves in their pores. For a proper understanding of how it can happen that a weaker magnet is at times attracted with a greater force than is a stronger one, or that bare unmagnetized iron is attracted with a stronger force than a reasonably strong magnet, let us suppose that we substitute for the magnet B an equal and similar magnet b, but one such that magnetic material moves much more easily in its pores. If we suppose that before this magnet is moved to A, magnetic fluid is already deficient in its part be and abundant in bf by a quantity$=\beta'$, and we apply here all the reasoning taken over from § 172, now calling the quantity of fluid which crosses from AD into AC, η', and that expelled from eb into bf, θ', we will obtain a formula similar to the one discovered above. The force by which the magnets attract one another will$=\frac{(\alpha+\eta')(\beta'+\theta')((r-r')-(\rho-\rho'))}{Qq}$.

174) Anyone who cares to give a little attention to this matter will easily perceive that the magnitude of η' depends on the magnitude of $\beta'+\theta'$, so that the greater this is, the greater η' ought to be also. From what has been dealt with above, it is sufficiently manifest that the force with which the magnet b attracts the fluid contained in A and forces it to cross from the part AD into the part AC increases while $\beta'+\theta'$ increases. If we admit this, it is readily apparent that

1) If $\beta'=\beta$, because of the greater facility with which we suppose the fluid contained in the magnet b to move, θ' would be$>\theta$, whence we obtain $\beta'+\theta'>\beta+\theta$ and hence, from what we have just proved, $\alpha+\eta'>\alpha+\eta$ also. It follows from this that the force by which the magnet A attracts the magnet b, which$=\frac{(\alpha+\eta')(\beta'+\theta')((r-r')-(\rho-\rho'))}{Qq}$, is greater than the force by which this same magnet attracts the magnet B, which has been discovered above to be equal to $\frac{(\alpha+\eta)(\beta+\theta)((r-r')-(\rho-\rho'))}{Qq}$. Then

2) Even if β' is supposed less than β, nevertheless on account of the reasoning we have just adduced, θ' will be$>\theta$, whence it can happen that $\beta'+\theta'>\beta+\theta$. But if this is so, $\alpha+\eta'$ will also be$>\alpha+\eta$,

---

[85] Musschenbroek, *Dissertatio* . . . *de magnete,* pp. 43–44.

so, of necessity (and this is clear in the same way as in the preceding number), the magnet b will be attracted by a greater force than the magnet B. Finally,

3) If β' is supposed to be exactly $=0$, in which case b possesses no magnetism, it is nevertheless clear that, if b consists of quite soft iron, but B of rather hard iron (in which case θ' will be quite large, but θ very small), it will be altogether possible that $\theta' > \beta + \theta$. Then because $\beta' + \theta' > \beta + \theta$, $\alpha + \eta'$ becomes $> \alpha + \eta$. So it follows once again that it is possible for an unmagnetized piece of iron b to be attracted by the magnet A with a greater force than is the magnet B.

Someone who has understood these things properly will be able to apply them without difficulty to the case where either a weaker magnet or a plain unmagnetized piece of iron is substituted for the magnet A, and he will find that exactly the same things happen. So it has been my lot again to give with the aid of our theory a simple explanation for a phenomenon which those who knew of it through experience alone could have found nothing but extremely difficult.

175) To complete our discussion about the action of a magnet on another magnet, it remains for us to consider the case where the magnets turn the same or like poles toward one another. Let us suppose first that the magnets AD, EH, Fig. XXXV, have turned positive poles toward one another, that is, poles in which magnetic fluid abounds. Therefore either of these magnets repels the magnetic fluid contained in the other. So the magnet AD forces some part of the magnetic material abundant round the end E of the other magnet to move toward the interior, and in turn the magnet EH exercises a similar action on the magnet AD. Let us call the quantity propelled in this way out of the end A, η, and that forced to cross toward the interior of the magnet from the end E, θ, and for the rest let us retain the notation used previously. But if we now imagine that each magnet is divided into three parts AB, BC, CD, and EF, FG, GH, and we say that the quantity of fluid abundant in $AB = \frac{\alpha}{2} - \eta$, in $BC = \frac{\alpha}{2} + \eta$,[86] in $CD = -\alpha$, and similarly the fluid abundant in $EF = \frac{\beta}{2} - \theta$, in $FG = \frac{\beta}{2} + \theta$, in $GH = -\beta$, and then in the general formula of § 43 we substitute the three former values for a, b and c, and the three latter values for e, f, g, we will get the formula expressing the force by which the magnets AD and EH attract each other.[87]

---

[86] The text has $-\frac{\beta}{2} + \eta$.

[87] Aepinus here assumes a highly artificial distribution of fluid for which he offers no justification at all!

ATTRACTION AND REPULSION   339

176) This comes to the same thing as if in the formula of § 43, c and g were first put negative, in which case there is obtained
$$\frac{-(ar+br'-cr'')e-(a\rho+b\rho'-c\rho'')f+(a\mathfrak{r}+b\mathfrak{r}'-c\mathfrak{r}'')g}{Qq},$$
and then $a=\frac{\alpha}{2}-\eta$, $b=\frac{\alpha}{2}+\eta$, $c=\alpha$, $e=\frac{\beta}{2}-\theta$, $f=\frac{\beta}{2}+\theta$, $g=\beta$. For different values of the quantities η and θ (and it is self-evident that they are variable), this formula will be at one time negative, at another zero, at another positive. To make this somewhat clearer, let us work at developing the cases in which each of them can happen. I say that as long as η and θ are not greater than $\frac{\alpha}{2}$ and $\frac{\beta}{2}$ respectively, that is, as long as the values of a and e are themselves positive, the formula must be constantly negative. For

1) r'' is $<r'$ and $c=a+b$, so cr'' is $<ar'+br'$. Further, since $r'<r$, and so $ar'<ar$, the conclusion a fortiori is that $cr''<ar+br'$ whence the quantity $ar+br'-cr''$ is always positive. In an exactly similar way we find that the quantities $a\rho+b\rho'-c\rho''$ and $a\mathfrak{r}+b\mathfrak{r}'-c\mathfrak{r}''$ are positive. Then

2) Since $\mathfrak{r}'$ and $\mathfrak{r}''$ pertain to distances, equally different, but greater than those to which pertain ρ' and ρ'' (§ 43), $\rho'-\rho''$ will be $>\mathfrak{r}'-\mathfrak{r}''$ (§ 140ff. and § 149). So because $a+b=c$, $a\rho'+b\rho'-c\rho''$ will be $>a\mathfrak{r}'+b\mathfrak{r}'-c\mathfrak{r}''$. Similarly $\rho-\rho'>\mathfrak{r}-\mathfrak{r}'$ and $a\rho-a\rho'>a\mathfrak{r}-a\mathfrak{r}$.
Therefore since we have $a\rho'+b\rho'-c\rho''>a\mathfrak{r}'+b\mathfrak{r}'-c\mathfrak{r}''$
and $a\rho-a\rho'\qquad>a\mathfrak{r}-a\mathfrak{r}'$,
by addition $a\rho+b\rho'-c\rho''>a\mathfrak{r}+b\mathfrak{r}'-c\mathfrak{r}''$.
By a similar method, it may be proved that $ar+br'-cr''$ is also $>a\mathfrak{r}+b\mathfrak{r}'-c\mathfrak{r}''$ and $>a\rho+b\rho'-c\rho''$. To express this more briefly, let $(ar+br'-cr'')=P$, $(a\rho+b\rho'-c\rho'')=Q$, and $(a\mathfrak{r}+b\mathfrak{r}'-c\mathfrak{r}'')=R$. Therefore, since from what has been proposed so far, $P>R$ and $Q>R$, Pe will be $>Re$ and $Qf>Rf$, so $Pe+Qf>Re+Rf$, or, because $e+f=g$, $Pe+Qf>Rg$. It is immediately obvious from this that under the assumed hypothesis the formula discovered above expressing the attractive magnetic force will always be negative.

177) If η and θ increase to the point that either a or e becomes negative, our formula can become positive. To make this clear, let e first of all be negative and $=-\gamma$; because a is still positive, what was demonstrated in nos. 1 and 2 of the previous § still holds good. Then because $e=\frac{\beta}{2}-\theta=-\gamma$, θ will be $=\frac{\beta}{2}+\gamma$, and $f=\beta+\gamma$, so the numerator of our formula, $-Pe-Qf+Rg$, becomes $+P\gamma-Q\beta-Q\gamma+R\beta$. Because $R<Q$, $R\beta-Q\beta$ is negative, and because $P>Q$, $P\gamma-Q\gamma$ is positive. So it is clear that γ can be so determined that $P\gamma-Q\gamma$ be-

comes $>Q\beta-R\beta$, and then our formula becomes positive. Again, our general formula (§ 176), by transposing the terms, can be expressed so that it becomes $= \dfrac{-(er+f\rho-g\mathfrak{r})a-(er'+f\rho'-g\mathfrak{r}')b+(er''+f\rho''-g\mathfrak{r}'')c}{Qq}$, [88] which for the sake of brevity we shall indicate by $\dfrac{-La-Mb+Nc}{Qq}$. It is readily obvious that by using an entirely similar procedure as in § 176, it can be demonstrated that as long as e is positive, the quantities L, M and N will be positive, and $L>M$ and $M>N$. So if $a=\dfrac{\alpha}{2}-\eta$ becomes negative $=-\delta$, η will be $=\dfrac{\alpha}{2}+\delta$ and $b=\alpha+\delta$, whence the numerator of the formula $=+L\delta-M\alpha-M\delta+N\alpha$. Because $L>M$ and $M>N$, $L\delta-M\delta$ is positive and $N\alpha-M\alpha$ is negative, whence it is self-evident that δ can be determined in such a way that $L\delta-M\delta>M\alpha-N\alpha$, and the formula expressing the attractive force of the magnets then becomes positive.

178) There is no advantage in proceeding further with this discussion, for I have done enough in showing generally that it is possible, under certain circumstances, for the value of the force by which the magnets attract each other to change from negative to positive. And so when two magnets turn positive poles toward one another, there are various possible outcomes of the experiment, for they may repel one another, but it is also possible that they will attract one another, and even that they will not act on one another at all. However, these latter cases do not take place unless η and θ are so great that a and e become negative (§ 176); and these quantities become greater if

1) The distance between them is diminished;

2) The force of the magnets acting on one another becomes greater;

3) If the magnetic fluid can yield more easily to repulsion, that is, if the magnets are made of rather soft iron.

It is therefore clear that the repulsion of the magnets can change into an attraction if

1) The magnets AD and EH are moved closer to one another;

2) We substitute for the magnet AD another better magnet of greater strength; [89]

3) If we substitute for the magnet EH another magnet made of rather softer iron.

[88] The text has $\dfrac{-(er+f\rho-g\mathfrak{r})a-(er'+f\rho'-g\mathfrak{r}')b+(er''+f\rho''-g\mathfrak{r}'')c}{Qq}$.

[89] In the text, these two cases are listed in the opposite order.

As a result we can easily see the reason for certain paradoxical phenomena, for it is readily clear why at times magnets repel each other when they are rather far removed from one another, but attract each other when they are brought closer (no. 1), or why at times a stronger magnet can attract another which it repels when it is weaker (nos. 2. 3).

179) All this can be applied to the case where magnets turn toward one another those extremities which have been emptied of magnetic fluid. In this case, each magnet attracts the fluid contained in the other; so in the magnet AD, as in the magnet EH, the magnetic matter from the remoter parts must move toward the ends A and E. And so in a manner similar to the previous case, we can suppose each magnet to be divided into three parts, and if in the general formula of § 43 we put $a=-\frac{\alpha}{2}+\eta$, $b=-\frac{\alpha}{2}-\eta$, $c=\alpha$, $e=-\frac{\beta}{2}+\theta$, $f=-\frac{\beta}{2}-\theta$, $g=\beta$, we will obtain the expression for the force by which each magnet attracts the other. It comes to the same thing if first of all in the general formula a, b, e and f are negative, in which case we get

$$\frac{-(a\mathfrak{r}+b\mathfrak{r}'-c\mathfrak{r}'')e - (a\rho+b\rho'-c\rho'')f + (a\mathfrak{r}+b\mathfrak{r}'-c\mathfrak{r}'')g}{Qq},$$

and then a becomes $=\frac{\alpha}{2}-\eta$, $b=\frac{\alpha}{2}+\eta$, $c=\alpha$, $e=\frac{\beta}{2}-\theta$, $f=\frac{\beta}{2}+\theta$, and $g=\beta$.

180) Since the formula evolved here coincides completely with the one evolved in § 176, everything said earlier (§ 176. 177. 178) can be applied here, and it would be superfluous to repeat it. So I add nothing further regarding this case. Writers who describe the properties of a magnet usually assume the two following rules, entirely similar to those used by writers dealing with electricity, and mentioned above in § 147: like poles of magnets repel one another, but unlike poles attract one another. As for the last rule, there is nothing which could prevent our accepting it as true, for it is in complete agreement both with the theory and with constant experience (§ 173). But it is sufficiently clear from § 177 that the former, as it is usually expressed, can be deceptive. Its converse must be fully recognized as true, that is, poles which repel each other are like. For only in this case is it possible for magnets to repel one another, hence it is sufficiently clear that in all other cases there is always an attraction between them.

181) That experience agrees most happily with our theory and with the consequences deduced from it can easily be tested. It is sufficiently, and more than sufficiently, known that unlike poles of magnets attract each other. But it will be worthwhile to adduce cer-

tain tests proving the truth of the propositions evolved in § 177. 178. Set a magnetic needle CD, whose pole C is north, and pole Ð south, close to the magnet AB, Fig. XXXVI. The needle at once takes up a position such that the south pole D faces directly toward the north pole A of the magnet. Then with the help of the wire CE fitted to the point C, or by any other means, rotate the needle round the point F and move the point C closer to the pole A. It will be observed initially that after the needle has been turned e.g. to the position cd, it still strives to recover its former position CD, which is an indication that the point C is still continuously repelled. But if the point C is brought perpetually closer to the pole A, it will come finally to a definite terminus, e.g. $\gamma\delta$, where the repulsion suddenly changes to an attraction, and the needle arranges itself of its own accord so that it adopts a position contrary to its former one, that is, the position where the point C faces the pole A. The experiment is also successful if the pole A is south, instead of north as we have supposed here. The magnet must be sufficiently powerful and the needle must be moved sufficiently close; if not, the repulsion cannot change into attraction, and the needle will of its own accord always revert to the position CD. The celebrated *Musschenbroek* observed this same alteration of the repelling force to an attractive force with the help of a very mobile balance, and was much struck by it.[90]

182) I have assumed so far that the positively magnetic part in any magnet is always of a magnitude equal to the negatively magnetic part, and that the fluid is distributed uniformly in each of these parts. But I have myself given warning already that I do not consider either hypothesis to be altogether true, and I suppose them only in order to facilitate the calculations I have introduced to avoid the long-windedness of the usual language. It is evident that the parts can be unequal, and the magnetic fluid distributed unequally. So far, I have been able to assume these hypotheses and the certain truth of the things deduced from them, even though they do not altogether agree with the hypothesis of nature, as long as I was concerned solely with the general laws of magnetic action and not with a determined quantity of this action. For it is readily clear that quite similar conclusions would emerge if the positive and negative parts of the magnet were unequal and the fluid was not distributed uniformly. But there remain certain experiments of mine which demand a closer look at the inequality of the parts and of the distribution of magnetic fluid, if we want to give a correct account of them.

[90] Musschenbroek, *Dissertatio . . . de magnete,* pp. 29–30. Musschenbroek attributed the discovery of this phenomenon to Polinière, citing his *Expériences de physique,* p. 279.

ATTRACTION AND REPULSION

183) Let us begin with a rather careful description of these experiments. To find what happens to magnets when they approach each other and act on one another, I tackled the tests in the following manner. I took a cylindrical iron wire CD, Fig. XXXVII, with a base diameter of about a line, and 10 in. 7 decimal lin. long, and magnetized it so that the magnetic center was at m, and Cm was the south pole, mD the north pole, the former being 5 in. 8 lin. long. Then opposite the north pole of the wire I placed the north pole A of a quite strong magnet AB at a distance AD. I observed that the magnetic center m moved appreciably toward C, as if fleeing from the pole A. Further, the closer I moved the magnet to A, the more the magnetic center was propelled toward C. So when AD=

1 in. 5 lin.,	Cm was = 5 in. 6½ lin.
0 – 5 –	– – 4 – 9½ –
0 – 1 –	– – 4 – 0 –
0 – 0 – that is in actual contact	– 3 – 0½ –

Similarly in another experiment, when the length of the wire was = 8 in. 6 lin. and Cm, before the magnet was moved up, = 4 in. 3 lin., I found that when

AD = 1 in. 5 lin.	Cm = 3 in. 9½ lin.
1 – 0 – –	– 3 – 7 –
0 – 5 – –	– 3 – 2 –
0 – 0 – –	– 2 – 1 –

I set up a great number of experiments of this kind, with varied circumstances too, using the south pole in place of the north. Since without exception they had the same success, they do not need to be described.

184) In these experiments, if the magnet AB was moved very close to the end of the wire D, or if it actually touched it, then after the removal of the magnet, the wire was found to have three poles and two magnetic centers, m and n. Cm was a south pole, mn north and nD south once more. Thus in the first of the experiments recounted in the preceding §, after the wire touched the magnet and the magnet was then removed, the south pole Cm was 3 in. ½ lin. long, the north mn was 6 in. 5 lin. long, and the south nD 1 in. 2 lin. long. In the second experiment, after contact, the south pole Cm was 2 in. 1 lin. long, followed by the north pole mn 5 in. 2 lin. long, and then the south pole nD 1 in. 1 lin. long. A large number of other experiments had completely similar results. In these tests I used rather long wires, but when I used shorter ones, the outcome was somewhat different. Thus e.g. when I used an iron wire 4 in. 5 lin. long and other shorter

ones, the magnetic center m, as in the previous experiments, moved closer toward the end C, the closer the magnet approached. But when I carried out a new test after the wire had touched the magnet, I did not find three poles and two magnetic centers as usually happened in the longer wires, but only a single center and two poles. But the poles were changed; for the extremity Cm which before had been south, was now north, and mD possessed the character of a south pole.

185) In this experiment, a remarkable thing happened whenever I used shorter wires, in that the place of the magnetic center remained where it was as long as the magnet was near the wire, but when a new test was made following the removal of the magnet, it was found to have moved spontaneously away from the end C somewhat toward D. Thus e.g. when I was using a wire 3 in. 2 lin. long, when the magnet was one inch distant from the end D, the distance of the magnetic center from the end C, or Cm, was $=9$ lin., but when the magnet was taken away, Cm immediately increased again to 1 in. 5½ lin. In another instance with a wire 2 in. 1 lin. long, when the magnet was brought up to a distance of 2 inches, Cm was $=7$ lin.; but when the magnet was taken away, it increased to a length of 1 in. ½ lin.

186) Since we are about to deduce the causes of these phenomena from our theory, we ought to warn in advance that we have been forced to refrain from an exact calculation, for this cannot be carried out as long as the function of the magnetic repulsion is unknown. Meanwhile it is within our power to show very clearly the agreement of these phenomena with the theory, so that the principles of the magnetic theory evolved here can be confirmed from this in no small degree. Therefore let us imagine a wire CD and a magnet A, Fig. XXXVIII, with the positive poles turned to one another first of all. Thus in Dm the magnetic fluid abounds beyond the natural quantity by a certain fixed quantity which we take to be uniformly distributed throughout the whole part mD. Let us suppose now that the cylinder of fluid mD is congealed and becomes solid, yet moves in the wire in exactly the same way with no greater difficulty than before.[91] So, even before the magnet A is moved up, the whole cylinder Dm will tend toward the evacuated part mC which attracts it, although in fact it will not cross toward the part mC because of the difficulty with which magnetic fluid moves through the pores of iron. But as the magnet A approaches it will repel the whole cylinder whence this will

---

[91] Newton had recourse to a similar device to demonstrate one of his theorems (*Mathematical Principles of Natural Philosophy*, Cambridge, 1934, pp. 394–95). So, too, did Henry Cavendish, in his paper on the theory of electricity (*Phil. Trans.*, LXI, 1771, 614–15).

be pressed toward C by a greater force than before, and will in fact make some progress. But finally this progress will stop. And since the force with which the magnet A propels the cylinder mD diminishes as the distance increases, the more the cylinder yields, the less is the force by which it is pressed, whence it ought to arrive finally at a place where the force propelling it toward C has decreased to the point where it cannot overcome the difficulty with which the fluid moves through the pores of the iron, and then its movement will stop completely.

187) Let us suppose that the cylinder Dm propelled in this manner, has been pushed into the position np, and let us consider that now it is instantaneously thawed out, so that any particle of the fluid can freely obey the force pressing on it. If then we look first at the particles clinging around p, it is clear that they must remain at rest, for they will be pressed on toward C by a force which is such that it cannot overcome the difficulty with which the fluid moves through the pores of iron. If this were not so, the whole cylinder would have been pushed somewhat further still toward C. The particles to be found around n, which will be pressed toward D by the attractive force of the evacuated part Dn, and by the filled part pn, will be able to flow toward D as far as is permitted by the repulsive force of the magnet A which is opposed to this flow. Anyone who cares to meditate on this himself a little more attentively will now be able to perceive that a state will be produced where in general the supply of fluid around p will be greater than that around D, and as circumstances are varied, the fluid around D can possibly at one time be abundant to a lesser degree than round p, and at another time it can actually be deficient. For the greater the repulsive force exercised by the magnet A, the less must be the supply of fluid around D.

188) From this it is immediately clear why, when the magnet A approaches closer to the end D, the magnetic center ought to be propelled so much the more toward the end C. But at the same time it is obvious that if the magnet possesses a sufficiently great force and is moved closer, it can happen that the end D of the wire will be emptied of fluid and become negatively magnetic; in this case the wire ought to gain two magnetic centers. As a result, a quite simple reason can be given for the phenomena recounted in § 183 and at the beginning of § 184. It may be readily imagined that, if the wire were shorter and the magnet were moved sufficiently close, it could happen that the force propelling the cylinder Dm would become so great that it would be pushed right to the very end C, in fact would be forced in part to issue forth from the end C. But if this happened, after the per-

formance of the experiment, the end C would not be negatively but positively magnetic, and it would not have three poles, as happens when the wire is longer, but only two, of which Cn would be positive and nD negative, Fig. XXXIX.[92] The experiments I described at the end of § 184 are pertinent here.

189) It seems possible that if the wire were quite short and the magnet of considerable strength, the whole cylinder nC would be expelled from the end C, in which case the whole wire CD would become negatively magnetic, yet in such a way that it was evacuated of fluid to a greater degree around D and to a less degree around C. Very many tests have taught me that this happens at times.[93] Thus, e.g. when I used a wire 2 in. 1 lin. long, which before the experiment had its magnetic center in m, such that Cm = 1 in. 1 lin., Fig. XXXVIII, when the magnet was moved to a distance AD

= 3 in. 0 lin.	Cm was	= 0 in. 9 lin.
2 – 0 –	–	– 0 – 7 –
1 – 0 –	–	– no center in the wire
0 – 5 –	–	– no center
0 – 2½ –	–	– no center
0 – 0 –	–	– no center

What ought to happen in this case when the magnet is taken away is easily judged from what I said at the conclusion of Chap. I., § 99.

190) If the cylinder np is propelled forward to the point where a space Dn is emptied of fluid and becomes of a significant size with respect to the emptied part Cm, Fig. XXXVIII, the cylinder will be pressed toward C by the attractive force of the part pC, which we may call m, and by the repulsive force of the magnet A, which we may call R, but toward D by the attractive force of the part Dn, which may be indicated by n. Therefore the whole force propelling the cylinder toward C = R + m − n; after the magnet has been removed this becomes m − n. If n is greater than m, which can quite easily happen, this force will be negative, and the cylinder np will attempt to return toward D when the magnet is removed. Therefore, if this force is strong enough to overcome the difficulty with which fluid moves through the pores of the wire, the cylinder will in fact proceed toward D after the magnet has been removed. This is what happened in the experiment described in § 185, and the reason for it is abundantly clear here. It is obvious at the same time that, for this phenomenon to occur, the part nD must significantly overcome the part mC, for

[92] The text has Fig. XLI, which is clearly wrong. For the purposes of the present discussion, Fig. XXXIX is perfectly appropriate, but see also the note at § 191 below.

[93] This claim seems to be quite unique in the history of magnetism.

ATTRACTION AND REPULSION

the force n has to be significantly greater than the force m to produce this result. It is clearly obvious from this why this phenomenon is not observed unless a magnet has been moved quite close and a rather short wire is used, for in a longer wire the evacuated part Cm remains too great for this effect to arise. Also I have never noticed with longer wires that when the magnet has been removed, the magnetic center has returned sensibly toward C; I have managed to observe this only in shorter wires.

191) If we now examine the second case, where the negative pole B of a magnet, Fig. XXXIX, is moved to the negatively magnetic end C,[94] it can easily be shown that similar things take place. For the particles existing in the replete part round m will be urged to cross toward C by the attractive force of the part mC and by the repulsive force of the part mD, even before the magnet is moved up. But since this force is not strong enough to overcome the difficulty with which the fluid moves through the poles, the fluid does not in fact cross over. But when the attractive force of the magnet B is added, together they overcome this resistance, and fluid ought actually to flow forward. As the particles get closer to the pole B, the attractive force always increases, and they cannot stop but must push on toward the end C. The magnet exerts a similar action also on the other particles constituting the cylinder Dm, but it solicits the more remote ones with less force. Thus e.g. a particle located near n will be urged toward C by a much smaller force than those found around m; it is pushed toward C by the attractive forces of the part mC and the magnet B, and by the repulsive force of the part nD. But it is evident that the two former forces, because of the increasing distance, and the third, because $Dn < Dm$,[95] are less than the three forces which propel a particle near m, and moreover the repelling force of the part nm reduces still further the solicitation toward C. Therefore, proceeding toward D, there will finally be a terminus such that particles situated beyond it will be urged toward C [96] by a force too small to overcome the difficulty of movement. Let that terminus be n, and the whole cylinder Dn ought to remain motionless, but nm will be as if broken

---

[94] The text has D. For the purposes of the present discussion, Fig. XXXIX is wrongly labeled. The labeling should be as in Fig. XL, viz.

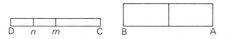

Figure 7. Corrected version of Aepinus's Fig. XXXIX.

[95] The text has $Dn > Dm$.

[96] The text has DC.

off, and the particles comprising it will be distributed throughout the whole part nC in such a way that fewer are found round n, and more round C.

192) It is clear that in this case the magnetic center ought to be pushed somewhat toward D. But since the particles of fluid in nC strive to move toward C, it can happen that if the attractive force of the magnet is sufficiently great, so great a quantity of fluid crosses toward this end that it becomes positively magnetic. But if the force is weaker, it can also happen that the end C remains negatively magnetic, yet in a weaker degree than it was before the approach of the magnet. This of course happens if the fluid which crosses toward C is not enough to overcome the deficiency. And so it is clear why in this case too the magnetic center flees, as it were, from the pole B, and if the magnet becomes sufficiently strong and it is moved sufficiently close, the end C becomes positively magnetic in such a way that the magnet gains three poles and two magnetic centers. It can happen further that the force of the magnet B is so great, and the wire CD so short, that the point n falls at D, and in this case the whole cylinder mD ought to cross toward C, so that the end D becomes empty of fluid and C becomes replete with it, in which case the wire will have not 3 but only 2 poles, and a single magnetic center. If finally the force of the magnet B becomes so great that particles of magnetic fluid clinging outside the wire near D are attracted by a force greater than the resistance which impedes their entry into that end D, it can happen that the whole wire becomes replete with magnetic fluid beyond the natural quantity and becomes positively magnetic, though in a greater degree around C than around D. What ought to happen if the magnet is then removed is readily clear from § 98.

193) I have supposed so far that the magnetic fluid is uniformly distributed in any part of the wire in which it either abounds or is deficient, but I have already warned on several occasions that this is a gratuitous hypothesis and that it cannot occur in nature. It can be tolerated as long as the conclusions deduced from it are such as are little changed even if the fluid is considered unequally distributed, as is really the case. But there remains a phenomenon which will be difficult to unravel unless we look to some extent at the unequal distribution of fluid. Let it be considered that in the case we have handled so far, the magnet B acts with such force and is so close to the wire that the latter gains three poles, a positive one Cm, Fig. XL, a negative one mn, and a positive one nD, and that the supply of fluid accumulated round nD is significantly less than that abounding round C. We can then represent the density of the fluid in any part of the

wire by the curve *fmng*, any ordinate of which indicates the extent to which fluid either abounds beyond the natural quantity, or is deficient, at the point to which it corresponds, ordinates falling above the axis being understood to be positive, and those below the axis negative. Any particle of fluid found in the part nm, e.g. p, is urged toward n by the repulsive force of the part mC, but it is pushed toward m by the part nD. Let the former of these forces $= m$, the latter $= n$, and the total force by which the particle p is enticed toward n will be $= m - n$. It is evident that this force will be positive if m is significantly greater than n, that is, if a much greater supply of fluid is found in the part mC than in nD. And so the force $m - n$ can be not only positive but also, as is self-evident, great enough to be able to overcome the difficulty with which fluid moves in the pores of iron. If this happens, the particle p ought to cross toward n, unless it is held by the attraction of the magnet B. Therefore if the magnet is taken away, the particle p, and other particles in its vicinity, will obey the force acting on them, and will in fact move toward n. So the density of fluid at the point n will be increased, and although there was no ordinate corresponding to this point before, there will now be a positive one. It is obvious that when this happens, the magnetic center ought to progress somewhat toward C.

CHAPTER III

Concerning the Communication of Electricity and Magnetism

194) It is abundantly established through experience that electrical and magnetic bodies can communicate their force to other bodies; in fact, in the magnetic case, I shall show later that all known methods of making a body magnetic can be reduced solely to communication, in such a way that unless there were some bodies endowed with magnetism of their own nature and without the help of art, it would be completely beyond our power to make any body magnetic. And so among the doctrines most worthy of examination must be those concerning the communication of the forces whose theory we are investigating here, and they merit closer discussion because, on the one hand, a very clear light can be cast on the similarity of the forces of electricity and magnetism, and on the other, the whole theory of artificial magnetism is to be found amongst them. I begin with an exposition of the laws of communication of the electrical force, and afterward I will proceed to draw out the laws which are valid for magnetism, for there is then an easy passage to them.

195) Let a certain body A, Fig. XLI, be positively electric, and consider it to be moved up to another body B surrounded on all sides by bodies electric *per se,* and suppose there is a certain cause impeding the transfer of fluid from one body into the other. If we examine what ought then to take place, it is clear from the things proposed above, which it would be superfluous to repeat, that the repulsive force of the fluid abounding in A would make the electric fluid in B leave the end C and accumulate toward the other end D. And so it is evident that a particle of electric fluid clinging at the very boundary between the bodies, e.g. m, will attempt to cross from the body A to the body B, and this attempt will succeed unless there is some cause preventing it. Therefore let us imagine the removal of the impediment we have so far supposed, and then we will need to distinguish various cases. Either both the body A and the body B are non-electric *per se,* or just one of them is, or finally both are electric *per se.* If the first is the case, it is clear that fluid ought actually to cross from A to B. It is readily clear that the fluid cannot stop and be brought to rest until it has been uniformly distributed throughout the whole combination of the bodies A and B in such a way that it has the same density in

every part.[97] For if the fluid is considered to have a greater density in one part than in a neighboring one, particles of fluid will be urged toward the latter part and there can be no equilibrium until the fluid has been reduced to a uniform density throughout.

196) Let the quantity of natural fluid pertaining to the body $A = q$, and that pertaining to $B = Q$, and let the fluid abounding in the body A, before it comes in contact with B, be α, and then the force by which the body A constituted in this state acts on particles of electric fluid by repelling then will be $= \frac{\alpha r}{q}$. But after the fluid has been uniformly distributed through both bodies, that force will be found $= \frac{\alpha r}{q+Q}$. It is clear from this that the electric force which is exercised by the combination after the uniform distribution of fluid has taken place is less than that which the body A alone exercises before it has come into contact with B, for $\frac{\alpha r}{q}$ is generally greater than $\frac{\alpha r}{q+Q}$, and the greater is Q, the greater is the extent to which the former quantity exceeds the latter. And so if we consider that the mass of the body B, to which the quantity Q is proportional, is infinite, or at least very large in comparison to the mass of the body A and to q, then after the uniform distribution has taken place, the force of the two combined will become evanescent and will be quite insensible. If therefore the body B touches other bodies not electric *per se,* and through these a connection is made with the terrestrial globe itself (which can be considered infinite in comparison to any body electrified by art), a single continuous body is formed and all the electricity of the body A is absorbed by the terraqueous globe, and will become vanishingly small. But if the body B is properly supported and is of a finite magnitude, it will become electric through contact with the body A, and indeed positively so, like the body from which it took its electricity.

197) All of my readers will be able to transfer these things to the contrary case without difficulty; that is, the case where the body A is negatively electric. Here, the body A, by attracting the electric matter contained in B, causes the end C to be filled with fluid and the end D to be evacuated. Then it is clear that the particle of fluid n, clinging to B at the point where the bodies make contact, must strive to enter

---

[97] See above, p. 125.

the body A. If therefore the bodies we are considering are not electric *per se,* and there is no cause impeding the actual transit, the particle n must actually cross from B to A. This flow cannot stop until the fluid has been uniformly distributed throughout the combination of the bodies. If the quantity of fluid deficient in A is called α, so that, before contact with B, the force by which the body A was attracting the particles of electric fluid was $=\dfrac{\alpha r}{q}$, the force of the whole combination after the equal distribution will be $=\dfrac{\alpha r}{q+Q}$. And so it is evident that things are the same here as in the previous case. If the body B is very large in comparison to the body A, or if B is connected with the globe of the earth through bodies non-electric *per se,* the electricity will be absorbed so that it will be almost evanescent to the senses. If the body B is properly supported, and is of finite magnitude, the combination of the bodies A and B will become negatively electric, as was A, but to a lesser degree the larger the body is, and to a greater degree the smaller it is.

198) These things happen if both bodies were non-electric *per se;* but a far different effect will be produced if either both or just one of them is electric *per se.* In this case, the fluid strives to cross from one body to the other, but because of the difficulty with which it both extricates itself from the pores of the body in which it abounds, and enters the pores of the body in which it is deficient, there is either no transit at all, or only a very slow one. So in these cases a uniform distribution of the same kind as in the preceding case must not be expected. It is sufficiently obvious from the preceding chapter what ought to happen in these cases and in all those where the body A has been sufficiently close to the body B so that, either repelling or attracting, it can act on the electric matter contained therein, and also what ought to happen when the actual transit of fluid from one body to the other is checked through some impediment, and there is no need to repeat it in detail and to make deductions from my theory once again. It is sufficient simply to recount the rules, since it is readily perceived from the previous Chapter how they flow from the theory.

199) I suppose it as sure and evident that

1) The body B always acquires electricity at the end C which is turned to the body A, but an electricity contrary to that possessed by the body A. If the latter is positively electric, the former becomes negatively electric at its end C; if A is negatively electric, C becomes positively electric (§ 108. 119).

2) This effect always takes place, but a weaker electricity is pro-

duced if the body B is electric *per se* than if it is non-electric *per se,* and it is always the weaker, the greater is the degree to which the body B is electric *per se* (§ 109. 118. 121).

3) At some distance from the end C, the body B gains the same electricity as the body A possesses; it there becomes positively electric if A is positive, negatively electric if A is negative (§ 108. 119).

4) All these things take place especially if the body B is surrounded on all sides by bodies electric *per se,* but bodies non-electric *per se* touched by other bodies non-electric *per se* must not be completely excluded from consideration. It is obvious from what has been expounded above (§ 115ff. 121) that the first of the rules adduced here quite evidently obtains in such cases, but that what was mentioned in the third does not, or, if so, only weakly.

200) These rules suffice for deciding the majority of cases, but there is one possible instance left that is well worth examining in particular detail. Suppose that a body B, Fig. XLII, is electric *per se* and of considerable length and that to begin with a positively electric body A is moved close to it. The electric fluid contained in the body B, yielding to the repulsion exerted by the body A, recedes from the end C toward the interior in such a way that it leaves a part CE empty and negatively electric while the fluid propelled from CE crosses into EF. This fluid will have to stop here for a moderate period of time at least, since because of the difficulty with which it moves through the pores of the body it cannot flow swiftly through the remaining parts of the body.[98] And so for this reason, once EF has become positively electric, it acts like a positively electric body, and by repelling the electric fluid contained in FG, the part of the body next to it, it makes this part empty of fluid in turn, and consequently GH, into which the propelled fluid retreats, becomes replete with it. Applying the same reasoning to the part GH, it is clear that the part HI must be emptied of fluid, ID filled, and so on. So it is clear that, through the approach of a positively electric body A, the body B can be reduced to a state where the end C becomes negative, but afterward the parts succeed each other and take over from one another, according to the circumstances and the varying length of the rod (at one time more, at another time fewer parts), alternately positively and negatively electric, after the fashion of the consequent points which, as I have already noted earlier, are usually seen in magnetized iron rods. It is further readily obvious that this alternating accumulation and evacuation of the parts ought to happen to a weaker degree, the further one proceeds from the end C toward D.

[98] A most implausible argument! Cf. above, p. 177.

201) Completely similar things ought to obtain in the second case, where a negatively electric body A is moved to the body B. It is established that, since little has been changed, all the reasoning produced earlier can be transferred to here. And so what I said in the previous § is valid in this case too, with the proviso that now the end C turned toward the body A becomes not negatively electric as before, but positively electric. I do not descend to calculation or a more detailed discussion at this point for these cannot be properly begun as long as the function of the electric repulsions is unknown, and no great benefit is to be expected from them. So I stop at the rather general statements offered to date, for they can be understood sufficiently clearly without entering on any calculations. Meanwhile I give the warning that it is impossible to determine the number of electric poles (so to speak) succeeding one another alternately produced in the manner described, and that in this much depends on the length of the body B apart from any other circumstances. It is readily clear that more poles will be found in a longer body and fewer in a shorter body, and that there can be lengths in which only two are produced.

202) I came upon the thoughts presented here simply through contemplating the theory, but keen to see the truth of my reasoning proved by the testimony of experience, I judged a test necessary; initially, however, my labors did not meet with success. I proceeded in the following way. I fitted a glass tube AB, Fig. XLIII, to a plate by one of its ends B, so that the other end A projected some feet beyond the plate. I then moved an electrified tube C to the end A; but when I took it away I did not discover in the tube AB the state I had hoped to induce in it. Not completely deterred by my lack of success, I sought a remedy for the matter, and began to think that simply moving the tube C to the tube AB could not impress enough electricity in the latter to produce the desired effect. So I changed the method of conducting the experiment somewhat. After everything had been done as just described, I not only moved the electrified tube C to the end A of the other tube, but I also rubbed the end A several times with the tube C, as is normally done if electricity has to be imparted by means of an electrificatory tube to a body suspended from silk threads or supported in some other way. After the test had been carried out I found that the end AE was positively electric for a length of 4 or 5 inches; beyond this, a part EF, about 2 inches long, was sensibly negatively electric, and beyond this the tube was again positively electric: weakly so, yet significantly enough. I repeated this test quite often and had the same success, and I was able to observe similar things if I used a solid cylinder of fused sulphur instead of either the

tube C or the tube AB. So although the test did not succeed in the way I had first hoped it would, nevertheless it is sufficiently established from this description of my tests that there is agreement between my reasoning and experience; for it cannot be doubted that the cause which prevented me from obtaining the desired effect simply by moving the tube C to the end A was none other than the weakness of the action exerted by the tube C under those circumstances.

203) Because of the similarity of the theories of magnetism and electricity, the reasoning presented so far can be transferred to magnetism, though only that concerning the case where the transit of electric fluid from one body to another is inhibited in some way. For since magnetic fluid moves with some difficulty in the pores of iron bodies, even if the magnet touches the iron, no detectable passage of magnetic fluid can be expected from the magnet into the iron or vice versa, in the same way as it also happens only with difficulty between bodies electric *per se*. So the communication of magnetism which takes place when one of the ends of a magnet is moved to an oblong piece of iron is reduced to the following rules.

1) If a piece of iron B, Fig. XLI, is moved to a short distance from the pole A of the magnet, or right up to contact with it, whether this is north or south, the end C turned to the magnet always acquires the opposite magnetism to that possessed by the pole A, north if it is south, and south if it is north.

2) Passing toward the end D, a point is reached where the same magnetism is found in the body as is possessed by the pole A (§ 148. 154).

3) If the rod B is longer, Fig. XLII, the result is that a part endowed with the opposite magnetism succeeds the part which has the same magnetism as the pole A, following this there is a part which has the same as A, then another with opposite magnetism, and so on. The force of these unlike poles succeeding one another alternately is found to be weaker the further we move away from the end C, and the number of these consequent points, apart from any other circumstances, depends largely on the length of the rod, so that ordinarily more are found in a longer rod and less in a shorter one (§ 200. 201).

204) Since it is well known that these laws deduced from the theory are in complete agreement with experience, there is no point in adducing special experiments to prove that agreement. So let us rather move on to a fuller consideration of certain special phenomena. First of all, it is self-evident that the magnetism acquired by the body B will be weaker, the harder is the iron from which it is made; but it is just as readily understood that the harder the steel or iron

into which it is impressed, the more durable the magnetism will be. It is generally established from what was said above, § 150, (by virtue of which we can admit this as a general rule, and, as is very well known, experience provides abundant confirmation) that magnetic force is induced in an iron body with the help of the repulsion or attraction exerted by a magnet with more difficulty, the harder the iron is, but that when magnetism has been produced in it, it lasts more tenaciously. To get a clear notion of the consequences of this, let us distinguish various possible cases. Readers will have to remember what I proposed previously about the degree of saturation of magnetic force.[99] It was by this term that I denoted the degree of magnetic force which can subsist for a reasonable length of time without sensible diminution in a magnetized body left to itself, and beyond which, if the magnetic force is increased, the excess is quickly destroyed. I proved that this degree of saturation depends on the difficulty with which magnetic fluid moves through the pores of the magnetized body, in such a way that it is always greater in more hardened bodies, and less in softer ones.

204A)[100] If now the body B is moved to the magnet A, Fig. XLI, the repulsive or attractive force of A, by means of which it arouses magnetism in the body B, is either so proportioned to the difficulty with which fluid moves through the pores of the body B that the magnetism which the magnet A produces in the body B is greater than this maximum degree, that is, the degree corresponding to the point of saturation, or it becomes equal to it, or less. In the former case, as soon as the magnet A is removed, the body B will straightway lose its excess of magnetic force beyond the saturation level; in the remaining cases, the degree of magnetism received from the magnet will be preserved unaltered.

205) If therefore two similar bodies B and b, Fig. XLIV, the former made from rather soft iron and the latter from more hardened iron, are put successively to the magnet A, generally speaking the magnet will induce a stronger magnetism in the former body and a weaker magnetism in the latter. If these bodies are separated from the magnet and are then examined, we get results containing something varied and paradoxical and unexpected. For either

α) The magnet A is so strong that it induces a degree of magnetism greater than the degree of saturation not only in the body B but also in the body b. In this case, both bodies when removed from the magnet immediately lose the excess beyond the saturation level.

---

[99] See above, pp. 291–92.
[100] Two successive paragraphs are numbered 204 in the text.

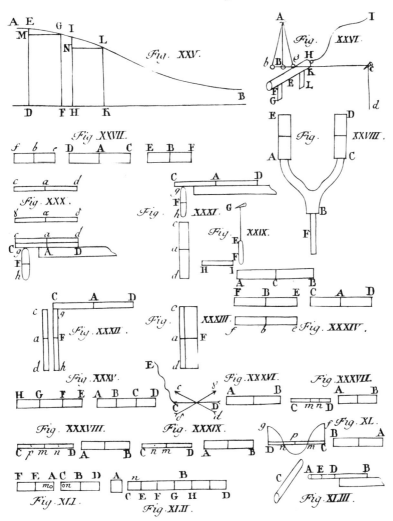

Figure 8. Facsimile Plate II from the *Essay*

Since there is a higher saturation level in the body b than in the body B, because b is harder, b will be found to have acquired a greater magnetism than B.

β) Or else the magnet A is so weak that the degree of magnetism which it induces in each body is less than saturation level. In this case, after the magnet is removed, both bodies preserve undiminished the magnetism they received from the magnet A. Since the magnet induces a lesser degree of magnetism in the harder body b than in the softer one, even after its removal, the body b will be more weakly magnetic than the body B.

γ) If finally the force of the magnet A is so adapted to the difficulty with which fluid moves through the pores of the bodies B and b that in the former a degree of magnetic force is induced either greater than or equal to the maximum, while in the latter it is less than the maximum, the former body loses the excess beyond the degree of saturation when the magnet is removed, but the latter preserves all the magnetism it received. It is possible then, from the varied degrees of hardness of the bodies B and b, that, when the magnet is removed, at one time the body B, and at another the body b, is found to have acquired the greater degree of magnetism; it is also possible for the bodies B and b to be of the same strength.

206) Further, since the saturation degree is the greatest a body can retain, magnetism is increased beyond it to no avail, since the body soon loses again the excess acquired beyond this degree. So if a weaker magnet A can induce in the body B a degree of magnetism equal to the saturation degree, no benefit will accrue from substituting another stronger magnet, and the body B will not receive a greater magnetism from the stronger magnet than from the weaker one. On the contrary though, if the magnet A cannot impress a degree of magnetism corresponding to the saturation level, the substitution of another stronger magnet for it brings a significant increase to the strength of the body B.

207) Anyone who has given attention to performing magnetic experiments ought to have observed the phenomena described in the preceding § rather often, so not only are we assured that experience is consonant with theory in these matters, but at the same time we know from § 205 why it is that the majority of writers have disagreed about the degree of magnetic force which iron bodies of different hardnesses can acquire, some saying that the hardest steel is suitable for receiving a more significant degree of magnetic force, others saying soft iron. We note generally that what is said here is equally as valid for other methods of magnetizing iron objects, soon to be described,

ELECTRICITY AND MAGNETISM 359

as for what has been considered so far. But before we pass on to these let us deal with some more considerations that are pertinent here. We have already noted above (§ 203) that if the method of magnetizing iron discussed here is employed, it is possible for the body B to acquire consequent points. But it is readily apparent that the harder the body B is, the more it is prone to receiving consequent points of this kind when the method of magnetizing considered so far is used. For since magnetic fluid moves with greater difficulty in hard steel than in soft, it is readily obvious that if, e.g., A, Fig. XLII, is the positive pole of a magnet, it will cause the end C of the body B to become negatively magnetic, but the part CE, which it empties of fluid, will be shorter if the body B is made from hard rather than soft iron. It is almost self-evident that if, e.g., the force of the magnet is strong enough to propel a particle of fluid situated at n still further if the iron in which it is contained is rather soft, it may be unequal to producing this effect if the resistance opposing the movement of the fluid increases with an increase in hardness. So it is quite possible that the magnet which made a part of the length CE empty in the soft iron is not able to evacuate the whole of this in a harder iron, but only the part Cn. Therefore since in a harder body fluid is propelled out of a shorter section, the following part, EF, replete with fluid, will also be shorter. It is easy to see that this argument can be similarly extended to the further parts FG, GH, etc. Even if the case is inverted and we assume that A is the negative pole of the magnet being used, it is readily evident that similar things take place. It is therefore clear that, even if all other things were equal, a longer rod made of harder iron would receive a greater number of consequent points than another similar rod made of soft iron. This property of hard bodies, that they are more prone to receiving consequent points than softer bodies, is valid only for the method of magnetizing described here. For I will soon demonstrate that, if other methods are used, the softer rather than the harder bodies are prone to contracting consequent points.

208) Physicists had already observed long ago [101] that with the communication of magnetism, a magnet such as we have considered here does not lose any of its strength, and they were much astonished at a thing which seems quite contrary to the analogy of the other operations of nature. Just as I have already managed several times to show full agreement between nature and phenomena of a kind that seemed to be strongly at variance with the custom of nature, I can

---

[101] Musschenbroek, *Dissertatio . . . de magnete,* pp. 104–105. Musschenbroek cites Norman, Gilbert and Hartsoeker as others who have remarked on this.

here prove the same once more. If a magnet A, Fig. XLI, imparts magnetism to the body B, and by repelling or attracting the magnetic fluid found in B makes the end C acquire an opposite magnetism to that of the pole A, and the end D the same magnetism as the pole A, there is no cause which could lessen the force of the magnet A, and if there is any change, it ought to consist in an increase rather than a decrease. For it is evident that the force of the magnet A cannot be lessened unless either the magnetic fluid crosses from A to B (or from B to A), or the fluid in the magnet A is forced to distribute itself more uniformly through it, and to cross from the part where fluid abounds to the other, evacuated part. But it is established that the former is prevented by the difficulty with which magnetic fluid moves through the pores of iron, even if the magnet touches the body B, while the latter is so far from being possible that we must rather expect the exact opposite to take place. If e.g. A is the positive pole of the magnet, since the part CB of the body B turned toward the magnet is empty of fluid, it attracts magnetic fluid and forces a greater accumulation of it than before toward the end A, and a greater evacuation of the remaining part of the magnet. So it tends to increase the force of the magnet A, not to diminish it. It is self-evident that the same also happens in the other case where A is the negative pole of the magnet. The arguments presented here can very easily be adapted to the remaining methods of magnetizing to be taught in what follows, for these too are such that, for similar reasons, they are completely harmless to the power of the magnets which are used for communicating the magnetic force.

209) Pertinent here, and easily explained from the principles presented so far, is the remarkable phenomenon that the strength of a magnet is increased slowly but significantly by iron placed close to it; the cause of this is immediately clear from what has just been discussed. For a body B placed against a magnet A becomes a magnet itself and, as has just been shown, it tends to make the magnet stronger: though this force is usually not large, after rather a long space of time it will be able to produce a noticeable effect. Similarly, the reason for another phenomenon very well known to those who have carried out magnetic experiments is to be found here. Hang an iron weight from a magnet (the biggest it can carry), and increase it little by little by successively adding small weights. After some weeks or months you will find that the magnet will carry a significantly greater weight than it used to pick up at first. Take this weight away from the magnet, then immediately put it back again. You will observe that the magnet is now unable to hold up the weight which it had

carried a moment before; it cannot be made to hold up the same weight unless you take part away and afterward add it again little by little. In this way, after some time, it becomes capable once again of carrying the same load as before.

210) If we consider what will happen to a magnet and to an iron suspended from it if they remain joined for rather a long time, it is clear from what has been discussed previously that if e.g. the pole A is positive, the part CB of the iron joined to it becomes negatively magnetic, and this effect will be stronger the longer the iron remains joined to the magnet. In a similar way, through the action of the body B which has become a magnet, the magnet A too becomes more strongly magnetic little by little, because of the continuous attraction which the iron B exerts on the magnetic matter included in the magnet A soliciting the fluid to move perpetually toward the end A. If the iron is now taken away from the magnet, not only does the iron (if it is soft, as the weights used for carrying by a magnet usually are) lose either all or at least the greater part of the magnetism it acquired, but also it is quite possible at times for the magnet itself to lose some of its strength, if through the long-protracted action of the iron B it has been made magnetic beyond the degree of saturation. If the weight B is once more joined to the magnet A, it cannot in a few moments recover the total degree of magnetism which it had acquired over a space of several weeks or months, so initially it should not be attracted by as great a force as before, but by one somewhat less. Further, if the last of the things mentioned also took place, that is, if the magnet itself lost some part of its strength after the separation of the iron, for this reason it too will initially attract the iron with a slightly weaker force than before, until after some time it completely recovers its former strength.

211) A slender oblong piece of iron EF, Fig. XLV, for example part of a rather slender iron bar, is touched at its mid-point by any pole of the magnet, first of all e.g. by the positive pole A. This acts on the fluid contained in the bar by repelling it. And so the fluid subsisting around the middle of the bar at DG, yielding to the repulsion, will be forced to leave the middle region of the wire and retreat toward the ends, some toward E, some toward F. So it is evident that the bar EF will become negatively magnetic around the middle DG, but at some distance from the middle, e.g. at DE and GF, it will become positively magnetic. It is just as obvious that if the bar were of a more considerable length it could acquire more consequent points both toward E and toward F. It is readily obvious that the same ought to happen if the pole of the magnet touching the middle of the bar is

negative, except in this latter case, DG, the middle of the bar touched by the magnet, will become positively, and the ends DE, GF negatively magnetic.

212) Experience testifies that the things I have said actually happen, though these are to be understood as applying only in the case where the bar or piece of iron used is not very thick. If the iron is thicker, and the positive pole A of the magnet approaches its center at HK, Fig. XLVI, the magnetic fluid, yielding to the repulsion exerted on it, and finding in the thicker iron another place to which it can retreat, crosses in part toward the sides E and D, but in part too, it moves toward FG, the surface of the iron opposite to the magnet, in such a way that the space HIK is emptied of fluid but DFGEKIHD becomes replete with it. Similarly, if the negative pole of the magnet is moved to the center, the magnetic matter will flow in toward the pole leaving the space DFGEKIHD empty, and will accumulate in the hemispherical space HIK.

213) From this can be found the reason for the quite remarkable phenomena that occur when an armature is joined in the customary manner to either an artificial or a natural magnet. Place a piece of iron GHI such as is shown in the figure against the magnet ABCD, Fig. XLVII, whose positive pole is BCEF and negative pole ADEF, and imagine that this iron is moved first to the positive pole BCEF. It is evident that the magnetic matter to be found in the part GH of the armature, or at least a great part of it, ought to retreat toward HI because of the repulsion exerted by the pole BCEF, so that the foot HI becomes positively magnetic. On the other hand if a similar iron bar MLK is joined to the negative pole ABEF, its foot LK will become negatively magnetic, the magnetic matter retreating from it into the other part ML. So the feet HI and LK of the two armatures become magnetic. According to differing circumstances, however, which it will be useful to consider in more detail, this force will be sometimes greater, sometimes less.

214) It is readily evident that, if the greatest force of which it is capable is to be acquired for the foot HI or the foot LK, the part of the armature GH or ML, which from now on I shall call *the wing of the armature* ought to be neither too thick nor too thin. If it is too thick, the magnetic matter propelled by the magnet will not all cross toward the foot HI, Fig. XLVIII, but part will remain around GH, and then the foot will not become magnetic to as great a degree as if the magnetic matter had all crossed into it. So the thickness of the wing must not be too great if all the magnetic matter propelled from the wing is to cross into the foot. Nor must the wing be made too

slender. If this happened, all the magnetic matter propelled from the wing would retreat to the foot; yet if the thickness of the wing were increased, the force of the pole would suffice to propel the magnetic matter contained in the additional part too into the foot, thus making it magnetic to a still greater degree. So the proper and most suitable thickness of the wing will be the greatest thickness allowing no accumulation of magnetic fluid toward GH.

215) All this can be applied to the armature MLK joined to the negative pole ADEF. If its wing ML were thicker, not all of it would become positively magnetic, but some of the fluid would cross from the surface ML toward the opposite one and leave the surface ML empty. Since the fluid accumulated in the wing has come from elsewhere than from the foot LK to fill the wing, the foot is evacuated to a smaller degree than at other times and becomes less magnetic. But if the thickness of the wing is imagined to be gradually reduced, it will finally reach a thickness such that the whole wing becomes positively magnetic, and this will be the thickness which gains the greatest force for the foot. If the wing is made still thinner, not as much fluid will be able to cross into it from the foot as would cross if the armature had been constructed with a thicker wing.

216) Further, because the fluid yields more easily if the armature is made of softer iron, its foot will become magnetic to a greater degree than it will if hardened steel is used, though some people have wrongly ordered this to be done. For the same reason, thicker wings can also be used if the armature is made of soft rather than of hardened iron, for then, for the same reason, the foot acquires a greater force in the former case than in the latter. So if the armature joined to the magnet is so constructed that the wings have the proper thickness as previously determined, and if the iron used for constructing it is as soft as can be obtained, the feet of the armature will acquire a significant magnetic force. All this is abundantly confirmed by experience, which long ago showed that the wings of the armature must be made not too thin and not too thick, and that the whole armature must be constructed from soft iron if one wishes to obtain a significant force in the feet.

217) An armature of this kind joined to a magnet not only conserves the strength of the magnet to which it is joined but indeed it slowly, yet significantly, increases it little by little. The cause of this is sufficiently evident from what has been discussed above (§ 209). For if the wing of the armature GH, Fig. XLVII, is joined to the positive pole it becomes negatively magnetic, but ML moved to the negative pole becomes positively magnetic, so both tend to propel the magnetic fluid more and more from the part ADEF into the part BCEF. A par-

ticularly noteworthy phenomenon concerning the effect of armatures joined to a magnet is this, that a magnet properly armed is found to lift a far greater weight than one not armed. In general terms the reason for this can easily be seen from the theory, so it is clear we ought to look here at the varying circumstances. First of all, imagine that a piece of iron G is moved to ABCD, Fig. XLIX, a bare magnet of parallelipiped shape, and it is obvious even to the eye that some parts of the pole CD situated to the side act quite obliquely on the body G, and are significantly distant from it, and it is self-evident that both these factors significantly reduce the force with which the magnet acts on the iron G. But if there is an armature joined to this same magnet, the force of the pole CD is completely collected as it were in the foot of the armature, which because of its narrow shape acts more directly and from a shorter distance on an iron body moved up to it, and so is able to exert a greater force than the magnet itself. Next, it is clear from the reasoning and the experiments presented in § 162 *et seq.* that if the iron G is moved to the bare magnet, the pole AEBF, possessing a contrary magnetism to the pole CDEF, significantly weakens the action which the pole CDEF exerts on the iron G, especially if the magnet ABCD is short, but that if the armature is joined to it, such a weakening will not take place, or not to such a degree; so for this reason too the foot of the armature can exert a greater force than the pole of the magnet itself.

218) We have so far considered the method of magnetizing which is achieved simply by the contact or approach of a piece of iron by a magnet. But since it is of great importance to give a strong force to magnetic needles, Physicists have tried various things and found various methods of giving a magnetic force to iron bars which it will be worthwhile to weigh against our theory. Stroke an oblong iron body DE, Fig. L, e.g. a piece of an iron bar, with A, one pole of a magnet, moving this pole over the whole length of the rod DE from D to E; repeat this operation several times. The end E where the stroking ended will be found to have acquired the opposite magnetism, and the end D where the stroking began, the same magnetism as the pole A.[102] If we consult our theory we very easily perceive that what experience shows happens ought to happen. For when the pole A is first applied to the end D, it imparts to it a magnetism contrary to that which it possesses, but afterwards when it is drawn from the end D toward E through the points F, H, G, etc. it successively induces in these points a magnetism opposite to its own, and meanwhile it takes away from

[102] The text has B.

the end D the magnetism which it had communicated to it initially, and imparts a contrary magnetism. Since it touches the end E last, this preserves the magnetism which it has had induced in it.

219) To clarify these things, let us distinguish the cases which can occur, and consider each of them separately. Suppose first that the positive pole of the magnet is used in the experiment, and it is evident that when it is brought up to the end D it forces the magnetic matter to leave this end and retreat toward more distant parts, e.g. toward F, while the end D becomes negatively magnetic. Then when the pole of the magnet is moved forward to F, it again propels the magnetic fluid accumulated there away from this point; so the rod is evacuated of fluid at that point, but is filled with it toward both D and H. Therefore when the magnet arrives at the point F, it destroys its own former effect and makes the point D positively magnetic (§ 211). The same happens throughout the whole length of the rod except at the end E which the magnet touches last. As is self-evident, it cannot again destroy the negative magnetism induced at this point and induce one contrary to it, so this preserves its negative magnetism. If the pole used is negative, by attracting the fluid toward the end D it initially makes this positively magnetic, and the more distant parts negatively magnetic. As in the preceding case, when moved along the length of the rod it similarly destroys its own effect and induces a negative magnetism in the end D, and a positive magnetism in the end E which it touches last.

220) Hardly more is to be expected from this method of magnetizing than can be produced simply by touching the end E with the magnet, as experience also testifies. But since this method is often employed to magnetize needles, instruments whose utility in common life is so great that nothing which assists in perfecting them should be overlooked, we note as well that it labors under the disability that the needle can easily acquire consequent points. Not only does experience confirm this, but it is also clear from what has been expounded above (§ 203. 200. 201) that it ought to happen, according to the way we understand it from the theory. For this method of magnetizing hardly differs in any way from that considered above which consisted solely in the contact of the end E with the magnet, since every effect the magnet produces while it is drawn over the length of the rod is destroyed, and it is only that generated when it touches the end E that remains.

221) Physicists have noted a certain effect of this method of magnetizing which I cannot pass over in silence since it is quite noteworthy in itself, and because of its excellent agreement with our theory

we must not neglect it. Imagine that the rod DE has been made magnetic with a strong magnet, by stroking it from D toward E. Take another weaker magnet, and stroke the rod as before with the like pole of this magnet in the same direction as before. It will be observed that not only has the magnetic force of the rod DE not been increased or remained the same, which one could expect, but rather it has been significantly lessened, and reduced to precisely that degree which the weaker magnet would have been able to induce in the rod if this had never been touched by the stronger one. The extremely accurate observations of the famous *Musschenbroek* on this phenomenon can be seen in his commentaries on the Essays of the Florentine *Accademia del Cimento* Part II, page 80.[103]

222) This phenomenon agrees so well with our theory that a greater congruence could not be sought. To make this clear, imagine that the rod DE has received from the positive pole of a stronger magnet a magnetism such that the end D has been made positively magnetic, and the end E negatively magnetic. Now stroke the rod with the positive pole of the weaker magnet, and it is evident not only that while this is applied to the end D it tends to make this end negatively magnetic, and to destroy the effect of the first magnet, but also that the same takes place perpetually as the magnet is drawn from the end D toward E: when it is a little distance from the end E, e.g. when it has reached the point G, it strives to induce in this end a magnetism contrary to the one it had received from the stronger magnet, that is, a positive one. Since completely similar things would take place if the negative poles of the magnets were used, it is clear generally that the effect of a weaker magnet consists in this, that it first of all destroys the effect of a stronger magnet and then it imparts a fresh magnetism to the rod. This phenomenon is only observed when the rod is stroked by a weaker magnet; it never happens if the end E is only touched by a weaker magnet, and the rod is not stroked by it over its whole length, for then the weaker magnet in no way tends to destroy the effect of the stronger, but it rather tends to increase it, as is shown not only by our theory but clearly also by experience.

223) What I have proposed here must not be understood without a caution, for I think there can be cases where the magnetism of a rod induced by a stronger magnet is not weakened but is rather increased when it is stroked by a weaker magnet. To understand this properly, imagine first that the rod DE, having been magnetized by a

[103] Accademia del Cimento, *Tentamina experimentorum naturalium captorum in Academia del Cimento,* trans. into Latin and edited, with notes, by P. van Musschenbroek, Leyden, 1731, Pt. II, p. 80.

ELECTRICITY AND MAGNETISM 367

stronger magnet, is stroked by a non-magnetic piece of soft iron AB. If the end D of the rod is a positive pole, the end A of the iron AB becomes negatively magnetic immediately it touches the end D, and as a result it tends to increase the positive magnetism of the point D. Similarly, when the iron AB is moved toward the end E, which is negatively magnetic, it likewise tends to increase the negative magnetism of the end E. So if the rod DE is stroked often and for a long time with a non-magnetic piece of soft iron, the force of the rod can eventually be increased appreciably.

224) If, in place of the soft iron, a magnet made of iron which is not completely hard is substituted, but the rod DE is quite hard, and of great strength, it can happen that when the magnet touches the positively magnetic end D with its own positively magnetic pole A, the latter end acquires the contrary magnetism, that is, negative, as a result of the action of the rod; this possibility is clearly established from what has been discussed previously. For the same reason, the same magnet can again acquire positive magnetism after it has been moved forward to the end E. So it is the same as in the case of stroking with a non-magnetic piece of soft iron, and thus the same consequence ought to follow from this operation if it is protracted for a long time, that is, an increase in the force of the rod, not a weakening.

225) There is another method of magnetizing sometimes used which hardly differs in effect from the one described so far. Usually a horseshoe shaped magnet is taken, either an armed natural magnet or an artificially curved one, whose pole A, Fig. LI, is moved to the end D, and then the magnet is drawn over the whole length of the rod in such a way that the pole A precedes and the pole B follows, and touches the end E [104] last, and this operation is repeated several times. As a result, the same effect usually follows as from the preceding method. That is, the end E receives a magnetism contrary to that which the pole B, which touched it last, possesses, and the end D acquires the same magnetism as this pole—unless perhaps the rod is so long that it gains consequent points.

226) It is clear at first sight that the following pole B destroys the whole of the effect produced by the preceding pole A, so it is evident that the same result ought to occur as when the rod DE is rubbed by the pole B alone. For since the pole B passes over all the points to which A has been applied, then necessarily, since it has a contrary force, it evacuates those places which had become replete through the

[104] The text has C.

action of the pole A, and fills those which had been evacuated. So a judgment is easily made about this method of magnetizing, for everything I taught about the one considered previously is valid for it. We must finally still warn that if this method of magnetizing, or that previously described, is used after the rod has been stroked in a similar way but in the contrary direction, all the magnetism previously produced will be destroyed, and the contrary state induced in the rod, so that the end D will acquire the magnetism which E possessed, and vice versa. How these things flow from the theory is so evident that no further exposition is needed.

227) Other methods of magnetizing are used at times, but I shall say nothing special about them since readers can decide about them from what has been expounded so far. But there is a remaining method which deserves a more careful discussion, since it helps to produce a far greater magnetism than others, and as a result it is used for making artificial magnets, and for animating magnetic needles. This method, which was given the name *"the double touch"* by its first inventor *Michell*,[105] is executed in the following manner. Two contrary magnetic poles, a positive one A and a negative one B, are placed at the middle of the iron rod DE, Fig. LII, a little apart; these poles, kept the same distance apart, are drawn backward and forward alternately over the rod DE, following the rule that the poles always remain the same distance apart and neither is moved beyond the end of the rod. Thus e.g. when the pole A has reached the end D, both are immediately drawn back and moved backward until the pole B has reached the end E, and then the poles are drawn back again. After this has been done several times and the poles are brought back to the position which they had at the beginning of the operation, near F, the middle of the rod, and they have been taken away in a line perpendicular to the rod DE, the rod will be found to be strongly magnetic, and the half FE will be homogeneous to the pole A and the other half FD will be homogeneous to the other pole B, unless perhaps the rod is quite long and has gained consequent points; but I shall not consider this case here, since I shall deal with it later.

228) There are different ways in which what I ordered in the previous § can be achieved. If artificial magnets parallelepiped in shape are available, join two of them, AD and BE, Fig. LIII, in such a way that they can be used in the manner indicated in the figure, so that their contrary poles adjoin one another; place between them a body G which prevents them from touching. A natural magnet AB, Fig. LIV,

[105] J. Michell, *A Treatise of Artificial Magnets,* Cambridge, 1750, p. 8.

armed in the usual fashion, or a curved artificial magnet AB, Fig. LV, of a horse-shoe or semi-circular shape, can also be used with a like success.

229) To see what ought to happen with the proposed method of magnetizing, let us imagine that the positive and negative poles A and B, Fig. LII, which have been moved to the rod, are of equal strength, and let us consider that a particle m of magnetic fluid located between the two poles and placed near the middle has been repelled from the positive pole by a force$=r$, and it is clear that the particle is urged first of all by the repelling force of the pole A from m toward q by a force$=r$. But it is similarly clear that the negative pole D, by attracting the particle m, solicits it in the same direction as the positive pole along the line mq with a force which likewise$=r$, because the forces of the two poles are equal. So it is manifest that the actions of the two poles conspire to produce the same effect, and the force which urges the particle m to cross from the area near the pole A to that close to the pole B is equal to the combined action of the two, that is$=2r$. Imagine further another particle n situated beyond the pole, and assume that the action of the positive pole on it again$=r$, and the force of the negative pole exerted on it$=\rho$. It is clear that ρ ought to be less than r, since the particle n is further from the pole B than from the pole A. Therefore the pole A repels this particle from n toward t, but the pole B attracts it in the contrary direction toward v, so the total force by which it is drawn towards t will be$=r-\rho$. In a similar fashion, a particle p is attracted by the negative pole in the direction ps [106] with a force$=r$, but is repelled by the positive pole in the contrary direction from p toward w by a force$=\rho$, and so the total force drawing it toward s will likewise$=r-\rho$.

230) It is clear then that the magnetic poles, when moved up to the rod DE in the described fashion, exert only a rather weak action on the particles of magnetic fluid enclosed in the parts of the rod AD and BE, so these parts must be expected to preserve the state in which they were constituted before the poles A and B were moved up to the rod, or at least to be very little disturbed from that state. But it is far different for the magnetic fluid in AB, the part of the rod located between the two poles A and B. For the particles in this section are urged by an appreciable force to cross from the region adjacent to the positive pole A to the region close to the negative pole B. It is clear that the force $2r$ which acts on the particles of fluid in this

---

[106] The text has qs.

central part of the rod is much greater than the force $r-\rho$ which urges the particles contained in the parts AD and BE. So the state to which the rod is reduced through the action of the poles A and B, Fig. LVI, will be that the part GF adjacent to the positive pole becomes strongly negative, but the part FH situated near the negative pole B becomes positively magnetic to a significant degree. AD and BE, the remaining parts of the rod not located between the poles, suffer hardly any change, or at least only one so small that it can be neglected. Thus they are not sensibly disturbed from the state they possessed before the poles A and B were moved up, whatever this state may have been, whether the natural state or one of either positive or negative magnetism.

231) If the poles A and B, Fig. LVII, are close to the end D of the rod, it is clear that the part DF of the rod lying close to the positive pole A ought to become negatively magnetic, and the fluid in it ought to be forced to leave the part DF and retreat toward the region near the pole B. But if the poles A and B are then moved forward over the length of the rod, and are drawn from D toward E, it is readily obvious to anyone who has been willing to consider the matter with only a little intelligence, that the end D [107] ought to preserve the negative magnetism it had received (for when the poles are somewhat removed from the end D, no force, or at least only a small one, is exerted to destroy again the magnetism impressed on the end D), but as the poles are moved forward over the whole length of the rod from D toward E, a constant effort is exerted by the poles to push the particles of magnetic fluid contained in the rod from D toward E. In a completely similar way, when the poles are close to the end E, the end *f*E near the negative pole B ought clearly to become positively magnetic, and magnetic fluid ought to cross to here from the part of the rod close to the positive pole A. It is just as clear that when the poles are moved from E toward D, not only does the end E retain undiminished the positive magnetism it had received, but also the action of the poles constantly tends to force the fluid to move from D toward E.

232) So it is clear in general terms that whether the poles are moved from D toward E or whether they are drawn back from E toward D, their action tends constantly to transfer the magnetic fluid contained in the rod DE from the end D toward the end E. It is also evident that this method of magnetizing ought to produce a much greater effect than either of those previously described and the remainder

[107] The text has E.

which are used at other times. For, because of the help which the poles A and B mutually offer to one another, the force which propels the magnetic fluid from D toward E is of a considerable size, and since their action tends constantly to produce the same effect whether the poles are drawn backward or forward, the force impressed on the rod DE will be greater, the longer the operation is continued, until finally the maximum level is reached. In part, the remaining methods of magnetizing do not exert as great a force to render the rod magnetic, and in part, they labor under the disability that when the operation is repeated rather often, an effect is produced hardly greater than when the operation is executed but once, since any operation that is repeated always destroys the effect of the preceding ones, and it induces new magnetism in the rod only as the previous magnetism is destroyed.

233) There is a further special benefit in this method, because with its help it can be arranged that the magnetic center falls precisely in the middle of a magnetized rod and this, as will be shown at length elsewhere,[108] is of great importance in constructing magnetic needles of better quality. For the magnetic center to be produced in exactly the middle of the rod in this method of the double touch, the only stipulation is that the half of the rod FE, Fig. LVI, is to be stroked precisely as many times as the other half FD. So if you have begun the stroking in the middle, as is usually the case, and have first of all drawn the poles toward D, when the operation is to be finished, you ought to proceed in such a way that last of all you bring the poles back from the end E to the middle, and then end the operation. If this is observed, it is evident that the half FE has been rubbed as many times as the other half FD. It is self-evident that, since the force which has evacuated the part FD of fluid and the force which has filled the other part FE with fluid are equal, and they have been applied an equal number of times, the magnetic center can fall nowhere but in the middle of the rod, for the action on the part FD is completely equal to that exerted on the part FE. So if the densities of fluid, or rather the differences between the density of the fluid at any point, and its natural density, are exhibited by the curve ACB, Fig. LVIII, the ordinates HI and GF, taken so that EH and DF, their distances from the ends, are equal, will likewise be equal. Since this curve must intersect the axis at some point, if it does so at C, the part of the curve DCB will be similar and equal to the part ABE. The consequence of this is that the point of intersection C, which is the magnetic center, must fall precisely in the middle of the rod.

[108] See Appendix, items **31A(i), 31B, 33.**

234) I said (§ 232) that with this method of the double touch, the longer the operation is continued, the greater is the increase in the magnetic force induced in the rod DE; but I have already added the limitation that it cannot be increased to infinity but only to a certain maximum degree. This requires some further discussion which must be added here. Imagine that the magnetization of the rod DE, Fig. LII, has already been continued to the fixed end-point; it is established from what has been discussed earlier that the greater the degree to which the part FD is evacuated of fluid and the part FE is replete with it, the greater is the force by which a particle of fluid m located near F is pushed toward D. Since this force is opposed to the crossing of the particle m from the part FD into the part FE, it is clear that at some stage the matter must be reduced to this, that as the magnetic poles A and B are drawn over the rod, they strive to propel the particle in the opposite direction from D to E, but they are incapable of producing this effect. Then no further increase of the force can be obtained even if the operation is repeated often. But various cases can occur here; for either this maximum degree of magnetism which, as we have just shown, the poles A and B can induce in the rod DE, is greater than the degree of saturation, or it is less than it. If the former, it immediately loses the excess degree of strength beyond saturation once the operation is finished and the rod is left to itself; but if the latter, it preserves unaltered the degree of magnetism it received. Further, in the former case, the rod receives as much magnetism as it can preserve, and so if stronger magnets were used to magnetize it, no benefit could be hoped for from this. But in the latter case, since the rod has acquired the greatest degree which the poles A and B are capable of producing, though not the greatest of which it was itself capable, were it stroked once more with stronger magnets, this action would not be without benefit, but rather a stronger magnetism would be induced in the rod as a result.

235) Let the rod DE, Fig. LIX, to be magnetized through the double touch, be located between two magnets placed at its ends D and E, in such a way that the pole G touching the end E is of the same kind as the pole B. It is easily understood that if the operation of the double touch is now begun, the poles G and H will significantly help the magnetization of the rod DE. For it is evident that these poles tend to induce in the rod the same state as the poles A and B themselves strive to impress, so that the magnetic fluid in the rod DE leaves the part DF and retreats toward the other part FE. But because of this help, it is clear that a smaller number of reciprocal strokings is needed to produce the same degree of magnetism as would have been neces-

sary had the poles G and H not been moved up to the rod. In the case where the poles A and B alone are not sufficient to make the rod magnetic right up to saturation level, it is evident that a greater degree of magnetism can be induced in the rod if the poles G and H are brought up to the ends than if this artifice is neglected.

236) If two parallelepipeds, made from soft iron and non-magnetic, are used in place of the magnetic poles G and H, we get almost the same result. For after the poles A and B have been drawn a few times backward and forward over the rod, and have made the part FD of the rod negatively magnetic, and the part FE positively magnetic, these affect the magnetic fluid contained in the iron parallelepipeds H and G, the latter by repelling, the former by attracting, so that G acquires a positive and H a negative magnetism. So this comes to the same thing as if two magnets had been brought up to the rod from the very beginning, and as a result the same benefit must be expected.

237) From the discussion so far, it is possible to give a reason for everything that usually happens when artificial magnets are prepared in the *Michellian* or *Cantonian* fashion; [109] so clearly indeed that this whole method could obviously have been developed *a priori* from our theory had it not already been noted previously by experiment. I have already resolved to follow up this matter, but I warn that I shall not give the full doctrine concerning artificial magnetism, but that I shall reserve part of it till the following chapter.[110] I shall not inquire here into how a magnetic force is to be induced in steel rods without the help of a magnet, but my intention is simply to discuss how, after they have already in some way or other acquired some degree of magnetic force, this can be increased in them right up to saturation level.

238) For artificial magnetism, pairs of steel rods are always used, e.g. 4 or 6 or 8 etc., since these experiments cannot be set up sufficiently well unless an even number is used. Although to lessen and abbreviate the task 4, 5 or 6 pairs of rods are usually constructed, my instructions here concern the method to be followed if only three pairs of rods are used, when their magnetism is to be increased to saturation level. Once the method of using 6 rods has been properly understood, it will easily be seen what is to be done if there are more than three pairs.

239) Suppose that, before the operation is carried out, some degree

[109] Michell and Canton discovered the "double touch" method independently of each other (Michell, *A Treatise of Artificial Magnets,* Cambridge, 1750; and Canton, *Phil. Trans.,* XLVII, 1751-52, 31-38). Cf. Clyde L. Hardin, "The scientific work of the Reverend John Michell," *Annals of Science,* XXII, 1966, 27-47.

[110] See below, pp. 444ff.

of magnetism has already been induced in all the rods intended to be used, even though this is weak and below saturation level. Mark one pole of each rod, e.g. the north, with a cross or in some other way, so that during the experiment the poles of each rod can always be distinguished easily. Finally, have ready at least two parallelepipeds of soft iron of the same thickness as the magnetic steel rods. If it has already been decided to increase the magnetism of the rods right to saturation level, the following method must be employed. First of all, place a pair of rods EF and GH, Fig. LX, on a wooden table between the soft iron parallelepipeds AB, CD, just mentioned, as indicated in the figure, and fix them in position so that they cannot easily move. Bind the remaining two pairs of rods into two little bundles as shown in Fig. LXI, and place a non-ferreous body K between them to keep the rods from contact and cause them to diverge appreciably lower down. Then magnetize the two rods EF and GH with these bundles by the method of the double touch described at length in § 227, by drawing the rods 7 or 9 times back and forth over each of them. After this operation has been completed, turn the rods EF and GH over so that the sides which touched the table now face upward, and using the bundles, stroke each rod again in exactly the same way as before.

240) Take away the second pair of rods from the bundles of magnets and substitute in its place the first pair of rods which has just been magnetized; place the second pair of rods between the parallelepipeds AB, CD, and magnetize them in exactly the same way as the first pair. When this operation is concluded, proceed to place the third pair of rods between the parallelepipeds AB, CD, and substitute the second pair in the bundles in its place; then magnetize the first pair again; and going round in a circle, repeat the operation until you no longer observe any increase of magnetism in the rubbed pair of rods. Since saturation level has then been reached, the operation must cease.

241) The method I have described is that which *Michell* and *Canton* used to make steel rods magnetic, and there is hardly anything in it not sufficiently and clearly explained from what has been discussed above. Yet there were certain circumstances in this method which I examined more fully and weighed against my theory. In doing so, I was led to a new method, both more efficacious and more convenient than that described so far. The considerations which prepared the way for the invention were the following. It was readily apparent from the theory that the perfection of the *Cantonian* method depended on the magnetic poles A and B, Fig. LII, exerting quite a strong action on the particles of magnetic fluid included in the central part of the rod, e.g. m, but only a quite weak action on the particles of fluid, e.g.

n and p, in the parts AD and BE of the rod. It is clear from the previous discussion that the action on the particles n and p is in a sense contrary to the action on the particle m, and it strives to destroy or at least to lessen the effect that the latter has produced. For the force exerted on n tends to propel fluid toward the evacuated end D, and similarly the force acting on p tries to draw the fluid accumulated near the end E back toward the interior of the rod. So the person who wants to perfect this method must direct his attention to increasing the action on the particle m as much as possible, and to lessening the actions on the particles n and p. At first sight, this seems easily managed, and no other artifice seems needed except moving the poles A and B as close to one another as possible. For the less is the distance between them, the closer each becomes to the particle m, and as a result each acts on it with a greater force. So the total force by which m is pushed from D toward E, which $=2r$, becomes greater, the more the poles A and B approach one another. Further, as the distance between the poles A and B decreases, it is clear that the distances of the particles n and p from the poles A and B become more nearly equal. It follows from this that the difference between r and ρ becomes less, so that if the poles are moved closer to one another, the force acting on the particles n and p, which $=r-\rho$, significantly decreases.

242) Whatever appearance of truth these reasonings possess, nevertheless I began a more diligent examination and readily saw that they contain a paralogism, or rather that they are constructed on a certain false principle which has been tacitly supposed. To make this clear, it must be remembered that the magnetic force of the steel rod, or that of any body, never acts only at a single point but is rather diffused and distributed throughout all the points of the rod. Thus if e.g. the end A of the rod AN, Fig. LII, were positively magnetic, and the end N negatively magnetic, these forces would by no means act only at the points A and N, but, rather, fluid abounds in the whole half AM of the rod, and in every point of it; similarly it is deficient in the whole of the other half MN. So the action which the positive pole AM and the negative pole BP exert on the particle m is made up of an infinite number of different actions. For it is evident that any point of the positive pole AM repels the particle m and similarly any point of the negative pole BP attracts it. But it is clear from statics that all these actions can be collected into a single mean equivalent to them. So if we suppose that these mean actions are exerted along the straight lines rm and sm, these lines will pass through certain points of the poles AM and BP, such as r and s. But it is readily apparent that these points, which I shall afterward call *the centers of mean directions,*

through which the mean directions of all the actions on the particle m pass, cannot coincide with the lowest points A and B of the poles AM and BP. Rather it is easily seen that they ought to be some distance from the rod DE and raised above it.[111]

243) Since the centers of mean directions r and s are not placed in a straight line with the particle m, it is clear that the whole force by which the particle m is urged from D toward E cannot be made equal to the sum of the actions which the negative and positive poles exert on it, as happened in § 241, but since the forces exerted by the positive pole A and the negative pole B include a certain angle rms, the magnitude of the force acting on the particle m must be judged in quite a different fashion. Take lines mt (in the direction of rm) and mp (in the direction of sm), and suppose that they indicate respectively the force by which the positive pole repels the particle m in the direction mt and that by which the negative pole attracts it in the direction mp; when the parallelogram $mpqt$ is completed, its diagonal mq will show the total force acting on the particle m, and it is self-evident that this will be less than the sum of the sides mt and mp, which is $2r$.

244) Other things being equal, the larger is the angle pmt, or the smaller is rms adjacent to it, the smaller will be the diagonal mq, and also the force exerted on the particle m represented by that diagonal. If the angle rms disappears completely (and this very nearly happens if the rods AN and BQ, Fig. LII, are moved right to contact with one another) the action on the particle m disappears simultaneously. If the poles A and B are removed further and further from one another, and are understood to be infinitely distant, the action on the particle m once more becomes zero, for because of the infinite distance the actions of both the positive pole A and the negative pole B on the particle m disappear. Anyone even a little used to mathematical disquisitions will easily be able to guess that there is a maximum between the two extreme cases previously indicated: that there is a distance between the poles A and B which makes the action on the particle m a maximum, and such that, when it is either increased or diminished, the action of the poles on the particle m is lessened.

245) To make this clearer in a special example, let us imagine that the actions of the poles A and B on the particle m, Fig. LXII, follow the law that they vary inversely with the distances rm and sm, and let us call the lines rf and $gs = a$, and the line $fg = x$. Calculation shows

---

[111] This notion, depending as it does on the idea that magnetic poles are centers of magnetic action, rather than merely the ends of a magnet's magnetic axis, is new with Aepinus.

that $rm = sm = \sqrt{a^2 + \dfrac{x^2}{4}}$, so the forces mt and $mp = \sqrt{a^2 + \dfrac{x^2}{4}}$, and since $mr : mf = mp : mb$, mq, or the force acting on the particle m, will be $\dfrac{4x}{4a^2 + x^2}$. If it is now assumed that its differential vanishes, or $16a^2 dx - 4x^2 dx = 0$, there will be a maximum force if $x = 2a$, or $fg = 2fr$. Although I am not proposing the law I assumed here as a true law of nature, it is nevertheless clear that, whatever law nature follows here, there can always be found, in a similar manner as in our hypothesis, a distance which makes the action on the particle m a maximum, and which cannot be increased or diminished without causing the force acting on the particle m to suffer loss as a result.

246) So although the method of the double touch is much superior to the rest, one must recognize that, as practiced by *Canton* and *Michell*, it labors under an inherent imperfection. To prevent the force acting on the particle m, Fig. LII, being weakened too much, care must be taken that the poles A and B do not come too close to one another, and be moved toward one another past the maximum point. Since the poles must be held apart with an appreciable distance between them, the forces acting on the particles n and p cannot be reduced either completely to vanishing point or to the point of being so small that they become undetectable. It is clear that major harm must be feared, since these forces, as we have shown before (and as is self-evident), tend to destroy the effect produced by the poles A and B. Although I thought initially it would be quite difficult to improve this method, I happened to find an artifice which overcame this inconvenience successfully. Before I explain it, I judge it best to discuss certain matters which it is opportune to explain at this point.

247) If magnetic poles, A positive and B negative, stand on a nonmagnetic rod of soft iron, DE, Fig. LVI,[112] it is established that the parts of the rod near the poles become magnetic according to the law that the part GF near the pole A becomes negatively magnetic and the part HF near the pole B positively magnetic. So the rod is attracted by the poles A and B, but, as is readily clear from the previous discussion, to a stronger degree the more strongly the parts GF and FH have become magnetic, and to a weaker degree the more weakly magnetic they have become. Since, as has just been demonstrated amply enough, the degree of magnetic force acquired by GF and FH depends

[112] The text has Fig. LIII.

in great part on the distance between the poles A and B, it is obvious that the magnitude of the force by which the iron DE is attracted depends on that same distance. If then the poles are so placed that their distance corresponds to the maximum, they attract the iron DE with the maximum force; but if they are moved either toward or away from one another, they attract it by a weaker force; in fact if the poles are thin and are moved into full contact with one another, they give no further sign of an attractive force.

248) Experience confirms all this extremely well, and if anyone wishes to see for himself he must proceed in the following manner. Choose a pair of thin magnetized steel rods and move them together first of all so they are completely touching; test the force by which they attract a given soft iron parallelepiped. Next move these rods gradually away from one another e.g. ½, 1, 1½, 2 inches etc. and as before check the amount of attraction exerted by the poles at each of these distances. When the rods touch, almost no attractive force will be detected; when the rods are moved away from one another, the attraction increases more and more until it reaches a certain fixed limit. If this limit is exceeded and the rods are moved further away from one another, the force will be found to decrease once more.

249) It seems that certain artifices can be sought from the things we have just discussed which will be quite useful in the construction of either horse-shoe or semicircular artificial magnets. It is obvious that the attracting force of magnets of this kind depends on the curvature induced in them, for they must not be too curved in case the legs, that is the poles AK and KB, get closer together than is proper, and the curvature must not be less than is proper or else the legs will be too far apart. The best curvature to choose is that which makes the distance between the poles correspond to the maximum. It is most desirable that the rule for this curvature should be discovered, but since this is impossible *a priori* because of the incompleteness of the theory, recourse must be had to experience; this can be done conveniently by preparing from the same steel, several equal and similar rods curved differently, some to a greater, some to a lesser degree, hardening them simultaneously and magnetizing them in the same way; then test the force by which each of them can pick up a given iron body. It will perhaps be found that the usual shape of such magnets is not the optimum, but that there are cases where the curvature is better if the magnet is bent into the shape of a half-Ellipse.

250) I think these things are well worth noting for armed natural magnets too. If a strong unarmed magnet of this kind has an axis that is longer or shorter than is proper, it can happen that, when the

ELECTRICITY AND MAGNETISM

armature is added to it, it does not acquire as great a strength as might reasonably have been hoped for. On the contrary, if another weaker magnet has the proper length, when it is armed it can carry a far greater weight than the former which was considered a much stronger magnet. Surely from this must be sought, at least in great part, the reason for the paradoxical phenomena usually observed when natural magnets are armed. We have read that after armatures have been added to them, the force of weaker magnets is at times significantly increased, but the force of stronger magnets is increased to a far less degree.

251) There are things discussed here which are illustrated in the *Michellian* or *Cantonian* method of magnetizing. Place the steel rods EF, GH, Fig. LX, between iron parallelepipeds AB, CD so that these parallelepipeds become magnetic and help the operation. The greater the degree of magnetic force they acquire, the greater their usefulness. But it is clear that this depends on the distance of the rods EF and GH from one another, so it is very important to choose the best distance, which is the one where the rods EF, GH impart the maximum degree of magnetic force to the parallelepipeds AB, CD. I dare not determine *a priori* what that distance is precisely; its discovery must be left to the help of experience.

252) I must finally give an exposition of the things I have thought of to perfect the method of the double touch; the following considerations have prepared the way for me. Suppose that the centers of mean directions r and s, Fig. LXIII, are situated in the same line as the particle of magnetic fluid m, and it is clear that the maximum condition which results in the case where the three points r, s and m are not in the same line, does not hold at all here, but rather that the closer the points r and s are moved to one another, the constantly greater and greater becomes the force which they exert on the particle m. So a ready conclusion is that, in the method of the double touch, the less the points of mean directions r and s are distant from the rod DE, the closer can the poles be safely moved toward one another and at the same time the greater can be made the force acting on the particle m. Thus e.g. assuming the hypothesis with which we made our calculations in § 245, it is clear that the more a, that is the distance rf, decreases, the more the distance between the poles corresponding to the maximum fg decreases at the same time, this being, of course, equal to twice the line rf; but at the same time there is an increase in the maximum force that can be acquired, this being equal to $\frac{1}{a}$. It is self-evident that similar things take place, whatever may be the true law followed here by nature.

253) It was not very difficult to imagine a way to bring the centers of mean directions r and s closer to the rod DE. If the rods AM and BN, Fig. LXIV, standing upright on the rod DE, are tilted as shown in the figure, so that they include quite acute angles with the rod DE, then by transposing the point r into R and s into S, it is clear that they become significantly closer to the rod DE. So the *Michellian* or *Cantonian* method can be changed with great benefit if the rods which are used in the magnetization are tilted as much as possible and lie almost horizontal, for because of the short distance of the centers of mean directions from the rod DE, the ends A and D of the rods AM, BN [113] can be moved toward one another without any resulting harm and with quite an appreciable increase in the force acting on the particle of fluid m.

254) So I have chanced to track down an artifice which helps to improve the method of the double touch; the force acting on the particle m is made much stronger by the method just described than if the operation is set up in the usual way. But it is clear from § 229 *et seq.* that in perfecting the method of the double touch, not only has the force on m to be increased, but also it is necessary for the forces pushing the particle n toward D and p toward E to be diminished as much as possible. Happily this requisite too happens to be satisfied by the same artifice that, as I have shown, helps to increase the force on m. This may be seen easily without a lengthy discussion. If e.g. the positive pole stands upright on the rod close to the end D, Fig. LXV, it is clear that it attempts to propel magnetic fluid from FG, the part of the rod near it, toward the end H and toward the end D alike; but if it is in a reclining position, Fig. LXVI, it is self-evident that it tends to expel the fluid from the whole part DF and from the end D itself. Since the same happens with the negative pole, it follows that the reclined position of the rods is to be preferred in this respect also to the upright position which is used in the common method.

255) Since what I have proposed here became known to me, I rarely use the *Michellian* or *Cantonian* method, but have got used to one, my own I rather think, which I call perhaps not incongruously *the improved method of the double touch*.[114] In this method the operation is performed in the following way. I get ready 4 rods already endowed with some magnetism, whose forces are to be increased right to saturation level. To achieve this, I place the first pair of them, EF,

---

[113] The text has BQ.

[114] The same improved method was discovered at about the same time by Antheaulme, but without the strong theoretical underpinning provided by Aepinus. See above, p. 181.

ELECTRICITY AND MAGNETISM 381

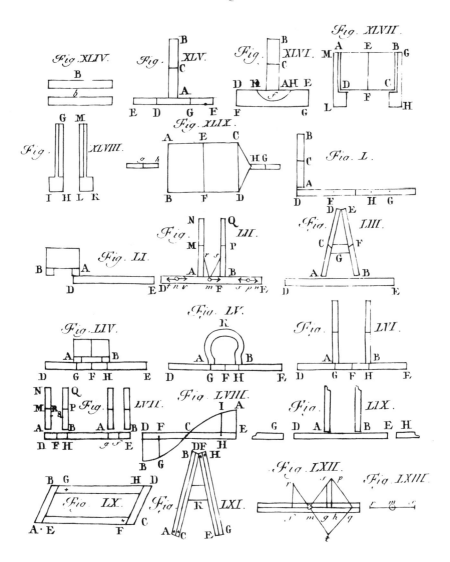

Figure 9. Facsimile Plate III from the *Essay*

GH, Fig. LX between iron parallelepipeds AB, CD exactly as in the *Michellian* method. Then I move the second pair of rods LM and IK to the middle of one of the rods e.g. EF, Fig. LXVII, as shown in the figure. I tilt them so that the angles LME and KIF included with the rod EF are sufficiently small, e.g. 15 or 20 degrees, but I move the ends of the rods M and I so close to one another that they are only a few lines apart. Then I draw them backward and forward over the rod EF, for the rest doing everything as in the usual *Michellian* method. After the rod FE has been rubbed on both sides in this way, I repeat the same operation on the other rod GH. Then I substitute the second pair of rods LM, IK, for the first, put them between the parallelepipeds AB, CD, and stroke them with the first pair of rods in the manner described above. Afterward I return to the first pair and, going in a circle, I repeat the operation until the force of the rods increases no further; at this point I finally stop the operation.

256) I have shown how, if there are two pairs of rods, their force can be increased to the maximum degree of which they are capable, but the whole operation succeeds similarly, though to a somewhat lesser degree, even if only three rods are used. Let us denote these three rods by the letters A, B, C, and let A be placed between the two parallelepipeds and stroked by the rods B and C according to my improved method of the double touch. After this, place the rod B between the iron parallelepipeds instead of A, and stroke it in a similar fashion with A and C, then do the same for C; then coming back to A, repeat the operation, going in a circle until all the rods are saturated with magnetic force.

257) If you want to use more pairs of rods e.g. 3 or 4 or 5, this whole operation is successful even then. If e.g. three pairs of rods are to be magnetized in this fashion let the first pair of them, EF, GH, Fig. LX, be placed in the usual way between the iron parallelepipeds AB, CD and then distribute the remaining two pairs of rods into two bundles; after putting them on the rod EF, Fig. LXVIII, in the manner indicated, stroke that rod and then the other rod GH with them according to the improved method of the double touch, and, going round in a circle as in the other cases, repeat the operation as many times as is needed to reach saturation level. Even though the operation succeeds well enough in this way (in fact magnetization is achieved perhaps more swiftly than if only 3 or 4 rods are used), nevertheless I do not suggest the use of more than four rods. Otherwise, there is an inconvenience one must be wary of, and in what follows I shall show what it is.

258) Frequent, almost daily, experience has shown me that this

method of mine is·more successful than the *Michellian* or *Cantonian* one, but if anyone still doubts the truth of this he will be able to test my assertion in the following way. He must prepare two similar and equal steel rods, hardened to an equal degree, and he must stroke the first of these with any pair of magnetic steel rods according to my method, and make the second pair magnetic with the same pair of rods, with an equal number of strokes after the *Cantonian* method; he will find that an appreciably greater degree of magnetism is always produced with my method. The reason for the superiority of my method over the *Cantonian* one is shown sufficiently by the previous discussion.

259) Apart from the fact that my method is more successful than the *Michellian* one, it also brings with it several advantages not found in the latter, and I think it worthwhile to give my readers an account of these. *Michell* and *Canton* always use several pairs of rods e.g. 3, 4, 5 and even more. But the drawback is that the greatest possible magnetism can be developed in only one pair of rods. For it is only the pair rubbed last that can be brought to saturation level, and the remaining pairs which were collected into bundles will have less than the saturation degree. If my method is used, employing not more than three or four rods, each pair of rods is always imbued with the maximum degree of magnetism of which it is capable. The reason for this is easily seen. In the *Michellian* method, the pairs of rods AB, CD collected into bundles are placed next to one another so that their like poles A and C, and B and D, mutually touch one another, and it is sufficiently self-evident that they then exert a mutual tendency to weaken the magnetic force which each possesses. It might occur to some that this harm could be corrected easily if in this case too only 3 or 4 rods were used, but for the rest the entire method were retained. But anyone willing to try this will find that the operation is tedious and it takes a long time before the saturation level can be attained. It is clear from this too why previously I was unwilling to advocate the use of more than three or four rods.

260) It is not very difficult to show that the method of the double touch administered in the usual way can err in that, even if the operation were repeated *ad infinitum,* there would be a possibility that the rods would never be brought to saturation level. Imagine that the two pairs of rods AB, CD, EF, GH, Fig. LXI,[115] have been gathered into bundles in the *Cantonian* fashion, and imagine that their angle of divergence CDFE is continuously diminished. It is clear from what

---

[115] The text has Fig. XLI.

has been discussed previously that in this way the force of the rods for magnetizing another rod can be lessened to the point where it becomes almost undetectable, or is at least made quite weak; this happens of course when the rods have been moved into full contact with each other. So it is possible for the angle of divergence to be made so small that the force corresponding to it is not great enough to impart the saturation degree to the rod being magnetized. Imagine therefore that you unfortunately happen to choose this angle, and it is evident that if you then stroke the rods EF, GH, even if the rods making up the bundles are saturated with magnetic force, nevertheless the degree of magnetism with which the rods EF, GH will be impregnated must remain below the saturation degree. It is clear *a fortiori* that this will be much more the case if the rods gathered into bundles do not enjoy the saturation degree. Therefore, if all the rods being used are endowed with but a weak magnetism at the beginning, and you toil at increasing their force in the *Cantonian* fashion, it is clear that the saturation degree can never be attained even though the operation is repeated innumerable times. But in my method there is no reason to fear this, for the closer the ends of the poles being used in the magnetization are brought to one another, the more the force is increased, and it is never diminished as in the common method. I can call on the celebrated *Canton* himself as a witness that what I present is consonant with experience, because he confesses that when he used only his own way of magnetizing, he could not impart to the rods the maximum degree of which they were capable. After the operation is over, he moves the two rods LM, IK, Fig. LXVII, in the tilted position to the middle of the rod EF which is to be magnetized, and, keeping the oblique position, he draws the rod ML toward E and the other rod IK toward F, and he affirms that the rod FE still increases in force.[116]

261) Anyone who has used the *Michellian* or *Cantonian* method in the preparation of artificial magnets will admit the experimenter's extreme fatigue during the operation, especially if he wanted to magnetize rods of rather large size and weight e.g. twenty or thirty pounds. If my method is used, the whole operation can be completed without difficulty, and experience itself testifies to this. A great convenience must be recognized in my method, because with a small and easily obtainable number of rods it can surpass everything the *Cantonian* method can only achieve with a great apparatus of rods.

262) I must not neglect to meet in advance a possible doubt concerning my improved method of the double touch. My instructions

[116] Canton, *Phil. Trans.*, XLVII, 1751–52, 37.

were that the experimenter should hold the rods LM, IK, with which he strokes the third rod EF, in such a way that they include with the rod reasonably acute angles LME, KIF, of approximately 15 or 20 degrees. But it may seem from what has been proposed above that it would be better to move the rods LM, IK to the rod EF in such a way that the angles LME, KIF completely vanish, that is, so that they completely touch the rod EF, and the reason I hesitate to allow this may not immediately strike the eye. Meanwhile I can easily demonstrate the reason for my decision. If the rods LM, IK, Fig. LXIX, are thought of as placed on the rod EF, and are then understood to be moved forward toward the other end, e.g. F, according to the method of the double touch, it is clear that the rod LM will come into complete contact with the rod EF over the whole of its length. Then since the like poles E and L of these rods will be immediately next to one another, it can only happen that the pole L will tend to weaken the magnetic force which the end E has received through the magnetization previously carried out. If my method is used, the ends L and E of the rods, Fig. LXX, remain more remote from one another and sensible harm need not be feared. The reason I adduce is valid, however, only for the case where the rod to be rubbed is of about the same length as those I use for magnetizing. So if it is intended to magnetize a rod, e.g. a magnetic needle, no more than half as long as the rods LM, IK, the rods serving to produce the magnetic force can with some benefit be moved to complete contact with it, so I usually use that method in this case.

263) The method of the double touch, whether it is administered after the *Cantonian* or *Michellian* fashion or after mine, tends constantly, as is clearly shown from the discussion so far, to propel the magnetic fluid, uniformly distributed through the rod DE, Fig. LXXI, as long as this was in the natural state, from the end D toward the other end E, and tends to make it rare near D, but more dense near E. So it is clear that this method never tends of itself to the production of consequent points in a rubbed rod, but rather, if there is no impediment, their production is powerfully prevented. It must not be thought, however, that it is in our power, using this particular method, to magnetize iron rods of any length in such a way that no consequent points occur. It can easily be shown that rods can be produced of such a length that even if it were supposed they had only 2 poles initially, yet several consequent points ought to arise spontaneously. The longer e.g. the positively magnetic part FG of the rod DE is imagined to be, the more its force repelling particles of magnetic fluid must be assumed to increase at the same time; so finally so great a

force ought to be reached that the fluid contained in the contiguous part DG is compelled to leave its place and move to more distant parts. When this happens, the parts of the rod close to G ought to be evacuated of fluid and to become negatively magnetic. In a completely similar way, if the negatively magnetic part FH of the rod becomes continually greater and greater, because its attractive force increases simultaneously it will finally grow to such a size that it will compel the fluid included in the further part of the rod HE to move toward the part near the point H and to be accumulated there.

264) So although neither the positively magnetic nor the negatively magnetic part of the rod can be extended to infinity without being destroyed by its own force, it is clear that each part can be extended to a greater length if the rod consists of strongly hardened steel than it can if this is of quite soft iron. For of course the greater the hardness of the iron, the greater the difficulty with which the magnetic fluid moves through its pores, and it is self-evident that, the greater this difficulty, the greater must be the force of the positive pole GF and the negative pole FH before the repulsive force of the former can evacuate the neighboring part G and before the attractive force of the latter can fill the part near H with fluid. It is clear then that, other things being equal, in a harder steel each of the positive and negative poles can be extended to a greater length without being destroyed by its own force than is possible in a softer steel. In a completely similar way, a wire of any metal can be extended to a length where it is destroyed by its own weight, but other things being equal, a harder metal supports a greater length than a softer one before this happens. Finally, readers will be able to see from this why I affirmed above that, if a suitable mode of magnetizing is used, soft iron is more suitable than hardened steel for receiving consequent points.

265) Eager to see whether experience agreed with my reasonings, I used my method to magnetize an iron wire of sufficiently soft iron, 4 feet long, and I found that it acquired four poles, a north DG, Fig. LXXI, a south GF, again a north FH, and a south HD. I then took a steel rod of the same length as the former, not hardened to the highest degree but enjoying the degree of hardness usually given by workmen to spring steel, and when I magnetized it in a completely similar fashion I found that it gained only two poles. I dare not assign the length to which hardened steel rods can be extended without contracting consequent points, for this, it seems, can only become known after repeated experiments, and I have so far been unable to do these. However, I think it is clear from the tests I have described that it is possible for strongly hardened steel rods appreciably longer than 4 feet

to be magnetized by this method without producing consequent points.

266) If anyone used to physical and geometrical disquisitions, and with a knowledge of how faithfully nature preserves the law of continuity, has been willing to ponder this somewhat, he will easily understand that it is sufficiently established that, in a long rod endowed with consequent points, magnetic fluid certainly cannot be uniformly distributed through the parts DG, GF, FH, HE in such a way that in every point of, say, the part GF it is condensed to the same degree as in every other, and the magnetism changes at G from the positive suddenly and sharply, as it were, to the negative, for this is completely contrary to the custom of nature. If the curve IGKFNHP is the curve of the densities (and what we understand under that name has been shown sufficiently above), this will be of such a nature that it will intersect the axis at G, F and H, but between pairs of intersections there will fall certain maximum ordinates such as KL,[117] MN here. In a similar fashion, it is easily seen that if the rod is too short to acquire consequent points, while it is then possible that the curve of densities LFM, Fig. LXXII, is such that the ordinates constantly increase as one proceeds from F toward D and E, and the very last of them DL and EN situated at the ends are the maxima, it can also be the case that the maximum ordinates are RQ, NP, Fig. LXXIII, pertaining to the points Q, P, somewhat removed from the ends; this last ought especially to take place in rods extended to a somewhat greater length.

267) It is easy to see that experience agrees excellently with these reasonings. If a magnetized rod endowed with two poles is tested, a fixed point will be found in either pole, from which it can carry a greater weight than from the rest, and which is therefore magnetic to a greater degree than the rest. When the test is made, it will be observed that these points possessing the greatest strength usually fall at the very ends of shorter rods but in longer rods they are ordinarily situated somewhat away from the ends toward the middle.

268) It is a matter of great moment, as much for magnetic needles (as I shall show elsewhere) [118] as for artificial magnets, that these points which are magnetic to the highest degree fall either at the very ends, if this is possible, or at least as close to the ends as possible. It is sufficiently clear that the method of the double touch, whether performed in the usual way or by my method, is useful in this respect too. For since this method of magnetizing constantly tends to propel all the fluid contained in the rod from one end to the other, it ought to make

---

[117] The text has GL.
[118] Aepinus, *Acta Acad. elect. Mogunt.*, II, 1761, 264–65.

the points of maximum repletion and maximum evacuation as close to the ends of the rod as possible. So the method of the double touch should be thought to surpass the others commonly in use in this respect too.

269) I could properly perhaps leave the discussion of the communication of magnetism at this point. For, though it is not a dearth of phenomena pertaining to it that forces me to do so (there are indeed innumerable things of this kind), I seem to have considered that which is special, and to have given my readers an explanation of my theory sufficient to enable them to carry it over easily to the remaining phenomena. Meanwhile I think it advisable to add certain remarks about the means of impregnating natural magnets with a strength greater than they have received from nature, and of making curved artificial magnets (they are usually made curved). I will say nothing here about the best method of animating magnetic needles, even though such a discussion is relevant at this point, since I shall treat it more fully elsewhere.[119] If, first of all, I intend to make curved artificial magnets, I usually proceed highly successfully in the following manner. I ensure that the highest possible degree of hardness is induced in a magnet of this kind, and when it is to be magnetized I place it on a firm wooden slab. From a number of magnetic rods I have ready, I select the biggest and strongest pair, and move these two rods to the extremities of the magnet as shown in Fig. LXXIV. I make sure that the rods AC, DB lie in a straight line and that the north pole of the rod AC touches that end of the magnet which is to become the south pole, and in turn the south pole of the rod BD is moved to the future north pole of the magnet. After this, as I have already instructed, I fix the magnet CKD and the rods AC, BD in position so they cannot easily be dislodged.

270) When this has been done I select another pair of very strong magnetic rods and I stroke the magnet CKD with them. I use the improved method of the double touch in this operation, and I tackle it in the following way. I move the rods FE, GH in the required manner to the middle of the magnet K, inclining them as much as possible so that they are almost parallel to the plane of the magnet, and holding the two of them in a straight line with their ends E, G a line or so apart. I then draw the two rods backward and forward over the magnet CKD according to the method of the double touch, taking care that the rods FE, GH always lie in a straight line and that they are parallel to the tangent of the curved magnet in the region near the poles E and G.

[119] See Appendix, items **31A(i), 31B, 33**.

After this has been done twenty or more times, I draw the rods back to the middle, K, and then I take them off. I also perform this operation on the other side of the magnet CKD, and then I always obtain a magnet of outstanding strength.

271) In this whole method there is nothing for which the reason is not either self-evident or apparent from the previous discussion, except the one artifice, my own I think, in that I place the rods AC, BD, in the one straight line and not parallel to each other as usually happens at other times; and I can affirm from frequent experience that this is beneficial. Imagine that the rods AC, BD, Fig. LXXV,[120] are put in a position parallel to the magnet CKD,[121] with e.g. the pole C positive, and D negative; and let the points of mean directions (a name I have previously explained) [122] fall at f and g. It is then obvious that m, a particle of magnetic fluid in the leg CK of the magnet, is repelled from the positive pole C toward the interior along the line fm,[123] but it is also clear that the negative pole D, if sufficiently close, exerts some action on the particle m, by attracting it. This latter force which acts along the line gm can be resolved into two forces, one mq parallel to the direction fm, the other mn normal to it; the former of these forces is opposed to the action of the pole C. Thus the negative pole D weakens the action exerted by the postive pole C and prevents it from achieving its full effect. But it is clear that the harm to be feared from this becomes less, the more the angle gmf which the directions of the forces include is increased toward a right angle, or (and this amounts to the same thing), the closer the point g becomes to the end D. All this is valid too for the negative pole D whose action is equally impeded by the positive pole C, the more the point f is distant from the end C. The easy remedy for this inconvenience is to arrange the rods AC, BD, so that they lie in a straight line, for it is evident that the directions according to which the actions of the poles C and D are exerted, that is fm and gm, are almost directly opposite to one another, so they cause one another either no harm or very little.

272) Curved artificial magnets are usually made so small that they cannot be conveniently stroked by steel rods, this usually being done precisely for its striking results. For such magnets of quite small mass can carry a very large weight compared to their own, even 50 times greater, and this is not the case with larger magnets. I apply such magnets, in the manner previously described, to two strongly magnetic

[120] The text has Fig. LXXIV.
[121] The text has CRD.
[122] See above, p. 375.
[123] The text has fn.

steel rods AC, BD, Fig. LXXVI, placed in the same straight line, and I leave them in that position for several hours or even days. To this I ordinarily add yet another artifice which I have found very useful in constructing magnets of this kind of small mass. In the manner described above, I move the end of the magnet to the rods AC, BD, but at the same time I move up to it another pair of rods FH, GI in such a way that the leg KC lies over the end F and the leg KD over the end G. I place a third pair of rods on the magnet in such a way that the leg KC is suitably covered by the pole of the former and the leg KD by the pole of the latter, and I leave the magnet at rest like this for some time. Similar artifices can be used, and anyone can easily think of things like this, if the pieces of iron to be magnetized cannot be stroked by rods according to the method of the double touch because of either their unusual shape or their small mass.

273) It is very well known that natural magnets can acquire artificially a greater degree of magnetism than they received from nature, for mention of methods of doing this is made on all sides by authors who deal with artificial magnetism.[124] Generally a parallelepiped shape is given to these magnets and very often they are too short to be magnetized according to the method of the double touch. Whether or not it is too short, the magnet must be taken from its own armature and placed between two strongly magnetic rods BE, CD, Fig. LXXVII, in such a way that the side B which is to become the north pole is touched by the south pole of the rod BE, and C which is to become the south is touched by the north pole of the other rod CD. If you do not possess rods of a mass sufficient to cover the whole polar regions B and C of the magnet, place a sufficient number of smaller rods next to one another as shown in Fig. LXXVIII. If then the length of the magnet permits, stroke it by using the improved method of the double touch. If it is too short to allow this conveniently, it is better if it is placed, as the figure shows, between several magnetic rods BE, GF, HI, Fig. LXXIX, and CD, MN, KL, and left at rest in that position for quite a long time. After this, the armature must be joined to the magnet again and the magnet moved so that each foot of its armature rests on one of the two strong rods BE and CD, Fig. LXXX, placed in a straight line, and it must be left in that position for several hours; for theory teaches and experience testifies that, as a result, its force is appreciably increased further. I have more things to say, both new and worthy of note, about the method of increasing the strength of natural magnets, but these can be discussed more conveniently below.

[124] Methods of increasing the strength of natural magnets had first been developed a few years earlier by Dr. Gowin Knight (*Phil. Trans.*, XLIII, 1744–45, 163).

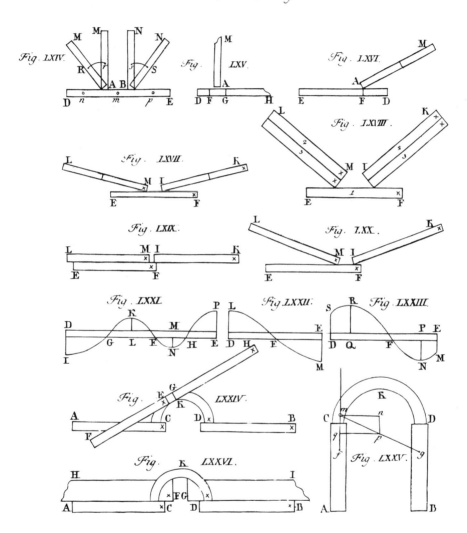

Figure 10. Facsimile Plate IV from the *Essay*

CHAPTER IV

Concerning Certain Phenomena of Bodies Immersed in Electric and Magnetic Vortices, and Concerning the Magnetism of the Earth

274) [125] Since I shall be using the terms vortex or electric atmosphere or magnetic atmosphere in what follows, it is advisable for me to declare at the very start the sense in which I employ these words, so that my readers may not attribute to them a significance foreign to my meaning. It is manifest from the preceding explanations of magnetic and electric phenomena that I never consider magnetic or electric matter as clinging outside the body or ambient to it, so it is clear that I am not using the words vortex or atmosphere in their proper sense. Whenever these words occur in what follows, nothing else must be thought to be denoted by them but what at other times is usually called the sphere of activity. For me these words designate nothing but the space to which the attraction and repulsion, electric or magnetic, sensibly extends itself in any direction around a given body.[126]

275) Some readers may be offended because with these words I deny the existence of any electric atmosphere, properly so called, for in general hardly anyone is to be found who doubts the existence of an atmosphere of this kind, and in fact the very celebrated *Franklin* himself, although his own theory suggests otherwise, declares himself to be of this opinion.[127] So it will be very proper for me to expound more clearly what I think about this subject. It is obvious from the previous discussion that I derive various phenomena of electricity, especially those which do not depend on the actual transit of matter from one body to another, from the action of electric matter enclosed in the pores of electrified bodies, and that I never have recourse to a vortex or atmosphere properly so called. I do not deny that the surrounding air close to a given body, either positively or negatively electric, being imperfectly electric *per se*, acquires the same electricity

---

[125] Many of the cross-references and paragraph numbers in this chapter are given wrongly. However, the error is always the same; the numbers given in such cases are consistently two smaller than they ought to be. This suggests that two of the early paragraphs in the chapter are late insertions into the text; §§ 275–76 seem the most likely candidates.

[126] See above, p. 109.

[127] *Benjamin Franklin's Experiments*, pp. 214–16.

CERTAIN PHENOMENA   393

as the body which it touches, though it does so slowly and to a weak degree, and only to a small distance from the body which it surrounds. So if anyone wanted to call this electrified air the "electric atmosphere," I would not quibble; but I warn that little or nothing in the production of many electric phenomena should be attributed to this atmosphere. This is clearly seen if a pair of bellows continuously whips up a wind which tears this atmosphere away from the electrified body, for nonetheless the phenomena take place as they usually do if the atmosphere surrounds the body undisturbed.

276) There will be some readers however who will think this hardly satisfactory; they will object that I am denying the existence of something which is directly apparent to the senses. For e.g. when the back of the hand is moved toward an electrified body or touches it, we are convinced of the existence of an electric atmosphere, it is believed, since we experience an entirely similar sensation as we do when a fine net, such as a spider's web, touches the hand. But must not this sensation be thought to be excited as much by the action of the electrified body on the electric fluid contained in the hand as on the parts of the flesh themselves? If that sensation arose from some atmosphere, it is evident that if you were to stand on pitch or some other suitable support and at the same time make yourself and another body electrified, and you were then to move your hands to that body, the same sensation ought to be produced as when you were standing on the ground. But experience teaches the contrary of this, for it shows that the experimenter, if he is electric to the same degree as the body to which he moves his hand, feels nothing at all. The arguments founded upon the electric odor observed in the vicinity of electrified bodies are not of great weight and are completely irrelevant at this point. For no matter to what degree any body is electrified, this odor is never noticed unless that body has one or two points or sharper, rather prominent parts from which or into which electric matter flows in the form of the well known electric brush. So this phenomenon must be referred to things which arise from the actual transit of electric fluid from one body to another.

277) Now that I have presented the warning which seemed necessary so that my terms would not be twisted in a sense foreign to my meaning, let us begin a more careful discussion of the phenomena to be investigated here. Imagine that a magnet EF, very freely mobile in any direction around the point D, is brought close to a rather large immobile magnet BC, Fig. LXXXI, and immersed in its vortex. Suppose further that AB is the positive pole of the latter and DF of the former, and similarly that AC and ED are the negative poles of

each. It is already clear from what has been discussed above that the part AB acts on the part ED by attracting it and on the part DF by repelling it, but on the contrary the part AC attracts DF and repels DE. And so it is evident that the magnet EF will not come to just any position, but will always take up the position in which the attractive and repulsive forces just enumerated are in equilibrium.

278) It is in general easily understood from this that, under the assumed circumstances, the mobile magnet EF will always take up various definite situations according to the various positions of the magnet BC with respect to EF. It will not be superfluous to inquire a little more exactly into the variation of that situation. Imagine first that the magnet EF is so arranged that its center d lies on the axis of the larger magnet BC, and it is readily clear that whatever position the magnet fe is imagined to have, there cannot be an equilibrium of the forces acting on it, unless the magnet fe coincides with the axis of the magnet BC, and is aligned with it. So, left to itself, it will be unable to come to rest until it has taken up a position where it is aligned with the magnet BC, and the unlike poles of the magnets BC and fe face one another.

279) I confess that the position I have described, in which the forces urging the magnet fe are in equilibrium, is not unique. For suppose that the magnet fe, Fig. LXXXII, is completely inverted so that it again coincides with the axis, but the like poles of the magnets BC and fe face each other, and it is readily clear from the principles of statics that, even in this position, the forces are in equilibrium. But I have not concerned myself with this position because it cannot be preserved. Suppose that the magnet ef is disturbed from its position for a short time, so that it includes with the axis an angle as small as you like. Since the magnet BC, Fig. LXXIII, repels the like pole of the other magnet with its own end C, and attracts the unlike one, it is clear that it repels the pole de in the direction mn and attracts the pole df in the direction pq. These forces do not tend to return the magnet fe to its former position, but act to turn it toward er and thus to give it a completely contrary position. So even if the magnet fe is imagined to have initially the position where the pole de faced the like pole of the magnet BC, it will hardly be able to preserve it for even a moment, since the slightest disturbance from its position will cause it to keep on moving more and more away from that position, until it acquires a position completely contrary to that which it had initially. So it is evident that although the position where the like poles of the magnets BC, fe are turned to one another is possible in principle, it is never found in practice, and we need not bother considering it.

280) The situation is far different in the case where the unlike poles of the magnets face one another. For suppose that the magnet *ef*, Fig. LXXIV, is again disturbed from the rest position; since its part *de* is already attracted by the pole of the magnet BC opposite to it in the direction *mn*, and the part *df* is repelled in the direction *pq*, and these forces try to rotate the magnet *ef* in the direction *er* about the point *d*, it is obvious that it will be drawn back to its own former position, which it will thus recover immediately. So this position is the only one which can be preserved and, as a result, it is the only one that can take place in practice. But let this suffice, for I do not want to be long-winded about a matter which is sufficiently obvious and clear to anyone who understands the elements of statics.

281) Consider now that from the center of the magnet BC, Fig. LXXXI, at the point A, a perpendicular Aδ of arbitrary length is erected to the axis, and the magnet EF or $\epsilon\phi$ is so placed that the point δ is found somewhere within its length. It is then very readily understood that the magnet $\epsilon\phi$ will not come to rest until it has achieved a position parallel to the axis CD, but in such a way that the unlike poles of the magnets face each other. For it is clear that only in this position is there an equilibrium of forces. It is true that here also there is still another rest position (that of course is the one where the position is directly contrary to the magnet $\epsilon\phi$), but it will readily strike anyone considering this matter that for this position too the same things are valid as were discussed in § 279.[128]

282) If finally the magnet EF is so arranged that its center is found neither on the axis nor on the perpendicular Aδ, but in some intermediate position, it is clear that it ought to assume a situation oblique to the axis BC such that it reclines a little from the pole AB from which it is a long way off, but inclines toward the other AC to which it is closer, in such a way that the negative pole of the magnet EF is still directed toward the region in which the positive pole of BC is found, and vice versa. Although in this case too there is another rest situation, directly contrary to the one just described, this situation, as is readily clear from the elements of statics, is such that it cannot be maintained, and it is obvious that everything said in the similar case in § 279 [129] can be applied to it.

283) Although the previous discussion makes it in general sufficiently apparent what situation ought to be assumed by a mobile magnet left to itself in the vicinity of a second magnet, it is impossible for us to determine exactly the law according to which these variations of situation occur. For here our theory deserts us, in part since the law

[128] The text has § 277.
[129] The text has § 277.

cannot be determined in the absence of a recognized function for the magnetic attraction and repulsion (which is so far unknown), and in part because numerous different cases can occur whose infinite variability prevents any general law from being established. For it is clear, as I argued above, that the positively magnetic part AB and the negatively magnetic part AC will not always be equal, and the magnetic fluid will not always be distributed through the parts AC and AB [130] in the same way, so now one and now another curve can serve as the scale of density. It is evident that with these variations the law describing the situation of the mobile magnet EF ought to vary at the same time, so it is clear that amid such a variability of circumstances we will toil in vain in seeking a constant law. In addition to this the magnetic body BC itself, which may be of innumerable different shapes, e.g. spherical, elliptical, or even of a completely irregular shape, is capable in such cases of producing now one, now another law for determining the situation of the mobile magnet. As a result it is also readily obvious that there is no hope that a constant law of this kind can be uncovered with the help of experience.

284) It is agreed that the magnetic needle (and no proof is needed for its recognition as a true magnet), brought to various points of the surface of the earth, successively assumes now one, now another position, and varies that position in a fashion quite similar to what happens when a mobile magnet immersed in the atmosphere of a larger immobile one is moved around the latter. To determine the position which the magnetic needle assumes of its own accord in any place, it is necessary to refer its position to two planes of given and constant position, whose location can easily be fixed. For this reason horizontal and meridional planes are chosen and the position of the magnetic needle is referred to them in such a way that an investigation is made of the angle at which the needle inclines to the horizontal plane, and similarly of the angle which the vertical plane passing through the needle includes with the meridional plane; the first of these angles is usually called the *inclination* of the magnetic needle, the latter its *declination*.

285) [131] It is sufficient for our purpose to expound only in general terms what has been observed about magnetic inclination and declination. The things to which we must especially turn our attention are the following:

1) The declination of a magnetic needle is nowhere very great, and has so far not been observed to rise beyond 30 degrees. So the

[130] The text has AC and CB.
[131] Incorrectly numbered 283 in the text.

magnetic needle does not stray far from the meridional line anywhere, though its declination varies from place to place. There are of course places where the needle falls in the meridian plane itself, and other places where it declines a few degrees or more to the east or to the west.

2) All the points for which the declination of the needle is zero are situated on a curve, an irregular and variously contorted line circling the globe of the earth, and the same may likewise be said of the loci to each of which there corresponds another declination, whether this be e.g. 5, 10, or 15 degrees.

3) This declination is mutable, and varies sensibly with the passing of time. But what law nature follows in that mutation is not yet understood. For it is so far not known whether the curves just mentioned only move over the surface of the earth very slowly, or whether these lines also experience another mutation in addition to the progressive movement, in such a way that they slowly acquire now one, now another curvature.

268) As for the magnetic inclination, it must be particularly noted that,

1) toward the equator it is usually zero, though this must not be taken as rigorously true. For the points at which the inclination is zero do not fall right on the equator of the terrestrial globe but only close to it.

2) Moving away from the equator toward the north pole of the earth, the point of the needle facing north is depressed more and more below the horizon, so that as the latitude increases, the inclination increases with it, and finally it arrives right at 90 degrees. It must not be thought however that this law is observed exactly. For the inclination is not the same at all points on the same parallel, and the point of maximum inclination, where the needle stands vertically to the horizon, does not coincide precisely with the pole of the earth, although it may be concluded from the way in which the inclination increases that it is not far from it.

3) The same is completely valid for the point of the needle facing south, as we progress from the equator toward the south pole.

4) The inclination too, just like the declination, does not remain constant at the one spot, but changes with the passing of time, according to some law that is so far completely unknown.

287) Anyone who is willing to weigh up these things a little will easily perceive that the situation of a magnetic needle brought to various places of the terrestrial globe changes in a way not unlike that in which it changes when the needle is led round an immobile

spherical magnet. As a result several of the old Physicists (and some of the more recent) have seized the opportunity to conclude that it is very probable and hardly open to doubt that the globe of the earth itself is endowed with magnetic force, and, just as is the case with an immobile magnet and a magnetic needle, the attraction and repulsion which it exerts on a smaller freely mobile magnet must be recognized as the sole and true cause of the directive force of a magnet, and of the magnetic needle.[132] So that we may admit the truth, the accuracy of this proposition seems clear from the phenomena just adduced, so that nothing further ought to be needed to prove it fully. Meanwhile readers will become more and more convinced of the truth of this matter in what follows, for I shall show that for the remainder of the magnetic phenomena too, the terrestrial globe shows exactly the same things as are usually shown by a magnet of smaller mass.

288) I have assumed therefore as an undoubted principle that there is included within the globe of the earth quite a large core made of matter which is ferreous in character, and imbued with magnetic force so that one of its hemispheres DHE, Fig. LXXXV, possesses a positive magnetism and the other DIE a negative magnetism. It must be noted however that the plane or surface DE which separates these hemispheres, though it differs little from the plane of the terrestrial equator BC, does not completely coincide with it, and that the shape of this core is in some way irregular, as is readily concluded from the laws of magnetic declination and inclination previously discussed and from the irregularities that occur in both. Certain questions arise from this supposition of ours and my readers will doubtless expect me, when writing about magnetic theory, to unravel them.

289) If it is conceded that the terrestrial core is made of some ferreous kind of matter and very probably of mineral iron, the question can be raised whence that core received the magnetic force it possesses, and what was the cause that forced the magnetic matter to be expelled from one hemisphere and retreat to the other. But I freely confess that I know no mechanical and efficient cause of this which I could put forward with any probability, for it is obvious enough that the daily movement of the terrestrial globe around its own axis, which might be suspected as the cause of this phenomenon, cannot be the cause of the propulsion of magnetic matter from one pole toward the other. But would it be altogether out of the question to say that this phenomenon is one of those for which we can recognize no efficient mechanical cause, and which must be derived from the immediate

---

[132] Gilbert, *On the Magnet*, esp. Book VI.

action of the creator of the world? For it cannot be denied that there are innumerable such phenomena, unless one dares to affirm that the whole world has been formed according to mechanical laws alone and without any concurrence of some intelligent being, or to take the position either that God does not exist at all, or that he has taken no part in the production of the world except as an inert spectator.[133] But not only *a priori* but also *a posteriori* we are certain that no cause of the phenomena can be recognized other than the action of the very wise creator of the world. Thus e.g. to be eager to investigate the mechanical reason for the terrestrial globe and the other planets revolving around the sun in a direction from east to west and not in the opposite direction is to undertake a vain task, or to investigate why the globe of the earth completes a full revolution round its own axis in exactly twenty-four hours, or why the earth has only a single companion or satellite while Saturn is attended by five secondary planets, is also useless, as is well known to anyone sufficiently skilled in the true system of Astronomical physics revealed by the great *Newton*.

290) But from it this new question arises, namely, what must be thought the final cause which could move God to impress magnetism on the terrestrial globe? For if we seek the cause of phenomena that do not depend on mechanism alone, recourse must be had to final causes, not to efficient courses. But here my readers will once again find me frankly confessing my ignorance. For although the magnetic force belonging to the globe of the earth is of great utility to the human race, in so far as it is the cause of the directive force of the magnetic needle which points the path over the seas just like, as *Gilbert* says, the finger of God,[134] yet I doubt that this could be recognized as its primary end. For in part the use of the magnetic needle was unknown for very many centuries, so it must be conceded that this remarkable property of the terrestrial globe was sterile for a long time; and in part, this opinion seems to rest on the belief that all things are created solely for the utility of man, and this indeed, when admitted without restriction, has stained with many errors that most noble part of Physics which deals with the ends of things.

291) Though I consider that to find the cause that first impressed magnetism into the terrestrial globe, recourse must be had to the very wise will of the creator and his most powerful action, yet it is different if a reply must be given to the question: why does neither the declination nor the magnetic inclination remain the same in the same place

---

[133] Aepinus thus rejects altogether the Cartesian ideal of ascribing a mechanical cause to every natural phenomenon.

[134] Gilbert, *On the Magnet*, p. 147.

but rather change slowly and perpetually? For it would be out of place to want to ascribe this to the immediate action of God. In general, it is obvious here that the magnetic core of the earth is subjected to a slow but continuous mutation, but I dare not determine whether that mutation consists in a variation of the shape of the core, or of the distribution of magnetic matter through each of its hemispheres, or of the position of the whole core in respect of the other parts of the globe of the earth; or what is the cause of these mutations themselves. The talented *Halley* is of the opinion that the magnetic core of the earth continually and slowly changes its position in respect of the globe of the earth,[135] and, in truth, that hypothesis seems to agree well enough with the observations that have been carried out until now. But I am unable to accede to *Halley's* opinion because I judge that insufficient observations have been carried out to date to remove every doubt completely. If it were definitely established that the curves about which I spoke in § 285 no. 2 [136] always remain the same, and are subject to no mutation, except only that they move in a slow motion around the surface of the earth, I would not hesitate to side with *Halley*, but I have already warned above that the truth of this is not properly established, and that perhaps the form of these curves is subject to variation also. But it is evident that, if this latter were the case, it could very probably be concluded that the core itself undergoes certain slow mutations either in shape or in the distribution of magnetic matter through it.

292) I think I will have chosen the better part in a matter so uncertain if I suspend judgment and leave it to coming ages to collect enough observations and then finally to put the decision on this matter beyond doubt. I have indeed considered at times that the motion of the earth around its own axis might be the cause of these mutations. For if GDHF, Fig. LXXXVI, is the hemisphere in which magnetic matter abounds, and GIEF on the other hand is that in which it is deficient, and if the terrestrial globe is supposed to perform its rotations round the axis HI, which is different from the magnetic axis, this is the same as if the terrestrial globe is thought to be at rest, but that any particle of magnetic fluid included in the core, e.g. G, strives to move in the parallel G*m* in a direction contrary to the motion of the earth. But if we suppose that all the particles of magnetic fluid in the core attempt to move in this way in circles parallel to the equator, it follows that the whole hemispheres GDHF and GIEF attain as it were a slow movement around the axis HI, and the mag-

[135] Halley, *Phil. Trans.*, No. 195, 1692, 563–78.
[136] The text has § 283 num. 2.

netic poles K and L revolve at a slow rate around the poles H and I of the diurnal motion in circles Kp, Ln, tending continuously from the east toward the west. Although initially this hypothesis seemed not inelegant, and probable enough, I abandoned it after a thorough examination. It is clear that it cannot be true unless it can be supposed that the particles of magnetic fluid are at rest and do not participate in the motion of the rest of the terrestrial globe. For it is obvious that, if they try to move with a velocity equal to the other constituent parts of the globe of the earth, in the same direction, they have no tendency to cross through the pores of the magnetic core in a direction contrary to the diurnal movement. But there can be no doubt that all the particles of magnetic fluid should tend from west to east with a velocity equal to the other parts of the terrestrial globe, for even if it had not pleased the very wise creator to impress in them the same motion as the rest of the terrestrial globe from the beginning, it is still clear from the elements of mechanics that long ago over the course of so many centuries, forthwith in fact, they ought to have completely acquired the common motion and velocity. For the same reason a very polished globe placed on a similarly highly polished horizontal plane is never seen to progress of its own accord from east to west, which ought to happen unless the globe tended to progress from west to east [137] with the same speed as the remaining points of the surface of the earth.

293) It is clear from this that, with the highest probability, the directive force of magnets and of magnetic needles should be considered as a secondary phenomenon, because it depends basically on a magnetic force inherent in the core of the terrestrial globe. If my readers still doubt the existence of a magnetic core of this kind, I trust that their doubts will be completely removed by what follows, for I am going to turn to other things. I have not so far discussed the phenomena of one electrified body close to another. Although I have little to say on this topic, I ought not completely neglect here a comparison of the electrical phenomena with the magnetic ones. Consider therefore that a body BC has been electrified so that the part AB is positively electric and the part AC negatively electric. Move close to it another electrified body EF whose part ED is negatively electric and part DF positively electric, and which is freely mobile round the point D. Since the laws which the electrical attraction and repulsion follow are completely similar to the laws of magnetic action, it is evident that all those things we previously considered in § 277–283 above [138] about

[137] The text has *ab ortu occasum versus*.
[138] The text has § 275–281.

the position which the body EF ought to assume when placed in various positions relative to the body BC, if the bodies BC and EF are magnetic, are valid also for the case we are considering here. There is no doubt that in this case too there is agreement between electrical and magnetic phenomena, and were the globe of the earth electric, as it is magnetic, there is no doubt that electrical needles could be constructed so as to take up a fixed direction, just as we are accustomed to construct magnetic needles.

294) These things are sufficiently self-evident and there is no need to test them by experiment. For experiments which illustrate only what is sufficiently established as true from elsewhere are empty and ludicrous. Yet anyone who wants to see the truth of the matter with his own eyes will be able to proceed in the following manner. Take a thin glass jar BCDEF, Fig. LXXXVII, of medium size with a narrow neck. Cover this externally up to CD with metal leaf and fill it with water up to the same height. Put an iron wire AB into the orifice of the bottle, B, projecting about a foot from the bottle, and fix on its point a lead globe A. Then on the exterior surface of the bottle fix a similar curved wire GH, reaching to the same height as the former wire, and likewise carrying a lead globe, H. Prepare another little machine FEAIKB, Fig. LXXXVIII, in the following manner. Glue metallic plates on each side of a square plate of thin glass AB, about 2 inches square; these metal plates CD, GH must be a little smaller than the plate AB, so they do not cover the whole of it, but leave a margin uncovered around the whole circumference. Finally fix metallic wires FE, IK to each plate and suspend this little machine from a silk thread in such a way that it can turn itself freely in any direction.

295) If it is desired to perform an experiment with the help of the instruments thus prepared, proceed in the following way. First fill the jar BCDEF with electricity in the usual way, then place it on a glass support and leave it to itself. It is established that then the wire BA and the globe A will be positively electric, and the globe H and the wire GH sustaining it will be negatively electric. Electrify the wire IK of the little machine FEAIKB while keeping the hand applied to the other wire FE, and it is obvious that in all essentials the little machine becomes a type of *Leyden* jar. When it is suspended it will be brought to the state where the wire IK gains a positive electricity and the wire FE a negative. But if this little machine, freely suspended, is moved to the jar BCDEF, or rather to the globes A and H which play the role of the poles here, it will take up completely similar positions to those a magnetic needle usually assumes next to another immobile magnet.

296) We have so far expounded what ought to happen to a magnet which is close to another magnet, but we need also to consider certain phenomena which occur if unmagnetized bodies, though ones capable of magnetism, e.g. soft iron, are moved to a magnet. From the preceding chapter, it is easy to make a judgment on this matter, for it is evident that a piece of iron FE, Fig. LXXXI, set up in the neighborhood of a magnet BC, ought soon to become magnetic itself according to the law that the end DE facing the positive pole AB of the magnet will become negatively magnetic, but the other end DF turned away from the negative pole AC of the magnet will become positively magnetic. So if the iron EF is suspended so that it can turn freely, everything is valid for it that I said previously about the magnetic needle and the position it assumes in the vicinity of another magnet.

297) Suppose now that a magnetic needle or alternatively an oblong iron bar EF, Fig. LXXXIX, has been set up so as to be freely mobile around a point D in the vicinity of the magnet BC, and imagine at the same time that the needle EF is quite short, so that its length can be taken to be zero. Let this needle, when its center is placed at D, assume a position such that it falls along the line EN. Suppose that the needle is moved forward an infinitely short distance toward N along the line EN so that its center comes to G, and it is clear that since its end F then approaches somewhat toward the pole AC and the end E recedes somewhat from the other pole AB, its position ought to change a little, in such a way that it now takes up a position HI and falls along the line HO which includes an infinitely small angle EGH with the line EN. If the center of the needle is moved along the line HO again so that its center arrives at K, the same happens once more and the needle takes up a position such that it falls along the line LP which includes an infinitely small angle HKL with the line HO. Since the same happens constantly whether the center of the needle is moved toward B or toward C, it is evident that the center will describe a curve QFGHR which enjoys the property that no matter where the center of the needle is located on this curve, the needle disposes itself of its own accord so as to lie on the tangent to this curve.[139]

298) It is difficult, in fact it is almost impossible, to determine what this curve is, partly because the function for the magnetic attractions and repulsions is unknown, and partly because the varying shapes of the magnet BC, and various distributions of magnetic matter through it, can give rise to innumerable irregularities in these curves. To illustrate rather than demonstrate the nature of the curve, imagine

---

[139] Aepinus thus formally defines a magnetic line of force in the way that has since become standard.

that the whole force of the positive pole AB, Fig. XC, and of the negative pole BC has been forced into the points M and N, and that the forces which these points exert, both attractive and repulsive, are equal, and let us investigate what the shape of our curve ought to be under this assumed hypothesis. If we examine the action which the point M, Fig. XCI,[140] exerts on the needle EF, it is clear that it attracts the end E of the needle, but repels the other end F, in the directions ME and MF, respectively. This double action conspires to rotate the needle in the direction Eg, whence we can substitute for this double action a single force acting on E and attracting the point E toward M. Exactly the same applies to the action of the point N, so this too may be considered as a single force applied to the point F and drawing it toward N. If we now consider that the point E of an infinitely small lever EF is attracted in the direction EM, and the point F in the direction FN, it is clear that these forces, which are oblique to the lever, can be resolved, the former into a force acting in a direction EI parallel to the lever, and another in the direction of the normal EL; and the latter into a force which draws the lever in the parallel direction FH and the normal direction FK. Of these forces, those which act along EI and FH are cancelled out, for we are supposing that the point D is immobile, so in determining the position of the lever we need consider only the normal forces acting along EL, FK. But it is evident that the lever ought to assume a position such that the force acting along EL is equal to the other acting along FK.

299) Imagine further that the forces exerted by the points M and N are reciprocally proportional to the distances ME, NF, and if we call $NS = x$,[141] $DS = y$, $MN = a$, the force acting along EM, because the

[140] The text has Fig. XLI. In any case, however, the reference would be better placed a few lines further on.

[141] The point S is clearly meant to be the foot of the perpendicular from D to MNG. Then the infinitesimals dx and dy introduced a little later will refer to the infinitesimal triangle, similar to the triangle DSG, which has EF as hypoteneuse, thus:

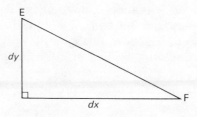

Figure 11. Diagram to clarify a mathematical argument of Aepinus

lever FE is infinitely small, will be $= \dfrac{1}{\sqrt{(a-x)^2+y^2}}$ and the force attracting the point F along FN $= \dfrac{1}{\sqrt{x^2+y^2}}$ Since further the force along FN is to the force along FK as FN is to NH, the force acting along FK $= \dfrac{\mathrm{NH}}{x^2+y^2}$. But since, as is clear from what has been expounded above, EF or IG is the tangent of the curve PDQ, DG : DS, that is EF : dy, will be $=$ NG : NH, whence we obtain NH $= \dfrac{y\,dx - x\,dy}{\mathrm{EF}}$, so that from this the force along FK [142] becomes $= \dfrac{y\,dx - x\,dy}{\mathrm{EF}(x^2+y^2)}$. Similarly the force acting along EL is found $= \dfrac{y\,dx - x\,dy + a\,dy}{\mathrm{EF}((a-x)^2+y^2)}$.

300) From this, for the curve PDQ that is to be discovered, we obtain the equation $\dfrac{y\,dx - x\,dy}{x^2+y^2} = \dfrac{y\,dx - x\,dy + a\,dy}{(a-x)^2+y^2}$, which when reduced gives
$$\begin{aligned}&+\ a y\,dx - a x\,dy\\&-2xy\,dx + x^2\,dy = 0.\\&\quad -\ y^2\,dy\ ^{143}\end{aligned}$$
To find the integral of this equation, let $x = zy$ and by substitution the equation will be transformed into
$$\begin{aligned}&+ a y^2\ dz - y^2 z^2\,dy = 0,\\&-2 y^3 z\,dz - y^2\quad dy\end{aligned}$$
which divided by y^2 reduces to
$$\begin{aligned}&+\ a\quad dz - z^2\,dy = 0,\\&-2yz\,dz - \quad dy\end{aligned}$$
whose integral is $az - yz^2 - y =$ Constant; this latter we indicate by β, and by restoring the value $x = zy$, we finally obtain $ax - x^2 - y^2 = \beta y$.

301) Thus under our assumed hypothesis, the curve PDQ will be a circle. Since, in the problem we have discussed, β is clearly not determined, this is an indication that any value can be assigned to it at will, so one circle after another will be generated perpetually, and any circle described with a radius not less than half MN, provided it passes through the points M and N, satisfies the problem. But it is clear that all these circles pass through M and N, since by making $y = 0$, x becomes either $= 0$ or $= a$.

302) Suppose therefore that some circle be drawn through the points M and N, Fig. XCII, and suppose that there are disposed on the periphery of this circle quite short magnetic needles ab, bc, cd, de etc., which touch one another at their extremities, and it is evident that

[142] The text has EK.
[143] The text has $- y^2 gy$.

each of them will be so disposed that it touches the circle or falls on its periphery. Since all these needles touch each other with their unlike poles, they will cling together and form a continuous curve. The same will happen if iron wires not imbued with magnetic force are substituted for magnetic needles, for when close to a magnet, these will immediately become true magnets or magnetic needles.

303) I have so far supposed that the centers of the needles or wires ab, bc, cd etc. are fixed and immobile, but provided they lie on a somewhat rough plane, so that they experience some friction, completely similar phenomena ought to follow. For because of the friction hindering the motion of a particle of this kind, e.g. de, ef etc., it cannot in fact be attracted or progress toward the points M and N. Nevertheless the force of the points M and N can be sufficiently great that it can force the particle to assume the position it would take up if it could move freely around its own center. This disposition of the particles is helped if the plane on which they lie is shaken gently, for then they jump off it, and, suspended in the air, they immediately acquire their proper position which they later preserve after falling back to the plane again.

304) Suppose therefore that the plane in which the points M, N are found is sprinkled with powdered iron or with iron filings, and that this plane is shaken gently, and it is evident that the filings will cling to one another, and form threads; since each particle acquires a position such that it falls on the periphery of some circle passing through the points M, N, these will form innumerable circular curves, all tending from the point M to N and returning into themselves. To illustrate somehow the kind of appearance this ought to produce, suppose the line MN, Fig. XCIII, is bisected at P, and a perpendicular of indefinite length is drawn through P. Then this line forms the locus of the centers of innumerable circles passing through the points M, N, and the iron filings will dispose themselves along innumerable peripheries such as are shown in the figure, the radii of which increase from MP all the way to infinity.

305) Since the force of the points M and N decreases perpetually as we recede from them, it is evident that the particles of iron some distance away can hardly be disposed by the force of the points M and N into their proper positions any more; it follows that the circles just mentioned, especially those of larger radius, cannot be shown whole; only their arcs close to the points M and N are formed sufficiently distinctly, since the more distant particles cannot be disposed by the force of the poles M, N, and so they ought to lie scattered indiscriminately.

306) The lines formed by powdered iron dispersed around a magnet do not completely coincide with the circles described here. For we have supposed that the force emanating from the points M and N decreases in inverse proportion of the distances, but I do not affirm that this is the law which magnetic actions follow. Anyone who has been willing to consider the matter with some attention will readily perceive that whatever the function according to which the magnetic actions are exercised, it does not coincide with the law assumed but differs from it; [144] the only consequence of this here is that the curves in which the iron filings arrange themselves are not exactly circular, but are of some other kind. For the rest, it is evident that everything else will apply, whatever the function for the magnetic actions. For the iron filings will form innumerable elliptic or oval curves or any other curves of a like kind which return into themselves, passing through the points M and N. It is now clear how from the laws of magnetic attraction and repulsion, there arises that regular arrangement of iron powder spread around a magnet which investigators of nature have so much wondered at, and which very many of them have claimed, erroneously in my opinion, as a manifest indication of the existence of a true magnetic vortex, or a river of magnetic matter moving continuously from one of the poles to the other.

307) To become surer through experience about the agreement between nature and the theory expounded so far, I fixed vertical iron wires AB, EF, Fig. XCIV, to a wooden plate at a distance of about two inches from each other. These wires had a base diameter of about one line and a length of half a foot. To B, F, the bottom ends of the wires, which were fixed to the plate, I moved two steel rods DC, GH enjoying, as equally as possible, a strong degree of magnetic force, in such a way that the north pole, C, of the former was touching the wire AB, and the south pole of the latter, G, was touching the wire EF. It is clear from the previous discussion that the end A in this case ought to acquire the force of a north pole and the end E that of a south pole. Afterward I touched the ends A, E, of the wires with a thin sheet of paper, and after spreading iron powder over it and tapping it lightly, I observed that the filings arranged themselves of their own accord exactly as the theory demands. For all the lines that they formed were oval, returning into themselves through the points of the paper touching the ends A, E, and, if the judgment of the eyes can be trusted, not very different from circles. But I draw no conclusions from this concerning the law of magnetic actions, for I do not deny that an estimate by eye can be quite faulty here.

[144] See above, pp. 115–17.

308) By setting up the experiment in the manner just described, I was able to obtain the desired agreement with the theory since I had satisfied sufficiently closely the conditions assumed in evolving the theory. For although the ends of the wires A, E cannot be considered with complete rigor as the points from which the magnetic force emanates, yet the magnetic force was here confined in a space sufficiently narrow that the phenomena could not differ very sensibly from those that take place if it spreads out from indivisible points. But as in magnetized steel rods the magnetic force is ordinarily diffused over a great area, the curves by dint of circumstances then vary, sometimes more, sometimes less, from the law which the theory has prescribed. Yet far from being contrary to our theory, when the cause of these aberrations is made clear, they tend to confirm it.

309) The reader who has understood properly what has been expounded so far will be able to understand not only the reason for the formation of the curves exhibited by iron filings sprinkled on a magnet, and able to transfer easily the principles which I have used here to other cases, but will at the same time perceive that these phenomena can be explained from magnetic attraction and repulsion alone, and that to save them we do not need, as very many have thought, a magnetic vortex properly so called, such as fluid which actually moves from one of the poles to the other. So let us pass now to a consideration of some other phenomena. It has often been noticed that any iron bodies placed in the vicinity of a magnet immediately become magnetic themselves, in such a way that either end gains a magnetism contrary to that possessed by the pole it is facing. Therefore if our supposition concerning the magnetic core enclosed within the globe of the earth is correct, this same phenomenon must necessarily take place also with the terrestrial globe, so that iron rods held in a fixed position should of their own accord acquire a magnetism produced by the action of the central magnet.

310) It is very well known that this is exactly how the matter is. For it was recognized long ago that if an unmagnetized iron rod is held in a position perpendicular or oblique to the horizontal, it acquires a magnetic force in such a way that if the experiment is carried out in the northern regions of the terrestrial globe the lower end of the rod repels the northern tip of a magnetic needle and attracts the south, but if the test is made in the southern hemisphere of the earth, the reverse happens. I shall not prolong my description of the experiments relating to this since the entire phenomenon is too well known to require it.

311) Experience shows that the amount of magnetic force which the

CERTAIN PHENOMENA 409

terrestrial globe induces in iron rods also varies with various positions of the rods in such a way that it is sometimes greater and sometimes smaller (in fact there is one position such that a rod held in it gains no force at all), but in general the reason for these phenomena is readily apparent from the theory. Yet the matter seems well worth a more careful examination. Before I begin this discussion, it will first be necesssary to take up certain considerations which will make our path easier. Consider that an iron rod AB is held in the position shown in Fig. XCV, and it is clear that any particle of magnetic fluid contained in it, such as C, is repelled by the positive pole of the terrestrial core and is attracted by the negative; let us suppose that the first of these actions is exercised along the line CE, the latter along the line CD. Suppose now that the rod AB is transferred to another place a thousand or even several thousand feet from the former, and that it varies its orientation so that it now takes up the position ab. Since the length of the rod AB, and the distance between the places AB and ab can be considered as infinitely small with respect to the immense magnitude of the terrestrial globe, it is clear that, after the rod has been transferred to the position ab, the forces acting on the particle c, repelling it in the direction ce and attracting it in the direction cd, will act along the lines cd, ce parallel to the lines CD, CE [145] and will be of the same magnitude as the corresponding forces in the former position AB. This is self-evident, for it is clear that a particle of magnetic fluid has changed its position with respect to the poles of the terrestrial core by only an infinitely small amount in moving from C to c, whence both the direction and the magnitude of the forces can likewise change only infinitely little.

312) Since therefore for any point on the surface of the earth and for a large enough space surrounding it, the forces acting on a particle of magnetic fluid located there are fixed in magnitude and in direction, let us suppose that C, Fig. XCVI, is such a particle, which the positive pole of the terrestrial core repels in the direction CE or CM, but which the negative pole attracts in the direction CD. Further, let GN be the horizontal line cutting the directions of the two forces at the points F and H, and let the magnitudes of the forces soliciting the particle C be shown by the lines CL, CM, and then not only will the lines CL, CM be given but also the angles CFG, CHG. Let us therefore put the lines $CM = M$, $CL = N$, and the angles $CFG = \mu$, $CHG = \nu$. Let us suppose further that the parallelogram CMKL is completed, and that its diagonal CK is drawn, which when produced meets the

[145] The text has CE, CD.

horizontal GN at the point N;[146] it is evident that the two forces acting along CM and CL coalesce, as it were, into a single force propelling the particle C in the direction CK with a force which is proportional to the diagonal CK. Call the line CK=A, and the angle CNG [147] enclosed with the horizontal by this line when produced=α. Since the angle KCL=$\pi-\nu+\alpha$, and HCF or KLC=$\nu-\mu$, and from that CKL=$\mu-\alpha$, we obtain from the triangle KCL the analogy sin $(\pi-\nu+\alpha)$: sin $(\mu-\alpha)$=M : N, or sin ν cos α − cos ν sin α : sin μ cos α − cos μ sin α = sin ν − cos ν tang α : sin μ − cos μ tang α = M : N, whence arises the equation N sin ν − N cos ν tang α = M sin μ − M cos μ tang α, which when reduced gives tang $\alpha = \dfrac{M\ sin\ \mu - N\ sin\ \nu}{M\ cos\ \mu - N\ cos\ \nu}$. If for the sake of brevity we put M sin μ − N sin ν = P, and M cos μ − N cos ν = Q, which makes tang $\alpha = \dfrac{P}{Q}$, we will have cos $\alpha = \dfrac{Q}{\sqrt{P^2+Q^2}}$ and sin $\alpha = \dfrac{P}{\sqrt{P^2+Q^2}}$. Since therefore sin $(\pi-\nu+\alpha)$: sin $(\nu-\mu)$=KL : KC=M : A, or sin ν cos α − cos ν sin α : sin $(\nu-\mu)$=M : A, by substituting the values of cos α and sin α discovered previously, $\dfrac{M\ sin\ (\nu-\mu)}{\sqrt{P^2+Q^2}}$: sin $(\nu-\mu)$=M : A, whence is obtained A = $\sqrt{P^2+Q^2}$.[148]

313) In place of the two actions exerted on the particle C by the twin poles of the terrestrial core, we can substitute a single one equivalent to them, whose magnitude and direction is given by a force A which = $\sqrt{P^2+Q^2}$ and which includes an angle α with the hori-

---

[146] The text has I.

[147] The text has CIG.

[148] In the text, the values of the various angles concerned were wrongly stated at the beginning of this argument, though not in a way which affected the final result. The initial values have here been corrected, and this has necessitated correcting the subsequent argument as well. The original is as follows:

Quoniam iam angulus KCL = $\nu - \alpha$, HCF vero, siue KLC = $\mu - \nu$, atque exinde CLK (sic) = $\pi - \mu + \alpha$, obtinemus ex triangulo KCL analogiam, sin $(\nu - \alpha)$: sin $(\pi - \mu + \alpha)$ = M : N, siue sin ν cos α − cos ν sin α :: sin μ cos α − cos μ sin α = sin ν − cos ν tang α : sin μ − cos μ tang α = M : N, vnde oritur aequatio N sin ν − N cos ν tang α = M sin μ − M cos μ tang α, quae reducta dat,

tang $\alpha = \dfrac{N\ sin\ \nu - M\ sin\ \mu}{N\ cos\ \nu - M\ cos\ \mu}$. Si iam porro brevitatis causa dicatur N sin ν − M sin μ = P, atque N cos ν − M cos μ = Q.

vt sit, tang $\alpha = \dfrac{P}{Q}$, erit cos $\alpha = \dfrac{Q}{\sqrt{P^2+Q^2}}$, et sin $\alpha = \dfrac{P}{\sqrt{P^2+Q^2}}$. Cum ergo sit sin $(\nu - \alpha)$: sin $(\mu - \nu)$ = KL : KC = M : A, siue sin ν cos α − cos ν sin α : sin $(\mu - \nu)$ = M : A, erit surrogatis valoribus ipsius cos α et sin α, antea repertis $\dfrac{M\ sin\ (\mu - \nu)}{\sqrt{P^2+Q^2}}$: sin $(\mu - \nu)$ = M : A, vnde obtinetur A = $\sqrt{P^2 + Q^2}$.

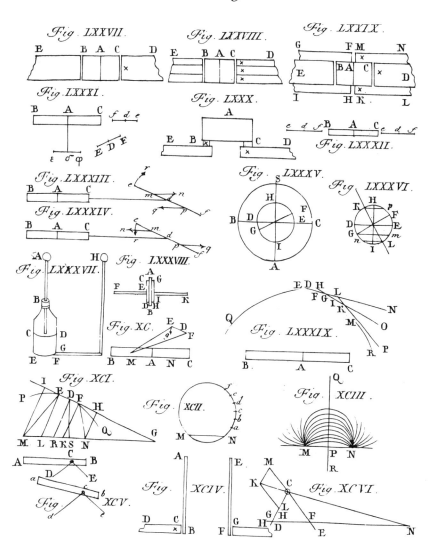

Figure 12. Facsimile Plate V from the *Essay*

zontal plane GN, such that tang $\alpha = \dfrac{M \sin \mu - N \sin \nu}{M \cos \mu - N \cos \nu}$.[149] Since it is equivalent to both forces, that exerted by the positive pole of the core as well as that exerted by the negative, we need consider only this force here. But before we ask what effect it can produce, it will be advisable to insert a remark at this point. Consider an iron rod AB, Fig. XCVII, exposed in a particular place to the action of the terrestrial core, and suppose that this is exerted along the line CK, driving the particle of magnetic fluid from C toward K, and it is clear that the force acting in the direction CK can be resolved into the force CF parallel to the length of the rod, and the force CG normal to the former. So a particle of magnetic fluid C included in the rod must be thought to be solicited by a double force, the former of which propels it along the length of the rod while the latter acts normally to the length of the rod. We can safely neglect the latter since there is no need for us to examine it. For if we suppose that it alone acts, its effect will be to propel the magnetic fluid from AB, Fig. XCVIII, toward HI, and to induce in the rod a state such that the part ABPQ is emptied of fluid, while the part PQHI becomes replete with it. But since the thickness of the rod AB is here supposed small, it is fairly obvious from the previous discussion in § 152. 160. 162 above that the magnetism arising from this action will be hardly, or not even hardly sensible, so there is no need to consider it, and the whole action of the normal force CG can be safely neglected in what follows; thus the whole business is reduced to tracking down the effect of the force CF acting along the length of the rod, and the quantity of magnetic force impressed on the rod can be safely assumed to be always proportional to this latter force.

314) If we now begin an inquiry concerning the quantity of the force which the iron rod AB gains from the action of the terrestrial core, it is abundantly clear that it depends on the position of the rod, so before anything else we ought to consider here how this can be determined. It can be done conveniently in the following manner. Through the line CK, Fig. XCIX, showing the direction along which the terrestrial core acts on the particles of magnetic fluid, which afterward I shall call the *magnetic direction,* let there pass a vertical plane DEFG, which is the same as that to which is given the name *the plane of the magnetic meridian,* and let it cut the horizontal in the line FG. Then let another plane HPLI, likewise vertical, be thought to pass through the rod AB exposed to the action of the core, cutting the hori-

[149] As before, the text has $\dfrac{N \sin \nu - M \sin \mu}{N \cos \nu - M \cos \mu}$.

zontal in the line IL, and let the length of the rod be supposed produced till it meets the horizontal line ILN at the point M. If first of all we determine the angle of inclination of the plane HPLI to the meridional plane DEFG, that is the angle GRL, which we shall call γ, and then also the angle AMR which the line of the rod includes with the horizontal line IRN, which we shall indicate by β, it is clear that the position of the rod has been completely determined.

315) Since the magnetic direction CK does not fall in the plane HPLI in which the rod AB lies, but is oblique to it, before the effects of the magnetic core on the rod can be determined, the force CK must first be reduced to the plane HPLI. Consider therefore that the magnitude of the force exerted by the core is exhibited by the line CK, Fig. C, and drop the normal KT from the point K to the plane QPRL; connect the points C, T, by the line CT, and from the points K and T draw the horizontals KS, TS which, it is agreed from Geometry, cross the line QR at the same point S. It is evident that thus the force CK is resolved into two forces CT, KT, the former of which lies in the plane QPLR, while the latter is normal to it. Since the latter is also normal to the rod AB located in the plane QPRL, it can be completely neglected for the reasons expounded above, so it can be supposed that only the force CT lying in the plane QPRL acts on the rod AB, and we must investigate the magnitude and direction of this. Retaining the denominations assumed above, and calling the angle CTS, δ, from the triangle CKS we will have CS = A sin α and KS = A cos α, whence is found from the triangle KST, ST = A cos α cos γ. From this it can be further deduced that $CT = A\sqrt{\sin \alpha^2 + \cos \alpha^2 \cos \gamma^2} = A \cos \alpha \sqrt{\tang \alpha^2 + \cos \gamma^2}$ and $\tang \delta = \dfrac{\tang \alpha}{\cos \gamma}$, whence $\sin \delta = \dfrac{\tang \alpha}{\sqrt{\tang \alpha^2 + \cos \gamma^2}}$ and $\cos \delta = \dfrac{\cos \gamma}{\sqrt{\tang \alpha^2 + \cos \gamma^2}}$.

316) If finally we want to solve our problem completely and determine the quantity of magnetic force which an iron rod held in any position gains from the action of the terrestrial core, the force CT, Fig. CI, must once more be resolved into two, the first, CZ, driving the particle C along the length of the rod AB, the other, CV, standing normally to the former; of these, the latter can be neglected for the reasons expounded above. It is found that since the angle ZCT or CTV = CWI − CMI = $\delta - \beta$, CZ = CT cos ($\delta - \beta$) = CT (cos δ cos β + sin δ sin β) = by substituting the values previously found for CT and sin δ and cos δ, A(cos α cos γ cos β + sin α sin β). Since this quantity expresses the total action exerted by the terrestrial core on the fluid enclosed in the rod along the very length of the rod, it also indicates

414 THEORY OF ELECTRICITY AND MAGNETISM

at the same time the magnitude of the magnetic force which the rod acquires, in that this is proportional to that force.

317) There are noteworthy conclusions flowing from this which deserve a brief mention. First of all, in any plane there is a particular position for the rod in which it acquires a greater magnetic force than if it is held in any other position. To determine this position corresponding to the maximum, the differential of the force CZ, or CT sin $(\delta-\beta)\,d\beta$, need only be put$=0$, from which we obtain $\delta=\beta$. So it is obvious that the rod AB acquires its maximum force when the angle CMI is equal to the angle CWI,[150] or, what comes to the same thing, if the rod AB is so held that its length falls along CT, the direction of the force of the terrestrial core when reduced to the vertical plane HPLI. Then there is another position of the rod where it acquires no magnetic force. That position is found by putting CT cos $(\delta-\beta)=0$ or cos $(\delta-\beta)=0$, whence we get $\delta-\beta=-\frac{\pi}{2}$, or $\beta=\frac{\pi}{2}+\delta$, and thus, if the rod AB is so placed that the angle CMI, Fig. CII, is equal to a right angle plus CWI, or so that it lies perpendicular to the line CT, it gains no magnetic force from the action of the terrestrial core.

318) As is clear from what has just been expounded, although in any vertical plane a position can be assigned such that a rod held there gains a greater degree of magnetism than in any other position, yet it is readily clear that the maximum forces which the rod gains in various planes are not equal but are at one time greater, at another smaller, according to the various positions of the plane in which it is located, in such a way that here there is a maximum of maxima. To find it, it is clear that, since $\beta=\delta$, the greatest force which the rod acquires when held in a given plane, Fig. XCIX, $=CT$[151] $=A\sqrt{\sin\alpha^2+\cos\alpha^2\cos\gamma^2}$, whence, by making its differential $\frac{-A\cos\alpha^2\cos\gamma\sin\gamma\,d\gamma}{\sqrt{\sin\alpha^2+\cos\alpha^2\cos\gamma^2}}=0$,[152] there is obtained either sin $\gamma=0$ or cos $\gamma=0$. In the former case, which pertains to the maximum, $\gamma=0$. With the angle GRL$=\gamma$ vanishing, it is clear that the rod obtains absolutely the maximum degree of magnetic force if it is held in the plane of the magnetic meridian itself, in a position such that its length coincides with the magnetic direction CK, in which case the force it acquires is$=A$. For since $\gamma=0$, cos $\gamma=1$, and hence the force which

---

[150] The text has CTW.

[151] The text has CD.

[152] The text has $\dfrac{-A\cos\alpha\cos\gamma\sin\gamma\,d\gamma}{\sqrt{\sin\alpha^2+\cos\alpha^2\cos\gamma^2}}=0$.

the rod acquires $=A\sqrt{\sin\alpha^2+\cos\alpha^2}=A$. The other case, where $\cos\gamma=0$, gives the minimum of the maxima. Since in this case $\gamma=\frac{\pi}{2}$, it is clear that the rod, so placed in the plane normal to the plane of the magnetic meridian that it acquires the greatest force which it can acquire in that plane, gains less force than if it had been situated in any other plane; this minimum force, because $\cos\gamma=0$, is equal to $A\sin\alpha$, and is induced in the rod while it stands vertically.

319) I leave these discussions about the various magnetic forces which an iron rod at a given point of the surface of the earth gains in different positions, yet I give warning that I shall present important matters in a special dissertation concerning the action of the magnetic needle, and I will in particular demonstrate that the position assumed of its own accord by a dipping needle, left to itself and freely mobile in any vertical plane, coincides with that in which the rod is to be placed if it is to gain the maximum degree of magnetic force.[153] But another question requiring a solution still remains. It is this: what force does a rod gain when it is brought to different places on the earth? Let AHGF, Fig. CIII,[154] be the terrestrial globe, and BEDC the core enclosed in it. Further, let EC be the plane which separates the part EBC endowed with positive magnetism from the negatively magnetic part EDC, and let the axis AG be perpendicular to this plane. Let us imagine further that all the force of the positive pole has been collected at the point B, and that of the negative one at the point D. If now we want to adapt the calculation and our discussions to some given point M of the surface of the earth, it is clear that by drawing lines MB, MD, from the point M, these will express the directions in which the poles B, D, of the terrestrial core act on the magnetic fluid enclosed in an iron rod in the vicinity of M. Suppose that through the point M a line NO is drawn perpendicular to the radius MK, touching the globe of the earth at M, which as a result will be the horizon. Now by producing the lines MB, MK, MD beyond M as far as Q, P, R, it is clear that the angle QMN will be $=\mu$, and the angle RMN $=\nu$ (§ 312 [155]). Suppose that the distance of the point M from the point A of the terrestrial globe which is vertically above the pole of the core, or the angle MKA, $=\phi$, the radius KM of the terrestrial globe $=r$, and the radius of the core $=a$, and let us meanwhile call the angle DMK $=\delta$ and KMB $=\gamma$. It is now clear that MK : KD $=$

---

[153] The promised dissertation seems never to have been published.
[154] No point G is shown on Fig. CIII. The point marked S has been wrongly labeled, and should be G.
[155] The text has § 310.

416   THEORY OF ELECTRICITY AND MAGNETISM

sin MDK : sin KMD = sin $(\phi - \delta)$: sin δ, whence is obtained $r \sin \delta =$ $a \sin \phi \cos \delta - a \cos \phi \sin \delta$, and tang $\delta = \dfrac{a \sin \phi}{r + a \cos \phi}$. In a completely similar way, tang γ is found $= \dfrac{a \sin \phi}{r - a \cos \phi}$, whence is further deduced $\cos \delta = \dfrac{r + a \cos \phi}{\sqrt{r^2 + 2 a r \cos \phi + a^2}}$, and $\cos \gamma = \dfrac{r - a \cos \phi}{\sqrt{r^2 - 2 a r \cos \phi + a^2}}$. But since RMN $= \nu =$ DMK + right angle $= \delta + \dfrac{\pi}{2}$, and QMN $= \mu =$ right angle $-$ BMK $= \dfrac{\pi}{2} - \gamma$, we will obviously have $\cos \delta = \sin \nu = \dfrac{r + a \cos \phi}{\sqrt{r^2 + 2 a r \cos \phi + a^2}}$ and $\cos \gamma = \sin \mu = \dfrac{r - a \cos \phi}{\sqrt{r^2 - 2 a r \cos \phi + a^2}}$.

320) Let us adapt our latter discussion to a certain case which deserves attention beyond the rest. Suppose that an iron rod is always held in a position such that it is perpendicular to the horizon, in which case the force which it gains $=$ A sin α (§ 318 [156]) $=$ P $=$ N sin $\nu -$ M sin μ (§ 312 [156]), where it must be noted that in this formula the quantities M, N are variables, and at the same time vary with the various positions of the point M.[157] If we substitute the values just discovered for sin ν and sin μ at some given point of the earth whose distance from the terrestrial magnetic pole is measured by the arc AM, the magnetic force which the iron rod gains when erected vertically will be $= \dfrac{N(r + a \cos \phi)}{\sqrt{r^2 + 2 a r \cos \phi + a^2}} - \dfrac{M(r - a \cos \phi)}{\sqrt{r^2 - 2 a r \cos \phi + a^2}}$.

321) If we consider that the rod is vertically above the magnetic equator at H, ϕ will be $=$ right angle $= \dfrac{\pi}{2}$, whence the force impressed on the rod will be $\dfrac{r(N - M)}{\sqrt{r^2 + a^2}}$. If the force of the positive magnetic pole B is assumed to be precisely as great as the force of the negative pole D, as we suppose here, then since in our case N $=$ M because of the equal distances HB, HD, the expression just found will be $= o$,

---

[156] The text has § 316 and § 310 respectively.

[157] In our corrected version of § 312, P was defined as M sin $\mu -$ N sin ν, not N sin $\nu -$ M sin μ as Aepinus once again writes it here and throughout the next few paragraphs. Since his inclusion of the additional minus sign has no effect on the argument (the direction of the induced magnetism always being inferred separately, and never being obtained directly from the algebraic formula), we have elected not to carry through all the amendments which would have been necessary to restore complete consistency.

CERTAIN PHENOMENA 417

whence an iron rod placed vertically over the magnetic equator will not be able to gain any magnetic force. But $\phi = o$ or $= \pi$, in which case cos ϕ becomes either $+1$ or -1, the magnetic force of the rod will be $= N - M$. When this happens the rod is above the magnetic poles B and D themselves. So if it is first located above the pole D, it is clear that, since the distance DG is less than the distance BG, N is greater than M; but if it moves to the point A,[158] for a similar reason, N will be less than M. Since further DG=BA and BG=DA, it is obvious that the rod, whether located at G or at A, acquires an equal force in either case, but at the point A [159] a contrary one to that which it gains at G.

322) Thus what ought to happen if a rod is placed vertically either above the equator or above the poles is already sufficiently clear. In the first case, it becomes completely non-magnetic, and in the latter it becomes magnetic in such a way that its lower end acquires a force contrary to that possessed by the pole in question. It is not yet determined what phenomena occur if the rod is taken to intermediate places. The determination of this matter depends on a knowledge not only of the law according to which the magnetic actions are exerted, but also on so many other circumstances that there is no possible hope of our solving it by means of theory alone. This will appear more plainly to anyone willing to give a little attention to the following discussion. I have demonstrated previously that the force acquired by a vertical iron rod generally $= N \sin \nu - M \sin \mu$, but when it is directly above the equator, $\sin \nu = \sin \mu$ as well as $N = M$, whence this force too vanishes. If we move from the equator toward the pole A, it is clear,

1) that each of the angles μ and ν increases; not at equal speeds, but unequally, in such a way that

2) the ratio of sin ν to sin μ departs from the ratio of equality; it first of all begins to increase and it becomes a maximum when ϕ has attained the value where cos ϕ becomes $= \dfrac{\sqrt{r^2 + a^2}}{\sqrt{3r^2 - a^2}}$.[160] But then,

3) it once more begins to decrease and again approaches more and more to the ratio of equality, until finally, after ϕ has vanished (when the pole has been reached), it attains the ratio of equality once more.

I shall not enter on a calculation demonstrating this since it is

---

[158] The text has H.
[159] The text has H.
[160] The text has $\dfrac{\sqrt{r^2 - a^2}}{\sqrt{3r^2 - a^2}}$.

obvious, and more prolix than difficult. But if we look at the variations which M and N undergo, it is evident that in passing from the equator toward the pole A, M continually increases, but N decreases, for we know for certain that the function of the magnetic actions, although unknown, is such that the force always decreases with increased distance, whence it is obvious that the ratio of M to N is increased as we move toward the pole A, and ought to depart continuously more and more from the ratio of equality.

323) There can be various outcomes for various laws of this increase. Imagine first of all that the ratio $\sin \nu : \sin \mu$ has increased by moving from the equator to the point M, and at the same time assume that meanwhile the ratio M : N has also increased, but to a lesser degree than the ratio $\sin \nu : \sin \mu$, and for the point M we shall have M : N $<$ $\sin \nu : \sin \mu$. So M $\sin \mu <$ N $\sin \nu$, whence N $\sin \nu -$ M $\sin \mu =$ a positive quantity. Since the ratio M : N constantly continues to increase, and the ratio $\sin \nu : \sin \mu$ decreases after passing the point where it is a maximum and once more tends toward a ratio of equality, it is clear that, in the latter part of the transition to the positive pole, we arrive at some stage at a point where M : N is $>$ $\sin \nu : \sin \mu$, and the force N $\sin \nu -$ M $\sin \mu$ is negative. It is obvious that, in moving from the equator toward the negative pole G, completely similar things take place. So under the hypothesis assumed at the beginning of the §, the following phenomena ought to occur in moving from the equator to the poles:

1) At the equator itself the rod acquires no force at all.

2) In moving toward the positive pole, the lower end of the rod becomes at first positively magnetic, and the upper end negatively magnetic; but finally

3) A point is reached between the equator and the pole where the force of a rod held vertically is again nothing; and then proceeding further to the pole,

4) the rod becomes magnetic once more, but in the inverse order, with the lower end now becoming negatively magnetic, and the upper end positive. In a similar fashion, by moving the rod from the equator toward the negative pole

1) initially it is negatively magnetic at its lower end, positively magnetic at its upper end. Then at some stage

2) a point is reached where its force is zero; and from then on

3) right to the pole the rod always becomes magnetic according to the law that its lower end is imbued with positive magnetism, its upper end with negative.

324) If, in moving from the equator to the pole, over the interval

where the ratio $\sin \nu : \sin \mu$ increases, the ratio M : N increases more rapidly, M : N will be constantly $> \sin \nu : \sin \mu$ from the equator to the pole, whence M $\sin \mu >$ N $\sin \nu$, and the force N $\sin \nu -$ M $\sin \mu$ will be constantly negative. In this case it is evident that when the rod is moved toward the positive pole, its lower end A will become constantly negatively magnetic and its upper end positively magnetic, and, similarly, in passing toward the negative pole G, the lower end will become positively and the upper one negatively magnetic. These things can be reached by theory alone, unaided by experiments, but which of these cases takes place can only be determined with the help of experiments.

325) Anyone willing to compare what has been proposed so far with commonly known experiments will discover outstanding agreement at every point. For it is very well known that if a rod is held in a horizontal position such that its length is perpendicular to the line in which the dipping magnetic needle disposes itself of its own accord, and is moved towards a magnetic needle poised on a stylus, it shows no power either to attract or repel the point of the needle, and this is what ought to happen according to the reasonings based on our theory presented above. But if the rod is held in a position inclined in any way to the horizon, it is known that, when brought close to the magnetic needle, its lower end repels the point of the needle facing north, and its upper end attracts it. But all this is well enough known. Finally, concerning a rod held vertically and carried to various areas of the terrestrial globe, experience has shown that the case which I expounded in § 322 [161] takes place truly in nature, as can be seen in the celebrated *Musschenbroek's* Dissertation on Magnetism p. 262, where he excerpted the tests pertinent to this from the English Transactions, Number 177.[162]

326) The phenomena we have considered so far supply a new and outstanding argument to prove the truth of the opinion introduced above, that there is a magnetic core enclosed in the terrestrial globe. For not only does the globe of the earth act on a magnetic needle or any other freely suspended magnet in exactly the same way as a magnet does, as we showed above, but, as is already obvious, it also does so

[161] The reference should be to § 324.
[162] Musschenbroek, in his *Dissertatio . . . de magnete,* pp. 262–63, sets out a series of observations taken from a report published in the *Philosophical Transactions* of the Royal Society of London (No. 177, for December 1685), under the heading "Several Observations of the Respect of the Needle to a piece of Iron held Perpendicular; made by a Master of a Ship Crossing the Æquinoctial Line, Anno 1684 . . ." (*Phil. Trans.,* xv, 1685, 1213–14).

in regard to the parts of a magnet, because it makes iron held in the proper position in its vicinity magnetic. Although these arguments might seem sufficient to win full conviction, yet from another phenomenon there arises hostile to my opinion a doubt which forced me to be undecided at first; after a more careful examination I was not only able to solve it, but I was also able to obtain from it a new argument for the truth of my opinion.

327) The matters I am discussing can be doubted for the following reason. If the terrestrial core possesses all the properties of a true magnet, it ought not only to act on a magnetic needle, and to make iron close to it magnetic, but it must also be expected to agree with the magnet in this phenomenon too, that it should attract non-magnetic iron, just as it should also attract or repel another magnet according as either the unlike or like poles are turned to one another.[163] It can easily be tested experimentally whether this is the case, and I shall briefly indicate the sort of test, and how it can be carried out.

328) Weigh a strongly hardened, non-magnetic steel rod on a very sensitive balance. Then imbue this rod with magnetic force, and weigh it again, first with the north pole downward, then the south pole. If the globe of the earth is a magnet, it ought to attract the rod in the former case, and repel it in the latter, whence it seems to follow that in the former case the weight of the rod ought to increase beyond the weight which it had before it was imbued with magnetic force, and in the latter it ought to be reduced below it. This outcome is all the more properly to be expected, since the phenomena adduced are always found to happen if the weighing is performed in the vicinity of an immobile magnet.

329) Place a strongly magnetic steel rod on a wooden boat, or push a magnetized iron wire through a little cork ball of convenient size. Let it float freely on the surface of water contained in a suitable vessel, and it is clear not only that the rod ought to turn itself in such a way that it lies in the magnetic meridian, with one end facing north and the other south, but also that, if what we have asserted about the magnetic force of the terrestrial globe is true, in our regions it should float to the northern rim of the vessel; it should never stay in the middle of the vessel, but should always move of its own accord toward the northern rim. Anyone can test how this actually happens when a magnet is held close to the vessel in the required manner.

330) Further, suspend a lead globe from a thin wire and let the posi-

---

[163] This objection seems to have originated with Athanasius Kircher, in his *Magnes, sive de arte magnetica,* published in 1654 (R. Taton, ed., *Histoire générale des sciences,* Paris, 1957–64, II, 333).

tion of the vertical line be determined with every possible care. Then take the lead weight away and in its place hang a magnetized steel rod on the wire. This, left freely to itself, will turn in such a way that one end faces north and the other south. If what I have been teaching is true, it seems that if the test is instituted in our regions, the rod, attracted by the north pole closer to it, ought to yield to this solicitation in such a way that the pendulum ought to deviate somewhat from the vertical toward the north, just as is seen when a magnet is brought close to the pendulum.

331) Finally, hang a soft iron rod from the arm of a very mobile balance in the position where, if retained, it acquires no magnetic force (§ 317 [164]). After noting its weight, suspend it (assuming that the test is set up in regions some distance from the equator) so that one end points directly downward or, better still, hang it in the position in which it gains the maximum magnetic force (§ 318 [165]), and weigh it once more. If the globe of the earth now attracts the rod in the latter position, as ought to be expected, its weight too should be found to be increased.

332) Whichever of these experiments you set up, you will always find that it does not succeed, and you will never discover any clear indication of an attraction or repulsion exerted by the terrestrial globe. So it is proper to doubt if ascribing a magnetic force to the terrestrial globe is accurate teaching, for it now seems clear that the phenomena are different from what ought to be expected on our hypothesis. The answer to this serious doubt might seem to be that the terrestrial globe is indeed a very great magnet but is at the same time very weak, and of so small a force that the attraction and repulsion which it exerts cannot be felt at all. But anyone who thinks about it properly will easily see that a response of this kind will be of no moment and can be properly rejected by an opponent. For tests show that if the globe of the earth is taken to be a magnet, it must be considered as a reasonably powerful one, and not as a very weak one whose force could hardly be felt. For both the quite notable directive force which it supplies to needles, and the sensible enough force which it induces in non-magnetic iron, are inconsistent with a very great weakness of the magnet enclosed in the bowels of the earth.

333) Yet, although this doubt is of rather great moment, after diligent examination, I have discovered that actually nothing should be feared from it; rather, it serves to prove particularly clearly the existence of a magnetic core in the earth, since it is in our power to

[164] The text has § 315.
[165] The text has § 316.

show that everything happens just as it ought under the hypothesis of a magnetic core. To make this clear, suppose that AB, *ab*, Fig. CIV, are two magnets, one of which, AB, is considerably larger than the other, *ab*. To each of them, to the former at a distance BC, to the latter at a distance *bc*, expose the completely similar and equal magnetic needles CE, *ce*, and I say that, if the needles CE, *ce* have been removed from the two magnets AB, *ab* to that distance at which the directive forces which those needles acquire from the action of the magnets are equal, then the force with which the larger magnet AB attracts the entire needle CE will be significantly less than the force with which the smaller magnet *ab* attracts the entire needle *ce*.

334) So that the truth of the matter might be seen more distinctly in a special example, suppose for the time being that magnetic actions are exerted according to the law that they are in the inverse ratio to the squares of the distances. To make our discussion easier, consider further that the whole force of the poles FB and *fb* has been collected at the points B and *b,* and that the force of the poles FA, *fa* has been collected at the points A and *a,* and imagine in a similar fashion that, in the needles CE, *ce,* the whole force of the poles CD and *cd* has been driven into the points C, *c,* and the force of the poles DE, *de* has been driven into the points E, *e,* so that thus the forces exerted by the poles B, *b,* of the magnet can be considered as if they were applied at the points C, *c* and E, *e*. With this supposition, let us express the distance BC by A, *bc* by *a,* and let us call the length of the needles CE and *ce, b*. Indicate by M the quantity of magnetic fluid in the part FB either above the natural quantity or below it, according as FB, the pole of the magnet AB, is positive or negative, and in a similar fashion let the quantity of fluid abounding or deficient in the part *fb* be called *m*.

335) It is clear first of all, partly as a result of our supposition, partly as a result of what we taught in Chap. 1, that the force by which the point B attracts the point of the needle C is to the force by which the point *b* of the other magnet attracts the point *c* of the needle exposed to it, as $\frac{M}{A^2}$ is to $\frac{m}{a^2}$. Further, since the magnetic matter expelled from the negatively magnetic part of the magnetic needles CE and *ce* is all propelled into the positively magnetic part, in such a way that in one part there abounds exactly as much magnetic fluid as is deficient in the other part, it is clear that the force by which the point C [166] is attracted by the pole B is to the force by which the point E is repelled

---

[166] The text has E.

CERTAIN PHENOMENA 423

from that same pole, as $\frac{1}{A^2}$ is to $\frac{1}{(A+b)^2}$,[167] whence if the former force is held to be expressed by $\frac{M}{A^2}$, the latter must be expressed by $\frac{M}{(A+b)^2}$. In a similar fashion, the force by which the point b repels the point e will be found $=\frac{m}{(a+b)^2}$. If we want now to discover the whole force by which the needle CE is drawn by the pole B, it is clear that since the point C is attracted by a force $=\frac{M}{A^2}$, but the point E is repelled by a force $=\frac{M}{(A+b)^2}$, the whole force soliciting the needle toward B $=\frac{M}{A^2}-\frac{M}{(A+b)^2}$, and the force by which the pole b attracts the needle ce is found by a similar argument to be $=\frac{m}{a^2}-\frac{m}{(a+b)^2}$.

336) Now that we have evolved the formulae which show the attractive force of the poles B and b, let us also inquire into the quantity of the directive force which the needles CE, ce acquire from the action of these poles. Accordingly let us suppose that the needles CE, ce are disturbed a little from their position so that they include angles CDG, cdg with the lines DB, db, and suppose that these angles are equal, and express them by ϕ. To discover the ratio between the forces which try to draw the needles CE, ce back to the lines DB, db, proceed as follows. It is clear from the previous discussion that the point G is attracted by the pole B in the direction GI. But because the angle CDG is quite small, the direction of the force GI can be taken to be parallel to the line DB and the force by which the point G is drawn can be taken to be equal to the force by which the point C is drawn. In a similar fashion, the point H is repelled by the pole B by the same force as that with which the point E is repelled, in a direction HM parallel to the line DB. By resolving these forces GI, HM into two forces, one parallel to the needle and the other normal to it, the former into the forces KG, PG, the latter into the forces HL, HN, it is clear that the parallel forces GK and HL are absorbed because of the immobility of the point D, but that the forces GP and HN conspire to draw the needle back to the line DB, so when they are added together they make up the total directive force of the needle. Because GI : GP = 1 : sin ϕ, the attracting force along GP $=\frac{M \sin \phi}{A^2}$, and for a similar reason the

[167] The text has $\frac{r}{(A+b)^2}$.

repelling force along $HN = \dfrac{M \sin \phi}{(A+b)^2}$. So the conclusion is finally reached that the total directive force of the needle is equal to $\left(\dfrac{M}{A^2} + \dfrac{M}{(A+b)^2}\right) \sin \phi$. The directive force of the needle ce exposed to the action of the smaller magnet ab is found, by the same arguments, $= \left(\dfrac{m}{a^2} + \dfrac{m}{(a+b)^2}\right) \sin \phi$.

337) It is therefore clear that the directive forces which the actions of the poles B and b exert on the needles CE, ce will be equal when $M\left(\dfrac{1}{A^2} + \dfrac{1}{(A+b)^2}\right) = m\left(\dfrac{1}{a^2} + \dfrac{1}{(a+b)^2}\right)$, or if $M : m = \left(\dfrac{1}{a^2} + \dfrac{1}{(a+b)^2}\right) : \left(\dfrac{1}{A^2} + \dfrac{1}{(A+b)^2}\right)$.[168] If we suppose that this ratio between M and m actually is the case, the directive forces on the needles will be equal, but nevertheless the force by which the pole B of the larger magnet attracts the whole needle CE will be considerably less than the force by which the pole b of the smaller magnet entices the needle ce.

338) For a clearer understanding of this matter, let us turn to a numerical example. Let us express $bc = a$ by 1, and let $CE = ce = b = 0{,}1$; then let $BC = A$ successively become 1, 10, 100, 1000, etc.,[169] and let the ratio between m (which we likewise express by unity) and M be assumed to be that for which the directive forces which the needles CE, ce gain from the poles B and b are equal. Then the values for the attractive forces are found to be as shown on the following table: [170]

A	1	10	100	1000
M	1	92	9141	913314
Directive force	1,8264462 × sin ϕ	1,8264462 × sin ϕ	1,8264462 × sin ϕ	1,8264462 × sin ϕ
Attractive force	0,1735537191	0,0181276432	0,0018263718	0,0001826628

If therefore there are four magnets whose absolute forces are as 1, 92, 9141, 913314, and there are similar magnetic needles at distances from these four magnets which are as 1, 10, 100, 1000, then the directive

---

[168] The text has $M : m = \left(\dfrac{1}{a^2} + \dfrac{1}{(a+b)^2}\right) : \left(\dfrac{1}{a^2} + \dfrac{1}{(A+b)^2}\right)$.

[169] The text has 10, 100, 1000, etc.

[170] Despite Aepinus's generally clear understanding of the proper role of mathematics in physics, it here becomes painfully apparent that the notion of "significant figures" was as foreign to him as it was to his contemporaries!

force on all the needles will be the same, but the force by which a larger magnet attracts the entire needle exposed to it will be significantly less than the force by which a smaller magnet entices its needle. Neglecting the last six decimal places, the forces are as 1736, 181, 18 and 2.

339) This makes the truth of the theorem proposed in § 333 [171] sufficiently clear, and at the same time it can be readily understood that if the magnet AB possesses as it were an infinitely greater force than the other magnet ab, and the needle CE is removed from the former to an enormous, or at least a very great distance, this will show the same directive force as will the needle ce located in the vicinity of the smaller magnet ab; but the force by which the larger magnet AB attracts the needle CE will in this case be as it were infinitely smaller than the force by which the needle is attracted by the magnet ab, and as a result it becomes completely insensible.

340) What we have shown in broad terms is that one cannot draw any conclusions about the strength of the attractive force exerted by a magnet from observations of the strength of its directive force. Rather, magnets of greater mass and greater strength can impart a sensible enough directive force into needles very much removed from them, even though, to the senses, their attractive force completely disappears. The same also happens in the case of the magnetic force which magnets usually induce in non-magnetic iron rods. For if such a rod ce is moved to the smaller magnet ab at a distance bc, and another similar and equal rod CE is exposed to the larger magnet AB at a distance BC, I say that although the distances bc, BC are so chosen that the same magnetism is induced in the two rods by their respective magnets, the larger magnet will attract the rod CE by a smaller force than that by which the smaller magnet attracts the rod ce. For the quantity of magnetic force induced in a rod depends on the force with which the magnet acts on the particles of magnetic fluid contained in the rod. So if we retain here the hypotheses and denominations used above, then in general the force which the smaller magnet ab exerts in order to magnetize the rod ce will be to the force exerted by the greater magnet AB as $\frac{m}{a^2}$ is [172] to $\frac{M}{A^2}$, and these forces, and the degrees of magnetism produced by them, will be equal if $M : m = A^2 : a^2$. Since the rods CE, ce gain the same degree of magnetism, the forces by

---

[171] The text has § 331.

[172] The text has $\frac{m}{A^2}$.

which they are attracted by the poles B and b will be as $M\left(\dfrac{1}{A^2}+\dfrac{1}{(A+b)^2}\right)$ is to $m\left(\dfrac{1}{a^2}+\dfrac{1}{(a+b)^2}\right)$.[173] So if we once more make $m=1$, $a=1$, and we assume that A becomes successively 1, 10, 100, 1000, etc.[174] but that M is always so determined that the forces tending to make the rods CE, ce magnetic become equal, the attractive forces will be found to be as shown in the following table:

A	1	10	100	1000
M	1	100	10000	1000000
Magnetic force of rod	1	1	1	1
Attractive force	0,1735537191	0,0197039600	0,0019980000	0,0002000000

which attractive forces, by neglecting the last six places, are approximately as 1736, 197, 20, 2. So it is clear in the same way as before that the larger magnet can impart the same degree of magnetism to a piece of iron reasonably far removed from it as a smaller magnet gives it at a smaller distance, but that the attractive force which the larger magnet then exerts is appreciably less than the attractive force of the smaller magnet: in fact it can happen that the very strongest magnet impresses a sensible enough degree of magnetism on a rod removed to a very great distance from it, yet nevertheless it does not sensibly attract it.

341) I am not afraid that readers will begin to doubt because I have assumed a fictitious hypothesis here, that the law of the magnetic actions is such that it follows the inverse ratio to the squares of the distances; for everyone will readily perceive that completely similar things take place whatever the true law of actions might be, provided it can be assumed that when the distance increases, the quantity of the action decreases. And I have not supposed one particular law here with the intention of putting it forward as the true law of nature, but so that in a determined example, everything might appear clearer visually with the help of specific numbers. For the rest, it is already quite apparent how we are to explain that paradoxical phenomenon which seemed completely contrary to my theory concerning the mag-

---

[173] The text has $m\left(\dfrac{1}{A^2}+\dfrac{1}{(a'+b)^2}\right)$.

[174] The text has 10, 100, 1000, etc.

netic core enclosed in the globe of the earth, because of course that core shows an outstanding directive force and a quite sensible power of making soft iron magnetic, but it betrays no indication of an attractive force on iron. If we assume that the terrestrial core is a very large magnet of great force, it is sufficiently clear that it is possible for it to extend its force sensibly to the immense distance to which we inhabitants on the earth's surface have been removed from it, and impart a significant directive force to magnetic needles, and magnetic force to iron rods held in the required position, notwithstanding the fact that it attracts or repels other magnets and non-magnetic iron so weakly that this action is completely imperceptible to our senses.

342) The reason for another phenomenon, which I recall has been noted by some rather diligent observers of the magnet, is clearly apparent from these same principles. Certain investigators of nature distinguished the *sphere of activity* of magnets from the *sphere of attraction*.[175] As the radius of the sphere of activity, they assume that at which the magnet sensibly acts on the magnetic versorium, but they fix the extension of the sphere of attraction from the distance at which a magnet exerts a sensible attraction on iron hung from a very mobile scale; and they further hold that experience teaches that generally the sphere of activity extends itself further than the sphere of attraction, but that the difference between the distances to which the two spheres extend is found to be greater, the larger is the magnet used in the experiment and the greater is its strength. So aptly do these things agree with my theory as expounded so far that I am confident that without further discussion on this matter, there will be hardly anyone who will doubt that I have found the true cause of the phenomenon.

343) We can now return to the path from which we were led in finding a solution of the doubt. There are very many phenomena which arise from the fact that the globe of the earth can communicate magnetism to iron implements and iron masses located near the surface of the earth, and which have always been considered by those skilled in nature to be marvelous and difficult to explain. But they flow so easily from our theory that they fully deserve a more careful treatment. First of all, in the spontaneous magnetism which iron rods gain from the action of the terrestrial core, a certain difference is observed in the experiment according as either a soft iron rod or a hardened one is used. For rods made of soft iron held in the required position gain a much more lively magnetism than hardened ones; and experience testifies that the degree of magnetism bears some kind of inverse ratio

[175] Gilbert was probably the first to do so (*On the Magnet*, pp. 103–104).

to the hardness, so that in general the magnetism is weaker the harder are the rods, and in rods endowed with the greatest degree of hardness possible, such as those which in hardness and fragility emulate glass (glatz=hart) it is hardly at all, or only very slightly sensible. However, we need not look very far for reasons for this phenomenon. Since magnetic fluid moves easily in soft iron and with difficulty in harder iron, and the communication of magnetism consists in the propulsion of magnetic fluid from one end of a rod toward the other, the particles of magnetic matter enclosed in softer iron will more easily than those enclosed in a harder steel obey the impulse exerted by the globe of the earth. For the same reason, as we have noted above, soft iron bodies in the vicinity of any magnet gain a greater magnetism than do hard bodies.

344) Hold the iron rod AB, Fig. CV, which was previously free of magnetism, in a vertical or any other convenient position. If then a magnetic needle is brought up to its lower end B (given that this experiment is instituted in the northern part of the earth), in agreement with what has been dealt with above, this repels the northern point of the needle and attracts the southern. If the needle is brought to the upper end A, the complete reverse obtains, for this end attracts the northern pole of the needle and repels the southern. Now turn the rod so that it comes to the position ba, such that the end A, previously the upper one, now becomes the lower, and it will be observed that both ends have immediately gained contrary forces to those they had before, for the end a now attracts the north point of the needle, and repels the south, but the end b repels the north and attracts the south. The sudden change experienced by the rod in this test has seemed marvelous to investigators of nature,[176] for the end B which was at first the north pole of the rod instantly becomes the south pole as soon as the rod is turned over quickly. But the reason for this is very easily deduced from my theory. For it is self-evident that when the rod is turned, the terrestrial pole toward which it is turned tries to induce in the rod a contrary state from that which it had previously, so it is by no means miraculous that the magnetism produced in a moment is also destroyed in a moment by an equal but contrary force, and a contrary state is induced in the rod.

345) I have noted above that rods made from rather hard steel, even if placed in the required position, nevertheless acquire only a weak magnetism from the action of the terrestrial core, and I have discussed the reason previously. But experience has revealed methods which

[176] Cf. Réaumur, *HAS,* 1723, Mém. pp. 146–47.

help even rather hard rods to acquire a greater and quite significant degree of magnetism. First of all, a longer stay in a convenient position can achieve this. For if hard iron rods or any other bodies suitable for receiving and conserving magnetism are properly placed and remain in that position for quite a long time—through whole years or even centuries—they are found to acquire a significant degree of magnetism; in fact we read that it is rather often observed that bodies of this kind, if they persist in the same position through several centuries, gain so great a force that they lack nothing by comparison with rather generous natural magnets dug up from pits. But it is apparent enough that this phenomenon presents no difficulties, for since the terrestrial globe continuously tends to make a rod magnetic as long as this remains in the required position, it is not greatly to be wondered at that its force achieves a greater effect than it can produce in a few moments if it continues to act constantly through several years or even centuries.

346) We must not be completely silent about the following phenomenon, although its cause is quite obvious. After an iron rod or any other ferreous body, following a long stay in the required position, has gained a significant degree of magnetism, turn it so that it takes up a position directly opposite to the former, and there will then be a different outcome from that which I described in the test in § 344.[177] For the magnetism of the rod is not immediately inverted as in that case, but rather, if the end A has gained south magnetism and B north magnetism, each end will retain the magnetism it possesses, even after the rod is turned. The reason for this phenomenon is readily discovered, for it is self-evident that the effect which the continual action of a force protracted over a long time has produced cannot be destroyed in a moment by an equal force tending to produce the contrary state. It is readily clear meanwhile that, if the rod is held for quite a long time in the inverted position, the force which it had gained over a long passage of time will be significantly weakened, and finally completely destroyed, in fact it can happen that, after all the magnetism which it formerly acquired has been wiped out, the rod is reduced to the opposite state. Yet it is obvious that this cannot happen unless the rod is held in the inverse position for a longer time than it stood upright. It is also easily understood from these principles why a magnet held for a long time in a position such that its south pole faces the north pole of the earth, and its north pole faces the south pole of the earth, will suffer a significant loss of its force, and, on the contrary, its vigor increases somewhat if its north pole is turned toward

[177] The text has § 342.

the north and its south pole toward the south, and the magnet is held in that position for a long time. But these things are sufficiently self-evident, so let us pass on to a treatment of phenomena of greater moment.

347) Consider a stratum *ef*, Fig. CVI, consisting of a rich ore of iron which has evolved sufficiently, to be immersed in the bowels of the earth but not deeply below the surface, and if *abcd* is the magnetic core of the earth it is clear from the previous discussion that its action will tend to make that ore magnetic. If therefore this action continues constantly through an immense time and for very many centuries, the stratum can become magnetic to such a great degree that parts of it torn out of a mine will show a powerful magnetic force, provided that the ore of which the stratum consists is sufficiently suitable for acquiring and conserving a strong magnetism. I think that this is the origin of the natural magnets which are dragged into the light every day from iron mines. For it seems lacking in all probability to suppose that our mines, which really descend a very small distance below the surface of the earth, are sufficiently deep down to touch the very core that we have discussed previously. It is not only evident that magnetic ore could arise in the fashion described, but also if, as many Physicists think, metals are still growing in the earth today,[178] it cannot be doubted that every day new magnets are still being generated and produced in the bowels of the earth in the said fashion. So my opinion that all the magnets which the labor of men has ever dragged from the abysses of the earth into the light were produced by nature in the described fashion does not seem open to doubt.

348) But my readers might share a doubt which occurred to me concerning that opinion. While I was trying to dispel it, I happened to reach not only a full solution, but also a not inelegant invention by which the strength of natural magnets can be increased to an unaccustomed degree. For if what I have asserted concerning the origin of natural magnets is true, it is possible to ask how it happens that not every iron ore torn from the bowels of the earth is found to be endowed with magnetic force. For it seems that my explanation of the origin of natural magnets is such that it becomes a consequence that no ferreous body can remain in the same position for long without acquiring a significant magnetic force. It is well known on the contrary that, among innumerable species of iron ore, there are only a

---

[178] This was still a commonly held opinion in the eighteenth century. Cf. Linnaeus: "*Minerals* grow; *Plants* grow and live; *Animals* grow, live and have feeling. Thus the limits between these kingdoms are constituted." (*Systema naturae* [1735], English trans. by M.S.J. Engel-Ledeboer and H. Engel, Nieuwkoop, 1964, p. 19).

CERTAIN PHENOMENA 431

few to which nature has imparted a magnetic force. So the theory seems to lead to consequences not consonant enough with experience.

349) Writers on mineralogy usually distinguish iron ores into *refractory* and *non-refractory;* by the latter term they designate those which are attracted by a magnet without further preparation, just as they are ripped from the bosom of the earth, either in great chunks or ground into a rather fine dust; whereas under the former term they include those ores on which a magnet exerts no action under the same circumstances.[179] Let us first of all consider very carefully the nature of the ores called refractory. Since I have stated above, and have proved with suitable arguments, that a magnet does not attract any body unless it is another magnet (i.e. a body which has gained magnetic force either from elsewhere or simply by approaching the magnet to whose action it is exposed), I have perceived that it is a necessary consequence of my theory that refractory iron ores should be completely unsuitable for receiving magnetic force. So I have easily understood the reason why there are so many kinds of iron ore which have received no magnetism from the prolonged action of the terrestrial core, even though they have lain buried in the bosom of the earth for very many centuries. For since all refractory iron ores are completely incapable of receiving magnetism, it is not surprising that those torn from the bowels of the earth are never found endowed with a power which they are not capable of receiving.

350) Believing that it is the duty of a circumspect investigator of nature not to trust too rashly in a theory, no matter the degree of probability with which it shines, lest he neglect to weigh the truth of its consequences on the balance of experience, I decided to inquire whether it is also proved by the testimony of experience that refractory iron ores completely reject any acquisition of magnetism. Among the refractory iron ores there is a certain species of *Haematite* called (Glaßfopf. From several crumbs of this mineral I produced a parallelepiped figure about one and a half inches long, 2 lines broad and $\frac{1}{4}$ line thick. With the help of very strong artificial magnets, and in the fashion described at length in the previous chapter, I worked at inducing some magnetic force in it, but no matter what I did, I found that my efforts came to nought. I had received from a friend a certain species of iron ore believed to have been brought from *Siberia,* which was naturally a rectangular parallelepiped in shape, and refractory, for neither great chunks of it nor even the finest dust was attracted by a magnet, even a very strong one. I placed a chunk of

---

[179] The terms "refractory" and "non-refractory" usually refer not to any magnetic property of an ore, but to its ability to resist fusion by heating.

it between two very strong artificial magnets in the manner previously described, and I held it in this position for several days, but again the attempt was fruitless, for it gave not even the slightest indication of having acquired any magnetism after this time had elapsed. Relying on these experiments, I consider myself sufficiently safe from error if for the future I assume the conclusion supplied by my theory to be true and in accord with experience: namely that all refractory iron ores are unsuitable for receiving magnetic force, and, as a result, though enclosed in the bosom of the earth for centuries, they never gain it.

351) Although the matter concerning refractory iron ore seems solved already, yet doubts can remain concerning the other kind of ores called non-refractory. For since these are attracted by magnets, theory bids us conclude that they can and in fact do gain magnetic force while they are being drawn by a magnet. Since they are suitable for receiving magnetism, it seems that they ought necessarily to have gained some magnetic force from the continuous action of the magnetic core while they were enclosed in the bosom of the earth. This seems the less subject to doubt since it can also easily be shown through experience that all the ores attracted by magnets can easily be imbued with magnetic force. But it is not too difficult to remove the doubts arising here, provided one is willing to attend a little to the following considerations. It is first of all self-evident that there can be a gradation in the property of iron ore to be either non-refractory or to obey a magnetic attraction. It is readily understood that there will be some ores which are non-refractory to a small degree, others to a larger degree, but generally, according to the principles of our theory, the less a body is drawn by a magnet, the smaller also is the magnetic force that it acquires from a magnetic action of a given magnitude. So among the number of non-refractory ores there will be certain ones which can acquire only a quite weak degree of magnetic force, and others which can gain a greater force and one more apparent to the senses. The magnetic force of the former cannot be observed in crude tests and cannot clearly strike the senses unless it is tested with great care and skill. So partly because these are ordinarily handled only by diggers of metal—a crude type of men intent on gain rather than on the disquisitions of physics—and partly because by chance it has not previously occurred to any Physicists to suspect that they are endowed with magnetic force, or to set up experiments for exploring this matter, they have been commonly held to be altogether inert and have been completely neglected. So the only ones commonly considered magnetic and sold under the name of magnets are those whose magnetic force is so great that it strikes even the more uncultivated senses.

CERTAIN PHENOMENA 433

352) In particular, some tests set up as opportunity offered on a Siberian iron ore showed me that this suspicion of mine is not basically empty. This ore was offered to me under the name Kammertz, and I suspect there is a certain affinity between it and haematite itself. It can be split into thin plates and is endowed with a blackish rusty metallic color, but when rubbed it yields a reddish dust which can hardly be distinguished from that rubbed from haematite. Yet it is quite different from haematite in that a magnet sensibly attracts its dust. I immediately suspected that this ore would be magnetic, and my suspicions did not deceive me. For when I tested it with a very mobile magnetic needle, I observed that any particle of it exerted a sensible enough attraction and repulsion after the fashion of a true magnet, even though as far as I know it had never even smelled a magnet. I ask those Physicists with the time and opportunity to carry out the same test on the other non-refractory iron ores which are not commonly held to be magnetic, and doubtless innumerable examples will convince them of the truth of my opinion.

353) There is another consideration regarding this matter which must not be overlooked. I have noted above that soft iron is suitable for receiving a significant degree of magnetism but not for preserving it. So perhaps there are iron ores too with the same character as soft iron, so that they gain a significant force when exposed to a magnet, but immediately lose either all or at least a very great part of it when they are taken away from the magnet. So these will be non-refractory and will be attracted by a magnet with a significant force. Doubtless too they possess a sufficiently great and sensible magnetic force in the bosom of the earth and when shut in mines, but when they are torn out and put in another position, they immediately lose that force. In this category seems to me to belong a type of black sand frequently found over the whole face of the globe and especially along seashores.[180] This sand is strongly attracted by a magnet, to at least the same degree as iron filings are, but when it is packed into a tube and magnetized, it is found to receive little magnetic force.

354) Although the discussion to date is enough to abolish all possible doubts completely, I could not rest here. I was eager to inquire into the source of the significant diversity between iron ores, when some do not obey a magnet at all, and others readily obey a magnetic attraction and freely take up a magnetic force. I soon understood that the answer was not, as had occurred to me, the poverty of the refractory iron ores, for some ores of this class are quite rich in metal, such as e.g. the species of haematite called Glatzkopf which I men-

[180] Cf. above, p. 242.

tioned before, a hundred pounds of which, when handled skilfully, usually provides up to eighty pounds of pure metal. I recalled that several metallurgical writers, and a good number of Physicists, hold that refractory iron ores are made obedient to a magnet if they are ignited in an open coal fire to white heat, which they call rösten, or if they are burned to white heat in a crucible with the addition of some inflammable material, or phlogiston.[181] So, thinking that this test ought to be carried out on the so-called refractory ores of § 350,[182] that is on haematite and on the particles of the Siberian ore, I brought them to white heat between coals, and when they had cooled down I observed that not only were they attracted by a magnet, but also, with the help of artificial magnets, I could give them a quite sensible magnetism using the methods described in the preceding Chapter. So it was obvious that ignition or burning with the addition of phlogiston makes refractory ores suitable for receiving magnetism. But it is not sufficiently agreed how this happens and what is that change induced in the ores by ignition which makes them obedient to a magnet. After considering various things, I came upon an hypothesis which I willingly submit to the judgment of men more expert than I in mineralogical matters.

355) To my mind the non-refractory iron ores are those which contain perfect iron in metallic form (gediegen Eisen), while refractory ores are those in which iron is found not perfect, but deprived of a certain constitutive part, or, so to speak, mineralized. I am not unaware of the opinion of several mineralogists that perfect iron endowed with metallic form is never extracted from mines, but I reject it since it does not seem to be confirmed either by experience or by the arguments ordinarily presented. First, as for experience; a pure and native particle of iron in the possession of *Master Marggraf*,[183] the celebrated chemist at the Royal Academy of Sciences in Berlin, places the possibility of the existence of native iron completely beyond doubt. If anyone thinks I am teaching things contrary to the testimony of the senses when I assert that natural magnets and other non-refractory iron ores contain perfect iron, I ask him to consider the following. It may be that in those minerals, very tiny particles of pure iron are so mixed and covered over with stony, earthy, and other foreign particles, that any single particle is so small and so hidden by foreign particles that it cannot easily be perceived either by the naked

---

[181] See above, p. 186.

[182] The text has § 348.

[183] Andreas Sigismund Marggraf (1709–82), one of the eighteenth century's foremost chemists, and a long-time (1738–82) member of the Berlin Academy of Sciences.

or the assisted eye. It is very well known that this happens at times with gold, a metal which is only found pure and native, and then they usually say that gold is sprinkled in the stone (eingesprengt). Next, as regards the arguments by which the possibility of extracting perfect native iron from mines is impugned, they are usually based on the high degree of corruptibility of iron. For with any solvent, no matter which, even common water, one can easily cause the destruction of pure iron and the disappearance of its metallic form; so it seems that even if perfect iron was supposed to exist in mines occasionally, it could not preserve this state for any significant length of time, but rather it would be destroyed again after a short time. Let me reiterate that it is quite possible for a particle of perfect iron enclosed in the ore to be so covered and hidden by particles of earth and stone that a solvent cannot penetrate to it, and so cannot destroy it. Thus, for example, if I assume that any grain of the magnetic sand described above, § 353,[184] contains a nucleus made from pure and perfect iron, but that it is surrounded on all sides by a stony, almost glassy matter, it may be readily understood how that iron particle can resist not only the rain water or sea water which moistens it every day, but, what is more important, can resist acids themselves (from which, as experience testifies, this kind of sand suffers no change). For the stony matter covering a particle of iron enclosed in it as a nucleus may defend it from the action of solvents in the same way as a hermetically sealed glass vessel can keep iron enclosed in it safe from the action of *aqua fortis,* if it is immersed in it.

356) These thoughts make it clear that I am defending an opinion which is not plainly foolish when I assert that there can be iron ores containing perfect iron endowed with metallic form. But the reason I think that non-refractory ores are of this nature will readily be understood from what follows. Experience teaches that refractory iron ores are made obedient to a magnet through ignition, and are thus reduced to the same state which non-refractory ores have received from nature itself without the help of our arts. If we weigh up the kind of change ignition induces in ores, it seems quite probable that it consists only in this, that with its help sulphureous and inflammable particles are combined with the particles of the metal contained in the ore. For as the ores grow white hot, the phlogiston of the coals penetrates them, and can join the metallic particles, and the resultant feeling that ores will be made obedient to a magnet to a still greater degree if they are strongly burned in a crucible with the copious addition of

[184] The text has § 351.

inflammable matter than if they merely grow white hot in the coals of an open fire, is so confirmed that it seems that scarcely anything more could be desired for full proof. If we ask further what phlogiston can achieve when entering into an ore and being intimately united with it through the strongest force of fire, it will be convenient to record that, although to date we know little about the nature of metals, it is nevertheless established by indubitable experiences that every metal is composed of a solid, friable part which Chemists usually call the *metallic earth,* and sulphur, or phlogiston, and we have discovered that it is to the latter that the metal owes its metallic character, that is, its ductility and that special sheen which we call metallic. So if I assert that refractory iron ores such as are dug from the bosom of the earth contain only a primogenital martial earth in which so far the phlogiston necessary to produce a metallic character is lacking; and that by subsequent ignition, or by cementation with inflammable bodies, phlogiston is united with that martial earth which had been deprived of it, so that consequently the mixed particles of the ore acquire the nature of a true metal; I think I am stating only what is quite conformable to the principles of metallurgy.

357) Since refractory iron ores are brought to the same state as non-refractory ones by ignition and cementation, and in all probability the change which ignition or cementation induces in the former consists entirely in this, that the martial earth contained in them is united with phlogiston and becomes a true perfect iron, we are led to conclude that non-refractory iron ores dug from mines already contain, without any help from our art, an iron that is perfect and endowed with metallic character. But a new question can be raised about the generation of those minerals, for it seems that their origin can be conceived in two ways. For nature either at the same time as it produces the ore, and in the same action, also produces perfect native iron in such a way that the ore already contains perfect metal right from its initial generation; or, on the contrary, while the ore was being formed it did not include the true metal but only a martial earth still deprived of the phlogiston needed before it could be called a perfect metal, but then this initially refractory ore was changed into a non-refractory one through some accident by which it was able to join phlogiston to its metallic earth and become true metal. Anyone who is aware of the changes which have been suffered by the terrestrial globe from the very oldest times will not deny that such accidents could have taken place. An accidentally lit subterranean fire could burn refractory ores still hidden in the bosom of the earth and change them into such as contain perfect iron and are attracted by a magnet. So although I dare not

CERTAIN PHENOMENA 437

defend the total impossibility of the former mode of generation of non-refractory iron ore, it still seems probable enough that it is the latter mode which nature normally uses, especially in producing magnetic iron ore, and I would not find it strange if anyone wanted to assert that all the magnets dug up from the earth, and all other non-refractory iron ores, have at some time been violently burned by a subterranean fire.

358) I have perhaps spoken here more boldly than is proper about matters that are in a way foreign to me. Therefore I leave the mature judgment to those who are more familiar with Mineralogy than I, and I pass on to more familiar disquisitions as I retreat to my own camp. I discovered from the literature that refractory ores such as we have received them from the hands of nature are completely unsuitable for receiving magnetism, but can be made to assume this force through ignition and cementation. So it was natural to suspect that if the artifice which can make ores that completely reject the acquisition of magnetic force suitable to assume that force were used on ores which of themselves do not refuse to take on a magnetic force, it would significantly increase their aptitude for receiving it. From this I concluded that natural magnets burned by a strong fire might be brought to the point where they would become capable of a far greater degree of magnetism than previously. This suspicion seemed to agree outstandingly with what I discussed previously concerning the reason for the difference between refractory and non-refractory ores. For, if, in agreement with my hypothesis, it is assumed that ignition intimately unites the martial earth contained in the ore with the inflammable principle and transmutes it into true iron, so to speak, the result that must be expected from the stronger ignition of a magnet is a greater evolution of the iron contained in it, and from this, a stone should become fit for taking on a greater magnetism through being fired. These were the considerations that led me to a not inelegant method of increasing the strength of natural magnets to an unaccustomed degree.

359) Repeated tests taught me that my suspicions were not in vain; I shall describe some of them rather carefully, partly to win support for my assertions, and partly for the purpose of clarifying the entire method I use to make natural magnets more powerful. The first magnet I tested had a regular parallelepiped shape, and was made from a blackish mass without metallic sheen, quite homogeneous and compact, such as magnets of the cheaper kind almost always are. This was of the weak class of magnets for, when bare, it weighed $25\tfrac{3}{64}$ ounces, and with its armature it was found to weigh $36\tfrac{2}{64}$ ounces, and it was

only able to carry a weight a little greater than its own, for it could only lift 4 ounces. I placed this magnet, devoid of armature, between two larger steel rods endowed with a considerable magnetic force, in the manner described in the previous Chapter, and after half an hour I found that its force had been increased to such an extent that after the armature had again been added it lifted $12\frac{1}{2}$ ounces, but no more. I exposed this magnet to an open coal fire and kept it very hot for half an hour. When it had cooled down I found that it had lost almost all its magnetic force. I placed it between the same rods for a quarter of an hour and then found that, fitted with the armature, it lifted $18\frac{1}{2}$ ounces. So after ignition the magnet gained a greater force in less time from the same magnetic rods than it had been previously able to acquire over a longer interval. It is evident then that the aptitude of this magnet for receiving magnetism increased in proportion to what it was previously by a factor of 37 to 25,[185] that is about 7 to 5, after I had employed this artifice.

360) There was another magnet, weighing $4\frac{1}{4}$ ounces without armature, and $5\frac{7}{8}$ ounces with armature, likewise made from a compact, uniform material, but richer in metal than the one previously described; with its armature, it carried $6\frac{3}{4}$ ounces. Placed for half an hour between the artificial magnets, before ignition, it could not carry more than $22\frac{3}{4}$ ounces. After being heated in a coal fire for half an hour and afterward cooled down, it had lost almost all its force, but placed for a quarter of an hour between the magnetized bars, it later easily carried $37\frac{1}{2}$ ounces, so its aptitude for receiving magnetic force had increased through my artifice in the ratio 150 to 91, that is in the ratio of about 5 to 3.[186] It seems then that, since I obtained a more significant increase in this case than in the magnet previously described, this method is more valuable, the more powerful is the magnet before being heated. The last point I shall mention here also showed me that the increase in force obtained in this fashion is stable enough, and the magnet does not easily lose it again provided it is looked after properly. A few days ago, after six months had passed, I tested it and found that it still picked up the same weight as at the beginning.

361) It seems that the artifice expounded here at some length can be brought to greater perfection, and is capable of further emendation. It is obvious in the first place that the evolution of the iron contained in the ore can be helped not only by ignition in an open fire, but also by cementation in a strong fire after the addition of a large supply of phlogiston. It is readily clear that cementation is

[185] The text has 37 to 27.
[186] The text has, respectively, 91 to 75, and 8 to 5.

without doubt more effective than mere ignition with coal in producing this effect. Then although ignition or cementation is quite conducive to the evolution of the iron, it is also very well known that iron ignited and submitted to a slow cooling becomes quite soft, and as a result is suitable for receiving magnetism, but unsuitable for retaining a very great degree of it. So it seems beneficial if the magnet, when sufficiently cemented and fiercely incandescent, is suddenly immersed in cold water. For in this way the iron evolved by cementation or ignition contracts and becomes harder and at the same time more suitable for conserving magnetism.

362) [187] I feel I should not overlook the opportunity offered me here of saying a few words about the best way of handling natural magnets. First of all I induce a regular parallelepiped shape in a magnet dug from a mine, by rubbing it on a sandy whetstone; taking care that it comes out as long as possible, for it has been shown above that, other things being equal, a longer magnet is capable of a greater force than a shorter one. In this operation there is no need for attention to be given in any way to the position of the natural poles of the magnet; it is enough to shape it as advantageously as possible, no matter where the natural poles may be. Cement the magnet thus prepared and dip it when white hot into cold water. Place it between two very strongly magnetic steel rods, or more if circumstances allow, and, if possible, stroke it with artificial magnets according to the method of the double touch, choosing as the polar faces those with the maximum distance between each other. Afterward, arm the magnet in the customary fashion, and you will obtain a magnet that is very powerful in proportion to the excellence of its ore and its shape, and endowed with as great a strength as possible.

363) Before I leave this discussion of natural magnets I have certain considerations I can add since they will fit in at this point. I greatly regret that I could find no trustworthy and diligent observations about the positions of magnets in the mines from which they were extracted. I should like skillful observations made to see whether the north pole of a magnet faces the earth's northern region before it is cut from the mine, and the south pole faces south, or whether the contrary is observed at times, so that the side of the magnet turned directly toward the north becomes the south pole of the magnet, and the side facing the south becomes the north pole. It seems that both can happen. For if AB, Fig. CVII, is the whole stratum of ore made magnetic through the action of the terrestrial core, it is easily understood that

---

[187] Incorrectly numbered 352 in the text.

440 THEORY OF ELECTRICITY AND MAGNETISM

the end B which faces north becomes the north pole, and A which faces south becomes the south pole; but it is obvious nevertheless that two cases can arise. For either the stratum AB acquires only two poles, or it has several succeeding one another in the manner of consequent points. Suppose it is the former, and let C be the magnetic center of the stratum. Anyone willing to remember what I remarked upon at the end of the first Chapter will readily perceive that whenever a piece MN or mn is cut from this stratum, it will be of such a character that its end M or m which had been facing south becomes the south pole, and the end N or n which had been turned north becomes its north pole. But if the stratum AE, Fig. CVIII, has consequent points, the result will be different according to the different places where a piece is cut from the ore. Let the curve of the densities of the magnetic fluid be LBKCHDF, whose ordinates become zero at the points B, C, D, and are a maximum at the points I, G. If A is the end facing the negative pole of the terrestrial core and E the end facing the positive pole of the core, it is clear that if the piece is cut out

between A and B, the piece MN will have M as its positive pole and N as the negative one
  – B – I – – OP – – O – – – – – P – – – –
  – I – C – – QR – – Q as its negative pole and R as the positive one
  – C – G – – ST – – S – – – – – T – – – –
  – G – D – – VW – – V as its positive pole and W as the negative one
  – D – E – – XY – – X – – – – – Y – – – –

and so on. But although skillful observations concerning this matter are very much to be desired, sufficiently trustworthy ones are hardly to be expected. For the opportunity of inquiring into these things is rarely given to Physicists with sufficient training, and the diggers of metal who usually handle ores either neglect such things, or deserve little trust.

364) At the beginning of this chapter, in dealing with the directive force of the magnetic needle, I indicated that experience shows that the declination and inclination of the nautical versorium is subjected to such great irregularities that it is reasonable to suspect that either the shape of the core is irregular, or the distribution of the magnetic fluid through it is irregular. A new cause has arisen here which could contribute to the production of these irregularities, in fact perhaps it produces them on its own. For the magnetic ores which lie buried under the surface of the earth without doubt occupy a large area dispersed irregularly through the entire globe of the earth, and by acting on a magnetic needle, they will disturb it from the direction it

would acquire if the magnetic core only was present. Since, as can very probably be assumed, new strata of ores are generated daily and ancient ones destroyed,[188] perhaps the principal cause of the variations observed in the position of the magnetic needle should be sought in the generation and destruction of magnetic ores, and in the other mutations which can occur to these. Things seem to agree outstandingly with this suspicion, because we read that at times after violent earthquakes the direction of the needle has suddenly suffered a considerable alteration; it is easily perceived that an earthquake will break the strata of the magnetic ores and tear them apart once they are broken, and so in this way it could produce significant changes in the direction of the versorium. Concerning this matter, it must be regretted that there have been no observations made with great enough skill. I should also like to test in the vicinity of volcanoes (especially those which, amongst other things, usually belch forth iron ore) whether the direction of a magnetic needle suffers any sensible change after violent eruptions. If these thoughts of mine are not altogether wide of the mark, it is clear how vain are the opinions of those who think that the law according to which the magnetic declination and inclination vary has already been discovered, or who hope that it will be found at some stage. For it may be seen that, in conformity with my idea that along with changes of the core (which are perhaps more regular), other changes take place in the ores hidden not far below the surface of the earth, and these latter ones will play some part in bringing about the changes in the direction of the needle. But it is clear that these changes in the ores cannot be reduced to a fixed and constant rule any more than can the changes in the weather during the year.

365) [189] We now have sufficiently considered the phenomena which result when the action of the terrestrial core induces a magnetism which is not easily destructible again in a short time into iron implements or ores containing a perfect iron, held for quite a long time in the required position. But there are other methods besides which can achieve the same thing in a few moments. For example, heat an iron rod which is not strongly imbued with magnetism, and while it is held in the required position for receiving magnetic force, allow it to cool, or better still, immerse it suddenly in cold water; it will acquire a significant degree of magnetic force, and this will not be immediately destroyed or reversed when the rod is turned over completely. It must be noted that if the test is set up in our regions, the lower end will gain a northern magnetism, and the upper end a southern magnetism.

[188] Cf. above, p. 430.
[189] Incorrectly numbered 363 in the text.

366) The test agrees outstandingly with our theory. For artisans who handle iron and steel know that iron is made softer by ignition, but grows hard once more with subsequent cooling, and that, other things being equal, it acquires a greater hardness the more swiftly and the greater the extent to which it is chilled. So since magnetic fluid moves more easily in softer iron than in hard, it is clear that the action of the terrestrial core will make a heated rod more strongly magnetic than a similar cold rod. So in an ignited rod AB, a greater quantity of fluid is propelled from one part BC into the other part AC than could have happened before the rod was heated. When the rod, after being endowed with a great degree of magnetic force while it is hot, is chilled, it immediately becomes harder again, so, since the difficulty with which magnetic fluid moves through its pores is thus increased, even if the force driving the fluid to cross from BC to AC should cease, nevertheless the fluid accumulated in the part AC will be forced to remain there and will not be able to return to the part BC. So it is clearly evident why the rod acquires a greater degree of magnetism in this test than it had unheated, and acquires so much that this cannot be immediately destroyed again when the rod is reversed.

367) While considering these things a thought came to my mind which is perhaps not unworthy of being presented. I noted in the preceding § that iron is made harder and more rigid with the help of cold. Common experiences testify that this is true, for they have shown that soft iron implements, or ones which are only moderately hard, become quite hard and fragile if exposed to extreme cold. So it seems that if a steel rod is magnetized, the degree of saturation will depend, to a large extent at least, on the degree of heat which it possesses. Thus a rod will not acquire as great a degree of magnetism during summer, when the air is warm, as it will if it is imbued with magnetic force in the winter, when it is extremely cold. The matter is worth testing experimentally, but I confess that these tests demand great skill if we want to avoid any mistakes.

368) There is a third method by which a rod can acquire a stable magnetism, and one not destructible simply by inverting the rod, from the action of the terrestrial core. While it is held in the required position for receiving magnetism, hit it hard, and rouse its smallest particles into a strong agitation, or use any other method which ultimately brings about the same effect. The rod will be found to have acquired a significant enough and greater degree of magnetism than it could have done had it not been struck, and if it is turned over, its magnetism will neither vanish nor be reversed. This phenomenon raises no difficulty, but rather can be happily explained, if we agree

that the motion of the magnetic material is freer when the rod is struck and the least parts are agitated by a strong tremor, than if those particles are completely at rest. For if this is assumed, it is immediately clear why a struck rod acquires a greater degree of magnetism than one that is not struck. For since the magnetic material can be more freely propelled through the pores of iron so long as the least particles making up the rod are trembling, it will be responsive to the force acting on it, soliciting it to cross from the part BC, Fig. CV, into the part AB, and fluid will be accumulated in AB and evacuated in BC to a greater degree, whence the whole rod will acquire a greater magnetism than it could were it not struck. If the rod is then reversed, the action of the terrestrial core tends to destroy the magnetism previously produced by inducing a contrary state, but that tendency will be ineffective since the fluid now moves with greater difficulty than before, and it will not be able to destroy the magnetism unless the rod is struck once again.

369) In assuming here that the magnetic fluid obeys the force propelling it from one end of the rod to the other more easily while the particles of the rod are trembling than if these particles are completely at rest, I think I am suggesting nothing improbable. For the reason fluid ought to move more freely in the pores of a body that has been struck violently and is trembling than it does if the least particles of the body are completely at rest can be found rather easily. Let another small body D touch the surface of the body AB, Fig. CIX, with D being attracted by the body AB by some force, and it is clear that if the particle D is driven by a certain force in the direction DE, that force will not produce any motion, and the particle D will not move from its position toward E, unless the force is greater than the friction which D experiences while it moves forward over the plane AB. If therefore this friction is called F and the force propelling the particle D toward E is represented by V, no movement can occur while $V<F$, if the particle D touches the plane AB. If the particle D is elevated a little and removed from contact, the friction either vanishes completely, or at least it becomes smaller, and the force V will then in fact be able to propel the particle. But it is clear that if the body AB is struck, its particles will be agitated, and the small body D will be removed from close contact with the surface AB and will, as a result, more readily obey the force acting on it. Therefore where there is reason to fear that friction will make our tests or experiments erroneous, we usually insure against the effects of friction by means of concussion. For this reason those performing meteorological observations tap the barometer before accepting the altitude indicated; astronomers

examining the positioning of their instruments with a level (which consists of an air bubble in a glass tube filled with spirits of wine) shake it so it trembles, and similarly those who perform observations on the magnetic needle tap the plane on which the needle stands.

370) When the experiment we are considering here is performed, it is self-evident enough that it does not matter how the least particles of the iron rod are incited to movement. Without doubt, electric lightning, whether given out by the *Leyden* jar on bursting from a cloud laden with thunder, is to be included among the things which can shake the least particles of iron very strongly, and when I assert this, I trust that all who have ever exposed their own bodies to the electrical commotion (as it is usually called) will agree with me. So it is easily understood how it happens that iron implements struck by lightning have quite often been made strongly magnetic, and magnetic needles have been found either to have lost their own force or to have exchanged it for a contrary one, and, according to *Franklin's* experiments, slender iron wires usually become magnetic after the electric lightning leaping from a *Leyden* jar has passed through them.[190]

371) This seems so obviously the reason that iron wires can be made magnetic with the aid of electricity that there is no need to look for another, and those who think that a hidden nexus of the causes of electricity and magnetism is revealed by this test seem to be mistaken. For nothing requires us to assume that the electric force acts as such here, since the whole of this effect is easily explained simply by the very strong concussion of the least particles, and the electrical concussion is found to produce here only what any other concussion could similarly have achieved. I admit that this deserves to be tested experimentally, either to place my opinion completely beyond doubt, or (though I do not fear this) to demonstrate its falsity, especially since to date the tests of the famous *Franklin* seem somehow doubtful. I have not been able myself to undertake it, for the disease which kept me confined to my bed for quite a long time has prevented me from adapting my electrical apparatus to experiments of this kind (which of course require a more violent degree of electricity than usual), or from carrying out these tests. But I shall work diligently to complete quickly what can still be demanded of me in this direction.[191]

372) There still remains a disquisition which must be numbered among the special cases in the total doctrine concerning magnetism. It concerns artificial magnetism, as it is called, or the method of producing a significant degree of magnetism in steel rods without using

[190] *Benjamin Franklin's Experiments*, p. 243.
[191] Aepinus never fulfilled this promise.

a natural magnet. But the whole doctrine concerning artificial magnets can be conveniently reduced to two heads. For on the one hand, such a disquisition must show how steel rods can be made magnetic without another magnet being used, and on the other it must give a method by which, once the rods have gained some degree of magnetic force, their force can afterward be increased right up to saturation level. I have said enough already about the latter question;[192] for not only have I considered at some length the methods used by others, but I have also supplied a more suitable and effective one than those discovered till now. But nothing has so far been said about the method of producing the first seeds, so to speak, of magnetic force in steel rods without the help of another magnet; which must be done if artificial magnets are to be achieved. The whole method depends on what has just been explained, and may easily be deduced from it.

373) A non-magnetic iron rod CD, Fig. CX, held erect, or in any other suitable position, immediately becomes magnetic from the action of the terrestrial core in such a way that in our region the lower end D acquires northern magnetism and the upper end C acquires southern magnetism, and the reverse occurs in the southern regions of the earth. Since the rod CD can be considered a true magnet so long as it is retained in the required position, it must be expected to show all the effects which ordinary magnets show. So if a wire or an iron rod AB is stroked with the rod CD, stroking it with its lower end D several times from A toward B, it is clear from the previous discussion that the end A ought to acquire a north magnetism and the end B a south magnetism.

374) These things agree with experience, as anyone can easily confirm. From this and from similar tests some Physicists have deduced conclusions which I think cannot be consistent with the truth. For some have been of the opinion that these tests show that magnetism can be produced by friction alone,[193] just as can electricity, and they feel they have found here an important analogy between the two forces. Others, going still further, consider that they can establish it as a general law that when iron is rubbed on another non-magnetic piece of iron, this becomes magnetic, and always according to the rule that the end from which the rubbing begins becomes the north pole, and that at which it ends, the south pole. I do not hesitate to pronounce each of these propositions as less than the truth. For the friction of iron on iron clearly cannot as such produce magnetism, and such magnetic force as is generated in the experiment is due altogether

---

[192] Cf. above, pp. 373ff.
[193] Musschenbroek, *Dissertatio . . . de magnete*, pp. 268–69.

to communication, not friction, in such a way that if this test were set up in a vacuum, far from the terrestrial globe, and removed from all other bodies rich in magnetic force, it could not produce even the slightest trace of magnetism, even though the labor was continued for centuries. This opinion of mine will be demonstrated completely in the subsequent discussion, from which it will become obvious that all the phenomena agree very well with the laws of communication of magnetism, but are completely contrary to those which should take place if friction alone, as such, were the cause of the magnetism.

375) I have an objection to fear at this point and I think it ought to be removed immediately, since it seems to militate on behalf of the opinion of those who propose the friction of iron on iron as the cause producing the magnetism. I have demonstrated above that in any given place on the earth, there are some positions such that if an iron rod is retained in them, it can gain no magnetic force from the action of the terrestrial core; such for example is the position parallel to the horizon at the magnetic equator. If the rod CD, Fig. CXI, is supposed to be held precisely in one of these positions and the rod AB is rubbed with it, it seems that if what I have supposed is true, it can gain no magnetism. For, under the circumstances previously described, the rod CD will not acquire any magnetic force from the action of the terrestrial core, and as a result, communication seems impossible. But something very different is taught by experience, which shows that the rod AB is made magnetic in this case too, although to a weaker degree than if the rod CD is held vertically upright. So it seems that in this test, the production of magnetism is the effect of friction alone, and not of communication.

376) This doubt is easily removed, provided what I presented earlier is duly pondered. For it does not follow from my assertions that the rod CD held in the manner previously described gains no magnetism at all, but it can only be concluded that the degree of magnetism it acquires ought to be significantly weaker than if the rod is held in the proper position. For I did not deny that the rod AB could be magnetized such that, for example, its upper half CDAB acquires southern magnetism and its lower half ABEF acquires northern magnetism; I merely held that ordinarily, because of the quite small thickness of the rod, the force produced in this way would only be small, so I was able to neglect it in the disquisitions presented earlier. It is clear then that even though it is held in the position which I described in the previous §, the rod CD will become truly magnetic, although weakly; so it is not surprising, or contrary to my doctrines, that when it is rubbed against the rod AB, the latter is made

magnetic. It is clear, according to the theory, that the magnetism produced in this fashion will be weaker than if the rod CD is employed erect in the test, and this we have found to agree completely with experience.

377) [194] As for the second rule which I previously pronounced false, namely that, when magnetism is produced by rubbing iron on iron, the end at which the rubbing begins becomes the north pole and that at which it ends becomes the south pole: there is no need for me to show its falsity in a special work, for what will be put forward shortly will disprove it sufficiently; so I consider it advisable to proceed rather toward bringing out certain things which follow from our theory. Hold the rod CD, Fig. CX, erect, or in some other convenient position, and if the test is carried out in our region, it is sufficiently well known already that its lower end becomes the north pole and its upper end the south pole. If therefore the rod ab is stroked by the upper end of the rod CD, exactly the same outcome is to be expected as is obtained when it is stroked by the south pole of some magnet. What was presented above, however, teaches that in this case the end a at which the rubbing begins becomes the south pole, and the end b at which it terminates, the north pole.

378) To become certain whether experience confirms what I had deduced as a consequence of the theory, I examined with the help of a very mobile magnet an iron forceps made of soft iron, to see whether it possessed a stable magnetism or not. After I had found that the latter was the case, and thus that the forceps was suitable for the test, I placed an iron wire with no magnetic force, about half a foot long and one line in diameter, on a table, and holding the forceps vertically upright I stroked this wire about twenty times with its lower end, beginning each rubbing at A and finishing at B. In this way the wire AB gained a sensible magnetic force, and the end A at which the rubbing had begun was its north pole, and the end B where it ended its south pole, as was shown by a very mobile magnetic needle.

379) This experiment completed, I once more stood the forceps vertically, but in the inverse position, so that the end which had previously been turned downward now faced up. Then in the same way as before I rubbed the wire ab, which had previously been examined carefully, with the end of the forceps which was now uppermost, following the rule that each stroke began at a and finished at b. When this operation had been completed about twenty times, the wire had been made extremely magnetic, but the end a at which the rubbing

---

[194] Incorrectly numbered 375 in the text.

had begun was now found to be the south pole and the end b where it ended the north pole.

380) These are the results of the experiment when it is performed in the northern hemisphere in which we live, but if it was repeated in the southern hemisphere, a completely opposite result would undoubtedly be observed. For there the lower end of an iron held vertically upright, or in some appropriate position, becomes the south pole, and its upper end the north pole. So we must consider as completely mistaken those who, from experiments performed only in the northern hemisphere of the earth (and not properly even then), have not hesitated to establish it as a rule that the end at which the stroking begins gains a north magnetism, and that at which it ends, a south magnetism.

381) To make it more evident that in this test, too, everything happens as it usually does in the vicinity of a magnet, I carried out the following experiments. On the plate GH, Fig. CXII, I placed a non-magnetic iron wire AB, and below the plate, at a distance of about half a foot, I placed a rather powerful artificial magnet K, with its north pole turned upward. Then taking a non-magnetic soft iron parallelepiped CD and holding it vertically upright so that its end D faced the pole K directly, I rubbed the wire AB about ten times, beginning each rubbing at A and finishing it at B, and the end A was found to have become a south pole and the end B a north pole. Keeping all the circumstances of the experiment the same, except that now the magnet was held above the plate with its north pole I turned downward, the end A where the rubbing began gained a northern force, and the end B where the rubbing ceased, a southern force. If I used the south pole of the magnet instead of the north pole used in this test, both experiments had a completely contrary result to the former case. It is sufficiently self-evident how similar these phenomena are to those which usually happen on the surface of the earth.

382) Methods have been developed from the experiments just described which help to produce the first traces of magnetism, when artificial magnets are to be prepared without the aid of natural magnets. It would be completely superfluous to recount these methods and apply our theory to them, since they are not difficult. But I trust it would not be unwelcome to my readers if I explain here some methods I have discovered which are more perfect and more successful than the common ones. I came across them when I saw that neither the *Michellian* nor the *Cantonian* method of the double touch can be applied in the operation where the first traces of magnetic force are produced. Since I recognized the superiority of the method of the

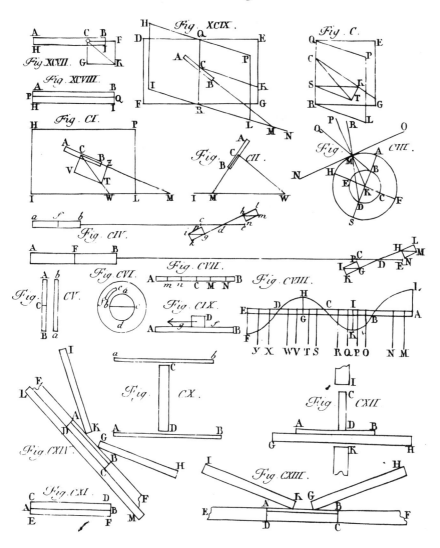

Figure 13. Facsimile Plate VI from the *Essay*

450   THEORY OF ELECTRICITY AND MAGNETISM

double touch over all others, I began to ponder a way of using it in this operation too, and I came upon two ways of doing it; both of them, I learned from experience, far excelled the usual ones.

383) To set up these tests, I used the following apparatus. First I had made four moderately hard iron rods, about 2 feet long, one and a half inches wide, and half an inch thick. Then I got twelve smaller steel rods, 6 inches long, 4 lines wide, and half a line thick; the first two pairs were made from soft steel, another four from steel hardened to the degree usual in steel springs, and the remaining four from steel hardened to the highest degree of brittleness. When all this was ready I tackled the matter in the following way.

384) I stood one of the larger iron rods mentioned at the beginning of the preceding § vertically and struck it a strong blow about two hundred times with a fairly large hammer. In this way it gained an appreciable magnetic force, so that it could easily pick up a small iron nail; the lower end of the rod was the north pole and the upper end the south pole. Handling the remaining three rods in the same way, I likewise imbued them with magnetic force.

385) Next, I placed AB, Fig. CXIII, one of the smaller rods made from soft iron, on a wooden plate between two large rods ED, CF, in the way required, and stroked it after the method of the double touch, described at length above, with a second pair of large rods, imbuing it with magnetic force; I performed the same operation on the remaining three smaller rods made of soft steel. Finally, after all these four rods had acquired some magnetic force in this way, I increased their force to saturation level by the method explained in the previous chapter.

386) I then proceeded to magnetize the four smaller rods enjoying the hardness of a spring. I placed two of them between two soft iron parallelepipeds in the way described in the previous Chapter, and divided the four previously magnetized soft rods into two bundles; I stroked the rods placed between the iron parallelepipeds with them after the method of the double touch, and performed exactly the same operation on the second pair of these rods. Putting aside the four soft rods, I increased the magnetism of the rods with the hardness of a spring to saturation level in the usual way.

387) To complete the operation, I repeated the whole method described in the previous §, using in place of the soft ones the rods with the hardness of a spring, and in place of the latter the rods endowed with the highest degree of hardness and brittleness. In this way I finally got four rods which were strongly magnetic in comparison to their mass, and with them I could transfer magnetism to larger and ever larger rods. There is no difficulty in the whole of this method,

so there is no need to explain the reasons for these phenomena according to the theory, for everything is sufficiently self-evident.

388) Let us move on therefore to the second method. To perform it I use exactly the same apparatus as for the preceding, except that now I take care that the larger iron rods mentioned in § 383 [195] are made from the softest iron. The operation itself is performed in the following manner. On a wooden plate LM, inclined to the horizontal at the angle taken up by the dipping magnetic needle, I place two of the larger iron rods in the same straight line, and one of the smaller rods made of soft steel between them, so that the three rods ED,[196] AB, CF, Fig. CXIV, lie in the straight line into which the dipping needle settles of its own accord. Then with the help of the second pair of larger iron rods (which I hold in such a way that they include with the rod AB either no angle at all or a very acute one), I make the rod AB magnetic using the method of the double touch, and in a completely similar way I imbue the remaining three smaller soft rods with magnetic force. When this is done I carry out the whole operation described at the end of § 385 and in § 386, 387,[197] and I obtain the desired result.

389) There are some things in this method which demand clarification in terms of the theory. It is obvious from the previous discussion that the rods DE, CF have been so placed that they immediately become true magnets from the action of the terrestrial core, according to the law that the end D of the upper rod becomes a north pole, and the end C of the lower rod becomes a south pole. So it is the same as if the rod AB had been placed between two magnets tending to induce in it a state such that the end B gains a north and the end A a south magnetism. Finally, the same thing happens for the rods KI, GH, for in the circumstances, these immediately turn into strong enough magnets, and the end K becomes the north pole of one and the end G the south pole of the other rod. As a result, the reason this method is more successful than the normal ones is readily clear, for everything else provides no difficulty.

390) It might seem that both methods recounted so far are unnecessarily complicated, for it is not immediately clear why I use quite small, thin rods, nor is it clear why I begin the magnetization on soft rods and ascend as if by degrees first to rods that are moderately hard, and finally to those enjoying the highest degree of hardness. For it might be thought that the method could be carried out equally

---

[195] The text has § 381.
[196] The text has EF.
[197] The text refers to § 383, 384 and 385, respectively.

happily, even if the operations through which the initial magnetism is won for the softer rods are performed immediately on the bigger, harder rods. But anyone willing to consult experience will observe that things either do not succeed, or only do so with some difficulty, unless we proceed along the indirect path I have described. The same thing is readily apparent from the theory, for since the difficulty which magnetic fluid has in passing through the pores of hardened steel is finite and has a definite magnitude, it is clear that even though we have four hardened steel rods endowed with magnetism, it is still possible that their force might be so weak that even when the method of the double touch is used, they will not be strong enough to propel the magnetic fluid and overcome the resistance to its motion. It is clear then that the magnetic force of the rods cannot be increased to the degree of saturation unless right from the beginning their primitive or original force transcends a given degree (which is to be defined through experience). Experience has shown, through the methods treated previously, that bigger, harder rods acquire a sensible degree of magnetic force, but not one that is sufficient.

391) Before I conclude my commentary, I judge it appropriate to note that it is quite obvious from what I have asserted above that there is no true and proper production of magnetism, but that all the methods by means of which iron bodies are made magnetic reduce to simple communication, in such a way that if there previously existed no body imbued with magnetic force, it would seem not to be in our power to produce this force. If anyone is willing duly to consider all the methods discussed so far by means of which iron bodies are imbued with a magnetic force they previously lack, he will easily see that they can be reduced to the three following classes. For manifestly we either seek this force from another magnet, for example when we draw iron over a magnet or we hold it in its vicinity; or iron seems to acquire magnetism spontaneously when we make it magnetic either by placing it in the required position or by heating or striking it; or finally this force is generated through the rubbing of two non-magnetic iron bodies. Concerning the former kind of phenomenon no question can arise, and concerning those belonging to the latter two classes, I have clearly demonstrated that this is due solely to the action of the magnet which occupies the core of the terrestrial globe. So these are without doubt specimens of the communication of magnetism, not the production of it.

AN EXPLANATION

of a Certain Phenomenon of the *Leyden* Jar Discovered by the Celebrated *Richmann*.

1) The celebrated *Richmann* first discovered a certain phenomenon of the *Leyden* jar which must it seems, at first sight, be numbered among the more difficult to explain.[198] The experiment is the following. From the two plates CD and IK, Fig. CXV, covering the sides of a *Leyden* jar, hang freely the flax threads Cπ and Iρ, 5 or 6 inches long. As the jar is filled with electricity in the usual way by electrifying the plate CD, it will be observed that the thread Cπ attached to the plate CD is raised, and, repelled by it, ascends to the position CP; the other thread Iρ will hang from the plate IK completely motionless in the vertical position. Now take away the chain LM and immediately the thread Cπ will swiftly descend from the position CP into the position Cp, about half way out, and the other thread Iρ will meanwhile ascend to approximately the position Ir. If then the jar is left freely to itself, both threads will descend equally (but slowly and with difficulty so that the motion will be sensible only after a long time) until, if the delay allowed is sufficiently long, both will return to the vertical positions Cπ, Iρ. But if, after the threads have reached the positions Cp, Ir, one of the plates, either CD or IK, is touched by the finger of a man not supported on pitch or glass, but standing on bodies non-electric *per se*, a weak spark leaps out, one hardly as violent as those that usually issue from bodies electrified without using the jar; but the thread hanging from the side which is touched descends completely and the one attached to the other side rises to the position CP or IR. This last experiment can be repeated very often, and without any sensible lessening of the electricity; it can indeed be repeated a hundred or two hundred times before the electricity is completely destroyed.

2) This fine experiment agrees so outstandingly with the principles evolved by me in Chap. I of my *Tentamen theoriae electr. et magnet.* that it could have been discovered through theory alone, had it not

---

[198] Richmann, *Novi commentarii Academiae Scientiarum Imperialis Petropolitanae*, IV, 1752–53, 324. The experiment here discussed is a slightly simplified version of the one described by Richmann.

already been noted previously through experience. So since a high degree of probability can accrue to the theory I have expounded if it is found to agree closely with more complex phenomena too, it seems that the causes of this experiment of *Richmann* seem to merit more careful examination. It is evident first of all that if the air surrounding the jar was perfectly electric *per se,* the phenomena recounted in the previous § could not take place. For when the chain LM was taken away, under the hypothesis of the jar being perfectly electrical *per se,* this should have remained in whatever state it was in when the chain LM was applied to the plate IK. Since in the case we are supposing the air can neither receive any electric fluid from the plate CD, nor permit any of it to enter the plate IK, exactly the same quantity of electric fluid ought to remain in each plate as there is at the moment the chain LM is taken away, so it is clear that the state of the jar can in no way be changed. Therefore the reason for the phenomena concomitant to *Richmann's* experiment can only be sought in the air being electrical *per se* to but an imperfect degree; it receives fluid into its pores, or emits it from them, with difficulty, but it does not completely impede this movement. That air is, as I suppose here, electrical *per se* only to an imperfect degree can be established already from the fact that the electricity in a body which is surrounded by air on every side does not last *in infinitum,* but is usually completely destroyed in a reasonably short time.

3) Since therefore the air continuously absorbs some electric fluid from the plate CD, let it be supposed that, after the jar has been electrified, the plate CD in a given time loses a quantity of fluid $=\epsilon$. Then α now changes to $\alpha-\epsilon$, and the force by which the plate CD repels the fluid, which is equal to $\dfrac{\alpha r - \beta r'}{Q}$ (§ 15. *Tent. Theor. Electr. et Magn.*), changes to $\dfrac{\alpha r - \beta r'}{Q} - \dfrac{\epsilon r}{Q}$, or, since $\beta = \dfrac{\alpha r'}{r}$, into $\dfrac{\alpha(r^2 - (r')^2)}{Qr} - \dfrac{\epsilon r}{Q}$. But from this, the force by which the plate IK attracts the fluid, which is equal to $\dfrac{\beta r - \alpha r'}{Q}$, also undergoes a change. For it becomes $\dfrac{\beta r - \alpha r'}{Q} + \dfrac{\epsilon r'}{Q}$, or, since $\beta = \dfrac{\alpha r'}{r}$, $+\dfrac{\epsilon r'}{Q}$. So the force of the plate IK, after CD has expelled the quantity ϵ, no longer remains zero as it was initially, but this plate immediately begins to attract fluid again. So while a quantity ϵ is expelled from the plate CD, the plate IK attracts some fluid: let us call this quantity γ. If we now suppose that as α changes into $\alpha - \epsilon$, β changes into $\beta - \gamma$, the force of the plate CD will

become $=\dfrac{\alpha(r^2-(r')^2)}{Qr}+\dfrac{\gamma r'-\epsilon r}{Q}$, and the force of the plate IK $=\dfrac{\epsilon r'-\gamma r}{Q}$.[199]

4) The whole matter now hinges on our determining the relation obtaining between ϵ and γ. As long as the chain LM remains joined to the plate IK, this relation can be investigated quite easily. For since fluid can then enter the plate IK very freely, so much of it will flow swiftly and in a very short time that the attracting force of the plate IK will be reduced again to o. Therefore we will immediately have $\dfrac{\epsilon r'-\gamma r}{Q}=o$ (§ 3), whence is obtained $\gamma=\dfrac{\epsilon r'}{r}$; by introducing this value into the formula expressing the force of the plate CD, this will become $=\dfrac{(\alpha-\epsilon)(r^2-(r')^2)}{Qr}$. So if the chain LM remains joined to the plate IK, the force of this plate will constantly remain zero, but the force of the other plate CD will slowly decrease, whence the thread Iρ will not change its position at all, but the thread Cπ will slowly descend from the position CP.

5) If we suppose that the chain LM is taken away, the relation between the quantities ϵ and γ is a little more difficult to discover. For it is evident that the force of the plate IK cannot then return immediately to o, since the air prevents the free influx of fluid into the plate IK. However the desired relation can be determined in the following way. Since the effects produced by different causes in equal times are, under the same circumstances, in the ratio of the forces by which they are produced, it is clear that the increments which ϵ and γ receive in the very small time dt ought to be proportional to the forces $\dfrac{\alpha(r^2-(r')^2)}{Qr}+\dfrac{\gamma r'-\epsilon r'}{Q}$ and $\dfrac{\epsilon r'-\gamma r}{Q}$. And so we will have $d\epsilon=\dfrac{dt}{Qr}(\alpha(r^2-(r')^2)+\gamma r'r-\epsilon r^2)$ and $d\gamma=\dfrac{dt}{Q}(\epsilon r'-\gamma r)$, whence is obtained by eliminating dt, $\dfrac{rd\epsilon}{\alpha(r^2-(r')^2)+\gamma r'r-\epsilon r^2}=\dfrac{d\gamma}{\epsilon r'-\gamma r}$, or $\alpha(r^2-(r')^2)\,d\gamma+rr'\gamma d\gamma-r^2\epsilon d\gamma-rr'\epsilon d\epsilon+r^2\gamma d\epsilon=0$.

Let this equation be multipled by [200]
$$\dfrac{2}{(r-r')\left(\dfrac{\alpha(r-r')}{r}+\gamma-\epsilon\right)\left(\dfrac{\alpha(r+r')}{r}-\gamma-\epsilon\right)},$$

[199] The text has $\dfrac{\epsilon r'-\gamma r}{Qr}$.

[200] One does not simply pluck an integrating factor such as this out of thin air! Aepinus undoubtedly arrived at this expression by a much more roundabout route than he here chooses to reveal.

and there will be obtained

$$\frac{d\gamma - d\epsilon}{\dfrac{\alpha(r-r')}{r}+\gamma-\epsilon} + \frac{r+r'}{r-r'} \times \frac{d\epsilon+d\gamma}{\dfrac{\alpha(r+r')}{r}-\gamma-\epsilon} = 0,$$

whence by integrating [201]

$$\log\left(\frac{\alpha(r-r')}{r}+\gamma-\epsilon\right) - \frac{r+r'}{r-r'}\log\left(\frac{\alpha(r+r')}{r}-\gamma-\epsilon\right) = \log.\text{Const.}$$

or proceeding to numbers

$$\frac{\dfrac{\alpha(r-r')}{r}+\gamma-\epsilon}{\left(\dfrac{\alpha(r+r')}{r}-\gamma-\epsilon\right)^{\frac{r+r'}{r-r'}}} = \text{Const.}$$

This constant can be determined from the fact that ϵ and γ disappear simultaneously at the beginning of the experiment; whence is finally deduced the equation expressing the common relation between ϵ and γ

$$\frac{\dfrac{\alpha(r-r')}{r}+\gamma-\epsilon}{\left(\dfrac{\alpha(r+r')}{r}-\gamma-\epsilon\right)^{\frac{r+r'}{r-r'}}} = \frac{\dfrac{\alpha(r-r')}{r}}{\left(\dfrac{\alpha(r+r')}{r}\right)^{\frac{r+r'}{r-r'}}}$$

6) Before we tackle the explanation of the phenomena of *Richmann's* experiment, it will be best to deal first with certain more general considerations. At the very beginning of the experiment, when the force of the plate IK is zero, and the repelling force of the plate CD $= \dfrac{\alpha(r^2-(r')^2)}{Qr}$ (§ 3),[202] it is self-evident that the force of the plate IK is less than the force of the other plate CD. I say that the same obtains throughout the whole course of the experiment, and that the force of the plate IK can never become greater than the force of the plate CD. For since the force of the plate IK, which we may call v, is initially less than the force of the plate CD, which we may indicate by V, the former can never overtake the latter, and cannot itself become the greater, unless they are at some stage equal to each other. If therefore we put $V=v$, we will have $\dfrac{\alpha(r^2-(r')^2)}{Qr}+\dfrac{\gamma r'-\epsilon r}{Q} = \dfrac{\epsilon r'-\gamma r}{Q},$

---

[201] In both this line of the calculation and the preceding one, the multiplying factor $\dfrac{r+r'}{r-r'}$ is given incorrectly in the text as $\dfrac{r+r'}{r-1}$.

[202] The text has $\dfrac{\alpha(r^2-(r')^2)}{\alpha r}$.

which equation gives $\alpha(r^2-(r')^2)+(\gamma-\epsilon)r(r+r')=0$, or $\alpha(r-r')+(\gamma-\epsilon)r=0$, whence is obtained $\epsilon=\dfrac{\alpha(r-r')}{r}+\gamma$. If this value is substituted in the equation derived in the preceding §,

$$\frac{\dfrac{\alpha(r-r')}{r}+\gamma-\epsilon}{\left(\dfrac{\alpha(r-r')}{r}-\gamma-\epsilon\right)^{\frac{r+r'}{r-r'}}}=\text{Const.}$$

there is obtained [203]

$$\frac{0}{\left(\dfrac{2\alpha r'}{r}-2\gamma\right)^{\frac{r+r'}{r-r'}}}=\text{Const.}$$

It is impossible for this equation to occur, unless the denominator of the fraction is $=0$, in which case γ becomes $=\dfrac{\alpha r'}{r}$. Therefore the force of the plate IK is constantly less than that of the plate CD until ϵ has become $=\alpha$ and $\gamma=\dfrac{\alpha r'}{r}$, that is, until the jar has returned to the natural state in each part. For since initially the quantity of fluid abounding in CD was $=\alpha$, and the quantity of fluid deficient in IK was $=\dfrac{\alpha r'}{r}$, it is evident that once ϵ has become $=\alpha$ and $\gamma=\dfrac{\alpha r'}{r}$, that is once V has become $=v$, the electricity of both plates has been completely destroyed.

7) Since V is always greater than v, $d\epsilon$ will always be greater than $d\gamma$. For these differentials are proportional to the forces V and v (§ 5). But since further dV is $=\dfrac{r'd\gamma-rd\epsilon}{Q}$ (§ 3), because $d\gamma<d\epsilon$ and likewise $r'<r$, dV will always be negative, whence the force of the plate CD ought to decrease continuously from the beginning of the experiment to the end. But things are different with the plate IK. Here, $dv=\dfrac{r'd\epsilon-rd\gamma}{Q}$, and even though $d\epsilon$ is constantly $>d\gamma$, because $r'<r$, this quantity can be now positive, now negative, and can in fact become equal to zero. But if $dv=0$, v will be a maximum. So it is evident that the force of the plate IK increases from o to a certain maximum value, and decreases from this to o again. To determine this maximum, let us suppose that $dv=\dfrac{r'd\epsilon-rd\gamma}{Q}=0$, and by substituting the

---

[203] The text has $\dfrac{0}{\left(\dfrac{2\alpha r'}{r}-\gamma\right)^{\frac{r+r'}{r-r'}}}=\text{Const.}$

458   THEORY OF ELECTRICITY AND MAGNETISM

values of the quantities $d\epsilon$ and $d\gamma$ found in § 5, after the required reductions we will have $\alpha r'(r^2-(r')^2)+\gamma r(r^2+(r')^2)-2\epsilon r^2 r'=0$, whence is obtained for the case of the maximum,
$$\gamma=\frac{2\epsilon r^2 r'-\alpha r'(r^2-(r')^2)}{r(r^2+(r')^2)} \text{ and } \epsilon=\frac{\gamma r(r^2+(r')^2)+\alpha r'(r^2-(r')^2)}{2r^2 r'}.$$

8) If we substitute the value of γ just discovered, corresponding to the maximum, into the equation evolved in § 5, containing the relation between γ and ϵ, after suitable reductions have been carried out it will give
$$\frac{\alpha(r-r')^{\frac{r-r'}{2r'}}(r^2+(r')^2)}{r(r+r')^{\frac{r+r'}{2r'}}}=\alpha-\epsilon.$$

But it is readily apparent that if r is little different from r', as is usually the case in the *Leyden* jar, this value of $\alpha-\epsilon$ is approximately equal to α. For if we express r and r' in such a way that the difference becomes $=1$ (which is always open to us, whatever ratio is supposed to hold between r and r'), $r-r'$ will be $=1$, or $r'=r-1$. If we substitute these values in the equation there will be obtained [204]
$$\frac{\alpha(2r^2-2r+1)}{r(2r-1)^{\frac{2r-1}{2r-2}}}=\alpha-\epsilon.$$

Since it is assumed that r and r' are little different from one another, our unit will be quite small with respect to the quantity r. So it is evident that $2r^2-2r+1$ is little different from $2r^2$, $2r-1$ from $2r$, and $\frac{2r-1}{2r-2}$ from 1. So the whole quantity $\frac{\alpha(2r^2-2r+1)}{r(2r-1)^{\frac{2r-1}{2r-2}}}$ will be approximately equal to $\frac{\alpha(2r^2)}{r(2r)}$, that is, almost $=\alpha$, whence it follows that ϵ will always be a quite small quantity with respect to α.

9) If this same value of γ which makes the force of the plate IK a maximum is substituted in the general formula expressing that force (§ 3), this maximum force will be found $=\frac{(\alpha-\epsilon)r'(r^2-(r')^2)}{Q(r^2+(r')^2)}$, but the force which the plate CD possesses, when the force of the plate IK is a maximum, will be $=\frac{(\alpha-\epsilon)r(r^2-(r')^2)}{Q(r^2+(r')^2)}$. Because ϵ is quite small and r is almost $=r'$, each of these forces differs very little from $\frac{\alpha(r^2-(r')^2)}{2Qr}$. Therefore since the force which the plate CD possessed initially

---

[204] The text has $2r^2+2r+1$, not $2r^2-2r+1$, both in this expression and twice more in the space of the next few lines.

is $=\dfrac{\alpha(r^2-(r')^2)}{Qr}$, each of these forces will be to it approximately as 1 is to 2.

10) It is easily understood that the situation reduces in a short time to the point where the force of the plate IK becomes the maximum. For to induce this maximum state in the jar, it is only required that a quite small quantity of fluid should be expelled from the plate CD, and this can in fact happen quickly (§ 8). Since the force of each plate in the meantime suffers a quite significant change, as I proved in the preceding §, it is clear why the thread $C\pi$ ought to descend swiftly and the thread $I\rho$ likewise ought to ascend at a considerable rate. This swift change in the forces (as I call it here) deserves still further consideration. I have already noted previously that as the jar is brought to the maximum state, the force of the plate CD decreases almost by half. It happens thus if the chain LM is taken away; but if we suppose this remains joined to the plate IK, we can easily understand that if the plate CD then expels the same quantity of fluid ϵ, the force V will experience in no way so significant a decrease as a result, but only a quite small one. For the force of the plate CD is then reduced to $\dfrac{(\alpha-\epsilon)(r^2-(r')^2)}{Qr}$ (§ 4), so the entire decrease is $=\dfrac{\epsilon(r^2-(r')^2)}{Qr}$, and because ϵ is quite small (§ 8), this quantity almost vanishes in comparison to the initial force $\dfrac{\alpha(r^2-(r')^2)}{Qr}$.

11) From this it is sufficiently clear why things take place as they do in *Richmann's* experiment as the jar is brought to the maximum state, and why they agree with the theory. It remains for us to investigate rather carefully the remaining phenomena also. This poses no difficulty, because once the jar has attained the maximum state, each force begins to decrease and each thread to descend, as is sufficiently clear from the previous discussion. But something significant occurs here, namely that, although the ascent and descent of threads to $I\rho$ and $C\pi$ was quite swift before the jar was brought to the maximum state, once it reaches this state, the movement of the threads is considerably retarded, and becomes so slow that any descent of the threads can only be observed to have taken place after a rather long time.

12) To shed full light on the reason for this, we will have to deal first with certain things which will prepare the way. Readers will recall from § 15 of the *Tent. Theor. Electr. et Magnet.* that if the quantity of electric fluid abounding in the plate CD is called α, and that deficient in IK, β, then the repelling force of the plate CD will

be $= V = \dfrac{\alpha r - \beta r'}{Q}$, but the attractive force of the plate IK, or v,
$= \dfrac{\beta r - \alpha r'}{Q}$. Let it be supposed that the quantity α is diminished by a quantity ζ, and β by a quantity $\zeta - \mu$; the decrease of the force of the plate CD will then be $= \dfrac{\zeta(r-r') + \mu r'}{Q}$,[205] and that of the plate IK will $= \dfrac{\zeta(r-r') - \mu r}{Q}$.[206] Thus if

1) $\mu = 0$, that is, if α and β decrease equally, the decreases in V and v will be equal and will $= \dfrac{\zeta(r-r')}{Q}$, whence if it is supposed that the forces V and v were equal at the beginning, even after the change of the quantities α and β their equality will be preserved.

2) if μ does not $= 0$ but is so small that it can be assumed to be evanescent, then the decreases in the forces V and v will be at least approximately equal. So if in this case the forces V and v were either exactly or approximately equal at the beginning, they also decrease almost equally, that is, they remain approximately equal.

3) if μ is of significant size with respect to ζ, that is, if the decreases in α and β differ significantly from one another, then the decreases in the forces V and v will also be significantly different. For the difference in the sizes of the decreases $= \dfrac{\mu(r+r')}{Q}$,[207] and this quantity is larger, the larger is μ.

4) further, if $\mu = 0$, the decrease which each force experiences will be quite small, and much less than if μ did not vanish and was of significant size. For in the former case this decrease of the forces $= \dfrac{\zeta(r-r')}{Q}$, and because r is little different from r', this quantity is quite small. If on the contrary μ is of significant size with respect to ζ, these decreases are almost $= \dfrac{+\mu r'}{Q}$ and $\dfrac{-\mu r}{Q}$,[208] and since these quantities are significantly greater than $\dfrac{\zeta(r-r')}{Q}$, a significant change ought to happen in each force, and the change will be greater, the greater is μ. If μ does

---

[205] The text has $\dfrac{\zeta(r-r') - \mu r'}{Q}$.

[206] The text has $\dfrac{\zeta(r-r') + \mu r}{Q}$.

[207] The text has $\dfrac{\mu(r-r')}{Q}$.

[208] The text has $\dfrac{-\mu r'}{Q}$ and $\dfrac{+\mu r}{Q}$, respectively.

not completely vanish but almost does so, the change that each force experiences will be small, for then it will be little different from $\frac{\zeta(r-r')}{Q}$.

13) If we transfer these generalities to the case we suggested ought to be considered here, it is clear that at the beginning of the experiment, when the forces V and v are quite unequal, the decreases in α and β, which are proportional to those forces (§ 5), ought to differ significantly from one another. But after the force of the plate IK has reached its maximum, it is established that the forces V and v have become approximately equal (§ 9). So the decrease of the quantities α and β will likewise be approximately equal, and the forces V and v will always remain approximately equal, from the point where the jar has acquired the maximum state, right to the end of the experiment and the complete destruction of the electricity in each plate.

14) It can easily be seen why the forces V and v change rapidly at the beginning of the experiment, but at a quite slow rate once the jar has attained the maximum state. It is clear that there can be two causes of this. If the value of ϵ, the decrease in α corresponding to the maximum force of the plate IK, is indicated by ζ,

1) At the beginning, this quantity ζ is expelled from the plate CD in a much shorter time, almost half the time, than after the maximum state has been induced in the jar. For the repelling force of the plate CD will initially be about twice as great as it is after the force of the plate IK has reached the maximum (§ 9).

2) If we imagine also that even when the jar has been reduced to the maximum state, the quantity of fluid expelled from the plate CD in the same time as before $=\zeta$, nevertheless the change of the forces V and v brought about in the time in which ζ is expelled will be much less than it was at the beginning before the jar had reached the maximum state. For since after the force of the plate IK has reached the maximum, the forces V and v are approximately equal (§ 9), and persist as such right to the end of the experiment, the decreases in α and β will also be approximately equal. We have proved in the preceding §, no. 4, that the forces V and v are changed only a little if the decreases in the quantities α and β are approximately equal, and the variations of the forces which are then produced are much less than they would be if the decreases in α and β were very unequal, which is the situation before the jar has attained the maximum state.

15) To make all this clearer, let us apply the generalities proposed so far to one or two special cases. Let us suppose that $r : r' = 100 : 99$, and let us express α by unity. At the beginning of the experiment,

therefore, the force of the plate CD will $= \frac{1,99}{Q}$ (§ 3). If we consider that the jar has attained the maximum state, ϵ, or the quantity of fluid expelled from the plate CD corresponding to the maximum state, $=0,03123$ (§ 8), and $\gamma=0,02128$ (§ 7). Hence the force of the plate CD will then $=\frac{0,97362}{Q}$, and the force of the plate IK $=\frac{0,96388}{Q}$, each of which values differs little from $\frac{0,995}{Q}$, that is, from half of the initial force of the plate CD. But if the chain LM had remained joined to the plate IK, after the plate CD expelled a quantity $\epsilon=0,03123$, the force of the plate CD would be $=\frac{1,9278}{Q}$, so it is clear that, unless the chain is removed, the force of the plate CD experiences very little loss.

16) If we suppose that after the jar has been reduced to the maximum state, the plate CD once more expels a quantity $\epsilon=0,03123$, so that the whole quantity of fluid expelled initially $=0,06246$, the γ corresponding to this, that is, the quantity of fluid which has as a result initially entered the plate IK, will be $=0,05246$, this value being easily elicited from the equation evolved in § 5 by the rule of false position, or by other normal methods of approximation. The force of the plate CD then becomes $=\frac{0,93754}{Q}$, and the force of the plate IK $=\frac{0,93754}{Q}$. If we now make a comparison of the forces which each plate gains successively, we shall obtain

Force of plate CD		Differ.	Force of plate IK	Differ.
initially	$=\frac{1,99}{Q}$	- - -	$=0$	
		$\frac{1,01638}{Q}$		$\frac{0,96388}{Q}$
After the jar has reached the maximum state.	$=\frac{0,97362}{Q}$	- - -	$=\frac{0,96388}{Q}$	
		$\frac{0,03608}{Q}$		$\frac{0,02634}{Q}$ [209]
After the plate CD has once more expelled a quantity $\epsilon=0,03123$.	$=\frac{0,93754}{Q}$	- - -	$=\frac{0,93754}{Q}$	

[209] The text has $\frac{0,02684}{Q}$.

DISSERTATION I

17) We have assumed so far, as is ordinarily the case in the *Leyden* jar, that r is little different from r'. So that readers can also see how the phenomena of *Richmann's* experiment alter if the quantities r and r' depart significantly from equality, let us subject still another case to our consideration. If $r : r' = 5 : 4$, then the initial force of the plate CD will be $= \frac{1,8}{Q}$, ϵ corresponding to the maximum will be $= 0,30770$, γ will be $= 0,12458$; hence when the jar is in the maximum state, the force of the plate $CD = \frac{0,75982}{Q}$, and the force of the plate $IK = \frac{0,60790}{Q}$. If after the jar has reached the maximum state, there is a further expulsion from the plate CD of a quantity $\epsilon = 0,30770$, the quantity of fluid expelled from the plate CD from the beginning of the experiment will be double this ϵ, $= 0,61540$, and γ corresponding to it $= 0,48738$, whence we obtain the force of the plate $CD = \frac{0,67252}{Q}$, and that of the plate $IK = \frac{0,02470}{Q}$. So we have

Force of plate CD		Differ.	Force of plate IK	Differ.
initially	$= \frac{1,8}{Q}$	- - -	$= 0$	
		$\frac{1,04018}{Q}$		$\frac{0,60790}{Q}$
After the jar has reached the maximum state.	$= \frac{0,75982}{Q}$	- - -	$= \frac{0,60790}{Q}$	
		$\frac{0,08730}{Q}$		$\frac{0,58320}{Q}$
After a quantity $\epsilon = 0,30770$ has once more been expelled from the plate CD.	$= \frac{0,67252}{Q}$	- - -	$= \frac{0,02470}{Q}$	

18) A look at the tables in § 16 and 17 shows that the particular differences in the phenomena that arise from a rather considerable difference in the quantities r and r' consist in the fact that if r is significantly greater than r',

1) It is necessary for a greater quantity of fluid to be expelled from the plate CD before the maximum state can be induced in the jar.

2) The forces of the plates CD and IK corresponding to the maximum state are significantly less than half of the initial force of the plate CD.

3) The above-mentioned forces also differ rather among themselves, in such a way that the force of the plate IK is considerably less than the force of the plate CD.

4) In the case of the force of the plate CD, after the jar has reached the maximum state, it decreases at a slow rate; and as for the force of the plate IK, the difference of the velocities with which it changes is not now as significant as it usually is, when r and r' approach nearly to equality.

19) All this is readily obvious from the general theory expounded above, even without the use of actual numbers. The truth can also be conveniently proved through experience in the following manner. Cover one side of a glass plate AB, Fig. CXVI, with a metal plate CD in such a way that a margin about one and a half inches wide around the periphery remains uncovered. Prepare a wooden plate IK, somewhat smaller than the glass plate AB; cover it with metal strips, and hang a thread $C\pi$ from the plate CD and a thread $I\rho$ from the plate IK. At first you apply the plate IK closely to the glass plate AB and perform *Richmann's* experiment, but afterward take the plate IK half an inch away from the glass plate AB and repeat the same experiment, and you will observe that everything occurs as I proved in the preceding §.

20) I still owe my readers an explanation of the last phenomenon I mentioned in § 1. For it must be shown why if one of the plates, either CD or IK, is touched by a finger after the jar has reached the maximum state, the thread which happens to be attached to that plate immediately descends, but the thread attached to the other plate ascends quickly; and there must be a similar inquiry into why this experiment can be repeated many times with a *Leyden* jar without the electricity of the jar suffering an appreciable diminution. For this reason let us consider that a quantity of electric fluid $= \alpha$ abounds in the plate CD of the jar, and a quantity $= \beta$ is deficient in the plate IK. Then the repelling force of the plate CD will be $= \dfrac{\alpha r - \beta r'}{Q}$, and the attractive force of the plate IK $= \dfrac{\beta r - \alpha r'}{Q}$. Suppose further that the jar has been established in a condition such that each of these forces is positive, i.e. the plate CD in fact repels the fluid and the plate IK at-

DISSERTATION I   465

tracts it. It is then evident that α ought to be $> \frac{\beta r'}{r}$[210] and $\beta > \frac{\alpha r'}{r}$. If therefore μ and ν are assumed to be positive numbers, let α be made $= \frac{\beta r'}{r} + \mu$, and $\beta = \frac{\alpha r'}{r} + \nu$, and then

1) by eliminating α

The force of the plate CD $= \frac{+\mu r}{Q}$.

The force of the plate IK $= \frac{+\beta(r^2 - (r')^2)}{Qr} - \frac{\mu r'}{Q}$.[211]

2) by eliminating β

The force of the plate CD $= \frac{+\alpha(r^2 - (r')^2)}{Qr} - \frac{\nu r'}{Q}$.[212]

The force of the plate IK $= \frac{\nu r}{Q}$.

21)

1) Let the plate CD be touched by a body non-electric *per se*, for example by a man not supported on bodies electric *per se*. Since the fluid repelled from the plate CD can now flow out freely, as much of it as is required for the repelling force of the plate CD to be reduced to 0 will immediately flow out, that is until α has become $= \frac{\beta r'}{r}$, and as a result μ will then $= 0$. The force of the plate CD is reduced to 0, but the force of the plate IK, which had been $= \frac{\beta(r^2 - (r')^2)}{Qr} - \frac{\mu r'}{Q}$, will become $\frac{\beta(r^2 - (r')^2)}{Qr}$,[213] and will increase by a quantity $\frac{\mu r'}{Q}$. But if

2) the plate IK is touched in a similar fashion, fluid is attracted by this plate and, being able to enter it freely, will flow into the plate until its attractive force has vanished, that is until $\beta = \frac{\alpha r'}{r}$ or $\nu = 0$. The force of the plate IK will therefore be reduced to 0, and the force of the plate CD, which had been $= \frac{\alpha(r^2 - (r')^2)}{Qr} - \frac{\nu r'}{Q}$, will be reduced to $\frac{\alpha(r^2 - (r')^2)}{Qr}$, and will increase by a quantity $\frac{\nu r'}{Q}$.

[210] The text has $\alpha > \frac{\beta r'}{r'}$.

[211] The text has $\frac{+\beta(r^2 - (r')^2)}{Qr} - \frac{\mu r'}{r}$.

[212] The text has $\frac{+\alpha(r^2 - (r'))}{Qr} - \frac{\nu r'}{Q}$.

[213] The text has $\frac{\beta(r^2 - (r')^2)}{Q}$.

22) If we now transfer this to the case of a jar constituted in the maximum state, and proceed to indicate the quantity of fluid abounding in CD by α, and that deficient in IK by β, from § 9 we obtain $\dfrac{\alpha r - \beta r'}{Q} : \dfrac{\beta r - \alpha r'}{Q} = r : r'$, whence we obtain

$\alpha = \dfrac{\beta(r^2 + (r')^2)}{2rr'}$, and $\beta = \dfrac{2\alpha rr'}{r^2 + (r')^2}$. And so because

$\alpha = \dfrac{\beta r'}{r} + \mu$, μ becomes $= \dfrac{\beta(r^2 - (r')^2)}{2rr'}$, and because

$\beta = \dfrac{\alpha r'}{r} + \nu$, $\nu = \dfrac{\alpha r'(r^2 - (r')^2)}{r(r^2 + (r')^2)}$. So if

1) We consider the case where the plate CD is touched, before it is touched the force of the plate $CD = \dfrac{+\mu r}{Q} = \dfrac{\beta(r^2 - (r')^2)}{2Qr'}$, and the force of the plate IK $= \dfrac{\beta(r^2 - (r')^2)}{Qr} - \dfrac{\mu r'}{Q} = \dfrac{\beta(r^2 - (r')^2)}{Qr} - \dfrac{\beta(r^2 - (r')^2)}{2Qr} = \dfrac{\beta(r^2 - (r')^2)}{2Qr}$. Now let the plate CD be touched, and because μ has become $=0$ the force of the plate CD will be $=0$, but the force of the plate IK will $= \dfrac{\beta(r^2 - (r')^2)}{Qr}$, so this force will increase to exactly double. But if

2) We adapt this to the case where the plate IK is touched, then initially the force of the plate IK will be $= \dfrac{\nu r}{Q} = \dfrac{\alpha r'(r^2 - (r')^2)}{Q(r^2 + (r')^2)}$,[214] but the force of the other plate $CD = \dfrac{\alpha(r^2 - (r')^2)}{Qr} - \dfrac{\nu r'}{Q} = \dfrac{\alpha(r^2 - (r')^2)}{Qr} - \dfrac{\alpha(r')^2(r^2 - (r')^2)}{Qr(r^2 + (r')^2)} = \dfrac{\alpha(r^2 - (r')^2)}{Qr} \times \dfrac{r^2}{r^2 + (r')^2}$. But when the plate IK is touched, its force will immediately become $=0$ because ν disappears, but the force of the plate CD becomes $= \dfrac{\alpha(r^2 - (r')^2)}{Qr}$, and because $\dfrac{r^2}{r^2 + (r')^2}$ almost $= \frac{1}{2}$ if r is little different from r', this force is approximately, though not exactly, double what the plate CD possessed initially.

23) Whether it is the plate CD or the plate IK which is touched, in each case the quantity of fluid abounding in CD and deficient in IK

---

[214] The text has $= \dfrac{\nu r'}{Q} = \dfrac{\alpha(r')^2(r^2 - (r')^2)}{Qr(r^2 + (r')^2)}$.

is little changed. For the quantities by which these are reduced are=
$\mu = \dfrac{\beta(r^2-(r')^2)}{2rr'}$ for the former, and $=\nu=\dfrac{\alpha r'(r^2-(r')^2)}{r(r^2+(r')^2)}$ for the latter,
and these are quite small so long as r differs little from r'.[215] Since the same things apply whenever the experiment is repeated, it is readily understood why the electricity still decreases little, even if the experiment is repeated many times. Further, since only a small quantity of electric fluid crosses to the finger touching the plate CD, and likewise only a small quantity crosses from the finger touching the plate IK into the plate itself, it is clear why the sparks which the jar gives off in each case are quite weak. For a large spark cannot be generated unless a great supply of electric fluid crosses from one body into another.

24) If we apply this to the special case treated above in § 15. 16, while the jar has been constituted in the maximum state, the force of the plate CD $= \dfrac{0{,}97362}{Q}$, and that of the plate IK $= \dfrac{0{,}96388}{Q}$,[216] but then $\alpha = 0{,}96877$, and $\beta = 0{,}96872$, whence $\mu = 0{,}00973$ and $\nu = 0{,}00963$. If therefore

1) the plate CD is touched, the force of the plate IK $= \dfrac{1{,}92776}{Q}$, which is double the initial force $\dfrac{0{,}96388}{Q}$. But if

2) the plate IK is touched, the force of the plate CD $= \dfrac{1{,}9278}{Q}$, which is almost double the initial $\dfrac{0{,}97362}{Q}$.

The quantities μ and ν are those by which α and β are reduced if either the plate CD or the plate IK is touched, and it is evident that here these quantities are quite small.

25) If r differs considerably from r', the things which have been deduced from the supposition of r and r' being approximately equal no longer apply. Then the quantities μ and ν, or the decreases in α and β, will be significantly greater; the force of the plate IK will still be doubled when CD is touched, but while the force of the plate CD will increase when IK is touched, it will remain significantly less than double the initial force. Thus if we take the case considered in § 17, for a jar constituted in the maximum state we shall have the force of

[215] The text has r.
[216] The text has $\dfrac{0{,}96888}{Q}$.

the plate $CD = \dfrac{0{,}75982}{Q}$,[217] the force of the plate $IK = \dfrac{0{,}60790}{Q}$, $\alpha = 0{,}69230$, $\beta = 0{,}67542$, whence is obtained $\mu = 0{,}15197$ and $\nu = 0{,}121575$. From this it emerges

1) if the plate CD is touched, the force of the plate $IK = \dfrac{1{,}21575}{Q}$, which is double the initial force $\dfrac{0{,}60790}{Q}$.

2) if the plate IK is touched, the force of the plate $CD = \dfrac{1{,}24614}{Q}$, which is significantly less than double the initial force $\dfrac{0{,}75982}{Q}$.

26) Everything I have proposed here is equally valid also for the case where from the beginning the plate CD is electrified negatively, not positively as I have supposed here. But since the whole method of reasoning used here is also valid with few changes for that case, I leave it to my readers to adapt to it for themselves what has been said so far. Some will perhaps be displeased that I have fallen into some rather lengthy mathematical disquisitions, for there are those who scarcely tolerate such things in physics. But I judge otherwise, and I consider this short dissertation recalling the phenomena of *Richmann's* experiment to their proper causes of the greatest value in proving the truth of the principles of the electrical theory I have evolved. For a great probability must be thought to accrue to a physical hypothesis if it is found to agree closely with more complicated phenomena which can be deduced from the theory only after a rather long series of reasonings. For, in part, if there had been any error in establishing the principles of the theory, it is to be hoped that such an error would be more readily obvious, the more lengthy is the series of reasonings by which a particular phenomenon is explained; and, in part, there can be no suspicion that the theory could have been accommodated to such rather intricate cases from the beginning, so they must be considered most suitable for setting up an examination of it.

---

[217] The text has $\dfrac{0{,}97362}{Q}$.

AN EXPLANATION
of a Certain Paradoxical Magnetic Phenomenon.

1) Recently a friend whose perspicacity in physics and mathematics I value highly reminded me of a certain quite paradoxical magnetic phenomenon of which I remember (though quite dimly) once reading a description. I have also completely forgotten where I found it mentioned, and my friend has no clear recollection of its source. As a result I am completely uncertain about the author of the experiment.[218]

2) The experiment I am talking about is the following. A thin piece of iron wire, *fe*, about one or at most two lines long, is placed on the flat plate AB, Fig. CXVII. Hold the magnet CD one or more inches above the plate, turning one of its poles C toward the plate AB: do not hold this pole directly above the wire *fe*, but a little on one side so that the point G, through which the vertical dropped from the pole of the magnet C to the plane AB passes, is one or two inches from the point *f*. The wire *fe* will then be observed to rise in such a way that it acquires the position indicated in the figure and includes with the plane an angle *ef*B, which is greater, the smaller is the distance *f*G. Now shake the plane gently so that the wire *ef* jumps a little, and it will be observed that the wire constantly approaches closer and closer to the vertical CG and finally reaches the point G itself.

3) So far nothing really remarkable or unexpected has happened. For the approach of the wire *fe* toward the point G seems readily understandable from the attraction which the pole C exerts on it. But change the manner of performing the test a little, and an unexpected result occurs. Do everything as before except hold the magnet below the plate, and once more the wire *ef*, Fig. CXVIII, will rise and include with the plane AB the angle *ef*B, which will be greater, and closer to a right angle, as the distance *f*G decreases. But if the plane is shaken, the wire *fe* does not as before approach the vertical CG, but rather it moves constantly away from it with a completely contrary motion, and thus the pole C now seems to be repelling the wire *fe*.

4) If you sprinkle the plate AB with iron filings and then hold the magnet above the plate, it is readily obvious that the same ought to

---

[218] Efforts to identify the source of this experiment have proved unavailing, as have attempts to discover the name of Aepinus's friend.

happen to any particle of iron dust as happens to the wire *fe*, so when the plate is shaken rather often, all the filings sprinkled here and there on the plate collect at last in a single heap directly below the pole C. But if you place a heap of filings on the plate near the point G when it lies directly above the pole C placed under the plate, when the plate is shaken, this heap simultaneously disperses and the particles of the filings, moving away in all directions from G, leave the space near the point G empty. The original experiment was first performed in this way; but it is clear that in § 2 and 3 I have changed nothing in the essentials of the test.

5) I am persuaded that I have uncovered the true cause of the phenomenon, but some general considerations will have to be inserted first so that I can expound it distinctly and duly apply it to all the circumstances of the phenomenon. Imagine that the iron wire *fe*, Fig. CXIX, is so fixed by an axis passing through its center of gravity *g*, or in some other way, that although the point *g* is immobile, the wire *fe* retains freedom of movement round the point *g*. Imagine that the force of the pole C of the magnet held on one side of AB has all been collected at the single point C, and that to the horizontal line C*m* passing through the point C the perpendicular *gm* is drawn from the point *g*; and I say that the wire *fe* will remain at rest only after it has acquired a position such that it lies along the line C*g* and faces the point C directly. To make this clear, recall first that because the axis passes through the center of gravity *g* the wire *fe* should be considered as having no gravity, whence we need look only at the action of the magnet on it. To determine this latter action, we ought to recall that from what has been explained extensively above, the wire *fe* close to the magnet will immediately become a magnet itself, following the rule that if C is a positive pole, the part *fg* turned toward it becomes negatively magnetic and the other part *ge* positively magnetic, but on the contrary if C is a negative pole, the former part of the wire becomes positively magnetic and the latter negatively magnetic. From this it follows that the part *fg* is attracted toward the point C, and the other part *ge* is repelled from it, and these forces are applied, let us suppose, at the mid-points of the parts *fg*, *ge*. Suppose now that initially the wire *fe*, Fig. CXX, does not have a position such that its direction passes through the point C, but includes an angle *fg*C with the line C*g*, and it is self-evident that the attractive force C*m* together with the repelling force C*n* conspire to rotate the wire *fe* in the direction *fr* round the point *g*, and to draw it back to the line *g*C. It is clear from this that the wire *fe* will not remain at rest in any position except when it lies in the line *g*C; therefore, even if it did not have

that position initially, or if for some reason it has been disturbed from it, it will soon acquire it, and will be forced to return to the line gC, just as we said was to be demonstrated here.

6) Not only ought the wire fe, fixed at the point g and freely mobile around it as if around a center, always acquire the position previously mentioned, but it ought also to take up the same position even if it is supposed that the wire fe is indeed freely mobile around the point g, but the point is not completely fixed but can freely ascend and descend in the vertical line qp. If the point g is so fixed to the line qp that it cannot be moved away from that line, but it either descends or ascends on the line when some force drives it, I say that nevertheless the wire fe will turn about the point g as it descends in such a way that its direction constantly passes through the point C. Assuming the conditions just mentioned, imagine that the wire fe, Fig. CXXI, does not have a position such that its direction passes through the point C, and it is evident that the part fg is attracted toward C in the direction mr, and that the part ge is repelled in the direction nk, and these directions can be assumed to be parallel to the line Cg if, as we have supposed here, the wire fe is quite short. Resolving these forces by the precepts of statics into two, one of which is parallel to the wire, the other normal to it, it is clear that the forces shown acting on the wire along the lines mr and nk can be resolved, the former into the forces ms and mt, the latter into forces nv and ni.[219] Of these forces, ms and nv, pulling in the direction of the wire, strive to remove the point g from the vertical qp, but the normals mt and ni conspire to draw the wire into the direction Cg. Therefore since *ex hypothesi* the former forces cannot achieve their effect, it is clear that we need look only at the latter ones, which, since nothing impedes them, cannot fail to produce the effect to which they tend, and cannot fail to bring the wire into a position such that its direction always passes through the point C.

7) This remains true as closely as possible, even if the point g is completely free so that it can freely obey any force soliciting it. If the wire is left freely to itself, or if it is tossed straight upward by some force, I say, whether it ascends or descends, that

1) it will always turn so that its direction passes through the point C.

2) its center of gravity, g, will either ascend or descend to all intents and purposes along the same vertical line pq, in such a way that it never sensibly deflects from it.

---

[219] The text has *ik*.

For the forces acting on the wire are first of all its gravity (which achieves nothing except that it urges the center of gravity directly downward, and as a result either retards the ascent of this point along the vertical line pq or accelerates its descent, and clearly does not tend to change the orientation of the wire fe, whatever this might be) and then the forces mr and nk due to the point C. When these last have been resolved as in the previous §, it is clear that each of the forces mt and ni strive to draw the wire back to the line Cg, whence it is clear that the wire ought to turn about the point g until its direction passes through the point C. Next the forces ms and nv strive to draw the point g away from the vertical line pq, the former by attracting it toward f, the latter by pressing it toward e. I say that the effect of these forces will be so small that they can be considered insignificant, and can be neglected. Since we suppose the wire fe to be quite short, and because the distances Cm, Cn are approximately equal, the force mr is only just greater than the force nk, and the force ms just greater than the force nv. So the difference of the forces $ms-nv$ by which the point g is urged to depart from the vertical line pq toward f will be quite small and can give rise to no sensible effect.

8) Let us pass now to the other case which we shall subject to examination. Suppose that, on the plane AB, Fig. CXXII, below which lies the magnetic pole C, the wire fe is placed to the side of the point C in such a way that the middle of the wire, g, is found not in the vertical line CD, but removed from the point D somewhat toward either the right or the left. I say that when the magnetic pole C is placed below the table, this wire fe will rise of its own accord so that the end e is somewhat raised and the wire includes an angle efB with the plane. I say further that this angle efB will always be less than the angle hfB which the wire would include with the plane if it acquired a position such that, when produced, its direction passed through C. To elucidate this theorem better, imagine that the wire fe is so fixed at the point f that while the point f remains immobile, the wire can rotate freely around it. Let the pole of the magnet be at C, and the horizontal line passing through the point f be AB. Now three forces act on the wire fe, which we consider as a lever whose point of rest is the point f. First of all gravity, or the weight of the wire, which we put $=p$, urges its center of gravity directly downward along the line gd; then the part fg is attracted toward the point C in the direction mr, and also the part ge is repelled in the direction nk from that point, and for the reasons indicated above it is proper to assume here that these directions are parallel to the line Cg. Let us suppose that the position of the wire shown in the figure is that where these forces are mutually

DISSERTATION II

in equilibrium with one another, and call the force attracting the part fg and repelling the part $ge = \alpha$. For the reasons adduced previously, we assume that these forces are equal. Finally let the angle $Bpg = \phi$, the angle $efB = \psi$, and the length of the wire $fe = m$. Since the three forces acting on the lever are supposed to be in equilibrium, suppose that they are resolved according to the precepts of statics into forces parallel to the lever and normal to it. If these forces are shown by the lines gd, mr and nk, when they are resolved, the lever will be drawn in a direction parallel to itself by forces gh, ms and nv, but in a direction normal to itself by the forces gc, mt [220] and ni. Because the angle $fgp = \phi - \psi$, mt [221] $= ni = \alpha \sin(\phi - \psi)$, and since $dgc = \psi$,[222] $gc = p \cos \psi$. And so because of the equality of the moments we shall obtain the equation $\dfrac{mp \cos \psi}{2} + \dfrac{m \alpha \sin(\phi - \psi)}{4} = \dfrac{3 m \alpha \sin(\phi - \psi)}{4}$, which when reduced gives $p \cos \psi = \alpha \sin(\phi - \psi) = \alpha \sin \phi \cos \psi - \alpha \cos \phi \sin \psi$, whence is obtained $\dfrac{\alpha \sin \phi - p}{\alpha \cos \phi} = \tang \psi$, or $\tang \psi = \tang \phi - \dfrac{p}{\alpha \cos \phi}$, and from this equation the position of the wire fe is determined.

9) This equation demands our developing somewhat more carefully the consequences which flow from it. It seems at first that there are two positions directly opposite to one another in which the lever fe can rest. For since $\tang \psi$, apart from the angle ψ, corresponds as well to an infinite number of other angles, $\pi + \psi$, $2\pi + \psi$, $3\pi + \psi$, $4\pi + \psi$, etc., if the wire fe rests in the position shown in the figure when the line fe is produced toward E,[223] it seems that it will also rest in a position fE directly opposite to the former, for the angle $BfE = \pi + \psi$. As for the remaining angles in the series of angles, continued either forward or backward, which have the same tangent as the angle ψ, there is no need to look at the angles $2\pi + \psi$, $3\pi + \psi$, $4\pi + \psi$, etc., or $-\pi + \psi$, $-2\pi + \psi$, $-3\pi + \psi$, etc., for obviously these repeat the positions of the lever fe and fE, but supply no new ones. In the actual case we readily understand that the things seemingly possible in abstract do not take place. What we have deduced from our equation would of course be true if the wire fe was converted to the position fE, and the part of it

---

[220] The text has ge, ml. In Fig. CXXII, the m has been wrongly placed. It should be at the foot of the perpendicular from t to fe.
[221] The text again has ml.
[222] The text has $dge = \psi$.
[223] Omitted from the figure.

Gf was still attracted toward C, and the part GE repelled from the point C, but according to the known laws of magnetic attraction, this does not happen in the position fE.[224] For the part EG will acquire a contrary magnetism to the pole C while the part fG will acquire the same magnetism as that of C, whence in this position the former part will be attracted, but the latter will be repelled. So it is clear that, in our case, there is only one position in which the wire fe can be at rest, and that is the one shown in the figure; and that of course is the only one which agrees with the circumstances of the case we are considering here.

10) If it is assumed that the angle gfB=right angle, or that $\phi=\frac{\pi}{2}$, sin ϕ will be=1, and cos $\phi = o$, whence tang $\psi = \infty$,[225] which is an indication that if the wire fe is placed in such a way that the point g is found in the vertical line CD, the wire fe ought to stand vertically. It is self-evident that the more the point g is removed to the right to the side of the line CD, the less the angle ϕ becomes, whence sin ϕ continuously decreases, but cos ϕ will increase. For both reasons the value of tang $\psi = \frac{\alpha \sin \phi - p}{\alpha \cos \phi}$ continuously decreases. So the angle ψ ought to become continuously less and less as we recede toward the right, until sin ϕ becomes $=\frac{p}{\alpha}$, when tang $\psi = o$, and the angle ψ also vanishes. But if then the wire fe is moved even further toward the right from D, because of the perpetual decrease in the force α, sin ϕ will eventually become $<\frac{p}{\alpha}$, and then tang ψ becomes negative. So the angle ψ itself will then have to be taken as negative, and the wire fe will descend below the horizontal AB.

11) If the wire fe is transferred from the line CD toward the left, it is readily obvious that the angle gPB is made obtuse, whence if pD=PD, our formula may be adapted to this case by making the angle ϕ change to $\pi - \phi$, in which case sin gPB=sin $(\pi - \phi)$=sin ϕ, but cos gPB=cos $(\pi - \phi)$= −cos ϕ. Substituting these values gives tang eFB=$-\frac{\alpha \sin \phi - p}{\alpha \cos \phi}$= −tang ψ, so the angle eFB=$\pi - \psi$ or eFA=ψ. It is clear therefore that, in receding from the vertical line CD toward the left, the angle eFA suffers the same change as is experienced by the angle eFB when the wire is moved toward the right.

[224] The text has FE.
[225] The text here uses the symbol $\sim$.

DISSERTATION II   475

12) If the magnetic pole C is held above the point f and above the horizontal line AB, Fig. CXXIII, passing through it, then retaining the denominations used previously and performing the calculation in exactly the same way, we obtain the equation tang $\psi = \dfrac{\alpha \sin \phi - p}{\alpha \cos \phi}$, which clearly coincides with the one discovered previously, and everything mentioned in § 9. 10 and 11 can be applied to it. Thus it is evident that if the wire fe is placed in a vertical line directly beneath the point C, it ought to stand vertically; if it is afterward transferred to the left, the angle gfB continuously decreases, until finally it becomes $= o$ and the wire fe rests on the horizontal line AB; but if we go beyond this point, it will descend below the horizontal. It is just as clear that by transferring the wire fe from the vertical line CD toward the right, the angle gFA experiences the same changes which the angle gfB was previously observed to undergo.

13) Now that the principles serving to explain our phenomenon have been sufficiently clarified, let us adapt them to the case which we set out to treat. It is readily obvious that if the wire fe is placed on the plate AB under which is the magnetic pole C, the point f can be considered as the point of rest of the lever fe. For the forces ms, gh, and nv, acting in a direction parallel to the wire fe, are absorbed by friction on the plane AB and its resistance, so the only forces needing to be considered are the normal forces mt, gc, ni.[226] So from § 9. 10 and 11 it is clear that if the wire is placed in the vertical line CD, it ought to stand perpendicularly to the plane of the plate AB, but if it is removed from the point D either to the right or to the left, the wire ought to be raised in such a way that it includes with the plane of the plate an angle efB, which is less than a right angle, and this ought to decrease as the distance from the point D increases until finally it ought to disappear completely. Let this terminus be at H; then since beyond this terminus the lever fe, to reach its rest position, has to descend below the plane AB, and this descent is prevented by the plane AB, it is clear that the wire removed away from D beyond the point H ought always to rest on the plane AB and not rise above it. Similar things happen if the magnetic pole C is held above the plate AB, for then too, while the pole is vertically above it, the wire stands perpendicularly to the plane, but when moved sideways it inclines less and less to the plate until it reaches the point where the angle which it includes with the plate becomes zero; and beyond this point, since to obtain the position of equilibrium it ought to descend below

[226] Again the text has *ml, ge, ni*.

the horizontal, and this in practice the plane AB does not allow, when the wire is further from the point D than H, it ought always to lie on the plane AB.

14) The truth of everything I asserted in § 8 is immediately clear, and it is evident that the angle $ef\mathrm{B}=\psi$ is always less than the angle $gp\mathrm{B}=\phi$. For since tang $\psi=$ tang $\phi-\dfrac{p}{a\cos\phi}$, it is clear that tang ψ is $<$ tang ϕ, so the angle ψ itself is $<\phi$, and this is valid too for the case where the pole C is held above the plate; for then the wire fe is raised a little, yet in such a way that the angle $ef\mathrm{B}$ which it includes with the plane, or ψ, is always less than the angle $gp\mathrm{B}$, or ϕ.

15) Hold the magnetic pole C below the plate AB, Fig. CXXIV, and let the iron wire fe be supposed to acquire the position $ef\mathrm{B}$. Shake the plate and the wire fe will leap; let us suppose that in this leap the point g is raised directly upward to t. It has been established already from the previous discussion that in its ascent and subsequent descent the point g does not depart sensibly from the line pq, but meanwhile the wire turns in such a way that its direction passes through the point C, or if the leap pt is quite small, as we suppose here, it acquires a position rts, parallel to the line Cg. When therefore on its descent the wire reaches the plate again, it will have a position FgE, so its lower end will be transferred from f to F. Since pf is to pF as cos ψ is to cos ϕ, it is clear that pF is less than pf, whence the point f and the whole wire fe ought to recede somewhat from the point D in its leap. After the wire fe descends to the point where it touches the plate AB again with its lower end F, it is clear that it will not retain its position parallel to the line Cg, but the end E ought to descend and the angle EFB decrease, in such a way that mFB becomes equal to the angle efB, or rather a little less than it. So every time the plate is shaken, the wire fe leaps and moves a small distance away from the point D. Therefore when the shaking of the plate is repeated frequently it ought to keep on moving further and further away from it.

16) Completely the contrary of this ought to happen if the magnetic pole C is placed above the plane AB, Fig. CXXV, for then when the plane is shaken, the wire fe ought to keep on moving closer and closer to the point D. Before the plate is shaken, the wire fe acquires a position such that the angle $ef\mathrm{A}=\psi$ is less than the angle $g\mathrm{RA}$ [227] $=\phi$. When the plane is shaken, the wire leaps upward in a vertical line so that the point g is transferred to t. During this ascent and the subsequent descent, the wire fe turns in such a way that its direction

[227] The text has gkA.

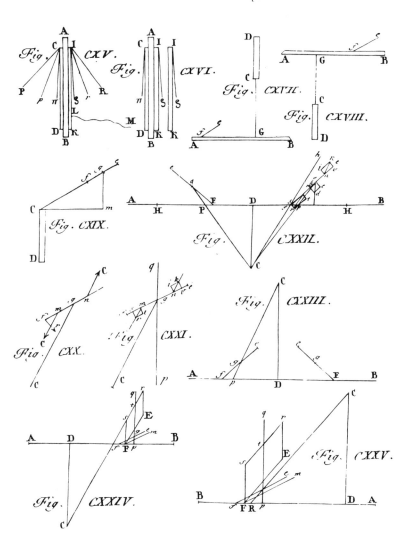

Figure 14. Facsimile Plate VII from the *Essay*

passes through C, or so that it acquires a position *str* approximately parallel to the line C*g*. So when the wire descends, retaining, approximately at least, its parallelism with C*g*, the end *f* is transferred to F, whence since pF is to pf as $\cos \phi$ is to $\cos \psi$, and hence $p\text{F} < pf$, the whole wire moves closer to the vertical line CD by the interval *f*F. After the wire *fe* has come to the position EF, its end E still descends a little so that it acquires a position *m*F, approximately parallel to the former position *ef*, or rather such that it takes up an angle *m*FA a little larger than the angle *ef*A. Therefore it is clear now why when the test is performed with the magnetic pole C held above the plate AB, the wire *fe* always approaches closer and closer to the vertical line CD, and why this outcome is altogether contrary to that which is observed when the magnet is placed below the plate.

APPENDIX

ANNOTATED BIBLIOGRAPHY OF AEPINUS'S PUBLISHED WRITINGS

A. COLLECTED WORKS

1. [Teoriya elektrichestva i magnetizma. Ed. Ya. G. Dorfman].
Ф. У. Т. ЭПИНУС / ТЕОРИЯ ЭЛЕКТРИЧЕСТВА / И МАГНЕТИЗМА / РЕДАКЦИЯ И ПРИМЕЧАНИЯ / ПРОФЕССОРА Я. Г. ДОРФМАНА // ИЗДАТЕЛЬСТВО АКАДЕМИИ НАУК СССР / 1951. 4to. 563p.
Contents: (1) ТЕОРИЯ ЭЛЕКТРИЧЕСТВА И МАГНЕТИЗМА. (Russian trans. of 11, omitting the dedication.) (2) РЕЧЬ О СХОДСТВЕ ЭЛЕКТРИЧЕСКОЙ СИЛЫ С МАГНИТНОЮ. (A reprint of **10B**, with spelling and punctuation modernized.) (3) МЕМУАР, СОДЕРЖАЩИЙ ТОЧНОЕ ОПИСАНИЕ ОПЫТОВ, ПРОВЕДЕННЫХ С ТУРМАЛИНОМ. (Russian trans. of **14A**(2).) (4) ДОПОЛНЕНИЕ К ПРЕДЫДУЩЕМУ МЕМУАРУ. (Russian trans. of **14A**(3).) (5) ЭПИНУС И ЕГО ТРАКТАТ О ТЕОРИИ ЭЛЕКТРИЧЕСТВА И МАГНЕТИЗМА. (Essay by Ya. G. Dorfman.) (6) Bibliography of works by or about Aepinus (incomplete, and with short titles only). (7) Index of names. (8) Table of Contents. (9) Errata.

B. WORKS PUBLISHED SEPARATELY

For each of Aepinus's separately-published works, at least one location is recorded wherever possible. Where locations in several countries are known, one location for each country is given. The following abbreviations are used:

BAN	Biblioteka Akademii Nauk S.S.S.R., Leningrad
BLM	Baillieu Library, University of Melbourne
BM	British Library, London
BN	Bibliothèque Nationale, Paris
BNR	Biblioteca Nazionale Centrale, Rome
CLE	Crawford Library, Royal Observatory, Edinburgh
CUT	Institute for the History of Electricity, Chalmers University of Technology, Göteborg
DSB	Deutsche Staatsbibliothek, Berlin
ESL	Engineering Societies Library, New York
GPIB	Gosudarstvennaya Publichnaya Istoricheskaya Biblioteka, Moscow
HU	Harvard University Library, Cambridge, Mass.
KVA	Kungl. Svenska Vetenskapsakademien, Stockholm
LC	Library of Congress, Washington
MSL	Medical Society of London (at Wellcome Institute for the History of Medicine, London)
NYP	New York Public Library

UA	Universiteits-Bibliotheek, Amsterdam
UG	Niedersächsische Staats- und Universitäts-Bibliothek, Göttingen
UM	University of Michigan Library, Ann Arbor
UR	Universitätsbibliothek, Rostock

—. Annvente Deo! / Consentiente Amplissimo / Philosophorvm Ordine, / de / Vulgarivm Opinionvm Vsv / Poetico, / in Academiae Patriae / Avditorio Majori, / d. IIII. April. A.O.R. CIƆ IƆ CCXXXXIIII. / Pvblice dispvtabvnt / Fratres Germani, / M [agister] Angelivs Joh. Dan. / Æpinvs, / et / Franc. Vdalric. Theodos. / Æpinvs, / Med. et Phil. Stvd. // *Rostochii,* / *Typis* Joh. Jac. Adleri, Sereniss. Princ. & Acad. / Typographi. 4to. 38 p.

Aepinus was to act as formal respondent at the public defense of this dissertation composed by his older brother. However, the brother fell ill, and had still not recovered by the time Aepinus had to leave for Jena. Hence the latter's response was never in fact delivered. The thesis was eventually defended in July 1744, with a new respondent in Aepinus's place.

2. De / Cvrvis, / in Qvibvs Corpora / Gravitate Natvrali / Agitata, / ea Lege Descendvnt, / vt / Qvantitatem Descensvs / Metiatvr / qvaevis Potestas Temporis. / Cvm Censvra Ampliss. Facvlt. Philos. // Anno MDCCXLVII. // *Rostochii,* / Typis Ioannis Iacobi Adleri, *Seren. Princ.* et *Academ.* / Typographi.

[2nd title] Qvo / ad Avdiendas Praelectiones / Mathematicas, in Celeb. / Segneri / Elementa Arithmetices et Geometriae, / et / Physicas, in Ill. Hambergeri / Elementa Physices / Nobilissimos atqve Clarissimos / Commilitones / officiose invitaret, / Tractationem hanc Typis / exscribendam cvravit / M[agister] F.V.T. Aepinus. 4to.

Title; 2nd title; pp. [3]–12; 1 plate.
Location: BAN.

3. Commentatio Mathematica, / de / Avgmento Sortis per / Anatocismvm. // ad / Virvm Excellentissimvm / atqve Consvltissimvm / HERMANNVM BECKERVM, / J[uris] V[trivsqve] Doctorem Celeberrimvm, / cvm / Professionem Jvris Pvblicam / ab / Amplissimo Senatv in ipsvm / delatam / prid. Calend. Septembr. CIƆ IƆ CCXXXXVII. / Solenni Ritv capesseret, / directa / a / M[agistro] F.V.T. Aepino. / Med. Cand. // *Rostochii,* / Typis Ioanni Iacobi Adleri, *Sereniss. Princ. et Acad.* Typographi. 4to.

Title; pp. [2]–8.
Locations: BAN, CLE, UR.

4. Q[vod] D[eus] B[ene] V[ertat]. / Meditationes / de / Cavsa et Indole Febrivm / Intermittentivm, / qvas / Consentiente Gratiosa Facvltate Medica, / Praeside, / *Viro* / Experientissimo et Excellentissimo, / Domino / GEORGIO CHRISTOPHORO / DETHARDINGIO, / Phil. et Med. Doctore, hvivsqve Professore / Dvc[ali] P[rofessore] O[rdinario] Celeberrimo, / Favtore ac Præceptore svo / Observantissime colendo / pro / *obtinendo Doctoris Medicinæ Gradv* / die . . . Aprilis Anni MDCCXLVIII. / pvblice tvebitvr / Franc. Vlr. Theod. Æpinus, / Rostoch. Phil. Mag[ister]. // *Rostochii,* / Typis Ioannis Iacobi Adleri, Academiæ Typographi. 4to.

Title; 1 blank p.; Dedication to Christian Ludwig, Duke of Mecklenburg, 2 p.; pp. [I]–XXXII.
Locations: BAN, MSL.

5. Rectore Acad. Rostoch. Magnificentissimo / Serenissimo Principe / ac Domino, / DOMINO FRIDERICO / Principe Haereditario Dvcatvs Meclebvrgici, / Principe Vandalorvm, Sverini, et Racebvrgi, / Comite Sverinensi, / Rostochii et Stargardiae Terrarvm Domino / Domino Nostro Clementissimo. / Demonstrationes / Primariarvm Qvarvndam / Aeqvationibvs Algebraicis / Competentivm Proprietatvm, / cvm Consensv / Amplissimi Philosophorvm Ordinis, / defendent / M[agister] Franc. Vdalr. Theod. Aepinvs / Medic. Doctorandvs / et / Christ. Lvdov. Wilhelm. a Flotav / Nob[ilis] Megapol[itensis]. // Rostochii / in *Avditorio Maiori* / nonis Ianvarii Anni CIƆDCCLII. // Litteris Adlerianis. 4to.

   Title; 1 blank p.; Dedication to the Royal Swedish Academy of Sciences, pp. [3]–[6]; pp. [7]–15.
   Locations: BAN, CLE, KVA, UM, UR.

6. Commentatio / de / Notione Qvantitatis / Negativae. / cvm / Praelectiones / per Instans Semestre Hybernvm habendas / indicaret, / edita / a / F. V. T. Aepino. / Ph[ilosoph.] D[oct.] et M[ed.] C[and.]. // Rostochii, / Typis Io. Iac. Adleri, Seren. Princ. et Acad. / Typographi. 4to.

   Title; 1 blank p.; pp. [3]–16. Signed at end, "Scr. Rostochii. A. CIƆ IƆ CCLIV. d. VI. Octob."
   Locations: BAN, CLE, UR.

7. De / Integratione, et Separatione / Variabilivm, / in Aeqvationibvs Differentialibvs, Dvas / Variabiles Continentibvs, / Commentatio // qvam / Rectore Academiae Rostochiensis Magnificentissimo / Serenissimo Principe ac Domino / DOMINO FRIDERICO, / Principe Haereditario Dvcatvs Mecklebvrgici, / Principe Vandalorvm, Sverini et Racebvrgi, / Comite Sverinensi / Rostochii et Stargardiae Domino, / Domino Nostro Clementissimo // cvm Consensv / *Amplissimae Facvltatis Philosophicae* / pvblice defendent / M[agister] FRANC. VDALR. THEOD. AEPINVS / in Acad. Reg. Scient. Berolinensi Astron. Prof. / et / HENR. JAC. BRANDT, / Ll. [=Legum?] Stud. Scharstorffio Megap[olitensis] / Rostochii. / In Avditorio Maiori / d. . . . Febr. A. CIƆIƆCCLV. // *Rostochii,* / Typis Johann. Jac. Adleri. 4to.

   Title; 1 blank p.; pp. III–XIX.
   Location: BAN, CLE.

   Pp. XVII–XIX, headed "Theses," are unrelated to the dissertation proper. On the Leningrad copy, the date on which the dissertation was to be defended has been altered by hand to "3. Mart."

8A. Calendarium / ad Annum Christi / MDCCLVI. / Bissextilem / et Meridianum / Berolinensem. // Jussu / Academiæ Scientiarum et Elegan- / tiorum Litterarum Borussicæ. 4to.

   Title; "Imprimatur. P.L.M. de Maupertuis, Academiæ Præses"; "Epochæ Celebriores," 1 p. unnumbered; "Cycli hujus Anni," 96 p., unnumbered;

"Explicatio hujus Calendarii," 19 p., unnumbered; "Etat de l'Academie des Sciences et Belles Lettres de Prusse," 8 p., unnumbered.
Location: UR.

The preparation of the Berlin Academy's annual astronomical calendar was for many years a task carried out by the Academy's official astronomer. One of Aepinus's successors in this position, Johann Bernoulli III, testifies explicitly to Aepinus's having continued the tradition during his term in office; see Bernoulli, *Reisen durch Brandenburg, Pommern, Preußen, Curland, Rußland und Pohlen, in den Jahren 1777 und 1778,* Leipzig, 1779–80, IV, 19–20. Hence this and the following three items (**8B, 9A** and **9B**) must be credited to him.

8B. According to both Houzeau and Lancaster, *Bibliographie générale de l'astronomie jusqu'en 1880,* I, 1587, and the records of the Berlin Academy itself, the above calendar was published in both Latin and German editions during this period. Houzeau and Lancaster give as a title for the German series, *Vollständiger astronomischer Calender auf das Jahr n auf den Berlinischen Mittagszirkel berechnet.* I have been unable to locate a copy of the German edition for 1756 to verify the reference.

9A, 9B. The same sources indicate that both Latin and German editions were published of an astronomical calendar for 1757, after which the series ceased. Once again Bernoulli is the authority for ascribing these to Aepinus. Presumably the titles were very similar to those of the preceding year. Once again, I have been unable to locate copies to verify the references.

10A(i). Sermo Academicvs / de / Similitvdine Vis Electricae / atqve / Magneticae / in Solenni Conventv Academiae / Imperialis Scientiarvm / A.O.R. MDCCLVIII. die VII. Septembris / cvm / Dies Nomini / Avgvstae / RVSSORVM IMPERATRICIS / ELISABETAE PRIMAE / Sacer / Celebraretvr, / pvblice praelectvs / a / F.V.T. Aepino. / In Academia Imperiali Scient. Petrop. Physices Professore, / Acad. Scient. et eleg. Litt. Regiae Berolinensis, et Electoralis / Mogunt. vtilium Scient. Societ. Erford. Membro. // Typis Academiae Scientiarum Petropolitanae. 4to.

Title; 1 blank p.; pp. [1]–32; 1 folding plate.

Locations: BAN, BM, BNR, KVA, LC, UR.

Reviewed in *Göttingische Anzeigen von gelehrten Sachen,* 1759, pp. 116–19.

10A(ii). Microcard edition of **10A(i)** by Readex Microprint, New York, 1976, as part of their "Landmarks of Science" collection.

10B. [Rech' o skhodstve elektricheskoĭ sily s magnitnoyu].
РѢЧЬ / О / СХОДСТВѢ / ЕЛЕКТРИЧРЕСКОЙ СИЛЫ / СЪ / МАГНИТНОЮ / ВЪ ПУБЛИЧНОМЪ СОБРАНІИ / ИМПЕРАТОРСКОЙ АКАДЕМІИ НАУКЪ / для торжественнаго дня / ВЫСОЧАЙШАГО ТЕЗОИМЕНИТСТВА / ЕЯ ИМПЕРАТОРСКАГО ВЕЛИЧЕСТВА / ВЕЛИКІЯ ГОСУДАРЫНИ ИМПЕРАТРИЦЫ / ЕЛИСАВЕТЫ ПЕТРОВНЫ / САМОДЕРЖИЦЫ ВСЕРОССІЙСКІЯ / ВЪ ДЕНЬ 7 СЕНТЯБРЯ 1758

ГОДА. / говоренная / АКАДЕМІИ НАУКЪ ПРОФЕССОРОМЪ ФИЗИКИ Ф.У.Ѳ. ЭПИНУСОМЪ. // ВЪ САНКТПЕТЕРБУРГѢ. / Печатана при Императорской Академіи Наукъ. 4to.

>Title; 1 blank p.; pp. [3]–36; 1 folding plate.
>*Location:* BAN.

>Russian trans. of **10A**, published simultaneously with this. The trans. was prepared by Mikhail Sofronov, and corrected by Stepan Rumovskiĭ before publication. Cf. also 1(2).

10C. Akademische Rede / von der / Aehnlichkeit der elektrischen / und magnetischen Kraft, / welche in der feyerlichen Versammlung der kai- / serlichen Akademie der Wissenschaften im Jahr 1758 / am 7ten Herbstmonats, als dem Nahmensfeste Ihro / Rußisch-Kaiserlichen Majestät Elisabeth der Ersten / öffentlich vorgelesen hat / Herr F.V.T. Aepinus, / der Naturlehre Professor bey der kaiserlichen Akademie der Wis- / senschaften zu St. Petersburg, der königl. Akademie der Wissen- / schaften und schönen Künste zu Berlin, und der churfürstl. mainzischen Akademie nützlicher Wissenschaften zu Erfurt Mitglied. // Leipzig, / in Joh. Friedr. Gleditschens Handlung. / 1760. 8vo.

>Title; 1 blank p.; pp. [3]–44; 1 plate.
>*Locations:* BAN, ESL, KVA, UR.

>German trans. of **10A**. Cf. also **18**. At least two other German editions were published in the periodical literature at the time, as follows (**10C** and **10E** have the same text; **10D** is clearly the work of a different translator).

10D. "Akademische Rede von der Aehnlichkeit der electrischen und magnetischen Kraft Bey der feyerlichen Versammlung der kaiserl. Akademie der Wissenschaften, am 7. Sept. 1758. An welchem das Namensfest der Allerdurchl. Kais. von Russland, Elisabeth der Ersten, gefeyret wurde, öffentlich vorgelesen von F.V.T. Aepinus, öffentl. Lehrer der Physik bey der kaiserl. Akademie der Wissensch. in Petersburg; der königl. preuss. Akad. der Wissensch. zu Berlin, und der churf. maynzischen Gesellschaft nützlicher Wissenschaften zu Erfurt Mitgliede. Aus dem Lateinischen übersetzt," *Hamburgisches Magazin, oder gesammlete Schriften, Aus der Naturforschung und den angenehmen Wissenschaften überhaupt* (Hamburg und Leipzig: bey Georg Christ. Grund und Adam Heinr. Holle), xxii, 1759, 227–72.

10E. "Akademische Rede von der Aehnlichkeit der elektrischen und magnetischen Kraft, welche in der feyerlichen Versammlung der kaiserlichen Akademie der Wissenschaften im Jahr 1758 am 7ten Herbstmonats öffentlich vorgelesen hat Herr F.V.T. Aepinus," *Allgemeines Magazin der Natur, Kunst und Wissenschaften* (Leipzig: in Gleditschens Handlung), xi, 1761, 90–131.

Extracts from **10C**, heavily edited and "brought up to date," have been published as follows:

10F(i). "Die Entdeckung der elektrischen Influenz und die Pyroelektricität. 1758. Aepinus, von der Ähnlichkeit der elektrischen und magnetischen Kraft." Friedrich Dannemann, *Grundriss einer Geschichte der Naturwissenschaften* zugleich eine Einführung in das Studium der naturwissenschaftlichen Litteratur. I. Band. Erläuterte Abschnitte aus den Werken hervorragender Naturforscher aller Völker und Zeiten, Leipzig, 1896, pp. 157–62.

10F(ii). 2nd ed., Leipzig, 1902, pp. 176–82.

10F(iii). (3rd ed.) *Aus der Werkstatt Grosser Forscher. Allgemeinverständliche erläuterte Abschnitte aus den Werken hervorragender Naturforscher aller Völker und Zeiten.* Bearbeitet von Dr. Friedrich Dannemann. Dritte Auflage des ersten Bandes des "Grundriss einer Geschichte der Naturwissenschaften." Leipzig, 1908. pp. 188–93.

10F(iv). (4th ed.) *Aus der Werkstatt Grosser Forscher. . . .* Von Dr. Friedrich Dannemann. Leipzig, 1922. pp. 182–87. (Further heavy editing has been done for this edition.)

11. Tentamen Theoriae / ELECTRICITATIS / et / MAGNETISMI. // Accedunt Dissertationes duae, quarum prior, / phaenomenon quoddam electricum, altera, / magneticum, explicat. // Avctore / F.V.T. Aepino / Acad. Scient. Imper. Petropolitanae, Regiae Berolinensis et / Elector. Mogunt. Erford. Membro. // Instar Supplementi Commentar. Acad. Imper. Petropolitanae. // Petropoli / Typis Academiae Scientiarvm. 4to.

Title; 1 blank p.; Dedication to Count Kiril Razumovskiĭ, 10 p., unnumbered; Praefatio, "scr. Petropoli A.R.S. MDCCLIX, m. Octobri," 8 p., unnumbered; half-title; 1 blank p.; Introduction, "scr. Petropoli. A.R.S. MDCCLVIII," pp. [1]–8; pp. [9]–354; half-title, "Explicatio phaenomeni cvivsdam lagenae lvgdvnensis a celeberrimo *Richmanno* inventi"; 1 blank p.; pp. 355–76; half-title, "Explicatio paradoxi cvivsdam phaenomeni magnetici"; 1 blank p.; pp. 377–90; seven folding plates.
Locations: BAN, BLM, BM, BN, BNR, HU, KVA, UR.

Published in an edition of 650 copies on 8 December 1759. Some copies include a French translation of the Dedication, dated "A St. Petersbourg, ce 3 Janvier 1760." For a Russian trans., cf. **1**(1). Reviewed in [*Rostock*] *Gelehrte Nachrichten*, 1760, 428–32 [in St. 38, dated 17 September 1760); in *Commentarii de rebus in scientia naturali et medicina gestis*, Supplement, Decade 1, 1763, pp. 681–93; and ibid., xxx, 1788, 313–23.

12A. [Izvestie o nastupayushchem prokhozhdenii Venery mezhdu Solntsem i Zemleyu]
ИЗВѢСТІЕ / о наступающемъ прохожде- / ніи Венеры между Солн- / цемъ и Землею. // Сочинено Ф.У.Т. Эпинусомъ, Членомъ Санкт -/ петербургской Императорской Академіи Наукъ. 8vo.
[St. Petersburg: Academy of Sciences Press, 1760]

Title; 1 blank p.; pp. [3]–[15]; two plates, one coloured.
Location: GPIB.

A separately-published version of the following, the same plates being used for printing the diagrams in each case:

12B(i). "ИЗВѢСТІЕ О НАСТУПАЮЩЕМЪ ПРОХОЖДЕНІИ ВЕНЕРЫ МЕЖДУ СОЛНЦЕМЪ И ЗЕМЛЕЮ," СОЧИНЕНІЯ И ПЕРЕВОДЫ, К ПОЛЬЗѢ И УВЕСЕЛЕНІЮ СЛУЖАЩІЯ (ВЪ САНКТПЕТЕРБУГѢ, ПРИ ИМПЕРАТОРСКОЙ АКАДЕМІИ НАУКЪ), ОКТЯБРЪ 1760 ГОДА, pp. 359–71 [in T. xii].

12B(ii). 2nd ed., St. Petersburg, 1801–1804. T. xii, pp. 359–71.

13A. Cogitationes / de Distribvtione Caloris / per Tellvrem, / in publico Academiae Imperialis / Petropolitanae conuentu, / NOMINI AVGVSTISSIMAE / IMPERATRICIS / ELISABETHAE, / Sacro / Die 6. Septembr. A. MDCCLXI. / praelectae, / a / Franc. Vlr. Theod. Aepino. / Sacr. Imp. Mai. Consiliario, Acad. Imper. Petropolitanae Physices / Professore, in Acad. Equestri studiorum Directore, / Acad. Berolinensis, Holmiensis et Erfordensis Membro. // *Petropoli* / Typis Acad. Imper. Scientiarum. 4to.

> Title; 1 blank p.; pp. [3]–23; Adnotationes ad Dissertationem Praecedentem, pp. 24–36.
> *Locations:* BAN, BM, HU, KVA.
>
> Reviewed in *Göttingische Anzeigen von gelehrten Sachen*, 1761–62, pp. 340–43.

13B. Reflexions / concernant / la Distribution de la Chaleur / sur le Globe de la Terre / lues dans l'Assemblée Publique / de l' Academie Impériale de St. Petersbourg / le 6. Septembre 1761. / Après / la Fête / DE SA MAJESTÉ IMPÉRIALE / de Toutes les Russies / etc. etc. etc. / par / Mr. Francois Ulric Theodose Aepinus / Conseiller de *Sa Majesté Impériale,* Professeur en Physique dans l'Aca- / demie Impériale de St. Petersbourg, Directeur des études du noble / Corps des Cadets, et Membre des Académies de Berlin, / de Stockholm et d'Erfurth. / Traduites par Mr. Raoult. // *à St. Petersbourg.* / de l'imprimerie de l'Academie Impériale des / Sciences. 4to.

> Title; 1 blank p.; pp. [3]–23; Remarques sur la dissertation precedente, pp. [24]–38.
> *Location:* BAN, NYP.
>
> French trans. of **13A**, published simultaneously with this.

13C. [Razsuzhdenie o razdelenii teploty po zemnomu sharu] РАЗСУЖДЕНІЕ / О РАЗДѢЛЕНІИ ТЕПЛОТЫ / ПО ЗЕМНОМУ ШАРУ / ДЛЯ / ДНЯ ТЕЗОИМЕНИТСТВА / ЕЯ ИМПЕРАТОРСКАГО ВЕЛИЧЕСТВА / ЕЛИСАВЕТЫ ПЕТРОВНЫ / САМОДЕРЖИЦЫ ВСЕРОССІЙСКОЙ / ВЪ СОБРАНІИ / САНКТПЕТЕРБУРГСКОЙ ИМПЕРАТОРСКОЙ / АКАДЕМІИ НАУКЪ / 1761 года Сентября 6 числа / читанное на Латинскомъ языкѣ / ФРАНЦ. УЛЬР. ѲЕОД. ЕПИНУСОМЪ, / Коллежскимъ Совѣтникомъ, Физики Профессоромъ, и Берлинской, / Стокгольмской и Эрфордской Академій Членомъ. // ПЕЧАТАНО ВЪ САНКТПЕТЕРБУРГѢ / при Императорской Академіи Наукъ 1761 года. 4to.

Title; 1 blank p.; pp. [3]-20; ПРИМѢЧАНІЯ НА СІЕ РАЗСУЖДЕНІЕ, pp. 21-32.
Location: BAN.

Russian trans. of **13A**, published simultaneously with this in an edition of 390 copies.

14A. Recueil / de / Differents Memoires / sur la / Tourmaline / publié par / Mr. Franc. Ulr. Theod. Aepinus, / Conseiller des Colleges de Sa Maj. Impériale, Professeur / en Physique de l'Academie Imp. des Sciences à St. / Petersbourg, Directeur des Etudes du Noble Corps / des Cadets, et Membre des Académies de Berlin, / de Stockholm et d'Erfurth. // *à St. Petersbourg* / de l'Imprimerie de l'Académie des Sciences / 1762. 8vo.

Title; 1 blank p.; Dedication to Catherine Alexievna, [Empress of Russia], 18 July 1762, 6 p.; Avant-propos, 8 p.; Table des pièces, 2 p.; pp. [1]-[193] (the latter wrongly numbered 179). Five plates.
Locations: BAN, BM, BN, KVA, NYP.

Contents: (1) Mémoire concernant quelques nouvelles Expériences électriques remarquables. (A slightly amended version of **27A**). (2) Memoire contenant une déscription exacte des experiences faites avec la Tourmaline. (Cf. 1(2)). (3) Supplement au Memoire précédent. (Cf. 1(3)). (4) Lettre du Duc de Noya Carafa sur la Tourmaline, à Monsieur de Buffon. Avec de Remarques de Monsieur Aepinus. (An annoted version of the Duke's pamphlet published under the same title in Paris in 1759). (5) Lettre de Mr. Aepinus, à Mr. le Duc de Noya Caraffa touchant la Tourmaline. (6) Experiences faites sur la Tourmaline par Mr. Benjamin Wilson . . . dans une Lettre à Mr. Guillaume Heberden. . . . (French trans. of Wilson's letter published in *Phil. Trans.*, LI, 1759, 308-39). (7) Remarques sur la Lettre de Mr. Wilson à Mr. Heberden et sur ses Expériences sur la Tourmaline par Mr. Aepinus.

14B. Verzameling / van / Verscheide Vertoogen / over den / Aschtrekker, / Uitgegeeven door den Heer / Franc. Ulr. Theod. Aepinus. / *Raad ter vergaderinge van Haare Keizerlijke / Majesteit, Hoogleeraar in de Natuurkunde / van de Keizerlijke Academie der Weeten- / schappen, enz. te Petersburg, Lid / der Academien van Berlin, / Stockholm en Erfürth.* Uit het Fransch vertaald door / Daniel Doornik. // Te Amsterdam, / By Gerardus Lequien, / Boekverkooper, MDCCLXXIII. 8vo. 268 p. 5 plates.

Location: UA.

Dutch trans. of **14A**. Johannes van Abkoude, *Naamregister van de bekendste en meest in gebruik zynde Nederduitsche Boeken* . . ., 2nd ed., Rotterdam, 1788, p. 9, lists what purports to be another Dutch edition published in Amsterdam in 1773 by H. Keizer, but I have been unable to find any trace of this.

15A. Abhandlung / von den / Luft-Erscheinungen, / in einer öffentlichen Versammlung / der Kayserlichen Akademie der Wissenschaften, / bey / Allerhöchster Gegenwart / Ihro Kayserlichen Majestät / **KATHARINA DER**

ZWEYTEN, / Kayserin und Selbstherrscherin / aller Reußen zc. zc. zc. / den 2. Julius 1763. / vorgelesen / von / Fr. Ulr. Theod. Aepinus, / Rußisch-Kayserl. Collegien-Rath, ordentl. Mitgliede der Kayserl. Akademie / der Wissenschaften, Ausseher der Ritter-Akademie in dem Hochadel. Land- / Cadetten Corps, der Akademien zu Berlin, Stockholm und Erfurt Mitglied. // St. Petersburg, / gedruckt bey der Kayserlichen Akademie der Wissenschaften. 4to.

Title; 1 blank p.; pp. [3]–16.
Locations: BAN, KVA, LC.

15B. [Razsuzhdenie o vozdushnykh yavleniyakh]
РАЗСУЖДЕНІЕ / О / ВОЗДУШНЫХЪ ЯВЛЕНІЯХЪ, / ПРИ ВЫСОЧАЙШЕМЪ ПРИСУТСТВІИ / ЕЯ ИМПЕРАТОРСКАГО ВЕЛИЧЕСТВА / ЕКАТЕРИНЫ ВТОРЫЯ / ИМПЕРАТРИЦЫ / И / САМОДЕРЖИЦЫ ВСЕРОССІЙСКІЯ / и прочая, и прочая, и прочая / читанное / въ публичномъ собраніи / Импера орской Академіи Наукъ / Іюля 2 дня 1763 года / ФР. УЛЬР. ТЕОД. ЭПИНУСОМЪ, / Коллежскиit Совѣтникомъ, ординарнымъ, оной Академіи Членомъ, Дирек- / торомъ Наукъ въ Сухопутномъ Шляхетномъ Кадетскомъ Корпусѣ, / и Членомъ Берлинской, Стокгольмской и Эрфуртской Академій. / переведено съ Нѣмецкаго языка. // ПЕЧАТАНО ВЪ САНКТПЕТЕРБУРГѢ / при Императорской Академіи Наукъ. 4to.

Title; 1 blank p.; pp. [3]–16.
Location: BAN.

Russian trans. of **15A**, published simultaneously with this.

16A. Hrn. Georg Wolfgang Kraffts / Kurze Einleitung / zur mathematischen und natürlichen / Geographie, / nebst dem Gebrauch / der / Erd-Kugeln / und / Land-Charten, / zum Nutzen der Rußischen studirenden / Jugend. // Zweyte mit Anmerckungen vermehrte Ausgabe / von / F.U.T. Aepinus. // St. Petersburg, / gedruckt bey der Kaiserl. Akademie der Wissenschaften. / 1764. 8vo. 390 p.

Location: BAN.

Reviewed in *Göttingische Anzeigen von gelehrten Sachen*, 1766, pp. 1014-15. The first edition had been published in St. Petersburg in 1738.

16B. [Rukovodstvo k matematicheskoĭ i fizicheskoĭ geografii]
(i) РУКОВОДСТВО / КЪ / МАТЕМАТИЧЕСКОЙ / И ФИЗИЧЕСКОЙ / ГЕОГРАФІИ, / со употребленіемъ / ЗЕМНАГО ГЛОБУСА / И / ЛАНДКАРТЪ, / вновь переведенное / съ примѣчаніями / ФР. УЛЬР. ТЕОД. ЭПИНУСА / изданіе второе. // ВЪ САНКТПЕТЕ-РБУРГѢ / при Императорской Академіи Наукъ / 1764 года. 8vo.

355 p. 5 plates.
Location: BAN.

Russian trans. of **16A**, published in an edition of 1200 copies. The translator was A. M. Razumov.

488 APPENDIX

(ii) [2nd ed.]

Reprint of **16B(i),** with an identical title and the same particulars on the title page concerning its publication. However, the watermark in the paper shows that it could not have been published earlier than 1803.

Location: BAN.

17A. Beschreibung / des / Welt-Gebäudes. // Voltaire. / Der Catechimus lehrt Gott, aber die Kinder: Newton / lehrt ihn die Weisen. // St. Petersburg, / gedruckt bey der Kayserl. Academie der Wissenschaften. / 1770. 8vo.

Title; 1 blank p.; Preface, pp. [3]–4; pp. [5]–56.
Location: UR.

In the unsigned Preface, this work is said to have been written in 1759, at the command of the then Imperial Grand Duchess, later to become the Empress Catherine II. For Aepinus's authorship of it, see Koppe, *Jetztlebendes gelehrtes Mecklenburg*, I, 14.

17B. [Razsuzhdenie o stroenii mira]

(i) РАЗСУЖДЕНІЕ / О / СТРОЕНІИ МІРА. / Переведено / СЪ НѢМЕЦКАГО ЯЗЫКА. // ВЪ САНКТПЕТЕРБУРГѢ / при Императорской Академіи Наукъ / 1770 года. 8vo.

Title; 1 blank p.; Preface, pp. [3]–4; [5]–56.
Location: BAN.

Russian translation of **17A**, published simultaneously with this in an edition of 600 copies. The translator was Grigoriĭ Vasil'evich Kozitskiĭ (1727–1775).

(ii) [2nd ed.]

РАЗСУЖДЕНІЕ / О / СТРОЕНІИ МІРА. / Изданіе второе. / Переведено / съ Нѣмецкаго языка. // ВЪ САНКТПЕТЕРБУРГѢ / при Императорской Академіи Наукъ / 1783 года. 8vo.

Title; 1 blank p.; Preface, pp. [3]–4; pp. [5]–58.
Location: BAN.

1200 copies of this edition were printed.

18. F.V.T. Aepinus / öffentlichen Lehrers der Physik bey der / kaiserlichen Akademie der Wissenschaften in / Petersburg, der königlichen preußischen Akademie der / Wissenschaften zu Berlin, und der churfürstlich- / maynzischen Gesellschaft nützlicher Wissen / schaften zu Erfurt Mitgliedes / ZWO SCHRIFTEN, / I. / Von der Aehnlichkeit der electrischen, und / magnetischen Kraft. / II. / Von den Eigenschaften des Tourmalins. / Aus dem Lateinischen übersetzt. // Grätz, / gedruckt bey den Widmanstätterischen Erben, / 1771. 8vo.

Title; 1 blank p.; Half-title, "I. Akademische Rede von der Aehnlichkeit der electrischen und magnetischen Kraft."; 1 blank p.; pp. 1–44; Half-title, "II. Abhandlung von einigen neuen Erfahrungen, die Electricität

des Tourmalins betreffend."; 1 blank p.; pp. 1–28 (new pagination); 1 plate.
Locations: CUT, DSB.

The first essay is a reprint of **10D**. The second is a German version of **27A** (of which, however, no Latin version is known, despite what is asserted here). Both known copies have bound between the title and the first half-title eight unrelated pages headed "Assertiones ex universa philosophia" advertising a disputation to be held at the local university in Graz in 1771.

19. Déscription des nouveaux Microscopes, / inventés par Mr. Aepinus, / Conseiller d'Etat actuel au Colleges des affaires étrangeres, Che- / valier de l'Ordre de Ste. Anne, Membre de differ. Académies. [St. Petersburg: Academy of Sciences Press, 1784]. 8 vo.

[No separate title p.]; I. Lettre à l'Académie des Sciences de St Pétersbourg ("Announce d'un Microscope achromatique d'une construction nouvelle, propre à voir les objets avec la lumière réflechie de leur surface"), $\frac{12}{23}$ April 1784, pp. [I]–VI; II. Réponse à Monsieur le Baron d'Asch, Conseiller d'Etat et premier Médecin de l'Etat Major des Armées Imper. Contenant des éclaircissements sur la construction des nouveaux Microscopes, $\frac{13}{24}$ June 1784, pp. VII–X; III. Enumeration des principaux avantages de mes nouveaux Microscopes, $\frac{10}{21}$ August 1784, pp. XI–XII.

Locations: BAN, BM, BN, UG.

Cf. **58, 60**. Reviewed in *Göttingische Anzeigen von gelehrten Sachen*, 1784, pp. 1692–94, and in [*Greifswald*] *neueste critische Nachrichten*, 1786, p. 232.

C. ARTICLES PUBLISHED IN THE PERIODICAL LITERATURE

20. ["Auflösung einer Aufgabe von den Logarithmen"], *Mecklenburgische Gelehrte Zeitungen. Auf das Jahr 1751*, Rostock und Wismar, bey Joh. Andreas Berger und Jacob Boedner, St. XIX, 12 May 1751, pp. 151–52.

21. "Demonstratio Theorematis Binomialis," *Gelehrte Nachrichten auf das Jahr 1752*, Rostock und Wismar, verlegts Joh. Andr. Berger und Jac. Boedner, 3 Beilage, 31 March 1752, p. 136.

22. ["Nachricht vom Durchgang des Merkur"], *Gelehrte Nachrichten auf das Jahr 1753*, Rostock und Wismar, St. XVI, 17 April 1753, pp. 170–73; St. XVII, 26 April 1753, pp. 182–84; St. XIX, 9 May 1753, pp. 210–13; St. XLIV, 31 October 1753, 489–92.

23. Review of J. N. De L'Isle, *Avertissement aux Astronomes sur le Passage de Mercure au devant du Soleil* . . . (Paris, 1753), in *Gelehrte Nachrichten auf das Jahr 1753*, Rostock und Wismar, St. XXVI, 27 June 1753, pp. 284–88; St. XXVIII, 10 July 1753, pp. 316–19.

24. ["Berechnung einer Mondfinsterniss nach Mayers Tafeln"], *Gelehrte Nachrichten auf das Jahr 1755*, Rostock und Wismar, St. X, 5 March 1755, pp. 108–109.

25. "Gedanken über die nahe bevorstehende Wiedererscheinung des Cometen vom J. 1682," *Gelehrte Nachrichten auf das Jahr 1755*, Rostock und Wismar, St. XLIV, 29 October 1755, pp. 499–504; St. XLV, 5 November 1755, pp. 525–28.

No indication of authorship is given anywhere in this article or in the *Register* of the volume in which it appears. Aepinus is identified as the author in G. C. Hamberger (rev. J. G. Meusel), *Das gelehrte Teutschland: oder, Lexikon der jetzt lebenden teutschen Schriftsteller*, Bd. IX, Lemgo, 1801, p. 13.

26. "De la figure des supports d'une voute," *Histoire de l'Académie Royale des Sciences et Belles Lettres [à Berlin], Année MDCCLV* (i.e. *Mém. de l'Acad.*, Tom. XI), Berlin, chez Haude et Spener, 1757, pp. 386–93; 1 plate.

The article is said to be "Traduit du Latin." Read 16 October 1755.

27A. "Mémoire concernant quelques nouvelles expériences électriques remarquables," *Histoire de l'Académie Royale des Sciences et Belles Lettres [à Berlin], Année MDCCLVI* (i.e. *Mém. de l'Acad.*, Tom. XII), Berlin, 1758, pp. 105–21.

"Traduit de l'Allemand." Read 10 March 1757. Reprinted in slightly amended form as **14A(1)**. Cf. also **14B, 18**. A version incorporating occasional alterations in the sentence structure was also included in each of the following compilations:

27B(i). *Mémoires de l'Académie Royale de Prusse.* . . . Par M. Paul. 7 vols., 8o, Avignon, 1768. Tom. V, pp. 369–400.

27B(ii). *Mémoires de l'Académie Royale de Prusse.* . . . Par M. Paul. 2 vols., 4o, Avignon, 1768. Tom. II, pp. 247–58.

27B(iii). *Mémoires de l'Académie Royale de Prusse.* . . . Par M. Paul. 7 vols., 8o, Paris, 1770. Tom. V, pp. 369–400.

A reprinting of the collection noted in **27B(i)**.

27B(iv). *Collection académique, composée des mémoires, actes ou journaux des plus célèbres Académies & Sociétés Littéraires.* . . . Par M. Paul. Paris, 1770. Tom. IX, pp. 247–58. (Vols. VIII and IX of this compilation are a reprinting of the collection noted in **27B(ii)**.)

A German translation of the article, different from the one which appeared in **18**, was published as follows:

27C. "Abhandlung von einigen neuen Erfahrungen die Electricität des Tourmalins betreffend," *Mineralogische Belustigungen, zum Behuf der Chymie und Naturgeschichte des Mineralreichs*, Erster Band, Leipzig, 1768, pp. 302–22.

In addition, a précis was published as follows:

27D. "Observations sur un Phenomene électrique," *Observations periodiques, sur la physique, l'histoire naturelle et les arts*, ed. F. V. Toussaint, Tom. II, Paris, 1757, pp. 341–45 (in the number for May 1757).

An extract has been published in English translation as follows:

27E. "Aepinus—Pyroelectricity," *A Source Book in Physics*, ed. William Francis Magie, New York/London, 1935, pp. 406–408.

28. "Recherches sur les inconvéniens qu'on a lieu de craindre dans l'usage du micrometre, surtout par rapport aux instrumens qu'on adapte au quart de cercle," *Histoire de l'Académie Royale des Sciences et Belles Lettres [à Berlin], Année MDCCLVI* (i.e. *Mém. de l'Acad.*, Tom. XII), Berlin, 1758, pp. 365–86; 2 plates.

"Traduit de l'Allemand." Read 2 December 1756. The original MS is in the archives of the Akademie der Wissenschaften der D.D.R., Berlin.

29. "Démonstration du theoreme de Harriot, avec une methode de chercher, si une équation algébrique a toutes les racines possibles, ou non?," *Histoire de l'Académie Royale des Sciences et Belles Lettres [à Berlin], Année MDCCLVIII* (i.e. *Mém. de l'Acad.*, Tom. XIV), Berlin, 1765, pp. 354–66; 1 plate.

"Traduit du Latin." The original MS is in the archives of the Akademie der Wissenschaften der D.D.R., Berlin.

30(i). "РАЗМЫШЛЕНІЯ О ВОЗВРАТѢ КОМЕТЪ, СЪ КРАТКИМЪ ИЗВѢСТІЕМЪ О НЫНѢ ЯВИВШЕЙСЯ КОМЕТѢ," ЕЖЕМѢСЯЧНЫЯ СОЧИНЕНІЯ КЪ ПОЛЬЗѢ И УВЕСЕЛЕНІЮ СЛУЖАЩІЯ, ВЪ САНКТПЕТЕРБУРГѢ, ПРИ ИМПЕРАТОРСКОЙ АКАДЕМІИ НАУКЪ, ОКТЯБРЬ 1757 ГОДА, pp. 329–48 [in Т. VI].

The article is dated 4 October 1757. It was translated into Russian by V. Lebedev.

30(ii). 2nd ed., St. Petersburg, 1801–1804. Т. VI, pp. 329–48.

31A(i). "РАЗСУЖДЕНІЕ О НѢКОТОРЫХЪ НОВЫХЪ СПОСОБАХЪ, ПРИНАДЛЕЖАЩИХЪ КЪ ПОПРАВЛЕНІЮ МАГНИТНЫХЪ СТРѢЛОКЪ И МОРСКАГО КОМПАСА," СОЧИНЕНІЯ И ПЕРЕВОДЫ, КЪ ПОЛЬЗѢ И УВЕСЕЛЕНІЮ СЛУЖАЩІЯ, ВЪ САНКТПЕТЕРБУРГѢ, ПРИ ИМПЕРАТОРСКОЙ АКАДЕМІИ НАЧКЪ, НОЯБРЬ 1758 ГОДА, pp. 483–506 [in Т. VIII].

31A(ii). 2nd ed., St. Petersburg, 1801–1804. Т. VIII, pp. 519–44.

A German version of this article was published, without the introductory eulogy of Peter the Great, as follows:

31B. "Abhandlung über einige neue Verbesserungen der Magnetnadel und des Seecompasses, aufgesetzt von F.U.T. Aepinus, Professor der Naturlehre zu St. Petersburg," *Hamburgisches Magazin, oder gesammlete Schriften, aus der Naturforschung und den angenehmen Wissenschaften überhaupt*, Bd. XXIV, Hamburg und Leipzig, 1760, pp. 563–84.

32(i). "КРАТКОЕ ИЗВѢСТІЕ О НОВОИЗОБРѢТЕННОМЪ СПОСОБѢ КЪ УМНОЖЕНІЮ СИЛЫ ВЪ НАТУРАЛЬНЫХЪ МАГНИТАХЪ," СОЧИНЕНІЯ И ПЕРЕВОДЫ, КЪ ПОДЬЗѢ И УВЕСЕЛЕНІЮ СЛУЖАЩІЯ, ВЪ САНКТПЕТЕРБУРГѢ, ПРИ ИМПЕРАТОРСКОЙ АКАДЕМІИ НАУКЪ, ГЕНВАРЬ 1759 ГОДА, pp. 48–53 [in Т. IX].

32(ii). 2nd ed., St. Petersburg, 1801–1804. Т. IX, pp. 51–56.

33. "Descriptio acuum magneticarum, noviter inventarum, quae vulgaribus praestantiores sunt, atque artificii, vires magnetum naturalium insigniter augendi. Autore F.V.T. Aepino. Prof. Petropolit. Acad. Imp. Petropol. et Elect. Scient. Utilium Sodali, A.R.S. 1759," *Acta Academiae electoralis Moguntinae scientiarum quae Erfurti est*, Tom. II, Erfordiae, apud Ioannem Fridericum Weberum, 1761, pp. 255–72.

34. "Descriptio ac explicatio novorum quorundam experimentorum electricorum," *Novi commentarii Academiae Scientiarum Imperialis Petropolitanae*, Tom. VII, 1758–59 (Petropoli, Typis Academiae Scientiarum, 1761), pp. 277–302; 1 plate.

Summary pp. 22–24. Submitted to the Academy, 1 December 1757; read 9 March 1758.

35. "Observatio optica de mutabilite diametri apparentis foraminis angusti, oculo propinque," *Novi commentarii Academiae Scientiarum Imperialis Petropolitanae*, Tom. VII, 1758–59 (Petropoli, 1761), pp. 303–308; 1 fig.

Summary pp. 24–26. Submitted 23 February 1758; read 11 May 1758.

36. "Demonstratio generalis theorematis Newtoniani de binomio ad potentiam indefinitam elevando," *Novi commentarii Academiae Scientiarum Imperialis Petropolitanae*, Tom. VIII, 1760–61 (Petropoli, 1763), pp. 169–80.

Summary pp. 27–29. Submitted 3 November 1757; read 19 December 1757.

37. "De functionum algebraicarum integrarum factoribus trinomialibus realibus commentatio," *Novi commentarii Academiae Scientiarum Imperialis Petropolitanae*, Tom. VIII, 1760–61 (Petropoli, 1763), pp. 181–88.

Summary pp. 29–30. Submitted 21 December 1758; read 24 May 1759.

38. "De nova quadam vectis proprietate dissertatio," *Novi commentarii Academiae Scientiarum Imperialis Petropolitanae*, Tom. VIII, 1760–61 (Petropoli, 1763), pp. 271–73; 1 fig.

Summary pp. 46–47. Submitted 4 June 1759; read 15 October 1759.

39. "Halonum extraordinariarum Petropoli visarum descriptio," *Novi commentarii Academiae Scientiarum Imperialis Petropolitanae*, Tom. VIII, 1760–61 (Petropoli, 1763), pp. 392–403; 2 plates.

Summary pp. 60–63. Submitted 20 December 1759; read 3 March 1760.

40A. "Experimenta Academ. Scientiar. Petropolitanae, quibus conglaciatio Mercurii vivi adprobatur," *Nova acta eruditorum* (Leipzig), February 1760, pp. 74–78.

For Aepinus's authorship of the report which served as the basis of this and the following item, cf. *Protokoly zasedanii konferentsii Imperatorskoĭ Akademii Nauk s 1725 po 1803 goda*, St. Petersburg, 1897–1911, vol. II, pp. 456–57.

40B. "Memoire sur la Congélation du Mercure dans un froid artificiel," *Journal des sçavans, combiné avec les mémoires de Trévoux* (Amsterdam), June 1760, pp. 259–67 [in Tom. LII, No. 7].

A slightly more elaborate account than in **40A**. The article falls in a section

ANNOTATED BIBLIOGRAPHY

headed "Additions de l'Editeur d'Hollande"; it did not appear in either the Paris edition of the *Journal des sçavans* or the *Journal de Trévoux*.

41. "Emendatio microscopii solaris," *Novi commentarii Academiae Scientiarum Imperialis Petropolitanae*, Tom. IX, 1762-63 (Petropoli, 1764), pp. 316-25; 1 plate.

   Summary pp. 33-34. Submitted 18 February 1760; read 5 May 1760.

42. "Dissertatio de experimento quodam magnetico, celeberr. domini Du Fay, descriptio in Commentariis Acad. scient. Paris a 1730," *Novi commentarii Academiae Scientiarum Imperialis Petropolitanae*, Tom. IX, 1762-63 (Petropoli, 1764), pp. 326-39; 1 plate.

   Combined summary of this and the following item, pp. 35-36. Submitted 22 December 1760; read 10 November 1761.

43. "Additamentum ad dissertationem de experimento magnetico celeber. do. Du Fay, continens nova experimenta magnetica detecta et explicata," *Novi commentarii Academiae Scientiarum Imperialis Petropolitanae*, Tom. IX, 1762-63 (Petropoli, 1764), pp. 340-51; 1 plate.

   Combined summary of this and the preceding item, pp. 35-36. Submitted 29 October 1761; read 8 March 1762.

44. "Instrumentorum astronomicorum, reticulo aut micrometro instructorum, nova emendatio," *Novi commentarii Academiae Scientiarum Imperialis Petropolitanae*, Tom. IX, 1762-63 (Petropoli, 1764), pp. 488-92; 1 plate.

   Summary pp. 46-47. Submitted 3 April 1760; read 22 May 1760.

45. "Observatio eclipseos lunaris, die 7 (18) maii 1761 habita in observatorio Imperiali Petropolitano," *Novi commentarii Academiae Scientiarum Imperialis Petropolitanae*, Tom. IX, 1762-63 (Petropoli, 1764), pp. 496-98; 1 fig.

   Summary pp. 48-49. Submitted 29 October 1761.

46A. "Observationes quaedam ad opticam pertinentes: Obs. I. De coloribus accidentalibus; II. De maculis nigris oculo obvolitantibus; III. De colore umbrarum," *Novi commentarii Academiae Scientiarum Imperialis Petropolitanae*, Tom. X, 1764 (Petropoli, 1766), pp. 282-95; 1 plate.

   Summary pp. 33-35. Read 8 May 1758. The different sections are often listed separately in bibliographies; sometimes not all three are recorded. A French translation of the first section was published as follows:

46B. "Observations sur les Couleurs accidentelles," *Observations sur la physique, sur l'histoire naturelle et sur les arts*. . . . Par M. l'Abbé Rozier . . . & par M. J.A. Mongez . . . Tome XXVI, Paris, 1785, pp. 291-94 (in the number for January 1785).

47. "Similitudinis effectuum vis magneticae et electricae novum specimen," *Novi commentarii Academiae Scientiarum Imperialis Petropolitanae*, Tom. X, 1764 (Petropoli, 1766), pp. 296-98; 2 figs.

   Summary pp. 35-36. Submitted 29 October 1761; read 8 February 1762.

48. "Dissertatio astronomica de effectu parallaxeos in transitu planetarum sub sole," *Novi commentarii Academiae Scientiarum Imperialis Petropolitanae*, Tom. x, 1764 (Petropoli, 1766), pp. 433–53; 3 plates.

Summary pp. 57–59. Submitted 4 December 1760; read 29 October 1761. Aepinus had previously read a paper with an almost identical title to the Berlin Academy on 17 March 1757.

49. "Dissertatio II. De effectu parallaxeos in transitu planetarum sub sole," *Novi commentarii Academiae Scientiarum Imperialis Petropolitanae*, Tom. x, 1764 (Petropoli, 1766), pp. 454–72.

Summary p. 60. Submitted 22 December 1760; read 2 November 1761.

50A. "De electricitate barometrorum disquisitio," *Novi commentarii Academiae Scientiarum Imperialis Petropolitanae*, Tom. xii, 1766–67 (Petropoli, 1768), pp. 303–24; 1 plate.

Summary pp. 28–31. Submitted 10 February 1763; read 9 June 1763. A German translation was published as follows:

50B. "Herrn F.V.T. Aepinus Abhandlung von der Electricität der Barometer . . . übersetzt von D. J.G. Krüniz," *Neue Hamburgisches Magazin, oder Fortsetzung gesammleter Schriften, aus der Naturforschung, der allgemeinen Stadt- und Land-Oekonomie, und den angenehmen Wissenschaften überhaupt*, Bd. ix, Leipzig, 1771, pp. 518–46.

51. "Examen theoriae magneticae a celeberr. Tob. Mayero propositae: Sect. I. De situ centri magnetici; II. De scala intensitatum vis magneticae. III. De functione attractionum et repulsionum magneticarum, et experimentis a Mayero pro confirmanda hypothesi sua institutis," *Novi commentarii Academiae Scientiarum Imperialis Petropolitanae*, Tom. xii, 1766–67 (Petropoli, 1768), pp. 325–50; 1 plate.

Summary pp. 31–34. Submitted 22 August 1763; read 12 September 1763.

52. "Descriptio novi phaenomeni electrici detecti in chrysolitho sive smaragdo Brasiliensi," *Novi commentarii Academiae Scientiarum Imperialis Petropolitanae*, Tom. xii, 1766–67 (Petropoli, 1768), pp. 351–55.

Summary pp. 34–35. Submitted 8 October 1764.

53A. "ПЕРЕВОДЪ РѢЧИ ЧИТАННОЙ КОЛЛЕЖСКИМЪ СОВѢТНИКОМЪ Г. ЭПИНУСОМЪ ПРИ ВСТУПЛЕНІИ ЕГО ВЪ ВОЛЬНОЕ ЭКОНОМ- ИЧЕСКОЕ ОБЩЕСТВО," ТРУДЫ ВОЛЬНАГО ЭКОНОМИЧЕСКАГО ОБЩЕСТВА КЪ ПООЩРЕНІЮ ВЪ РОССІИ ЗЕМЛЕДѢЛІЯ И ДОМОСТРОИТЕЛЬСТВА, 1766 ГОДА, ЧАСТЬ iv, ВЪ САНКТПЕТЕ- РБУРГѢ, ПРИ ИМПЕРАТОРСКОЙ АКАДЕМІИ НАУКЪ, pp. 197–200.

A German translation of this was published as follows:

53B. "Rede des Herrn Collegienraths Aepinus, bey seiner Aufnahme in die freye ökonomische Gesellschaft," *Abhandlungen der freyen ökonomischen Gesellschaft* in *St. Petersburg zur Aufmunterung des Ackerbaues und der Hauswirthschaft in Rußland*, Bd. IV, 1766 (Aus dem Rußischen übersetzt. St. Petersburg, Riga und Leipzig, 1774), pp. 308–10.

54A. "РАЗСУЖДЕНІЕ О СРЕДСТВАХЪ СЛУЖАЩИХЪ КЪ ОТВРАЩЕНІЮ ПОВРЕЖДЕНІЯ ПРОИЗВОДИМАГО ГРОМОВЫМЪ УДАРОМЪ И МОЛНІЕЮ ДОМАМЪ И ДРУГИМЪ СТРОЕНІЯМЪ," ТРУДЫ ВОЛЬНАГО ЭКОНОМИЧЕСКАГО ОБЩЕСТВА, КЪ ПООЩРЕНІЮ ВЪ РОССІИ ЗЕМЛЕДѢЛІЯ И ДОМОСТРОИТЕЛЬСТВА, 1770 ГОДА, ЧАСТЬ XVI, ВЪ САНКТПЕТЕРБУРГѢ, ПРИ МОРСКОМЪ ШЛЯХЕТНОМЪ КАДЕТСКОМЪ КОРПУСѢ, pp. 257–85.

An edited version of this was published as follows:

54B(i). ["НАСТАВЛЕНІЕ ОТВРАЩАТЬ ГРОМОВЫЕ УДАРЫ"], МѢСЯЦОСЛОВЪ ВЪ ПОЛЬЗУ ДОМОСТРОИТЕЛЬСТВА НА 1773 ГОДЪ, ed. Я.Я. ШТЕЛИН, ВЪ САНКТПЕТЕРБУРГѢ, 1772, pp. 29–51.

54B(ii). This was reprinted in СОБРАНІЕ СОЧИНЕНІЙ, ВЫБРАННЫХЪ ИЗЪ МѢСЯЦОСЛОВОВЪ НА РАЗНЫЕ ГОДЫ, ЧАСТЬ VII, ВЪ САНКТПЕТЕРБУРГѢ, 1791, pp. 394–412.

55A. "Ueber den Bau der Mondfläche, und den vulcanischen Ursprung ihrer Ungleichheiten, von F.V.T. Aepinus, würklichen Staats-Rath beym Collegio der auswärtigen Affairen zu St. Petersburg, Mitglied verschiedener Akademien," *Schriften der Berlinischen Gesellschaft Naturforschender Freunde*, Bd. II, Berlin, im Verlage der Buchhandlung der Realschule, 1781, pp. 1–40.

Noted in *Göttingische Anzeigen von gelehrten Sachen*, 1782, pp. 698–99. The original MS is in the archives of the Museum für Naturkunde, East Berlin. A Russian translation of the first part of the article was published as follows:

55B. "О СТРОЕНІИ ПОВЕРЬХНОСТИ ЛУНЫ И О ПРОИЗХОЖДЕНІИ НЕРОВНОСТЕЙ ОНЫЯ ОТЪ ВНУТРЕННЯГО ОГНЯ," АКАДЕМИЧЕСКІЯ ИЗВѢСТІЯ НА 1781 ГОДА, ЧАСТЬ VIII, St. Petersburg, 1781, pp. 918–33 (in the number for May-July 1781).

The remainder of the article was scheduled to appear in succeeding numbers of this journal, but it ceased publication at this point. Documents in the archives of the Soviet Academy of Sciences suggest that Aepinus may then have arranged to have 100 copies of the translation printed at his own expense, but no trace of any such publication has been found.

56(i). "Plan des wirklichen StatsRats Hrn. Aepinus," *Stats-Anzeigen, gesammelt und zum Druck befördert von August Ludwig Schlözer* . . ., Bd. III (Heft 11), Göttingen, in der Vandenhoekschen Buchhandlung, 1783, pp. 260–78.

The article forms the major part of a report headed "Rußische Schulprojecte," dated Mainz, 14 January 1783.

56(ii). Aepinus's plan was reprinted as an appendix (pp. 134–47) to Д. А. ТОЛСТОЙ, "ГОРОДСКІЯ УЧИЛИЩА ВЪ ЦАРСТВОВАНІЕ ИМПЕРАТРИЦЫ

APPENDIX

ЕКАТЕРИИЫ II," ЗАПИСКИ ИМПЕРАТОРСКОЙ АКАДЕМІИ НАУКЪ, T. LIV, St. Petersburg, 1887, Supplement I.

57. ["Letter concerning the Russian school project"], *Stats-Anzeigen, gesammelt . . . von August Ludwig Schlözer . . .*, Bd. v (Heft 17), Göttingen, 1783, pp. 3–7.

    The letter is dated 22 June 1783 (O.S.).

58. ["Vorschlag ein Fernrohr als Mikroscop zu gebrauchen"], *Göttingische Anzeigen von gelehrten Sachen unter der Aussicht der Königl. Gesellschaft der Wissenschaften*, St. 125, 5 Aug. 1784, pp. 1249–52.

    A report based upon **19**, Pt. II.

59. ["Letter concerning the new microscope"], *Göttingische Anzeigen von gelehrten Sachen unter der Aussicht der Königl. Gesellschaft der Wissenschaften*, St. 65, 25 April 1785, pp. 654–56.

    The letter is a response to a review of **19** that had appeared in the 1784 volume of the same journal (St. 169, 21 October 1784), pp. 1692–94.

60. "Lettre de S.E. Mr. le Conseiller d'Etat actuel & Chevalier Aepinus, à Messieurs de l'Académie Impériale des Sciences de St. Pétersbourg: Annonce d'un Microscope achromatique, d'une nouvelle construction, propre à voir les objets avec la lumiere réflechie de leur surface," *Nova acta Academiae Scientiarum Imperialis Petropolitanae*, Tom. II, 1784 (Petropoli, 1788), Histoire pp. 41–49.

    Letter dated St. Petersburg, 20 March 1784; read to the Academy 8 April 1784. Cf. **19**, Pt. I, which differs by only a few very minor changes in phraseology.

61A. "Lettre de S.E. Mr. le Conseiller d'Etat actuel & Chevalier Aepinus, à M. le Conseiller de Collèges Pallas: sur les Volcans de la Lune," *Nova acta Academiae Scientiarum Imperialis Petropolitanae*, Tom. II, 1784 (Petropoli, 1788), Histoire pp. 50–52.

    Letter dated St. Petersburg, 18 June 1784; read to the Academy 21 June 1784. An English translation, with explanatory notes by J.H. de Magellan, was published as follows:

61B. "Letter from Mr. Aepinus, Counsellor of State in Russia, to Mr. Pallas, Counsellor of the Imperial Colleges at St. Petersburg, in Consequence of a Communication made to the Imperial Academy of Sciences by Mr. de Magellan, Member of the same Academy, concerning a Volcano discovered in the Moon, on the 4th of May, 1783, by Mr. Herschel, F.R.S.," *Gentleman's Magazine*, Vol. LIV, Part II, London, 1784, pp. 563–64 (in the number for August 1784).

62. "NormalSchulen in Rußland," *Stats-Anzeigen, gesammelt . . . von August Ludwig Schlözer . . .*, Bd. VII (Heft 25), Göttingen, 1785, pp. 82–92.

    A letter from Aepinus dated St. Petersburg, 13 June 1784.

63. ["Letter concerning the Russian school project"], *Stats-Anzeigen, gesammelt . . . von August Ludwig Schlözer . . .*, Bd. IX (Heft 36), Göttingen, 1786, pp. 408–21.

    The letter is dated St. Petersburg, 22 October 1786.

64. "Two letters on Electrical and other Phenomena; addressed to Dr. Matthew Guthrie . . . by his Excellency M. Æpinus," *Transactions of the Royal Society of Edinburgh*, Vol. II, Edinburgh, printed for T. Cadell and J. Dickson, 1790, Pt. II, Sect. I, pp. 234–44.

   Noted in *Göttingische Anzeigen von gelehrten Sachen*, 1791, p. 828. The letters are dated St. Petersburg, 7/18 January 1789, and 23 January 1789.

65. "ПЕРЕВОД ЛАТИНСКОГО МЕМУАРА Ф. У. Т. ЭПИНУСА «Additamentum ad demonstrationem impossibilitatis tubi nyctoptici»," pp. 82–92 in ЛОМОНОСОВ : СБОРНИК СТАТЕЙ И МАТЕРИАЛОВ, Т. II, РЕД. С. И. ВАВИЛОВ, МОСКВА/ЛЕНИНГРАД, 1946.

   Aepinus's memoir is published as an appendix to Vavilov's article, "НОЧЕЗРИТЕЛЬНАЯ ТРУБА М. В. ЛОМОНОСОВА."

D. EXPOSITIONS OF AEPINUS'S WORK

66A(i). Exposition/raisonnée/ de la théorie/ de/ L'Électricité/ et du Magnétisme,/ D'après les principes de M. Æpinus, des / Académics de Pétersbourg, de Turin [i.e. Berlin], &c./ Par M. l'Abbé Haüy,/ De l'Académie Royale des Sciences, Professeur Émérite/ de l'Université.// A Paris,/ Chez la Veuve Desaint, Libraire, rue du Foin- / Saint-Jacques. // M. DCC.LXXXVII./ Avec Approbation, & Privilége du Roi. 8vo. xxxii + 238 p. 4 plates.

   Noted in *Observations sur la physique, sur l'histoire naturelle et sur les arts*, Tome XXXI, Paris, 1787, pp. 234–35 (in the number for September 1787); reviewed by "A.-M. V." in *Biblioteca oltremontana ad uso d'Italia*, 1788 (v.2), pp. 139–60, continued in ibid., 1788 (v.3), pp. 221–38.

66A(ii). Microcard edition of 66A(i) by Readex Microprint, New York, 1969, as part of their "Landmarks of Science" collection.

Extracts from Haüy's book were published as follows:

66B. "Des principes généraux de la théorie de l'électricité, par M. Æpinus: Extrait de l'Ouvrage de M. l'Abbé Haüy," *Observations sur la physique, sur l'histoire naturelle et sur les arts*, Tome XXXI, Paris, 1787, pp. 401–17 (in the number for December 1787).

A German translation of the book was also published, as follows:

66C. Darstellung/ der Theorie/ der Elektricität/ und des/ Magnetismus/ Nach den Grundsätzen des Herrn Aepinus/ vom/ Bürger Hauy/ Aus dem Französischen übersetzt/ und/ mit Anmerkungen begleitet/ von/ D. Karl Murhard./ Mit 7. Kupfertafeln.// Altenburg,/ bei Rinck und Schnuphase./1801. 8vo. xxxiv + 310 p.

   Both C. G. Kayser, *Vollständiges Bücher-Lexikon enthaltend alle von 1750 bis zu Ende des Jahres 1832 in Deutschland und in den angrenzenden Ländern gedruckten Bücher*, Leipzig, 1834–36, Th. III, p. 70, and H. Fromm, *Bibliographie deutscher Übersetzungen aus den Französischen 1700–1948*, Baden-Baden, 1950–

53, Bd. III, p. 311, list a second edition of this translation, published in Leipzig in 1809, but I have been unable to trace a copy.

NOTES:

1. According to Koppe, *Jetztlebendes gelehrtes Mecklenburg*, I, 13, Aepinus contributed various items either unsigned or under other names to publications in Rostock and elsewhere. With the probable exception of 25 above, however, it has proved impossible to identify any of these.
2. The following work has been attributed by some to Aepinus:

   КРАТКОЕ ПОНЯТІЕ О ФИЗИКѢ, ДЛЯ УПОТРЕБЛЕНІЯ . . . ВЕЛИКАГО КНЯЗЯ ПАВЛА ПЕТРОВИЧА, St. Petersburg, Academy of Sciences Press, 1760. 12 mo. ii + 65 p.

   Others, however, have disputed the attribution. I know of no evidence which might settle the matter one way or the other.
3. V. I. Vsevolodov, in his standard compilation, *Alfavitnyĭ ukazatel' stateĭ, napechatannykh v Trudakh i drugikh periodicheskikh izdaniyakh Imperatorskago Vol'nago Ekonomicheskago Obshchestva*, St. Petersburg, 1849, p. 427, attributes the following work (and also, therefore, the corresponding German version that was published a little later) to Aepinus:

   "ПРИМѢЧАНІЯ О РЫБНОЙ ЛОВЛѢ НА ВЗМОРЬѢ КРОНШТАТСКОМЪ," ТРУДЫ ВОЛЬНАГО ЭКОНОМИЧЕСКАГО ОБЩЕСТВА, 1767 ГОДА, ЧАСТЬ VI, pp. 145–49.

   On internal grounds, however, the attribution cannot be correct; the author of the article in question remarks at one point that at the time of writing he had been living in St. Petersburg for twenty-six years, whereas Aepinus at that stage had been there for no more than ten years.
4. In the Vorontsov Archives (P. I. Bartenev, ed., *Arkhiv Knyazya Vorontsova . . . Bumagi Grafa M. L. Vorontsova*, Moscow, 1870–95, Vol. XXIX, p. 271) there is a letter dated 24 July 1789 from F. H. La Fermière to Count S. R. Vorontsov, in which reference is made, *inter alia*, to an "atlas compilé par monsieur Oepinus." I have been unable to find any record, however, of such an atlas ever having been published.
5. J. C. Houzeau and A. Lancaster, in their *Bibliographie générale de l'astronomie*, Bruxelles, 1880–89; 2nd ed. rev. by D. W. Dewhirst, London, 1964, Vol. II, col. 1226, attribute the following article to Aepinus: ["Sur la constitution physique de la lune"], *Philosophical Transactions of the Royal Society of London*, 1794, pp. 429ff. However, no such article exists!

BIBLIOGRAPHY OF SECONDARY WORKS CITED

Allgemeine deutsche Biographie, 56 vols., Leipzig, 1875–1912.

Arnault, A. V. et al., *Biographie nouvelle des contemporains, ou dictionnaire historique et raisonné* . . . , 20 vols., Paris, 1820–25.

Barrière, Pierre, *L'Académie de Bordeaux, centre de culture internationale au XVIIIe siècle*, Bordeaux, 1951.

Bartenev, P. I., ed., *Arkhiv Knyazya Vorontsova*, 40 vols., Moscow, 1870–95.

Becker, Hermann, ed., *Sacrum piis manibus et exequiis viri illustris atque summe reverendi Domini Franc. Alberti/Æpini* . . . sanctum, Rostock, 1750.

Bernoulli, Johann III, *Reisen durch Brandenburg, Pommern, Preußen, Curland, Rußland und Pohlen, in den Jahren 1777 und 1778*, 6 vols., Leipzig 1779–80.

Berthelot, A., ed., *La Grande encyclopédie: inventaire raisonné des sciences, des lettres et des arts*, 31 vols., Paris, 1882–1902.

Bilyarskiĭ, P. S., *Materialy dlya biografii Lomonosova*, St. Petersburg, 1865.

Bolkhovitinov, N. N., "V arkhivakh i bibliotekakh S.SH.A.: Nakhodki, vstrechi, vpechatleniya," *Amerikanskiĭ ezhegodnik*, 1971, pp. 329–41.

———, "1783 god. Peterburg-Filadel'fiya," *Nauka i zhizn'*, 1976 (no. 7), pp. 65–67.

Boss, Valentin, *Newton and Russia: The Early Influence, 1698–1796*, Cambridge, Mass., 1972.

Bowles, Geoffrey, "John Harris and the powers of matter," *Ambix*, XXII, 1975, 21–38.

Briggs, J. Morton, Jr., "Aurora and Enlightenment: eighteenth-century explanations of the aurora borealis," *Isis*, LVIII, 1967, 491–503.

Buchholz, A., *Die Göttinger Rußlandsammlungen Georgs von Asch*, Giessen, 1961.

Burgmann, Johann Christian, *Memoriae posteritatis perennaturae, insigne gloriae monimentum, quod sibi, dum vixit, statuit ipse, Vir maxime venerabilis . . . Dominus Franc. Albert. Aepinus . . .* , Rostock, 1750.

Cantor, Geoffrey, "The changing role of Young's ether," *British Journal for the History of Science*, v, 1970, 44–62.

Carlid, G. and J. Nordström, eds., *Torbern Bergman's Foreign Correspondence*, Stockholm, 1965.

Castéra, J., *History of Catharine II. Empress of Russia*, London, 1800.

Chenakal, V. L., "Problema metallicheskoga termometra v fizika XVIII v. i

Lomonosov," pp. 96–135, in *Lomonosov: sbornik stateĭ i materialov*, VI, Moscow/Leningrad, 1965.

Cohen, I. B., ed., *Benjamin Franklin's Experiments*, Cambridge, Mass., 1941.

——, *Franklin and Newton: An Inquiry into Speculative Newtonian Experimental Science and Franklin's Work in Electricity as an Example Thereof*, Philadelphia, 1956.

——, ed., *Isaac Newton's Papers and Letters on Natural Philosophy*, Cambridge, 1958.

Coxe, William, *Travels into Poland, Russia, Sweden and Denmark*, 3 vols., London, 1784–90.

Dahlgren, E. W., *Kungl. Svenska Vetenskapsakademien Personförteckningar 1739–1915*, Stockholm, 1915.

Daujat, Jean, *Origines et formation de la théorie des phénomènes électriques et magnétiques*, Paris, 1945.

Delambre, "Eloge historique de M. Coulomb," *Mémoires de l'Institut National des Sciences et Arts (sciences mathématiques et physiques)*, VII, 1806, Hist. pp. 206–23.

Demkov, M. I., *Istoriya russkoĭ pedagogii, Chast' II: Novaya russkaya pedagogiya (XVIII-ĭ vek)*, 2nd ed., Moscow, 1910.

Dictionary of Scientific Biography, 15 vols., New York, 1970–78.

Dorfman, Ya. G., "Epinus i ego traktat o teorii electrichestva i magnetizma," pp. 461–538 in F.U.T. Aepinus, *Teoriya elektrichestva i magnetizma*, Moscow/Leningrad, 1951.

Eschenbach, J. C., *Annalen der Rostockschen Academie*, 13 vols., Rostock, 1796–1807.

Finn, Bernard S., "An appraisal of the origins of Franklin's electrical theory," *Isis*, LX, 1969, 362–69.

Forbes, Eric G., ed., *The Euler-Mayer Correspondence*, London, 1971.

——, *The Unpublished Writings of Tobias Mayer*, 3 vols., Göttingen, 1972.

Gillmor, C. S., *Coulomb and the Evolution of Physics and Engineering in Eighteenth-Century France*, Princeton, 1971.

Gliozzi, Mario, "Beccaria nella storia dell'elettricità," *Archeion*, XVII, 1935, 15–47.

Gnucheva, V. F., *Geograficheskiĭ Departament Akademii Nauk XVIII veka*, Moscow/Leningrad, 1946.

Guerlac, Henry, "Newton's optical aether: his draft of a proposed addition to his *Opticks*," *Notes and Records of the Royal Society of London*, XXII, 1967, 45–57.

Hackman, W. D., "Electrical Researches," in R. J. Forbes et al., eds., *Martinus Van Marum: Life and Work*, 6 vols., Haarlem, 1969–74.
Hall, A. R., and M. B. Hall, eds., *Unpublished Scientific Papers of Isaac Newton*, Cambridge, 1962.
Hans, N., *History of Russian Educational Policy 1701–1917*, New York, 1964.
Hansteen, Christopher, *Untersuchungen über den Magnetismus der Erde*, Christiania, 1819.
Hardin, Clyde L., "The scientific work on the Reverend John Michell," *Annals of Science*, XXII, 1966, 24–47.
Hawes, Joan L., "Newton's two electricities," *Annals of Science*, XXVII, 1971, 95–103.
Heathcote, N. H. de V., "The early meaning of *Electricity:* some *pseudodoxia epidemica*," *Annals of Science*, XXIII, 1967, 261–75.
Heilbron, J. L., "Robert Symmer and the two electricities," *Isis*, LXVII, 1976, 7–20.
———, "Franklin, Haller, and Franklinist history," *Isis*, LXVIII, 1977, 539–49.
Hobson, E. W., *"Squaring the Circle": A History of the Problem*, Cambridge, 1913.
Hofmeister, A., *Die Matrikel des Universität Rostock*, 4 vols., Rostock, 1904.
Home, R. W., "Franklin's electrical atmospheres," *British Journal for the History of Science*, VI, 1972, 131–51.
———, "Aepinus and the British electricians: the dissemination of a scientific theory," *Isis*, LXIII, 1972, 190–204.
———, "The origin of the lunar craters: an eighteenth-century view," *Journal for the History of Astronomy*, III, 1972, 1–10.
———, "Electricity in France in the post-Franklin era," *Actes du XIVe Congrès International d'Histoire des Sciences, Tokyo-Kyoto 1974*, Tokyo, 1974–75, II, 269–72.
———, "On two supposed works by Leonhard Euler on electricity," *Archives internationales d'histoire des sciences*, XXV, 1975, 3–7.
———, "The scientific education of Catherine the Great," *Melbourne Slavonic Studies*, no. 11, 1976, 18–22.
Hoskin, M. A., "'Mining all within': Clarke's notes to Rohault's *Traité de physique*," *The Thomist*, XXIV, 1961, 353–63.

Institut de France, *Index biographique des membres et correspondants de l'Académie des Sciences . . .* , Paris, 1968.

Kohfeldt, G., *Rostocker Professoren und Studenten im 18. Jahrhundert*, Rostock, 1919.
Koppe, J. C., *Jetztlebendes gelehrtes Mecklenburg*, 3 vols., Rostock/Leipzig, 1783–84.
Koyré, A., *From the Closed World to the Infinite Universe*, Baltimore, 1957.
———, and I. B. Cohen, "Newton and the Leibniz-Clarke correspondence," *Archives internationales d'histoire des sciences*, XV, 1962, 63–126.

―――, eds., *Isaac Newton's Philosophiae naturalis principia mathematica: the Third Edition (1726) with Variant Readings*, 2 vols., Cambridge, 1972.

Labaree, Leonard W., ed., *The Papers of Benjamin Franklin*, New Haven, 1959–.
Lauch, A., *Wissenschaft und kulturelle Beziehungen in der russischen Aufklärung zum Wirken H.L.Ch. Bacmeisters*, Berlin, 1969.
Lawrence, George H. M., ed., *Adanson: The Bicentennial of Michel Adanson's "Familles des plantes,"* 2 vols., Pittsburgh, 1963–64.
Leighton, Albert C., "Some examples of historical cryptanalysis," *Historia mathematica*, IV, 1977, 319–37.
Lemay, J. A. Leo, *Ebenezer Kinnersley: Franklin's Friend*, Philadelphia, 1964.
Le Sueur, A., ed., *Maupertius et ses correspondants*, Montreuil-sur-Mer, 1896.
Love, Rosaleen, "Revisions of Descartes's matter theory in *Le Monde*," *British Journal for the History of Science*, VIII, 1975, 127–37.
Lyubimenko, I. I., ed., *Uchenaya korrespondentsiya Akademii Nauk XVIII veka*, Moscow, 1937.

McCormmach, Russell, "Henry Cavendish: a study of rational empiricism in eighteenth-century natural philosophy," *Isis*, LX, 1969, 293–306.
McGuire, J. E., "Force, active principles and Newton's invisible realm," *Ambix*, XV, 1968, 154–208.
Madariaga, I. de, *Britain, Russia and the Armed Neutrality of 1780*, London, 1962.
Massardi, Francesco, ed., *Epistolario di Alessandro Volta: Edizione Nazionale*, 5 vols., Bologna, 1949–55.
Maxwell, J. Clerk, ed., *The Electrical Researches of the Honourable Henry Cavendish, F.R.S.*, Cambridge, 1879.
May, W. E., *A History of Marine Navigation*, Henley-on-Thames, 1973.
Modzalevskiĭ, B. L., *Spisok chlenov Imperatorskoĭ Akademii Nauk 1725–1907*, St. Petersburg, 1908.
Mohrmann, H., *Studien über russisch-deutsche Begegnungen in der Wirtschaftswissenschaft (1750–1825)*, Berlin, 1959.
Morrell, J. B., "The University of Edinburgh in the late eighteenth century: its scientific eminence and academic structure," *Isis*, LXII, 1971, 158–71.

Nugent, Thomas, *Travels through Germany . . . With a Particular Account of the Courts of Mecklenburg*, 2 vols., London, 1768.

Oseen, C. W., *Johan Carl Wilcke, Experimental-Fysiker*, Uppsala, 1939.
Ostrovityanov, K. V., ed., *Istoriya Akademii Nauk S.S.S.R.*, Moscow/Leningrad, 1958–.

Pace, Antonio, *Benjamin Franklin and Italy*, Philadelphia, 1958.
Palter, Robert, "Early measurements of magnetic force," *Isis*, LXIII, 1972, 544–58.
Pekarskiĭ, P. P., *Istoriya Imperatorskoĭ Akademii Nauk v Peterburge*, 2 vols., St. Petersburg, 1870–73.
Perevalov, V. A., *Lomonosov i Arktika*, Moscow/Leningrad, 1949.
Poggendorff, J. C., *Biographisch-Literarisches Handwörterbuch zur Geschichte der exacten Wissenschaften*, vol. I, Leipzig, 1863.
Poroshin, S. A., *Zapiski, sluzhashchiya k istorii Ego Imperatorskago Vysochestva . . . Pavla Petrovicha*, 2nd ed., St. Petersburg, 1881.
Priestley, Joseph, *The History and Present State of Electricity*, London, 1767; and 2nd ed., London, 1769.
Protokoly zasedanii konferentsii Imperatorskoĭ Akademii Nauk s 1725 po 1803 goda, 4 vols., St. Petersburg, 1897–1911.
Pupke, H., "Franz Ulrich Theodosius Aepinus. Zur 225. Wiederkehr seines Geburtstages," *Die Naturwissenschaften*, XXXVII, 1950, 49–52.

Radovskiĭ, M. I., "Issledovaniya Epinusa v oblasti elektromagnetizma," *Elektrichestvo*, 1940, 67–70.
Raskin, N. M., *Khimicheskaya laboratoriya M. V. Lomonosova*, Leningrad, 1962.
Robinson, Eric, and Douglas McKie, eds., *Partners in Science: Letters of James Watt and Joseph Black*, London, 1970.
Rozhdestvenskiĭ, S. V., *Istoricheskiĭ obzor deyatel'nosti Ministerstva Narodnago Prosveshcheniya 1802–1902*, St. Petersburg, 1902.
Russkiĭ biograficheskiĭ slovar', 25 vols., St. Petersburg, 1896–1918.

Schofield, Robert E., ed., *A Scientific Autobiography of Joseph Priestley (1733–1840)*, Cambridge, Mass., 1966.
——, *Mechanism and Materialism: British Natural Philosophy in an Age of Reason*, Princeton, 1970.
Shields, M. C., "The early history of graph in physical literature," *American Physics Teacher*, V, 1937, 68–71.
Sobol', S. L., *Istoriya mikroskopa i mikroskopicheskikh issledovaniĭ v Rossii v XVIII veke*, Moscow/Leningrad, 1949.

Taton, René, *L'oeuvre scientifique de Monge*, Paris, 1951.
——, ed., *Histoire générale des sciences*, 4 vols., Paris, 1957–64.
Tilling, L., "The Interpretation of Observational Errors in the Eighteenth and Early Nineteenth Centuries" (unpublished Ph.D. thesis, University of London, 1973).
Tolstoĭ, D. A., "Gorodskiya uchilishcha v tsarstvovanie imperatritsy Ekateriny II," *Zapiski Imperatorskoĭ Akademii Nauk*, LIV, 1887, Supplement I.
Torlais, J., *L'Abbé Nollet 1700–1770: un physicien au siècle des lumières*, Paris, 1954.

Tunbridge, Paul A., "Faraday's Genevese friends," *Notes and Records of the Royal Society of London,* XXVII, 1973, 263–98.

Turnbull, H. W., et al., eds., *The Correspondence of Isaac Newton,* Cambridge, 1959–.

Vavilov, S. I., "Nochezritel'naya truba M. V. Lomonosova," pp. 71–92 in *Lomonosov: Sbornik stateĭ i materialov,* II, Moscow/Leningrad, 1946.

Wehnert, J. M., "Franz Ulrich Theodosius Aepinus," pp. 99–105 in *Mecklenburgische Gemeinnützige Blätter,* Bd. VII, Parchim, 1803.

Westfall, R. S., *Force in Newton's Physics,* London, 1971.

Whittaker, E. T., *A History of the Theories of Aether and Electricity,* London, 1910.

Winter, E., ed., *Die Registres der Berliner Akademie der Wissenschaften 1746–1766,* Berlin, 1957.

Wonzel, P. van, *Etat présent de la Russie,* St. Petersburg/Leipzig, 1783.

Yushkevich, A. P., et al., eds., *Die Berliner und die Petersburger Akademie der Wissenschaften im Briefwechsel Leonhard Eulers,* 3 vols., Berlin, 1959–76.

———, *Leonhardi Euleri Opera omnia, Series Quarta A: Commercium epistolicum,* vol. I, Basel, 1975.

———, and E. Winter, eds., *Leonhard Euler und Christian Goldbach: Briefwechsel 1729–1764,* Berlin, 1965.

Zedler, J. H., ed., *Grosses vollständiges Universal-Lexicon aller Wissenschaften und Künste,* 64 vols., Leipzig/Halle, 1732–50.

INDEX

Académie royale des sciences, Paris, *see* Paris, Royal Academy of Sciences
Académie royale des sciences et belles lettres de Berlin, *see* Berlin, Royal Academy of Sciences and Letters
Accademia del Cimento, Florence, 92, 366
action at a distance, 86, 196, 216; in Aepinus's theories, 13, 15, 109, 111, 114, 117, 121–22, 187; his attitude toward, 111–12, 240, 259
Adams, George, the younger (1750–1795), 203, 205
Adanson, Michel (1727–1806), 94
Aepinus, Angelius Johann Daniel (1718–1784), 6, 8, 9, 18, 19–20, 62–63, 91n, 480
Aepinus, Francisca Agnesa Beata (1719–?), 6, 20
Aepinus, Franz Albert (1673–1750), 3–7, 8, 9, 19, 62, 63
Aepinus, Franz Ulrich Theodosius (1724–1802)
  Career: early life, 3–7; scientific education, 7–17; graduation, 18, 19; teacher at Rostock University, 6, 18–20; academician in Berlin, 22–28; academician in St. Petersburg, 26, 27, 28, 29, 30, 31–53, 63, 64; courtier and civil servant, 36, 47–48, 53–61, 113; last years, 61–62
  astronomical work, 20–23, 43–47, 50, 51, 482; attitude toward hypotheses, 112, 239; correspondence, 63, *with* Delisle 20, *with* Euler, 14, 20, 22, 190, *with* Franklin, 57–58, 192, *with* Göttingen, 21, 22, *with* Swedish Royal Academy of Sciences, 21, *with* Wilson, 19, 94–95, 191–92, 202, 204; cryptography, 54, 62n; dispute with Lomonosov, 36–47, 48, 62; early work in mathematics, 7–8, 14, 15–17, 18, 20, 33–34; *Essay*, publication of, 26, 35–36, 39, 106, 189, 484; final causes, 399; and Franklin, 34, 55, 57–58, 65–66, 73, 91–92, 93, 95, 97–98, 106, 107–10, 114, 117–24, 135–36, 164, 166, 175, 192, 194–95, 200, 202, 203–204, 214, 230, 234, 237, 241–42, 282–88; friendship, *with* Catherine the Great, 47–48, 50, 53, 63, *with* Euler, 14, 17, 24–25, *with* Robison, 222, *with* Rumovskiĭ, 24, 43, 52, *with* Wilcke, 18–22, 25, 79, 89–92, 129–30, 281n; geography, interest in, 50; graphs, use of, 321–23, 326; laws of electrical and magnetic action unknown to, 115–17, 168, 218, 255, 258, 306, 320–21, 323, 327, 335, 344, 354, 396, 403, 407, 417, 418, lecture notes, 9–10; mathematical approach to physics, 113–15, 126, 127, 136, 168, 187, 201–203, 211, 215, 219, 221–22, 224, 468; note book, 32, 72, 131–35, 187–88; optical research, 23, 33, 56–57; religious convictions, 63; repulsion between particles of ordinary matter, 118–21, 198–200, 212, 219, 258–59
Aepinus, Johann (1499–1553), 5
aether, 13–14, 66, 139, 148–49, 165–66, 224n; Aepinus's views on, 72–73, 112, 243; the Eulers and, 15, 66, 68–71, 72, 78, 145–48; discussed by Newton, 13, 82, 154, 155
air, attempts to explain its elasticity, 82, 115, 117, 149; dielectric in Leyden experiment, 96–98, 108, 234, 281–85. *See also* "air condenser"
"air condenser," 123, 125, 214. *See also* air, dielectric in Leyden experiment
analogy between electricity and magnetism, 164–66, 210–11; advanced by Aepinus, 26, 32, 34, 107, 111, 112, 164, 167, 187, 230, 237–39, 243–44, 350, 355
Antheaulme, *Syndic des Tontines* in Paris, 36, 159, 180, 215; method of magnetizing iron, 181–82, 234n, 380n
armature, magnetic, 138, 362–64, 378–79, 390

Armed Neutrality, Treaty of (1780), 57
Asch, Georg von (1727–1807), 57, 489
astronomy, Aepinus's work in, 20–23, 43–47, 50, 51, 482
atmosphere, Aepinus's use of the term, 392
atmosphere, electrical, 66, 75, 87, 218, 224n; Franklin's doctrine, 83–85, 97, 98–102, 135; elaborated by his supporters, 98, 101–106, 135, 195–98, 210n, 212; rejected by Aepinus, 98, 109–10, 114, 121–22, 134, 136, 195, 214, 392–93
attraction, Newtonian, 110, 119–20, 152–53, 155, 201, 258–59. *See also* action at a distance *and* gravity
attraction and repulsion, electrical, 70–71, 75, 109, 133, 135, 198; in Franklin's theory, 80, 82–83, 91, 101; re-assessed by Aepinus, 108, 120–21, 262–68; law of variation with distance, 65, 119, 133, 217–19, 224; not assumed by Aepinus, 115–17, 255, 258, 306, 320–21, 323, 354. *See also* atmosphere, electrical
attraction and repulsion, magnetic, 138, 140, 142–43, 144, 147, 150, 153, 162, 188n, 263–68, 329, 341; law of variation with distance, 162, 168–73, 206–208, 217–19, 224; not assumed by Aepinus, 168, 255, 306, 327, 335, 344, 396, 403, 407, 417, 418
aurora borealis, Halley's explanation of, 157
"avenging electricity" (*elettricità vindice*), 131

balance, torsion, invented by Coulomb, 217
Banks, Sir Joseph (1743–1820), 220
Barbeu Dubourg, Jacques (1709–1779), 203, 205n, 213–14
Barletti, Carlo (1735–1800), 193
barometer, electricity associated with, *see* "lucent barometer"
Bauer, General F. W. von (1731–1783), 61
Beccaria, Giambatista (1716–1781), 86, 89, 103–105, 110, 131, 135, 193, 196, 197–98, 212
Becker, Johann Peter (?–1757), 6, 8n

Becker, Peter (1672–1753), 6, 8, 16, 20, 22
Béraud, Laurent (1703–1777), 68n, 165, 237n
Bergman, Torbern (1735–1784), 96, 191n
Berlin, Gesellschaft Naturforschender Freunde, 56, 495
Berlin, Royal Academy of Sciences and Letters, 21, 25, 26, 27, 46, 49, 90, 96, 113n, 148, 208, 238, 282, 494; Aepinus's membership of, 14, 22–23; receives a copy of Aepinus's essay, 190, 191, 208
Bernoulli, Daniel (1700–1782), 17, 145, 148–50, 218
Bernoulli, Jean (Johann I) (1667–1748), 145, 148–50
Bernoulli, Johann III (1744–1807), 55, 482
Bertrand, Louis (1731–1812), 24
binomial theorem, 33–34, 489, 492
Black, Joseph (1728–1799), 220, 222n
Blumenbach, Johann Friedrich (1752–1840), 57
Boerhaave, Herman (1668–1738), 9–10, 19, 74, 78, 159
Bohnenberger, Gottlieb Christian (1732–1807), 212n
Bordeaux, Royal Academy of Sciences, Letters, and Arts, 78n, 165, 237
Bose, Georg Matthias (1710–1761), 67
Bowdoin, James (1726–1790), 192
Boyle, Robert (1627–1691), 10, 12, 92, 157. *See also* Boyle's law
Boyle's law, Newton's explanation of, 82, 115, 117
Braun, Josef Adam (1712–1768), 41, 42
Brisson, Mathurin-Jacques (1723–1806), 194n
Brugmans, Anton (1732–1789), 209, 210, 216
Buffon, Georges-Louis Leclerc, comte de (1707–1788), 79, 219–20, 486

Calandrini, Jean-Louis (1703–1758), 173n
calculus, dispute over invention of, 16, 157
Canton, John (1718–1772), 92, 102, 105, 122, 132, 135, 191n, 192; experiments on tourmaline, 93, 95; ideas on electrical atmospheres, 103, 195–98, 210n;

INDEX

induction experiments, 92, 97, 98, 195; method of magnetizing iron, 159, 180, 181, 182, 301, 448; method analyzed by Aepinus, 373–85

capacitance, electrical, 125

Carafa, Giovanni, Duc de Noya (1715–1768), 94, 486

Catherine the Great, Empress of Russia, 49, 52, 55, 57, 58–60, 61, 486–87, 488, 496; shows favor to Aepinus, 47–48, 50, 53, 63

Cavallo, Tiberius (1749–1809), 192, 205, 222, 223

Cavendish, Henry (1731–1810), 125–26, 191, 197, 201–202, 218, 344n

cementation, of iron ores, 437–39

Chambers, Ephraim, (c. 1680–1740), 158, 205

Chappe d'Auteroche, Jean-Baptiste (1728–1769), 52n

Chichagov, Vasilii Yakovlevich (1726–1809), 50

Cigna, Gian Francesco (1734–1790), 200, 210–11, 214

circle, quadrature of, 111, 240

circulation theory of magnetism, *see* magnetism, "Cartesian" or circulation theory

Clairaut, Alexis-Claude (1713–1765), 17

Clarke, Samuel (1675–1729), 156–57

Colden, Cadwallader, (1688–1776), 192, 203

College of Foreign Affairs, Russian, 53–54, 57, 62n

Collinson, Peter (1694–1768), 78, 81n

comet, Halley's prediction concerning, 32–33, 490, 491

communication, of electricity, 350ff; of magnetism, 187n, 295, 297, 350ff, 452. *See also* induction

compass, mariner's, 139, 159. *See also* needle, magnetic

consequent points, 177, 182, 187n, 207, 294–95, 343, 355, 359, 365, 368, 385–87; electrical analogue, 134, 353

continuity, law of, 135, 387

"contrary electricities" 89, 90–91, 92, 93; Wilcke's thesis on, 25, 89, 90, 91–92, 113n, 129–30. *See also* Franklin, theory of electricity

Cotes, Roger (1682–1716), 111

Coulomb, Charles Augustin (1736–1806), 186, 204, 214–15, 216–17, 219, 220, 221, 224; establishes force laws, 65, 168, 169, 217–19; molecular theory of magnetism, 143, 175, 217; prefers "two-fluid" theories, 198–200

Cousin, Jacques-Antoine-Joseph (1739–1800), 220

Coxe, William (1747–1828), 61

cryptography, 54, 62n

d'Alembert, Jean le Rond (1717–1783), 163, 224

d'Arcy, Patrick (1725–1779), 133–34

Dashkova, Princess Ekaterina Romanovna (1744–1810), 61

declination, magnetic, 138, 139–40, 148, 150, 185, 396–97, 399–401, 440–41

Delisle, Joseph Nicolas (1688 1768), 20, 489

De Luc, Jean André (1727–1817), 219n

Derham, William (1657–1735), 157

Desaguliers, John Theophilus (1683–1744), 12, 77n, 158, 161

Descartes, René (1596–1650), 11, 13, 15, 17, 399n; his theory of magnetism, 139–43, 145, 154, 156, 158

Detharding, Georg Christoph (1699–1784), 8, 480

Diderot, Denis (1713–1784), 163

dipping needle, 415, 419, 451. *See also* inclination, magnetic

discharge, electrical, 131, 281; oscillatory nature of, 125, 280

Dollond, John (1706–1761), 38n

Domashnev, Sergeï Gerasimovich (1743–1795), 60

Dorfman, Yakov Grigor'evich, 27n, 113n, 115–16, 125, 479

"double touch" method of magnetization, 180, 182, 387–88, 390, 448–50, 452; Aepinus's analysis of, 368–77; improved by Antheaulme and Aepinus, 181, 182, 214, 234, 379–85

Dubourg, Jacques Barbeu (1709–1779), 203, 205n, 213–14

Du Fay, Charles François de Cisternai (1698–1739), 67, 73, 75, 92, 126, 132n, 135, 325, 493; theory of magnetism,

143–44, 150, 158; two electricities, 87–89, 91, 199n
Duhamel du Monceau, Henri-Louis (1700–1782), 159, 179–80, 181, 215
Du Tour, Etienne-François (1711–1789), 145, 150

Earth, magnetic character of, 138, 139; due to magnetic core, 183–86, 231, 397–401, 408–30, 439–43, 451, 452; surface temperature distribution, 49–50, 485
Edinburgh, Royal Society of, 61, 497; University of, 116, 223
Education Commission, Russian, 59–60; Aepinus's recommendations concerning, 58–60, 495–96
electric fluid, condensed in Leyden experiment, 116, 125, 272, 280–81; equilibrium distribution of, 115–17, 125–26, 350–51, 352; natural quantity of, 78, 101, 118, 245–48, 252, 257
electrical atmosphere, see atmosphere, electrical
electrical attraction and repulsion, see attraction and repulsion, electrical, and repulsion, electrical
electrical discharge, see discharge, electrical
electrical induction, see induction, electrical
electrical potential, 125
electricities, contrary, see "contrary electricities"
electricity, analogy with magnetism, see analogy between electricity and magnetism; "avenging" (*elettricità vindice*), 131; of barometers, see "lucent barometer"; J. A. Euler's theory, 24, 68–72, 91, 101; L. Euler's interest in, 24, 68; Franklin's theory, 32, 34, 73, 77–89, 91, 93, 135, 164, 166, 202, 214, 281; improved by Aepinus, 65–66, 97–98, 106, 107–10, 114, 117–24, 136, 194–95, 200, 241–42, 282–88; Hamberger's discussion, 25, 66–67; importance in the natural economy, 229–30, 277; and light, 70, 76, 133; and magnetism, differentiated by Gilbert, 138, 164; Newton's experiment, 85; Nollet's theory, 65–66, 67–68, 73–78, 88, 89, 94, 97, 98, 101, 102, 165, 213; "two-fluid" theory, 199–200, 209, 219–20; Wilcke's thesis on, see "contrary electricities"
electrometer, d'Arcy's, 133–34; Richmann's, 113n
electrophorous, 129–30, 131, 210, 212
elettricità vindice, 131
Elisabeth, Empress of Russia, 34, 47, 49, 229, 482, 483, 485
Ellicott, John (1706–1772), 92, 110
Encyclopédie, of Diderot and d'Alembert, 163
Erfurt, 9; Society of Sciences, 35, 492
Erxleben, Johann Christian Polykarp (1744–1777), 212
Euler, Johann Albrecht (1734–1800), 24–25, 31n, 47n, 53n, 63n, 66, 73, 74, 77–78, 90–91; theory of electricity, 24, 68–72, 91, 101
Euler, Leonhard (1707–1783), 20, 21n, 22, 23, 27, 28, 31n, 33, 34, 43n, 46, 48, 63, 190, 191, 208–209, 224; influence on Aepinus, 14–15, 17, 24–25, 68, 113n; interest in electricity, 24, 68; theory of magnetism, 145–48, 149, 164, 185, 204–205, 216; on unexplained forces, 15, 68, 110, 112, 166, 174; views on aether, 15, 66, 68–69, 72

final causes, Aepinus's views on, 399
fire, Boerhaave's theory of, 74, 78; "common" and "electrical," distinguished by Franklin, 82; subterranean, 436–37
Florence, Accademia del Cimento, 92, 366
Fontenelle, Bernard le Bovier de (1657–1757), 17, 143, 161
force, magnetic, lines of, 184, 403–408; short-range, 110. See also action at a distance *and* attraction *and* gravity
Franklin, Benjamin (1706–1790), 18, 55, 57–58, 63, 73, 74, 91, 93, 97, 110, 123, 186, 198n, 202, 214, 219, 221, 223, 246, 312, 444; Aepinus's debt to, 34, 91, 93, 107–108, 114, 117–21, 135, 164, 214, 230, 234, 237, 241; explanation of Leyden experiment, 80–81, 85–87, 97, 122–25, 272, 282–88; lightning an electrical discharge, 24–25, 79; reception of Aepinus's work, 109, 175, 192, 194–

95, 203–204; theory of electricity, 32, 34, 65–66, 68, 77–87, 88–89, 91, 93, 97, 98–102, 122, 241–42, 258, 260–61, 281, 392. *See also* atmosphere, electrical, *and* Franklinists

Franklinists, 89, 105; elaborate concept of electrical atmosphere, 98, 101–106, 135, 195–98, 210n, 212

friction, as cause of magnetism, denied, 445

Frisi, Paolo (1728–1784), 68n, 116n, 193

Gehler, Johann Samuel Traugott (1751–1795), 212

Gilbert, William (1544–1603), 74n, 92, 137–39, 162, 170, 182, 183n, 298n, 359n, 398n, 399, 427n; distinguishes between electricity and magnetism, 138, 164

glass, its role in Leyden experiment, according to Franklin, 80, 85–87, 97, 127; re-assessed by Aepinus, 97, 108–109, 194, 241–42, 282–90

Goldbach, Christian (1690–1764), 54

Goldsmith, Oliver (1730–1774), 205

Gordon, Andreas (1712–1751), 67

Göttingen, 56, 60; Royal Society of Sciences, 57, 205, 496; University, 20, 21, 22, 52,

Graham, George (1675–1751), 157

Gralath, Daniel (1739–1809), 67, 92

Grandi, Guido (1671–1742), 17

graphs, Aepinus's use of, 321–23, 326

'sGravesande, Willem Jacob (1688–1742), 12

gravity, universal, 13 119–20, 201, 258–59; according to Newton, 111–12, 152, 155, 240; Euler's attitude toward, 110, 112, 147; explained by Descartes, 139, 141

Gray, Stephen (1666–1736), 67, 92, 279n

Green, George (1793–1841), 219

Gregory, David (1659–1708), 154

Gregory of St. Vincent (1584–1667), 16

Grischow, August Nathaniel (1726–1760), 28, 31, 43, 50

Guericke, Otto von (1602–1686), 92

Guthrie, Matthew (1743–1807), 61, 497

Haller, Albrecht von (1708–1777), 9–10, 78n

Halle University, 9, 10, 67

Halley, Edmond (1656–1742), 32–33, 138, 157, 185, 400

Hamberger, Georg Erhard (1697–1755), 9–15, 16, 25, 66–68, 84, 164

Hamilton, Sir William (1730–1803), 55–56

Harris, John (1666–1719), 157

Hartsoeker, Nicolaas (1656–1725), 12, 164, 359n

Hauksbee, Francis, the elder (?–1713), 66, 85, 92, 157, 169–70

Haüy, René-Just (1743–1822), 220–24, 497

Helmholtz, Hermann von (1821–1894), 280n

Helsham, Richard (?1682–1738), 171n, 173n

Hemmer, Johann Jakob (1733–1790), 212

Herschel, William (1738–1822), 496

Hire, Gabriel-Philippe de la (1677–1719), 43

Hire, Philippe de la (1640–1718), 143

Huber, Johann Jakob (1733–1798), 24

Hübner, Lorenz (1753–1807), 210

Huygens, Christiaan (1629–1695), 142–43, 146, 147

hypotheses, Aepinus's attitude towards, 112, 239

Imperial Academy of Sciences of St. Petersburg, *see* St. Petersburg, Imperial Academy of Sciences

Imperial Corps of Noble Cadets, St. Petersburg, 48, 50, 51, 53

inclination, magnetic, 396–97, 399–401, 440–41. *See also* dipping needle

induction, electrical, 130, 197; Aepinus's understanding of, 108, 121–22, 126, 129, 131, 136, 212, 261, 304–25; Canton's experiments on, 92, 97, 98, 195. *See also* communication

induction, magnetic, 138–39, 183, 207, 421; Aepinus's analysis of, 177–79, 181, 183, 184–85, 325–49. *See also* communication

iron, magnetization of, 144, 179–82, 185–86, 355–59, 364–90, 408–19, 425–31, 441–52; mineralogy of, according to Aepinus, 186, 433–37 iron filings, patterns formed by, 158–59, 161, 163, 176, 184, 301n, 406–408, 469–78

Jacquier, François (1711–1788), 173
Janković de Mirievo, Teodor (1741–1814), 59–60
Jena University, 5, 9–10, 16, 18, 19, 480
Jesuit missionaries in Peking, 130, 200

Karl Leopold, Duke of Mecklenburg-Schwerin, 3, 4
Keill, John (1671–1721), 156
Kinnersley, Ebenezer (1711–1778), 81n, 87–89, 92, 100, 194, 203
Kircher, Athanasius (1602–1680), 420n
Kleist, Ewald Georg von (c1700–1748), 71
Klingenstierna, Samuel (1698–1765), 91, 92
Knight, Gowin (1713–1772), 158–60, 179, 390n
Kotel'nikov, Semyon (1731–1806), 24, 27
Krafft, Georg Wolfgang (1701–1754), 50, 73n, 173n, 487
Krasil'nikov, Andreĭ (1705–1773), 46, 47, 52
Kratzenstein, Christian Gottlieb (1723–1795), 67
Kruse, Karl Theodor, fl. c1750, 41
Kühn, Karl Gottlob (1754–1840), 212n
Kurganov, Nikolaĭ Gavrilovich (1725–1796), 46, 47, 52

La Caille, Nicolas-Louis de (1713–1762), 32
La Fermière, F. H. (1737–1796), 61, 62n, 63n, 498
Lambert, Johann Heinrich (1728–1777), 208–209, 217
Laplace, Pierre-Simon de (1749–1827), 220
Legendre, Adrien-Marie (1752–1833), 220
Lehmann, Johann Gottlob (1719–1767), 24, 25, 26n, 52, 92–93
Leibniz, Gottfried Wilhelm (1646–1716), 11–12, 14, 16, 156, 157
Le Maire, Pierre, 159, 179–80
Lémery, Louis (1677–1743), 93, 143
Le Monnier, Louis-Guillaume (1717–1799), 163

Le Monnier, Pierre (1675–1757), 161, 163
Le Roy, Jean-Baptiste (1720–1800), 89, 213, 214
Le Suer, Thomas (1703–1770), 173
Leyden experiment, 71–72, 76–77, 83, 91, 101, 106, 132, 135; Aepinus's discussion of, 122–25, 129, 194–95, 212, 268–90; condensation of electric fluid in, 116, 125, 272, 280–81; Franklin's explanation of, 80–81, 85–87, 272, 283–84, 286–88; function of glass in, 80, 85–87, 97, 108–109, 127, 194, 241–42, 285; Richmann's adaptation, 127–29, 453–68, 484. *See also* air, dielectric in Leyden experiment *and* Leyden jar
Leyden jar, 174, 186, 311, 402, 444; analogy with magnet, 167, 238, 293–96. *See also* Leyden experiment
L'Hospital, Guillaume-François-Antoine, Marquis de (1661–1704), 17
Lichtenberg, Georg Christoph (1742–1799), 212n
Lieberkühn, Johann Nathanael (1711–1756), 33
light, electrical, 70, 76, 133
lightning, an electrical discharge, 24–25, 79, 285; renders iron magnetic, 186, 444
lightning conductors, 55, 495
lines of force, magnetic, 184, 403–408
Linnaeus, Carl (1707–1778), 93n, 430n
Lomonosov, Mikhail Vasil'evich (1711–1765), 28n, 29–31, 51, 52, 66, 113n; dispute with Aepinus, 36–47, 48, 62; "night-vision" telescope, 37–38, 39, 497
London, Royal Society of, 21, 78–79, 94n, 151, 153–54, 158–60, 169, 191, 419n
"lucent barometer," 67, 69, 131, 494
lunar craters, 55–56, 495, 496, 498

Magalhães, João Jacinto de [Jean Hyacinthe de Magellan] (1722–1790), 496
magnet, analogy with Leyden jar, 167, 238, 293–96; armed, 138, 362–64, 378–79, 390; artificial, 158–59, 179–82, 231, 292, 327, 373, 378, 387–90, 432, 434, 444–52;

directive force of, *see* needle, magnetic; natural, 186, 231, 378–79, 430,

434, 437–39; result of breaking in two, 175–76, 216–17, 297–303; sphere of activity and sphere of attraction distinguished, 170, 420–27; "unipolar," possibility of, 167, 176, 290, 294, 296–97, 346–48. See also iron, magnetization of

magnetic center, 176–77, 207, 299–303, 343–49, 371

magnetic declination, see declination, magnetic

magnetic fluid, natural quantity of, 167, 245–48, 252, 257; non-uniform distribution of, 255, 298, 342, 348–49

magnetic induction, see induction, magnetic

magnetic lines of force, 184, 403–408

magnetic saturation, see saturation, magnetic

magnetism, Aepinus's theory, 164, 166–68, 174–75, 177–79, 187, 203–205, 214, 216, 219, 223, 239-40, 242–43, 290–303, 325–49, 355ff; analogy with electricity, see analogy between electricity and magnetism; "Cartesian" or circulation theory, 139–51, 153–64, 165, 167, 180–81, 186, 204–205, 208, 209, 210n, 215–16, 407–408; distinguished from electricity, 138, 164; and from gravity, 152; molecular theory, 143, 175, 217; not caused by friction alone, 445; terrestrial, see Earth; "two-fluid" theory, 209, 216, 219–20

magnetization, methods of, 144, 179–82, 185–86, 355–59, 364–90, 408–19, 425–31, 441–52. See also "double touch"

Mahon, Charles (third Earl Stanhope) (1753–1816), 198

Marggraf, Andreas Sigismond (1709–1782), 434

Mariotte, Edmé (?–1684), 12

Martin, Benjamin (1704–1782), 33, 158, 171n

mathematics, Aepinus's early work in, 7–8, 14, 15–17, 18, 20, 33–34; application to physics, 113–15, 126, 127, 136, 168, 187, 201–203, 206–208, 211, 213, 215, 218–24, 468

Maupertuis, Pierre Louis Moreau de (1698–1759), 17, 22–23, 28

Maxwell, James Clerk (1831–1879), 125, 165, 218n

Mayer, Johann Tobias (1723–1762), 21, 22, 23, 46, 489; magnetic researches, 185, 205–208, 217, 494

Mecklenberg-Schwerin, Karl Leopold, Duke of, 3, 4

mercurial phosphorus, see "lucent barometer"

mercury, congelation of, 41–42, 61, 492

Mercury, transit of, observed by Aepinus, 20, 489

Merian, Johann Bernard (1723–1807), 24

Mesmer, Franz Anton (1734–1815), 210

metal ores, origin of, 430

Mézières, Ecole du Génie, 215

Michell, John (1724–1793), 163, 173, 187; and "double touch" method of magnetization, 180–82, 234, 368, 373–74, 377, 379–85, 448

microscope, achromatic, devised by Aepinus, 56–57, 489, 496

microscope, solar, improved by Aepinus, 33, 493

Model, Johann Georg (1711–1775), 41, 52

molecular theory of magnetism, 143, 175, 217

Monge, Gaspard (1746–1818), 219

moon, craters on, 55–56, 495, 496, 498

Müller, Georg Friedrich (1705–1783), 15n, 24n, 27, 28n, 33, 35, 42, 43n, 48n, 52, 189–90

Munich, Bavarian Academy of Sciences, 209–11

Musschenbroek, Pieter van (1692–1761), 12, 63, 179, 242n, 335n, 337n, 359n, 419n, 445n; attempted determination of magnetic force law, 170–73, 327; an authority on experimental magnetism, 147–48, 177, 192–93, 219–20, 237, 294, 342, 366, 419; discovers Leyden experiment, 71, 76, 282; on the theory of magnetism, 161–65

Musschenbroekian jar, see Leyden jar

Musschenbroek's apparatus, see Leyden jar

Musschenbroek's experiment, see Leyden experiment

needle, dipping, 415, 419, 451. *See also* inclination, magnetic needle, electric, analogous to magnetic needle, 402

needle, magnetic, 387; directive force of, 147, 184–85, 396–402, 403, 415, 419–20, 421–25, 440–41; method of magnetizing improved, 182, 364–65, 371, 388, 491–92; used to test magnetism of iron, 179, 333, 433

Newton, Isaac (1642–1727), 8, 11, 12, 17, 34n, 63, 78n, 81n, 104–105, 151, 157, 160, 173, 229, 239, 344n, 399, 488, 492; attempted determination of magnetic force law, 168–70; calculus controversy, 16, 157; distinguishes between magnetism and gravity, 152; doctrine of universal gravity, 119–20, 201, 240, 258–59; electrical experiment, 85; explanation of Boyle's law, 82, 115, 117; ideas concerning aether, 13, 82, 154–55; unexplained forces used by, 13, 15, 110, 111–12, 152–53, 155, 184, 240; views on magnetism, 151–56

"night-vision" telescope, 37–38, 39, 497

Nikolaĭ, Baron Andreĭ L'vovich (1737–1820), 61

Nollet, Jean Antoine (1700–1770), 67, 84, 87–88, 92, 97, 101, 102, 103, 132n, 165, 194n, 201, 283n, 285; predominance of his ideas in France, 65–66, 213; theory of electricity, 73–78, 89, 94, 98; views on magnetism, 215–16

Norman, Robert (fl.late 16th cent.) 359n

Noya Carafa, Giovanni, Duc de (1715–1768), 94, 486

Nugent, Thomas (1700–1772), 3–4, 62

Oersted, Hans Christian (1777–1851), 166n

Osterman, Ivan Andreevich (1725–1811), 54, 62n

Osterwald, Dietrich (1729–c.1790), 59

Pallas, Peter Simon (1741–1811), 221, 496
Panin, Nikita Ivanovich (1718–1783), 54
Paris, Royal Academy of Sciences, 21, 38, 42, 73, 76, 89, 134, 142, 143, 151, 161, 179, 213; Aepinus proposed as Foreign Associate, 220–21; magnetism prize competitions, 144–51, 159, 163, 214

Parrot, Georg Friedrich (1767–1852), 62
Pastukhov, Peter Ivanovich (1732–1799), 59
Paul Petrovich, Tsar of Russia (1754–1801), 61, 62, 498; Aepinus tutor to, 53
Paulian, Aimé-Henri (1722–1801), 216
Peking, Jesuit missionaries in, 130, 200
phlogiston, 10, 186, 434, 435–36, 437, 438
points, consequent, *see* consequent points
Poisonnier, Pierre Isaac (1720–1798), 42
Poisson, Siméon-Denis (1781–1840), 217, 219
Polinière, Pierre (1671–1734), 179n, 342n
Popov, Nikita (1720–1782), 45, 50
potential, electrical, 125
Prévost, Pierre (1751–1839), 221, 223
Price, Richard (1723–1791), 204
Priestley, Joseph (1733–1804), 95–96, 97, 103, 105n, 129n, 192, 193, 194–97, 200–202, 204, 214, 223; criticizes Aepinus's theory, 115–17, 201; infers laws of electrical force, 218

quadrature of the circle, 111, 240

Razumovskiĭ, Count Kiril Grigor'evich (1728–1803), 43, 228–31, 484
Réaumur, René Antoine Ferchault de (1683–1757), 12, 73; on magnetism, 143, 181, 334, 428n
repulsion between particles of ordinary matter, 118–21, 198–201, 212, 219, 258–59
repulsion, electrical, 66–67, 102; between two negatively electrified bodies, 83, 100, 119, 262–63. *See also* attraction and repulsion, electrical
Richman, Georg Wilhelm (1711–1753), 27, 28n, 31, 92, 113n, 115; Aepinus's discussion of his experiment, 127–29, 234–35, 453–68, 484
Rive, Gaspard de la (1770–1834), 223
Robison, John (1739–1805), 116, 218, 222–24
Rohault, Jacques (1620–1675), 156
Rostock, 3–6, 9, 11, 15–16, 18, 19, 22, 28, 36n, 49, 62, 498; *Gelehrte Nachrichten*, 20, 72, 484, 489–90; University, 3, 4–9, 14, 18–20, 89

INDEX

Rotterdam, Batavian Society for Experimental Philosophy, 212
Royal Society of London, see London, Royal Society of
Rozier, François (1734–1793), 214, 493
Rumovskiĭ, Stepan (1734–1815), 24, 27, 31, 43, 45, 47n, 50, 52, 53n, 63n, 483
Russian educational reform, Aepinus's recommendations, 58–60, 495–96

St. Petersburg, Free Economic Society, 54–55, 62n, 494–95, 498
St. Petersburg, Imperial Academy of Sciences, 17n, 21, 28–31, 60–61, 62n, 127, 130; Aepinus's career in, 26–27, 31–53, 131, 189–90; distribution of publications, 189–91; geography department, 50n, 51, 53; observatory, 43, 45–46, 50, 51; prize competitions, 24, 68, 181, 233; university, 30, 39–41, 51
St. Petersburg, Imperial Corps of Noble Cadets, 48, 50, 51, 53
St. Vincent, Gregory of (1584–1667), 16
saturation, magnetic, 168, 291–92, 356–58, 361, 372, 382, 383, 445, 452
Savery, Servington (?–1744), 157–58, 160–61
Schlözer, August Ludwig (1735–1809), 52, 58n, 60n, 495, 496
Schumacher, Johann Daniel (1690–1761), 31
Segner, Johann Andreas von (1704–1777), 16, 21, 22, 27
Shuvalov, Ivan Ivanovich (1727–1797), 37
Sigaud de la Fond, Joseph-Aignan (1717–1799), 161n, 213, 214–15
Socin, Abel (1729–1808), 212n, 218n
Stahl, Georg Ernst (?1660–1734), 10
Stählin, Jakob von (1709–1785), 52, 190n
Stanhope, Charles Mahon, third Earl (1753–1816), 198
Steiglehner, Georg Christoph (1738–1819), 210–11
Stiles, Ezra (1727–1795), 192, 202, 203
Stockholm, Royal Swedish Academy of Science, 21, 49, 62, 191, 481
Strömer, Mårten (1707–1770), 92
Sturm, Johann Christoph (1635–1703), 12, 164

Symmer, Robert (1707–1763), 199–200, 209

Taylor, Brook (1685–1731), 157, 177n, 294n; attempted determination of magnetic force law, 169–71, 173
telescope, achromatic, 38n, 55–56
telescope, "night-vision," 37–38, 39, 497
Thomson, William (Lord Kelvin) (1824–1907), 217
Tiedemann, J. H., *instrument maker in Leipzig*, 62n
torsion balance, invented by Coulomb, 217
tourmaline, 49, 113n, 282, 479, 486, 488–89, 490; Aepinus's investigation of, 25–26, 28, 32, 34, 89–90, 91, 92–96, 132, 213–14; analogy with magnet, 26, 164, 187, 238
transit, of Mercury, 20, 489; of Venus, 32, 43–47, 50, 53, 484, 485, 494
Treaty of Armed Neutrality (1780), 57
Troostwyk, Adriaen Paets van (1752–1837), 212
Turin, Royal Academy of Sciences, 200
"two-fluid" theory, of electricity, 199–201, 209, 219–20; of magnetism, 209, 216, 219–20

Van Marum, Martinus (1750–1837), 212
Van Swinden, Jan Hendrik, (1746–1823), 193, 210–11, 212, 217
Varignon, Pierre (1654–1722), 17
Venus, transit of, 32, 43–47, 50, 53, 484, 485, 494
volcanoes on the moon, 496. See also lunar craters
Volta, Alessandro (1745–1827), 129–31, 193, 219, 276n
Voltaire, François-Marie Arouet de (1694–1778), 63, 488
Vorontsov, Count Semyon Romanovich (1744–1832), 61, 62n, 63n, 498
vortex, magnetic, see magnetism, "Cartesian" or circulation theory

Waitz, Jacob Seigismund von (1698–1777), 67, 92
Wallis, John (1616–1703), 17

Wargentin, Pehr (1717-1783), 21n
Watson, William (1715-1787), 77-78, 81, 92, 110
Watt, James (1736-1819), 222n
Weiss, L. F., *Magister* at Rostock, 8, 15
Whiston, William (1667-1752), 160-61, 170
Whittaker, Edmund T. (1873-1956), 86, 120n
Wilcke, Johan Carl (1732-1796), 21-22, 26n, 49, 79, 96, 105, 108, 113n, 191, 195-96; Aepinus's student, 16n, 18, 19-20; collaborates with Aepinus, 25, 32, 89-92, 129-30, 281n; favors "two-fluid" theories, 200, 216
Wilcke, Samuel (1704-1773), 19

Wilson, Benjamin (1721-1788), 195, 213, 214; correspondence with Aepinus, 19, 94, 191-92, 202, 204; experiments on tourmaline, 93-96, 486
Winckler, Johann Heinrich (1703-1770), 67
Winthrop, John (1714-1779), 109n, 192, 194, 202, 203
Wolff, Christian (1679-1754), 5, 7-8, 16, 30, 63

Young, Thomas (1773-1829), 120n, 223-24

Zavadovskiĭ, Peter Vasil'evich (1739-1812), 59
Zeiher, Johann Ernst (?-1794), 41-42, 51

Library of Congress Cataloging in Publication Data

Aepinus, Franz Ulrich Theodor, 1724-1802.
   Aepinus's Essay on the theory of electricity and magnetism.

   Translation of Tentamen theoriae electricitatis et magnetismi.
   "Annotated bibliography of Aepinus's published writings": p.
   Includes index.
   1. Electricity—Early works to 1850. 2. Magnetism—Early works to 1800. I. Home, Roderick Weir. II. Connor, Peter James. III. Title. IV. Title: Essay on electricity and magnetism.

QC517.A2813   537   78-10105
ISBN 0-691-08222-7